Laser Material Processing

T0178036

William M. Steen · Jyotirmoy Mazumder

Laser Material Processing

4th Edition

William M. Steen, em. Univ. Prof. Dr.
Greenacre
Old Wimpole Road
Arrington
Herts, SG8 0BX
UK
w.steen@btinternet.com

Jyotirmoy Mazumder, Prof.
G.G. Brown Laboratory Center for Laser-aided
Intelligent Manufacturing
Department of Mechanical Engineering
and Material Science and Engineering
University of Michigan
College of Engineering
2350 Hayward Street
Ann Arbor, MI 48109-2125
USA
mazumder@umich.edu

ISBN 978-1-84996-061-8   e-ISBN 978-1-84996-062-5
DOI 10.1007/978-1-84996-062-5
Springer London Dordrecht Heidelberg New York

British Library Cataloguing in Publication Data
A catalogue record for this book is available from the British Library

Library of Congress Control Number: 2010932029

Cover design: eStudioCalamar, Girona/Berlin

Printed on acid-free paper

Springer is part of Springer Science+Business Media (www.springer.com)

This emission is dedicated to stimulating friendships

Acknowledgements

Since the first edition of this book was written in 1991 the subject of laser material processing has grown not only in the scope of its technology and industrial application but also in the growing need for industry to be staffed by competent engineers who have some level of fluency in the application of optical energy. Thus, the aim of this edition is to be a definitive text on the basic science underlying the interaction of laser beams with matter and their application to industrial processes.

This book is the product of courses given by the authors at Liverpool University, UK, and University of Michigan, USA. The authors would like to acknowledge the support they have gained from the enthusiasm of the many students who have passed through the research schools they have led at Imperial College and Liverpool University for Bill Steen and at University of Illinois and University of Michigan for Jyoti Mazumder. Many of these now run their own laser businesses, teach the subject or have found other ways of making money from laser material processing. To witness this activity is one of the greatest pleasures an academic can have.

We are particularly grateful to Prof. Ken Watkins, who now leads the Liverpool University laser group, for writing Chapter 10 on laser cleaning and to Dr. Geoff Dearden, his second in command, for helping to update Chapter 9 on bending; to Dr. Bill O'Neill, who now runs his own laser group at Cambridge University, for his help in editing the original versions of Chapters 1 and 2 and some of the material for Chapter 7; to Prof. Lin Li, who now runs his own laser group at Manchester University, for his help in editing the original version of Chapter 12 on automation; to Dr. Neville Krasner, one of the pioneers in photodynamic therapy work, for reading and checking the draft of Chapter 11 on medical uses; to Dr. Rehan Akhter, who works with lasers in Pakistan, for his help in editing and assembling the illustrations for the original version of Chapter 4; to Dr. Guru Dinda of Focus Hope and Dong Kam of University of Michigan for helping write the sections on scaffold fabrication for biomedical application and application of micromachining for implantable lungs in Chapter 11. Some of the present students at University of Michigan, including Dallas Manning, Joohan Shin and Seung Lee, helped to collect and draw figures, and so the message spreads! Both of us would finally like to acknowledge with thanks the support of our wives, Margaret and Aparajita, who created an ambience in which to live and complete this fourth edition. Jyoti is also thankful to his boys Debashis and Debayan for tolerating their father's prolonged absence on many a day.

On the principle of a picture being worth a thousand words and a spoonful of sugar helping the medicine go down, the authors are also most grateful to the artistry of the two cartoonists Patrick Wright and Noel Ford.

The cover picture is an SEM photograph of a crater in a metal target caused by the impact of 1 mJ pulses at 6 kHz from a copper vapour laser operating at 511 nm wavelength. The photograph was taken by Dr. David Coutts of Macquarie University.

Contents

Prologue

It has been true throughout history that every time mankind has mastered a new form of energy there has been a significant, if not massive, step forward in our quality of life. Owing to the discovery of the laser in 1960, optical energy in large quantities and in a controlled form is now available as a new form of energy for the civilised world. It is therefore reasonable to have great expectations.

Consider the analogy with other forms of energy. The start of civilisation is identified with the ability to make tools – by the application of mechanical energy. The Lower Palaeolithic of some 1.75 million years ago produced ancient crude stone or bone tools. Finer stone tools were produced through the Middle Palaeolithic and Upper Palaeolithic period to reach pinnacles of excellence over a development period of around 1.5 million years. We might grumble about technology transfer being slow today, but they really had an argument then! This simple technology based on the application of mechanical energy caused a major change in our quality of life – it took us out of the trees and converted us from animals to human beings.

Centuries later the control of chemical energy in the form of organised fires and convectively blown furnaces was achieved and the Bronze Age (around 6500 BC) and the Iron Age (around 1500 BC) resulted in superior tools. Owing to the increased productivity of agriculture using these superior tools and the improved security afforded from swords and chariots, stable political groupings formed, the Greek and Roman empires were born, the arts flourished and again a major step forward in our quality of life resulted.

The ability to harness wind and water energy in the form of windmills and waterwheels started industrialisation, whereas in the form of sailing ships it opened up the world to international trade. The great navigators discovered new worlds by applying wind energy with the help of the now-sophisticated product of mechanical energy, the chronometer.

In 1701 Newcomen built the first working steam engine for pumping water at Dudley Castle in Chester, UK. By 1790 the Industrial Revolution was in full swing with steam engines doing the back-breaking work of previous ages. The quality and the speed of life both increased.

In 1831 Michael Faraday invented the dynamo and, after the improvements of Thomas Edison in 1878, electricity became available in a controllable form and in large quantities. The electric motor is the heart of many domestic machines and industrial plant. Arc welding, electric heating, radios, TV (there is a slight overlap with electromagnetic or optical energy here), telephones, lighting, computers and refrigerators are more examples of the dramatic effect the mastery of electrical energy has had on our quality of life.

Nikolaus Otto and Eugen Langen, working with the designs of Alphonse Beau de Rochas, started production of the first four-stroke internal combustion engine in Deutz, Germany, in 1867 and so found a new way of harnessing chemical energy from oil and petrol. Personalised transport and flight became a reality. Although some argue this has brought no advance in our quality of life, there is hardly a soul who would do without such means of transport. Swift travel has begun to make a true world community. International trade allows one to have fresh vegetables all the year around and the benefits of many cultures can be shared.

Nuclear energy became available when Enrico Fermi built the first atomic pile, which went critical on 2 December 1942 in the squash court at the University of Chicago, USA. Atomic energy has been used directly only as a bomb or for medical radiation treatment. As such, it has altered world politics. It has questioned the wisdom of settling arguments by fighting and so far has thus resulted in peace between the superpowers – though there may be a problem with other countries. The current direct application of atomic power is only as a heating system in power stations, as a form of coal substitute. Thus, it seems that a further invention is needed before we have truly mastered this form of energy.

Arthur Schawlow and Charles Towns first published the concept of the laser in 1958. In 1960 Theodore Harold "Ted" Maiman (1927–2007) [1] invented the first working ruby laser, shown in Figure 0.1. It was not a surprise but the result of considerable investment following Einstein's papers in 1916 [2] and 1917 [3] in which he showed that "stimulated emission" – the basic physics generating laser radiation – was an everyday occurrence. His argument ran along the lines of a heat balance on a hot object. The radiation from the object, I, would be proportional to the number of excited species, N_2, decaying to give the radiation and the loss of that radiation by reabsorption from the unexcited species, N_1, each according to some probability factor, A and B, respectively.

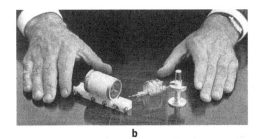

a   b

Figure 0.1 **a** Theodore Maiman and one of his first ruby lasers as shown in Hughes publicity shots, and **b** Theodore Maiman's first laser, which is less photogenic

In brief the radiation coming from the hot object would be

$$AN_2 - BIN_1 = I ,$$

Rearranging, we get

$$I = \frac{AN_2}{1 + BN_1} \approx \frac{AN_2}{BN_1} .$$

However, the ratio N_2/N_1 is known from the Boltzmann distribution:

$$\frac{N_2}{N_1} = \exp\left[\frac{-hc}{\lambda k T}\right] .$$

The result was something that was too small to satisfy Planck's law:

$$I(\lambda) = \frac{2\pi c^2 h}{\lambda^5}\left[\frac{1}{e^{(hc/(\lambda k T))} - 1}\right] .$$

The energy balance could only be achieved if there was a further term based on a stimulated emission of the form CIN_1, where C is a further probability factor. So just as nuclear energy was something of a surprise to us even though it had been staring us in the face most days, so it was with stimulated emission as we had been sitting in front of a fire. In both cases they are a phenomenon of great importance but are only "visible" by mathematics – a thought for our educationalists!

With the great stories of H.G. Wells, *e.g.*, "War of the Worlds" written in 1890, as a guide, the military soon realised that a death ray would be handy on any battlefield. There resulted an avalanche of research funding – the only time I have heard of one laboratory requesting a grant for laser development and being awarded three times as much as was asked for [4]! This was the TRG proposal for $300,000 by Gordon Gould for which the Defence Advanced Research Project Agency (DARPA) awarded $1,000,000. However Maiman won this race by a few months with his solid-state ruby laser. In the months and years following it seemed that almost anything could be made to lase. Table 0.1 shows the wavelength bands covered by commercial lasers of today together with the active species.

The history of the invention of the laser makes fascinating reading. The outline is shown in Table 0.2 based on data taken from the book "Laser Pioneers" by Jeff Hecht [5]. It includes the curious tale of Gould's laboratory notebook being notarised by a candy-store owner as having legal precedence over a "Physical Review" paper by Townes and awarded patent rights; even though neither of them actually invented a working laser! It also includes the period of great excitement just after Maiman's invention when the Optical Society of America put on four optical maser symposia on 2–4 March 1961 in Pittsburgh. The ballroom was filled for the presentation of the keynote paper by Townes and overflowed into the balconies. In the words of the reporter Bromberg [6]:

"Normally if you had a paper at the Optical Society, you might draw a hundred people. There might be two or three cameras taking pictures of the slides. These halls were packed, the ballroom was packed, for these papers. I remember as a high point

Table 0.1 Range of wavelengths for current commercial lasers

Laser type	Lasing species	Principle wavelength (µm)	Region	Date invented/commercialised
Excimer	F_2	0.157	UV	1975/1976
	ArF	0.193	UV	
	KrF	0.248	UV	
Nd:YAG frequency-quadrupled	$Nd^{3+} \times 4$	0.266	UV	
	XeCl	0.308	UV	
	XeF	0.351	UV	
Nitrogen	N_2	0.337	UV	1966/1969
AlGaN diode	Band gap	0.38–0.45 (tunable)	Blue	
Helium–cadmium	Cd^+	0.4416	Blue	1968/1970
Argon	Ar^+	0.4880	Blue	1964/1966
	Ar^+	0.5145	Green	
Copper vapour	Cu^*	0.5106	Blue-green	1966/1981
	Cu^*	0.5782	Yellow	
Nd:YAG frequency-doubled	$Nd^{3+} \times 2$	0.532	Green	
Helium–neon	Ne^*	0.6328	Red	1962
Ruby	Cr^{3+}	0.6943	Red	1960/1963
Alexandrite	Cr^{3+}	0.700–0.820 (tunable)	IR	1977/1981
Ti:sapphire	Ti^{3+}	0.670–1.100 (tunable)	IR	
AlGaAs diode	Band gap	0.7–0.9 (tunable)	IR	1962/1965
Nd:YAG or Nd:glass	Nd^{3+}	1.064	IR	1964/1966
Yb:YAG or Yb:glass	Yb^{3+}	1.030	IR	1990s
Chemical oxygen–iodine	Chemical $(O_2 + I_2)$	1.3	IR	1964/1983
Er:YAG	Er^{3+}	1.5	IR	
Hydrogen fluoride	Chemical $(H_2 + F_2)$	2.6–3.0	IR	1967/1977
Helium–neon	Ne^*	3.39	IR	
Carbon monoxide	CO vibration	5.4	IR	
Carbon dioxide	CO_2 vibration	9.4	IR	
		10.64	IR	1964/1966
Dye	Fluorescence	1.1–0.3 (tunable)	IR–UV	1962/1965
Free electron	Electron vibration	12.0–0.1 (tunable)	IR–UV	1963/1969

Art Schawlow getting up to give a talk. Every slide he projected, there was a veritable staccato machine gun fire of Minoltas going off. It was unbelievable! Panicsville."

In material processing the laser must be reasonably powerful, which reduces the number of eligible lasers to only a few – essentially the CO_2, rare-earth solid-state [(Nd:YAG), Yb:YAG, Er:YAG, Nd:glass], diode (GaAs, GaN) and excimer lasers, with some peripheral interest in CO, copper vapour and free-electron lasers and synchrotron radiation.

Table 0.2 Outline history of the development of the laser [5]

Date	Name	Achievement	References
1916	Albert Einstein	Theory of light emission. Concept of stimulated emission	[2, 3]
1928	Rudolph W. Ladenburg	Confirmed existence of stimulated emission and negative absorption	[17]
1940	Valentin A. Fabrikant	Noted possibility of population inversion	[18]
1947	Willis E. Lamb, R.C. Retherford	Induced emission suspected in hydrogen spectra. First demonstration of stimulated emission	[19]
1951	Charles H. Townes	The inventor of the maser at Columbia University. First device based on stimulated emission. Awarded the Nobel prize in physics in 1964	[20]
1951	Joseph Weber	Independent inventor of the maser at University of Maryland	[21]
1951	Alexander Prokhorov, Nikolai G. Basov	Independent inventors of the maser at Lebedev Laboratories, Moscow. Awarded the Nobel prize in physics in 1964	[22]
1954	Robert H. Dicke	"Optical bomb" patent based on pulsed population inversion for superradiance and a separate Fabry–Perot resonant chamber for a "molecular amplification and generation system"	[23]
1956	Nicolaas Bloembergen	First proposal for a three-level, solid-state maser at Harvard University	[24]
1957	Gordon Gould	First document defining a laser; notarised by a candy-store owner. Credited with patent rights in the 1970s	[25]
1958	Arthur L. Schawlow, Charles H. Townes	First detailed paper describing an "optical maser". Credited with the invention of the laser; from Columbia University	[26]
1960	Arthur L. Schawlow, Charles H. Townes	Laser patent no. 2,929,922	[27]
1960	Theodore Maiman	Invented the first working laser based on ruby, 16 May 1960. Hughes Research Laboratories	[1]
1960	Peter P. Sorokin, Mirek Stevenson	First uranium laser – second laser overall, November 1960. IBM Laboratories	[28]
1961	A.G. Fox, T. Li	Theoretical analysis of optical resonators at Bell Laboratories	[29]
1961	Ali Javan, William Bennett Jr., Donald Herriott	Invented the helium–neon laser at Bell Laboratories, Murray Hill, New Jersey	[30]
1962	Robert Hall	Invention of the semiconductor laser at General Electric Laboratory followed swiftly by others	[10]
1964	J.E. Geusic, H.M. Marcos, L.G. Van Uitert	Inventor of the first working Nd:YAG laser at Bell Laboratories	[8]
1964	Kumar N. Patel	Invention of the CO_2 laser at Bell Laboratories, Murray Hill, New Jersey	[7]
1964	William Bridges	Invention of the argon ion laser at Hughes Laboratories	[31]
1965	George Pimentel, J.V.V. Kasper	First chemical laser at University of California Berkley	[32]
1966	William Silvast, Grant Fowles, B.D. Hopkins	First metal vapour laser, Zn–Cd, at University of Utah	[33]

Table 0.2 Outline history of the development of the laser [5]

Date	Name	Achievement	References
1966	Peter Sorokin, John Lankard	First dye laser action demonstrated at IBM Laboratories	[34]
1969	G.M. Delco	First industrial installation of three lasers for automobile application	(D. Roessler, private communication, 1995)
1970	Nicholai Basov's group	First excimer laser at Lebedev Laboratory, Moscow. Based on Xe only	[15]
1970	Zh.I. Alferov *et al.*	Invention of double heterostructure for laser diodes	[35]
1974	J.J. Ewing, Charles Brau	First inert gas halide excimer laser at AVCO Everet Laboratories	[14]
1977	John M.J. Madey's group	First free-electron laser at Stanford University	[16]
1980	Geoffrey Pert's group	First report of X-ray lasing action, Hull University, UK	[36]
1981	Arthur L. Schawlow, Nicholaas Bloembergen	Awarded the Nobel prize in physics for work in non-linear optics and spectroscopy	
1984	Dennis Matthews's group	First reported demonstration of a "laboratory" X-ray laser; from Lawrence Livermore National Laboratory	[37]
2000	Zh.I Alferov, H. Kroemer	Awarded the Nobel prize in physics for heterostucture invention	

The CO_2 laser was invented by Patel [7] in 1964 working at Bell Laboratories. His first laser used pure CO_2 and produced 1 mW of power with an efficiency of 0.0001 %. Addition of nitrogen improved the power to 200 mW and when helium was added the power jumped to 100 W with an efficiency of 6 % – all this within 1 year! Today all CO_2 lasers have a gas mixture of approximately 0.8:1:7 $CO_2/N_2/He$. The commercial potential for this sort of laser was immediately perceived [4]. Spectra-Physics worked on developing the technology from 1965. It stopped a year later but the team working on the laser went on to found their own company – Coherent. They marketed the first CO_2 laser at 100 W in 1966 and a 250-W version in 1968. The need to cool the gas was soon understood and methods of convectively cooling lasers were designed. AVCO came out in the early 1970s with a 15-kW CO_2 transverse flow laser. It was an anachronism and the market was not ready for such a powerful laser; however, much early work on this laser gave insight into the potential for material processing even though the mode was poor and so the focus was never very fine. Today the CO_2 laser remains the workhorse for material processing, with slow flow, fast axial flow, transverse flow, sealed and waveguide styled lasers operating at average powers up to 25 kW or even 100 kW for military-funded laboratories. High-powered pulsed CO_2 lasers have also been developed, such as the transverse excitation atmospheric pressure lasers. These have megawatts of power and operate in a pulsed mode with up to 10 J per pulse and 1-ms pulse lengths. Another growth direction has been into sealed low-power units, sometimes as waveguide lasers, for medical and guidance uses which are now available with up to several kilowatts or so output and a lifetime of around 4,000 h. CO_2 lasers fitted with a suitable grating as a rear mirror have an output which can be a single spectral line. This spectral line can be tuned between 8- and 11-μm wavebands, which is of use in processing certain plastics. The tuning can be continuous if the pressure of the tube is increased. This has allowed growth into the communications markets.

The Nd:YAG laser was invented at Bell Laboratories in 1964 [8]. Quantronix, Holobeam, Control Laser (which later bought Holobeam) and Coherent were quick to enter the market since the application for resistance trimming of electronic circuits was soon appreciated as a large potential market. However, the market stayed at less than $1 million for many years since the lasers were of poor quality. In 1976 the market changed abruptly when Quanta-Ray introduced the first reliable high-performance YAG laser. The design used an unstable resonator and was robust. It gave 1-J pulses at half the price of its rivals. Since then the Nd:YAG lasers have progressed with the introduction of sophisticated pulse shaping such as those for the petawatt (10^{15} W) laser at Rutherford Appleton Laboratory in the UK or the commercial lasers developed by JK Lasers (which became Lumonics, and was then bought by Sumitomo and then by General Scanning to become GSI-Lumonics and finally GSI Group – a tale similar to many as the industry crystallises into larger, better capitalised units) and others with powers ranging up to many kilowatts continuous wave or even terawatts (10^{12} W) pulsed when used in a master oscillator power amplifier arrangement. Beam expansion and compression techniques have allowed extremely short pulses down into the femtosecond range (10^{-15} s). This is an area of physics yet to be explored – megawatts arriving in less than a wavelength are leading to solid-state plasmas, whatever they are. The host material for the neodymium has also developed into several varieties: still the YAG crystal, but now a variety of glass materials as well as yttrium lanthanum fluoride, yttrium aluminium phosphide and others. The most exciting are the ytterbium-doped fibre lasers or disc lasers giving high brightness and potentially very high powers in a compact, robust form.

Diode lasers were late starters but are fast becoming the laser considered to have the brightest future. The first suggestion of a possible diode laser came from Basov et al. [9] in 1961. Soon after that, in 1962, the first laser of this kind was demonstrated in three laboratories at nearly the same time: by Hall et al. [10] at General Electric, Nathan et al. [11] at IBM and Quist et al. [12] at MIT Lincoln Laboratory. These early diode lasers were more of a curiosity than useful, since they needed cryogenic temperatures and could only be pulsed. The idea of a heterostructure [13] that introduced a higher bandgap layer to keep the carriers in certain areas of the device reduced the large current densities that had been required. This was the breakthrough needed, and in 1969 the first modern diode laser was operated. Since then unprecedented improvements have been made and today the diodes can be arranged as arrays of almost any power needed up to multi kilowatts, but still with a problem with beam quality partly solved by using fibre delivery. Nevertheless diodes are small, robust and cheap. Clearly they will radically alter the laser market in years to come, making optical energy cheap and conveniently available from small optical generators.

The excimer laser, working on the improbable chemistry of noble gas halides, e.g., KrF, was invented in 1974 at AVCO Everett Laboratories by Ewing and Brau [14] after the Russians in Basov's group at Lebedev Laboratories in Moscow had demonstrated ultraviolet radiation from pure xenon gas lasers [15]. These ultraviolet lasers seemed to have potential for military purposes owing to the low reflectivity of most materials to short-wavelength radiation. Considerable money has been put into developing them and as a result the material processing industry now has some robust ultraviolet lasers capable of "cold cutting" with immense prospects for the electronic industry, microlithography processing, weird chemistries and 9-ns suntans! However, other sources of ultraviolet radiation can be obtained from frequency-quadrupling Nd:YAG radiation, which may prove to be competitive, the Nd:YAG laser being more user-friendly.

Another source of ultraviolet radiation, or almost any other wavelength, is the free-electron laser first demonstrated by Madey's group [16] at Stanford University in 1977. Sending electrons at high speed through an undulating magnetic field causes them to oscillate and thus radiate at the frequency defined by the spacing of the "wiggler" and their speed through it. Almost any wavelength can be generated, from infrared to soft X-ray. However, owing to the velocity profile of the moving electrons in a vacuum under an electrostatic field, the actual output is more like a rainbow, with the shorter wavelengths coming from the higher velocity regions, such as the centre of the electron flow. The free-electron laser is a compact form of synchrotron radiation. The application of soft X-rays is mainly for very fine lithography.

As a result of this activity we now have optical energy in a controllable form. The question is: Will the mastery of this form of energy also give a massive boost to our standard of living?

In the previous application of new forms of energy there were people who got hurt and many who did well from these changes. That is the law of natural selection which has been the theme underlying all changes in this world. It is not something to control so much as something to note as a lesson from life. It is also the hard reality associated with progress. We and our politicians might try to mitigate the suffering, yes; but fight the changes, no! The application of optical energy will be no exception – why should it be? Is it possible to stop the clock and pretend something has not been invented? The winners in the past have always been those who see change as an opportunity, not a threat. I hope the reader is of that mind and that this book may help to open new opportunities for him or her.

References

[1] Maiman TH (1960) Stimulated optical radiation in ruby. Nature 6 August
[2] Einstein A (1917) Zur Quantentheorie der Strahlung. Phys Z 18:121
[3] Einstein A (1916) Strahlungs-emission und -absorption nach der Quantentheorie. Verh Dtsch Phys Ges 18:318–332
[4] Klauminzer GK (1984) Twenty years of commercial lasers – a capsule history. Laser Focus/Electrooptics Dec 54–60
[5] Hecht J (1985) Laser pioneers. Academic, London
[6] Bromberg JL (1991) The laser in America 1950–1970. MIT Press, Cambridge
[7] Patel CKN (1964) Continuous wave laser action on vibrational–rotational transitions of CO_2. Phys Rev A 136:1187
[8] Geusic JE, Marcos HM, Van Uitert LG (1964) Laser oscillations in Nd doped yttrium aluminium, yttrium gallium and gadolinium garnets. Appl Phys Lett 4:182
[9] Basov NG, Kroklin ON, Popov YM (1961) Production of negative temperature state p–n junctions of degenerate semiconductors. Sov Phys JETP 13:1320

[10] Hall RN, Fenner GE, Kingsley JD, Soltys TJ, Carlson RO (1962) Coherent light emission from GaAs junctions. Phys Rev Lett 9:366

[11] Nathan MI et al (1962) Stimulated emission of radiation from GaAs p–n junctions. Appl Phys Lett 1:62

[12] Quist TM et al (1962) Semiconductor maser of GaAs. Appl Phys Lett 1:91

[13] Kroemer H (1963) A proposed class of heterojunction injection lasers. Proc IEEE 51:1782

[14] Ewing JJ, Brau CA (1975) Laser action on the bands of KrF and XeCl. Appl Phys Lett 27:350

[15] Basov NG, Danilychev VA, Yu MP, Khodkevich DD (1970) Laser operating in the vacuum region of the spectrum by excitation of liquid Xe with an electron beam. JETP Lett 12:329

[16] Madey JMJ (1971) Stimulated emission of bremsstrahlung in a periodic magnetic field. J Appl Phys 42:1906

[17] Landenburg R (1928) Research on the anomalous dispersion of gases. Phys Z 48:15–25

[18] Bertolotti M (1983) Masers and lasers: an historical approach. Hilger, Bristol

[19] Lamb WE Jr, Retherford RC (1947) Fine structure of hydrogen by a microwave method. Phys Rev 72:241

[20] Gordon JP, Zeiger HJ, Townes CH (1954) Molecular microwave oscillator and new hyperfine structure in the microwave spectrum of NH_3. Phys Rev 95:282

[21] Weber J (1953) Amplification of microwave radiation by substances not in thermal equilibrium. Trans IRE Prof Group Electron Devices PGED-3:1–4

[22] Basov NG, Prokhorov AM (1954) 3-level gas oscillator. Zh Eksp Teor Fiz (JETP) 27:431

[23] Dicke RH (1954) Coherence in spontaneous radiation processes. Phys Rev 93:99

[24] Bloembergen N (1956) Proposal for a new type of solid state maser. Phys Rev 104:324

[25] Hecht J (1994) Winning the laser patent war. Laser Focus World Dec 49–51

[26] Schawlow AL, Townes HC (1958) Infrared and optical masers. Phys Rev 112:1940

[27] Schawlow AL, Townes HC (1960) A medium in which a condition of population inversion exists. US Patent 2,929,922, 22 March

[28] Sorokin PP, Stevenson MJ (1960) Stimulated infrared emission from trivalent uranium. Phys Rev Lett 5:557

[29] Fox AG, Li T (1961) Resonant modes in a maser interferometer. Bell Syst Tech J 40:453

[30] Javan A, Bennett WR Jr, Herriott DR (1961) Population inversion and continuous optical maser oscillation in a gas discharge containing a He–Ne mixture. Phys Rev Lett 6:106

[31] Bridges WB (1964) Laser oscillations in singly ionised argon in the visible spectrum. Appl Phys Lett 4 128–130; Erratum (1964) Appl Phys Lett 5:39

[32] Kasper JVV, Pimentel GC (1965) HCl chemical laser. Phys Rev Lett 14:352

[33] Silfvast WT, Fowles GR, Hopkins BD (1966) Laser action in singly ionised Ge, Sn, Pb, In, Cd, Zn. Appl Phys Lett 8:318–319

[34] Sorokin PP, Lankard JR (1967) Flashlamp excitation of organic dye lasers – a short communication. IBM J Res Dev 11(2):148

[35] Alferov ZhI, Andreev VM, Garbuzov DZ, Zhilyaev YV, Morosov EP, Portnoi EL, Trofim VG (1970) Fiz Tekh Poluprovodn 4:1826–1829. English translation (1971) Sov Phys Semicond 4:1573–1575

[36] Jacoby D, Pert GJ, Ramsden SA, Shorrock LD, Tallents GJ (1981) Observation of gain in a possible extreme ultraviolet lasing system. Opt Commun 37(3):193–196

[37] Matthews DL et al (1985) Demonstration of a soft X-ray amplifier. Phys Rev Lett 54:110–113

"Perhaps when you have finished messing around,
you could give me a hand with these wheels."

1 Background to Laser Design and General Applications

To have begun is half the job: be bold and be sensible

Horace (65–68 BC), Epistles I ii 40

The first part of this chapter briefly describes the basic principles of the physics and the construction of a laser. The second part of the chapter gives a sketch of the numerous ways in which the laser can be used other than as a material processing tool. The whole chapter is aimed at providing a review of the overall state of laser science and applications, which should be useful for an engineer of laser material processing. There are several textbooks available on laser physics which deal with the subject in detail [1–5].

1.1 Basic Principles of Lasers

The word "laser" is an acronym for light amplification by stimulated emission of radiation. It was first proposed by Schawlow and Townes [6].

1.1.1 Stimulated Emission Phenomenon

The mystical part of a laser is that it works at all. This is entirely due to the stimulated emission phenomenon, which was predicted by Einstein [7] in 1916 using a mathematical argument (see the Prologue, page 1). It may thus appear to be an esoteric phenomenon, not of everyday experience. In fact it was not until 1928 that Ladenburg first confirmed it by observing negative absorption in his spectroscopic work (see Table 0.2). However, we now realise it is commonplace, contributing to the radiation from every radiating object.

Through Einstein's analysis of radiation from hot objects, he postulated that there must be a radiant term based on a photon of radiation striking an excited species and causing it to release the energy of excitation. This has since been shown to be true. The stimulated photons are found to be in phase and travelling in the same direction as the stimulating photons. A photon originates from the energy change between an excited state and a lower state. It is thus usually spectrally pure if the change is between electronic or vibrational quantum states, but may have a range of wavelengths if the change is from an energy well to a lower state (as with excimer lasers, diode lasers, titanium sapphire lasers and others; see Table 0.1).

Many materials can be made to show this stimulated emission phenomenon, but only a few have significant power capability, since a further condition is that a population inversion is necessary, whereby there are more atoms or molecules in the excited state than in the lower-energy state, so as to allow amplification as opposed to absorption. To achieve this, the lifetime of the excited species has to be longer than that of the lower-energy state.

The main lasers used in material processing are carbon dioxide (CO_2), carbon monoxide (CO), neodymium-doped yttrium aluminium garnet (YAG; Nd:YAG), neodymium glass (Nd:glass), ytterbium-doped YAG (Yb:YAG), erbium-doped YAG (Er:YAG), excimer (KrF, ArF, XeCl) and diode (GaAs, GaAlAs, InGaAs, GaN and others being developed) lasers.

1.1.2 Basic Components of a Laser

A laser must have the following three basic components (see also Section 1.2 on laser construction):

1. active medium, which serves as a means to amplify light (see Figure 1.1);
2. pumping source, which is a means to excite the active medium to the amplifying state: and
3. optical resonator, which is a means to provide optical feedback.

The active medium may be any material that is solid, liquid, gas or plasma. It is believed that something in everything will lase if the substance is hit hard enough. In fact, even an edible laser was reported by Hänsch *et al.* [8]. They made a laser out of a Jell-O®[1] dessert. The common laser mediums include ruby, Nd:YAG and Nd:glass (solids); organic dyes, such as rhodamine 60, coumarin 2 and coumarin 30, dissolved in solvents such as alcohol or water (liquids); and He–Ne, CO_2, argon and nitrogen (gases).

Any energy source can be used as a pumping source. The common pumping sources include flash lamps (incoherent light), lasers (coherent light), electrons (DC, RF or pulsed gas discharge, electron beam), chemical reactions, ion beams and X-ray sources.

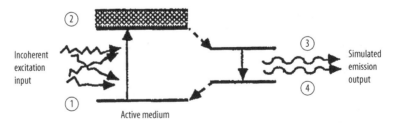

Figure 1.1 Energy level diagram for a four-level laser system showing the means to amplify light

---

[1] Jell-O® is a registered trademark of Kraft Foods Holdings Inc., Three Lakes Drive, Northfield, Illinois, 60093, USA. www.kraftfoods.com

Output power is proportional to the power of the pumping source and the amount of active medium; therefore, power can be controlled by controlling either the pumping source or the active medium, *e.g.*, fast axial flow CO_2 provides $0.7\,kW\,m^{-1}$, whereas a slow flow or sealed tube design provides $0.05\,kW\,m^{-1}$ (see Figure 1.2).

An optical resonator causes the light generated by the active medium, parallel to its axis, to be reflected back and forth through the medium. If the light is amplified owing to this action, and if the gain equals the round trip losses in the resonator, then the combination of the amplifier and the resonator is at the threshold for lasing. Light in the excited resonator travelling parallel to the axis is amplified several times. Only part of it is released in each pass through a partially transmitting window as a laser beam (Figure 1.3). Therefore, the optical resonator is a cavity defined by a 100% reflective mirror at one end and a partially transmitting mirror at the other end. For a laser, the cavity is filled with an active medium and a pumping source, such as an electromagnetic field, is provided to excite the active medium. The optical resonator is also responsible for making the laser monochromatic and unidirectional, and it also imposes the spatial distribution.

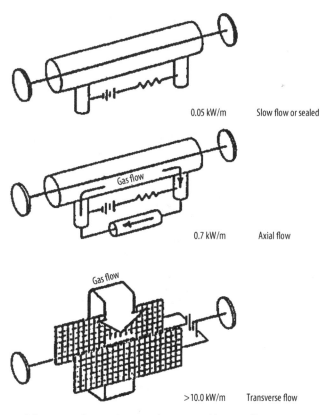

0.05 kW/m   Slow flow or sealed

0.7 kW/m   Axial flow

>10.0 kW/m   Transverse flow

Figure 1.2 Types of electric gas lasers, showing the practical limits of laser output power per metre of amplifying medium (from AVCO circular no. 7705)

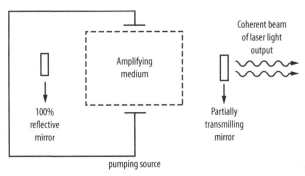

Figure 1.3 Optical resonator for a laser

1.1.3 Physics of the Generation of Laser Light

Clearly, the amplification of light and the laser output are achieved by the interaction of the atoms and molecules of the active medium with the electromagnetic field of the pumping source. Atomic systems, such as atoms, ions and molecules, can exist in certain states, each of which is characterised by a definite (excitation) energy. In an atom (Figure 1.4), each orbit occupies certain energy levels. A transition between states of energy E_1 and E_2 results in the emission or absorption of a photon, the frequency v_{12} of which is given by

$$h v_{12} = |E_1 - E_2| , \qquad (1.1)$$

where h is Planck's constant.

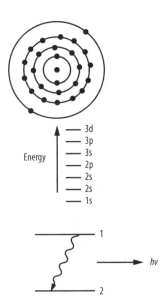

Figure 1.4 Energy levels in an atom

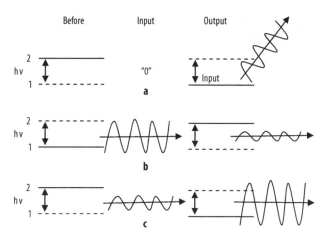

Figure 1.5 Radiative processes in a two-level system **a** spontaneous emission; **b** absorption; **c** stimulated emission

A system in the lower (ground) state may absorb photons of the appropriate frequency to rise to a higher state.

Einstein identified three major processes by which the atoms can interact with an electromagnetic field:

1. spontaneous emission (A_{21});
2. absorption (B_{12}); and
3. stimulated emission (B_{21})

A_{ij} and B_{ij} are constants for a given transition and are called Einstein coefficients.

These three processes are illustrated for a two-level system separated by $h v$ energy, where h is Planck's constant and v is the frequency of the wave (Figure 1.5). The viability of a laser and light amplification by stimulated emission can be demonstrated by developing a relationship between the three above-mentioned processes, the Einstein coefficients, the necessary boundary conditions of black-body radiation and lifetime broadening [2].

1.1.3.1 Spontaneous Emission

In this process, when an atom in level 2 decays spontaneously to level 1, it adds excess energy in the form of a photon (where each photon has energy equivalent to $h v$) to the cavity. But, by definition, for spontaneous emission, the photon comes out in a random direction (Figure 1.5a). That means photons can radiate randomly into anywhere in the 4π steradians without any polarisation contributing to the field.

If the population density in level 2 is N_2 the rate equation for decay of this state is given by

$$\frac{dN_2}{dt}\bigg|_{\text{spontaneous}} = -A_{21} N_2 , \qquad (1.2)$$

where A_{21} is the Einstein constant for this process.

If no other process takes place in the cavity, the atomic population will steadily shift to level 1 with time constant τ.

$$\tau = (A_{21})^{-1} \text{ radiative lifetime} \tag{1.3}$$

Obviously, the population density of the bottom level (N_1) must increase just as fast as the population density of the top layer decreases.

1.1.3.2 Absorption

In this process, an atom in state 1 absorbs a photon from the field and thus converts the atom into state 2. The rate at which this process takes place must depend upon the number of absorbing atoms and the field from which they extract the energy. Thus, we have

$$\left.\frac{dN_2}{dt}\right|_{\text{absorption}} = +B_{12}N_1\rho(v) = -\left.\frac{dN_1}{dt}\right|_{\text{absorption}}, \tag{1.4}$$

where $\rho(v)$ is the energy density of the field or the electromagnetic energy density (J m^{-3}) and B_{12} is the Einstein coefficient for this process. Note from Figure 1.5b, in this process, that the wave decreases in amplitude and the atom in state 1 is converted to state 2 and the part of the wave not absorbed continues along its path.

1.1.3.3 Stimulated Emission

This process is the reverse of absorption. The atom gives up its excess energy, hv, to the field, adding coherently to the intensity. The added photon has the *same frequency, same phase, same sense of polarisation and propagates in the same direction* as the wave that induced the atom to undergo this type of transition.

Obviously, the rate is dependent upon the number of atoms to be stimulated and the strength of the stimulating field:

$$\left.\frac{dN_2}{dt}\right|_{\substack{\text{stimulated}\\\text{emission}}} = -B_{21}N_2\rho(v) = -\left.\frac{dN_1}{dt}\right|_{\substack{\text{stimulated}\\\text{emission}}}. \tag{1.5}$$

A relationship between the three Einstein coefficient can be established by adding Equations 1.2, 1.4 and 1.5 and introducing Boltzmann statistics and necessary boundary conditions. At this point, one should note that besides radiation, collision with another atom, an electron or a lattice vibration (a phonon) can also cause a transition to take place.

However, stimulated emission is the process which is most significant for a laser since it provides energy, or photons, with the same frequency, same phase, same sense of polarisation and that propagates in the same direction. This makes amplification of light possible and produces coherent laser light. But for stimulated emission to occur, it is evident from Figure 1.5c that one needs population inversion, *i.e.*, the number of atoms in level 2 has to be higher than that in level 1, which is an unstable situation for nature.

We will now examine the rate equations and the relationship between the Einstein coefficients to determine the necessary theoretical conditions for a viable laser. For simplicity, only classical mechanics will be used. The objective is only to identify the important components of a laser.

1.1.4 Relationship Between the Einstein Coefficients

The total rate of change of the population density in state 2 (or 1) due to the radiative processes can be obtained by adding Equations 1.2, 1.4 and 1.5:

$$\frac{dN_2}{dt} = -A_{21}N_2 + B_{12}N_1\rho(v) - B_{21}N_2\rho(v) = -\frac{dN_1}{dt} . \tag{1.6}$$

At equilibrium, the time rate of change must be zero. Thus,

$$-A_{21}N_2 + B_{12}N_1\rho(v) - B_{21}N_2\rho(v) = -\frac{dN_1}{dt} = 0 \tag{1.7a}$$

$$B_{12}N_1\rho(v) = A_{21}N_2 + B_{21}N_2\rho(v) \tag{1.7b}$$

or

$$\frac{N_2}{N_1} = \frac{B_{12}\rho(v)}{A_{21} + B_{21}\rho(v)} . \tag{1.8}$$

Einstein also used Boltzmann statistics for the ratio of populations in the two energy states at thermodynamic equilibrium:

$$\frac{N_2}{N_1} = \frac{g_2}{g_1}\exp\left[-\Delta E/kT\right] = \frac{g_2}{g_1}\exp\left[-hv/kT\right] , \tag{1.9}$$

where g_1 and g_2 are the degeneracy of states, *i.e.*, the number of independent ways in which the atom can have the same energy, and where k is the Boltzmann constant.

Now combining Equations 1.8 and 1.9, we have

$$\frac{g_2}{g_1}\exp\left(-\frac{hv}{kT}\right) = \frac{B_{12}\rho(v)}{A_{21} + B_{21}\rho(v)} . \tag{1.10}$$

For extreme temperature, $kT \gg h$ the left-hand side of Equation 1.10 approaches g_2/g_1 and the right-hand side of Equation 1.10 will be dominated by $\rho(v)$. Then, one can approximate Equation 1.10 as

$$\frac{B_{12}}{B_{21}} = \frac{g_2}{g_1} \text{ or } g_2B_{21} = g_1B_{12} . \tag{1.11}$$

Equation 1.11 provides an important relationship between absorption and stimulated emission coefficients.

Now, we shall solve Equation 1.10 for $\rho(v)$ using Equation 1.11:

$$A_{21}\frac{g_2}{g_1}e^{-hv/kT} + B_{21}\frac{g_2}{g_1}e^{-hv/kT}\rho(v) = \frac{g_2}{g_1}B_{21}\rho(v) \tag{1.12}$$

or

$$A_{21}e^{-hv/kT} = B_{21}\rho(v)\left[1 - e^{-hv/kT}\right]$$

or

$$\rho(v) = \frac{A_{21}}{B_{21}}e^{-hv/kT}\Big/\left(1 - e^{-hv/kT}\right) = \frac{A_{21}}{B_{21}}1\Big/\left(e^{hv/kT} - 1\right) \qquad (1.13)$$

Now, from black body radiation theory, by integrating Planck's law over the range of wavelengths, we have

$$\rho(v) = \frac{8\pi v^2}{c^3}\frac{hv}{e^{hv/kT} - 1}$$

$$\text{and} \quad A_{21} = B_{21}\frac{8\pi hv^3}{c^3}. \qquad (1.14)$$

Note that since $B_{12} = (g_2/g_1)\,B_{21}$ all three coefficients are interrelated. It is most important to realise that these coefficients are characteristic of the atom.

Radiation is not the only thing that can affect an excited atom. The atoms can undergo a *collision* with *another atom, an electron* or a lattice *vibration* (a phonon), which can also cause a transition to take place.

1.1.5 Lifetime Broadening

So far, in considering atomic transitions, we have only considered discrete energy levels, as shown in Figure 1.6a, where all energy is concentrated in one frequency. But in real life, energy is distributed over a range of frequencies owing to energy fluctuations such as Doppler effects.

Figure 1.6b greatly exaggerates the smearing, but shows the fact that (1) radiation can and does appear on either side of the line centre; (2) different states have different broadening and (3) the band of frequencies emitted by the transition, $2 \rightarrow 1$, will reflect the smearing of both upper and lower states.

One can interpret these bell-shaped curves as being the relative probability of an atom being found in a band, dE_2, around energy E_2 given that the atom is in state 2.

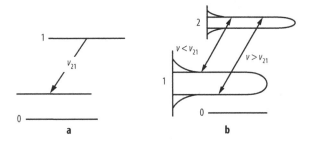

Figure 1.6 a Elementary energy level diagram, and **b** energy level diagram with lifetime broadening

The same applies for the atom in state 1. Since a transition can take place between dE_2 and dE_1, emitted radiation is also distributed over a range of frequencies, with the most intense part appearing between the bands with the greatest occupation probability at v_{21}.

To explain this, a function, $g(v)$, known as the "line shape" is defined such that $\{g(v)dv\}$ is the probability of emission of a photon with frequency between v and $v \pm dv$.

Obviously, if the atom emits a photon, it has to appear somewhere since energy is conserved.

$$\int_0^\infty g(v)dv = 1 \qquad (1.15)$$

Even though the limit is between 0 and ∞ the main contribution will be around v_{21}.

Note that the ground state is still sharp, since an atom in the ground state is, by definition, at the end of the line, its lifetime is infinite and its energy is well defined.

The idea of a line shape is most important, quite general and independent of the maze of mathematics surrounding its development. It can be summarised as follows.

The "line-shape function", $g(v)dv$, is the relative probability that:

1. a photon emitted by a "spontaneous" transition will appear between v and $v \pm dv$;
2. radiation in the frequency interval from v to $v \pm dv$ can be absorbed by the atoms in state 1; and
3. radiation in this interval will stimulate the atoms in state 2 to give up their internal energy.

The same line-shape function can be applied to all three processes (spontaneous emission, absorption and stimulated emission).

Many of the real-life line-shape functions are asymmetrical and are mathematically intractable.

1.1.6 Transition Rates for Monochromatic Waves

For the discussion of Einstein coefficients, we were concerned with the interaction of a continuous radiation spectrum, $\rho(v)$ with the "discrete" energy levels of a group of atoms. We assumed that the bandwidth of the radiation (*i.e.*, energy spread) was much larger than the bandwidth of emission or absorption by the atoms.

In the case of lasers, the situation is reversed. Here, one has a finite amount of radiant energy (per unit volume) in a bandwidth that is much smaller than the corresponding spread expressed by the line shape of the transition. Therefore, the arithmetic of Equations 1.4 and 1.5 has to be changed accordingly.

Thus, the rate of change of the population of state 2, due to a monochromatic wave at a frequency v with energy density ρ_v in joules per cubic metre is

$$\left.\frac{dN_2}{dt}\right|_{\substack{\text{absorption}\\ \text{stimulated}\\ \text{emission}}} = -N_2 B_{21} g(v)\rho_v + N_1 B_{12} g(v)\rho_v . \qquad (1.16)$$

Note: that $\rho(v)$ is a continuous distribution of frequencies and has units of joules per unit volume per unit frequency.

For an ideal monochromatic wave, all its energy is of one frequency, but the atom reacts only according to $g(v)$ [the dimension of $g(v)$ is (*frequency*)$^{-1}$] and thus the units of ρ_v are joules per cubic metre.

It is useful to convert energy density into intensity, I_v (watts per unit area), by recognising that electromagnetic energy travels at the velocity of light, c. Thus,

$$I_v = c\rho_v . \tag{1.17}$$

Then, Equation 1.16 can be modified using Equations 1.11, 1.14 and 1.17:
From Equation 1.11

$$B_{12} = \frac{g_2}{g_1} B_{21} .$$

From Equation 1.14, $B_{21} = \left(c^3/8\pi h^3\right) A_{21}$.
Substituting for B_{12} and B_{21} in Equation 1.16, we have

$$\frac{dN_2}{dt} = -N_2 \frac{c^3}{8\pi h v^3} A_{21} g(v)\rho_v + N_1 \frac{g_2}{g_1} B_{21} g(v)\rho(v)$$

Hence

$$\frac{dN_2}{dt} = -N_2 \frac{c^3}{8\pi h v^3} A_{21} g(v)\rho_v + N_1 \frac{g_2}{g_1} \frac{c^3}{8\pi h v^3} A_{21} g(v)\rho_v$$

and using Equation 1.17 to substitute for ρ_v

$$\frac{dN_2}{dt}_{\substack{\text{spontaneous} \\ \text{stimulated} \\ \text{absorbed}}} = -A_{21} \frac{c^3}{8\pi h v^3} g(v)\frac{I_v}{c}\left[N_2 - \frac{g_2}{g_1}N_1\right]$$

which is

$$\frac{dN_2}{dt}_{\substack{\text{spontaneous} \\ \text{stimulated} \\ \text{absorbed}}} = -A_{21} \frac{c^2}{8\pi v^2} g(v)\frac{I_v}{h v}\left[N_2 - \frac{g_2}{g_1}N_1\right] .$$

Now $c = v\lambda$ where λ = wavelength.
Hence

$$\frac{dN_2}{dt}_{\substack{\text{spontaneous} \\ \text{stimulated} \\ \text{absorbed}}} = -A_{21} \frac{\lambda^2}{8\pi} g(v)\frac{I_v}{h v}\left[N_2 - \frac{g_2}{g_1}N_1\right] . \tag{1.18}$$

Showing the need for a population inversion.

1.1.7 Amplification by an Atomic System

Verdeyen [2] examined the concept of amplification of electromagnetic energy by its interaction with atoms. He used Einstein's view of interaction but did *not* assume thermodynamic equilibrium.

Let us imagine an experiment in which a slab of these atoms, ΔZ long, is being irradiated by a polarised electromagnetic wave of intensity I_v (W/m^2), which after amplification (or attenuation) is received by a detector (Figure 1.7).

The detector system does not distinguish between the photons generated by different processes. Photons from stimulated emission and spontaneous emission will have the same response. Thus, spontaneous emission will contribute noise to the system.

Noise can be minimised using:

1. A *filter* with a passband Δv around frequency, v of the source.
2. A *polariser* to reject half of the spontaneous power orthogonal to the source.
3. A *field of view* with an acceptance cone of $d\Omega$ to match the incoming beam. The beam energy is distributed uniformly over 4π steradians and the detector will accept a fraction, $d\Omega/4\pi$.

Therefore,

$$\text{output} = \text{input intensity} + \text{stimulated radiation}$$
$$- \text{amount of radiation absorbed}$$
$$+ \text{spontaneous radiation}.$$

Then, the intensity balance is

$$
\begin{array}{ccccccc}
 & 1 & 2 & 3 & 4 & 5 & 6 \\
\Delta I_v = & hv \times B_{21} & \frac{I_v}{c} & \times g(v) & \times 1 & \times 1 & \times N_2 \Delta Z \\
 & -hv \times B_{12} & \frac{I_v}{c} & \times g(v) & \times 1 & \times 1 & \times N_1 \Delta Z \\
 & +hv \times A_{21} & \Delta v & \times g(v) & \times 1/2 & \times \frac{d\Omega}{4\pi} & \times N_2 \Delta Z,
\end{array}
\tag{1.19}
$$

where column 1 represents the package of energy contributed by each transition, column 2 represents the rate per atom 1×2 (equivalent to the power contributed by each atom), column 3 represents the power contribution per atom scaled by the line shape, column 4 represents the probability the photon involved has the proper polarisation, column 5 represents the probability that the photon is within the solid angle specified and column 6 represents the number of atoms involved in the interactions.

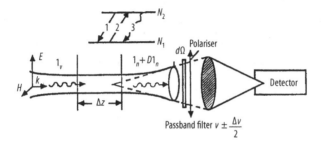

Figure 1.7 Measurement of the gain of an atomic system

Now, manipulating Equation 1.19, we obtain

$$\frac{\Delta I_v}{\Delta Z} \to \frac{dI_\lambda}{dz} = \left[\frac{hv}{c} (B_{21} N_2 - B_{12} N_1) \, g(v) \right] I_v + \frac{1}{2} \left[h(v) A_{21} N_2 g(v) \Delta v \frac{d\Omega}{4\pi} \right] . \quad (1.20)$$

The last term is a noise term since it will be present in the detector even without input intensity because it represents the spontaneous emissions; thus, we can neglect it. Now using Equations 1.11 and 1.14, we obtain

$$\frac{dI_v}{dz} = \left[\frac{hv}{c} \left(A_{21} \frac{c^3}{8\pi h v^3} N_2 - A_{21} \frac{g_2}{g_1} \frac{c^3}{8\pi h v^3} N_1 \right) \right] I_v g(v)$$

or

$$\frac{dI_v}{dz} = \left[\frac{hv}{c} A_{21} \frac{c^3}{8\pi h v^3} \left(N_2 - \frac{g_2}{g_1} N_1 \right) \right] I_v g(v)$$

using $c = v\lambda$ and Equation 1.18

$$\frac{dI_v}{dz} = \left[A_{21} \frac{\lambda^2}{8\pi} \left(N_2 - \frac{g_2}{g_1} N_1 \right) \right] I_v g(v) = \frac{dN_2}{dt} I_v \quad (1.21)$$

and

$$\frac{dI_v}{dz} = \Delta v_0(v) I_v ,$$

where $v_o(v)$ is defined by this equation and is the gain coefficient (m^{-1}), with subscript o, indicating that the incoming intensity is sufficiently small as to cause negligible perturbation on populations N_2 and N_1. Equation 1.21 is a very important central equation for laser theory.

The values of the gain coefficient varies with the material; for example, for ruby a value of 0.23 cm^{-1} was found, whereas for diode-pumped Nd:YAG a value of 10 m^{-1} is reported in the literature [9, 10].

Factors in the gain coefficient can also be expressed in terms of a stimulated emission or absorption cross-section:

$$\sigma_{SE} = A_{21} \frac{\lambda^2}{8\pi} g(v) , \qquad \sigma_{AB} = A_{21} \frac{\lambda^2}{8\pi} g(v) \frac{g_2}{g_1} . \quad (1.22)$$

Therefore, the gain coefficients can be written as the product of the stimulated emission cross-section and the population inversion, ΔN:

$$v_o(v) = \Delta N \sigma_{SE}(v) , \quad (1.23)$$

where

$$\Delta N = N_2 - \frac{g_2}{g_1} N_1 . \quad (1.24)$$

Having found the differential form of the gain, we can integrate Equation 1.21 to obtain the power gain

$$I_v(z) = I_v(0) \exp\left[v_o(v)Z\right] \qquad (1.25)$$
$$G(v)I_v(0) \, ,$$

where

$$G = \exp\left[v_o(v)d\right] \, ,$$

the power gain of an amplifier of length d.

For a rod-shaped Nd:YAG laser, d will be the length of the rod, whereas for the disc laser, it is the thickness of the disc.

1.1.8 The Laser: Oscillation and Amplification

A laser is essentially an amplifier with positive feedback, such that the loop gain provided by stimulated emission exceeds unity. The simplest form of a laser consisted of a medium with a population inversion, which is placed in an optical resonator as shown in Figure 1.8. It is obvious from the laser gain, Equation 1.21, that it is necessary to have $N_2 > (g_2/g_1)N_1$ to have gain.

Now, to construct a simple laser, we have to construct an amplifier, where the threshold for oscillation is determined by the requirement that the round trip gain exceeds 1.

Ignoring losses such as scattering, diffraction, *etc.*, and only considering mirror losses, we can define the threshold value for the gain coefficient $v_o(v)$ for oscillation as

$$R_1 R_2 e^{(2v_o(v)L)} \geq 1 \, , \qquad (1.26)$$

where L is the length of the medium.

$$G = \exp\left[v_o(v)L\right] \, ,$$

the total power gain of length L or

$$v_0(v) \geq \frac{1}{2L} \ln\left(\frac{1}{R_1 R_2}\right) = \alpha \qquad (1.27)$$

where α is the gain per unit length.

In other words, the gain per unit length $v_o(v)$ must exceed the loss when that loss is prorated on a per unit length basis. Equations 1.25–1.27 are very useful for quick estimation of the gain of any laser system and, thus, determination of the feasibility of any laser.

Example 1.1. If we consider a 30-cm Nd:YAG laser cavity which includes a mirror with 98 % reflectivity and an output window with 65 % reflectivity, what should be the minimum gain coefficient for the laser to be viable?

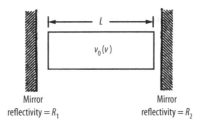

Figure 1.8 Simple laser

The gain coefficient is $v_o(v)$; its minimum value is given by Equation 1.26

$$e^{2v_0(v)L} \geq \frac{1}{R_1 R_2}$$

or

$$2v_0(v)L \geq \ln \frac{1}{R_1 R_2}$$

or

$$v_0(v) \geq \frac{1}{2 \times 30} \ln \left(\frac{1}{0.98 \times 0.65} \right)$$
$$\geq 0.007516 \, \text{cm}^{-1}$$
$$\geq 0.7516 \, \text{m}^{-1}$$

The gain coefficient for Nd:YAG [10] is $10 \, \text{m}^{-1}$; therefore, a Nd:YAG laser will be viable under this design.

1.2 Laser Construction Concepts

1.2.1 Overall Design

The basic laser consists of two mirrors which are placed parallel to each other to form an optical oscillator, that is, a chamber in which light travelling down the optic axis between the mirrors would oscillate back and forth between the mirrors forever if it is not prevented by some mechanism such as absorption. Between the mirrors is an active medium which is capable of amplifying the light oscillations by the mechanism of stimulated emission as just described. There is also some system for pumping the active medium so that it has the energy to become active. This is usually a DC or RF power supply, for gas lasers such as CO_2, excimer and He–Ne lasers, or a focused pulse of light for the Nd:YAG and solid-state lasers or an electric current for a semiconductor or free-electron lasers or a chemical reaction, as with the iodine laser. The optical arrangement is shown in Figures 1.9. One of the two mirrors is partially transparent to allow some

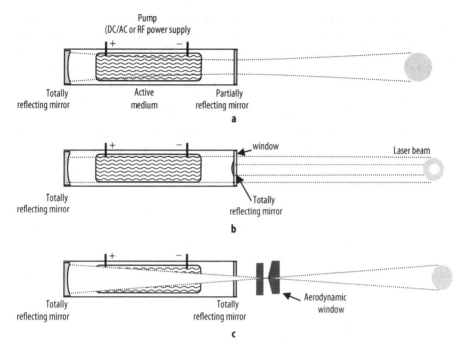

Figure 1.9 Basic construction of a laser cavity: **a** stable, **b** unstable, and **c** stable cavity with an aerodynamic window

of the oscillating power to emerge as the operating beam. The other mirror is totally reflecting to the best that can be achieved (99.999 % or some such figure). This mirror is also usually curved to reduce the diffraction losses of the oscillating power and to make it possible to align both mirrors without undue difficulty – this would be the case if both mirrors were flat. The design of the laser cavity hinges on the shape of these mirrors, including any others in a folded system, the dimensions of the cavity and the cooling of the active medium so that it can quickly return to the ground state.

1.2.1.1 Cavity Mirror Design

Kogelnik and Li [11] wrote one of the fundamental papers on cavity design. They showed by geometric arguments that the mirror curvatures at either end of the cavity could only fall within certain values or the cavity would become "unstable" by losing the power around the edge of the output mirror. Cavities can be identified as "stable" or "unstable" depending upon whether they make the oscillating beam converge into the cavity or spread out from the cavity. Most lasers up to 2 kW use stable cavity designs (Figure 1.9a) because it is safe to transmit that level of power through the output mirror without risk of breakage. The output mirror is made partially transparent to the laser radiation; for CO_2 lasers operating at 10.6-μm infrared radiation the output mirror is made of zinc selenide (ZnSe), gallium arsenide (GaAs) or cadmium telluride

(CdTe); for Nd:YAG lasers with 1.06-μm radiation the output mirror is made of BK7 fused-silica glass or the like. In all cases the output mirror is carefully coated to give the required level of reflection into the cavity (typically a CO_2 laser output window would have 35 % reflectivity for feedback into the cavity). Breakage of the output window is serious from the implosion aspect for CO_2 lasers, which have a low-pressure cavity at around 54 Torr, explosion for an excimer laser, which has a high-pressure cavity at around 4 atm and for all the problems associated with the costs of replacement and shut down of production. Thus, for higher-powered lasers it is not uncommon to find that the cavity is designed as an unstable cavity (Figure 1.9b) taking the power from around the edge of the output mirror, which is in this case a totally reflecting metal optic. The larger ring-shaped beam thus passed means reduced power density on the window sealing the cavity. An alternative, for low-pressure systems such as the CO_2 laser, is to have an aerodynamic window, in which a venturi arrangement ensures that the vacuum is held while the beam passes through the high-velocity, low-pressure gas to the atmosphere (Figure 1.9c). The shape of a stable beam is determined by the shape of the output aperture, whereas the shape of the unstable beam is the same as the edge of the output reflecting mirror. It is usually round or ring-shaped but may be square as with some excimer lasers and slab lasers.

1.2.1.2 Cavity Dimensions – the Fresnel Number

When the laser is working, the radiation within the cavity oscillates back and forth as a resonant standing wave. The ratio of the length of the cavity to the width of the output aperture determines the number of off-axis directions or modes (see Section 2.7.3) which fit an exact number of half wavelengths between the two mirrors. This number is described in the Fresnel number, $N = a^2/\lambda L$, a dimensionless group, where a is the radius of the output aperture, L is the length of the cavity and λ is the wavelength of the laser radiation. The Fresnel number, N, equals the number of fringes which would be seen at the output aperture if the back end mirror was uniformly illuminated.

To give some idea of where this term comes from, consider a cylindrical cavity of length L and radius a, as illustrated in Figure 1.10. When the path differences between the direct and off-axis beams differ by an integral number of wavelengths, there will be a bright ring or fringe, due to constructive interference.

Thus, the number of fringes, or off-axis modes, which can oscillate in the cavity, will be, by Pythagoras' theorem,

$$a^2 + L^2 = (L + n\lambda)^2 .$$

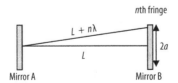

Figure 1.10 The meaning of the Fresnel number

Table 1.1 Dimensions of typical laser types and their Fresnel number. Note that Fresnel numbers less than 1 indicate some waveguiding is likely

Type of laser	Power (W)	λ (µm)	Cavity radius (mm)	Cavity length (m)	Fresnel number	Mode
CO_2 slow flow (MF600)	600	10.6	7.7	14.4	0.4	TEM00/TEM01*
CO_2 fast axial flow (PRC3000)	3,000	10.6	11	5.2	2.2	TEM00/low
Nd:YAG (JK401)	400	1.06	2.4	0.55	9.8	Multimode
Fibre laser Yb:YAG	500	1.03	0.1	10	0.00097	TEM00

Therefore, neglecting the $n^2\lambda^2$ term,

$$a^2 = 2Ln\lambda,$$

and hence

$$n = a^2/(2L\lambda) = (N)/2. \tag{1.28}$$

Thus, a low Fresnel number gives a low-order mode. Off-axis oscillations are lost by diffraction and hence will not occur in an amplifying cavity. Table 1.1 lists some of the Fresnel numbers associated with some current industrial lasers. A higher Fresnel number cavity may be controlled to give lower-order beam modes by controlling the mirror design – the flatter the mirror, the lower the mode order; an example is the flexible mirror design of the Laser Ecosse (now Ferranti Photonics) AF5 laser. This laser can be engineered to give any mode from TEM00 to a multimode beam by altering the curvature of the fully reflecting mirror by a centrally applied stress.

The off-axis modes describe the transverse electromagnetic mode (TEM) structure of the power distribution across the beam, which is essentially a standing wave across the beam formed from the interference between these various longitudinal standing waves. This mode structure is discussed in Section 2.7.3. The mode coming from a given laser can be modified by inserting apertures into the cavity, to alter the cavity Fresnel number. This will, of course, reduce the power output but may give a more easily focused beam.

1.2.1.3 Cooling Considerations

A significant part in the design of lasers is the cooling of the active medium. In the case of gas lasers this is done by conduction cooling as in slow flow lasers, or convective-cooling as in fast axial flow lasers or transverse flow lasers. For solid-state lasers it is by conduction from different-shaped active media supports, such as rod, disc or fibre shapes. For semiconductor lasers it is done via conduction from a support, sometimes, as in space applications, by using the Peltier effect – the opposite to a thermocouple, when a current is passed one junction heats up and the other cools (first observed by Jean Peltier in 1834).

1.2.1.3.1 Cylindrical Conduction Cooling

This group includes slow flow gas lasers, sealed lasers, rod solid-state lasers and fibre lasers.

The power per unit length depends upon the cooling efficiency as well as the physics described in section 1.1.7. There is a temperature above which lasing action will not occur because at higher temperatures the lower energy level cannot be emptied fast enough to the ground state. Thus, for maximum power we are interested in the temperature at the centre of the cavity in a slow flow system or a solid-state rod, since this is the region of highest temperature that must not exceed the critical temperature for lasing action to occur [12].

If the waste heat from the pumping action is uniformly generated within the cavity space at, say, Q W m^{-3}, then by considering a heat balance on a cylindrical element in that space, we have (see Figure 1.11.)

Heat generated = $Q\pi r^2$ per unit length,

Heat removed = $-2\pi rk(\mathrm{d}T/\mathrm{d}r)$ per unit length (from Fourier's first law).

For continuous output – continuous wave (CW) – operation there will be equilibrium and these two terms will be equal. Hence,

$$\frac{\mathrm{d}T}{\mathrm{d}r} = \frac{Qr}{2k},$$

with the boundary condition $T = T_c$ at $r = a$, where T_c is the coolant temperature.

If the thermal conductivity, k, is assumed constant, this differential equation can be solved by separating the variables:

$$\int_T^{T_c} \mathrm{d}T = \int_r^a -\frac{Qr}{2k}\mathrm{d}r ,$$

$$T - T_c = \frac{Qa^2}{4k}\left[1 - \frac{r^2}{a^2}\right] .$$

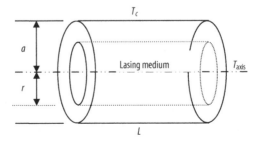

Figure 1.11 Conduction model for slow flow or rod geometries

The maximum temperature $T_{max} = T_{axis}$ when $r = 0$, where

$$T_{axis} = \frac{Qa^2}{4k} + T_c \,.$$

The maximum power per unit length is when $T_{axis} = T_{lim}$, the highest temperature at which lasing can be achieved. Therefore,

$$Q = \frac{4k}{a^2} \left[T_{lim} - T_c \right] \,.$$

Now

$$\eta Q = \frac{P}{\pi a^2 L} \,. \tag{1.29}$$

(If E is the overall energy supplied then ηE is P, the laser beam power, by definition; but $QV + P = E$, where V is volume; hence $P \approx \eta(QV)$, assuming $1 - \eta \approx 1$.)

Therefore, maximum laser power

$$P = 4\pi\eta kL \left(T_{lim} - T_c \right) \,, \tag{1.30}$$

i.e., only the length affects the power in the slow flow or solid-state rod design of lasers.

Putting reasonable figures into this equation, $\eta = 0.12$ (12 % efficiency conversion to optical energy, the rest is the waste heat to be removed), $k = 0.14\,\mathrm{W\,m\,K^{-1}}$ (He at $0\,^\circ\mathrm{C}$), $T_{lim} = 250\,^\circ\mathrm{C}$ (from statistical mechanics), $T_c = 10\,^\circ\mathrm{C}$ (refrigerated cooling water), produces power

$$P = 4\pi(0.12)(0.14)(L)(240) = 50\,\mathrm{W\,m^{-1}} \,.$$

The gain per metre is relatively small and so these lasers are either not very powerful, up to 2 kW, or very long, as with the first lasers at Essex University in the 1960s, which ran to some 70 m in a straight line. In any case, to get power, the cavity length will be long but may be subtly folded. These long cavities mean a low Fresnel number and hence a low-order mode, which is the most suitable form for focusing – see Chapter 2. The relatively smooth plasma formed in the slow flowing gases also ensures a good mode by giving uniform gain across and along the cavity. These lasers are amongst the best for cutting owing to this superior mode.

1.2.1.3.2 One-dimensional Area Cooling:

This group includes *waveguide or diffusion-cooled lasers and disc lasers.*

Waveguide lasers are made with a very thin slit between the electrodes, which are pumped with RF power, to avoid sparking. The generated beam is reflected from the electrode faces as it oscillates within the cavity. The beam is thus waveguided within the narrow passage.

A disc laser consists of a small coin-shaped active medium cooled from one side (see Figure 1.20).

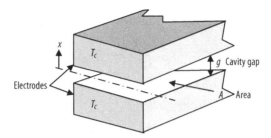

Figure 1.12 A waveguide cavity

In these cases (see Figure 1.12) the heat generated in the space of area A and thickness x is QAx. The heat lost by conduction from the surface of this space is $-2kA(\mathrm{d}T/\mathrm{d}x)$. Solving as before with $\eta Q = P/Ag$, we get

$$P = \eta \frac{A}{g} 4k \left(T_{\mathrm{lim}} - T_{\mathrm{c}} \right) . \qquad (1.31)$$

This arrangement is no longer dependent only on length for generating power. For a 1-m-long cavity with a 2-mm electrode separation and 10-mm width, this gives 80 W m^{-1} per centimetre-wide channel.

Lasers of this style can be compact and powerful but generate elliptical beams, which may need further optical manipulation. They may also be designed to give arrays of beams [13].

The calculation for cooling disc lasers is similar and shows that large-area and thin discs would give the most power – but note from Equation 1.29 that there is more power available from a thicker active volume, but cooling consideration do not allow that.

1.2.1.3.3 Convective Cooling

This group includes *fast axial flow lasers and transverse flow lasers*.

Fast axial flow lasers achieve their cooling by convection of the gas through the discharge zone. The general arrangement is shown in Figure 1.13. Typically the gases flow at 300–500 m s^{-1} through the discharge zone. Control of the gas mixture and avoidance of any leakage allows smooth plasmas to be produced. The axial nature of the flow, discharge and optical oscillation favours an axially symmetric power distribution in the beam. The cavity length is usually of a fairly low Fresnel number, and so the beam mode is of a low order and thus more easily focused to a small point. The gain is typically 500 W m^{-1}, some 10 times more than for a slow flow system, and hence compact high-powered units have been made this way. Once again the output power is a function of the cooling efficiency. In this case the fluid enters the cavity cool and leaves hot. The limit to lasing action is when the output temperature exceeds T_{lim}. The heat acquired comes from the waste heat, Q, which heats up the flowing gas during its residence in the discharge zone. The time in the heating zone is given by the distance travelled divided by the speed. If conduction is neglected, a heat balance on the gas

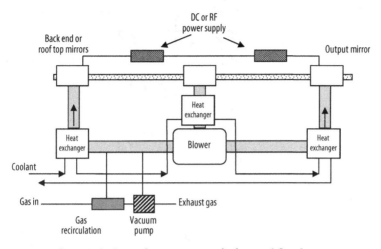

Figure 1.13 General construction of a fast axial flow laser

gives

$$\rho C \left(T - T_{\mathrm{c}} \right) = Q \left(\frac{x}{V} \right).$$

The maximum temperature at the end of the linear temperature rise is

$$T_{\max} = \frac{QL}{V\rho C} + T_{\mathrm{c}}.$$

Once more $\eta Q = P/AL$ and hence the maximum laser power, P, is given by

$$P = \eta A V \rho C \left(T_{\mathrm{lim}} - T_{\mathrm{c}} \right). \tag{1.32}$$

The power is proportional to the velocity and the cross-sectional area. The problem with the area route to designing this form of high-powered laser is the Fresnel number, $a^2/L\lambda$, which is approximately $A/L\lambda$. Many short fat laser cavities have been tried which produce the power, but they also produce a beam which cannot be focused very finely because of the high mode number, or M^2 value (see Section 2.8.1).

Putting typical values into the equation for the power gives 650 W m^{-1} using $\eta = 0.12$ as before, $\rho = 0.05$ kg m^{-3} (low-pressure He), $C = 5$ kJ kg^{-1} (approximate value), $T_{\mathrm{lim}} = 250\,^{\circ}$C (statistical mechanics), $T_{\mathrm{c}} = 10\,^{\circ}$C (refrigerated water), $V = 300$ m s^{-1} and $A = 3 \times 10^{-4}$ m^2 (20-mm-diameter tube).

Transverse flow lasers they are once more convectively cooled, but this time the flow is transverse to the discharge. Cooling is thus more effective and very compact high-powered lasers have been built this way. The main disadvantage with these lasers lies in their lack of flow symmetry. For example, the gas enters the cavity cold and becomes heated as it traverses the lasing space. Thus the gain, which is a function of temperature, falls across the cavity and an asymmetric power distribution across the beam results.

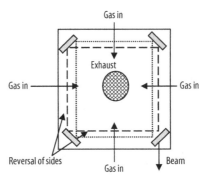

Figure 1.14 A square cavity for more uniform amplification across the beam

The MLI laser has a square cavity which attempts to smooth this effect. The design is shown in Figure 1.14.

1.2.1.3.4 Other Designs

Various flow patterns on these general themes have been and are being explored. One was the Photon Sources Turbo laser using a spiral flow which rotated in and out of the discharge zone.

1.3 Types of Laser

The wavelength, λ, of a laser is decided by the energy difference as the excited species is stimulated to a lower energy level [$E = hc/\lambda$, where h is Planck's constant (6.626×10^{-34} J s) and c is the velocity of light (3×10^8 m s^{-1})]. In general, the quantum states refer to molecular vibration levels for long-wavelength lasers, to electron orbit levels for visible laser radiation and to ionisation effects with ultraviolet lasers.

For material processing, CO_2, Nd:YAG and fibre lasers are the most popular systems. Excimer and diode lasers are also fast appearing on the scene. Performance characteristics of the commercially available lasers for material processing are provided in Tables 1.2–1.4. Figure 1.15 shows that a single laser can perform several processes if the power density and interaction time are manipulated. This is one of the main reasons why lasers are such popular candidates for flexible manufacturing systems.

1.3.1 Gas Lasers

1.3.1.1 Carbon Dioxide Lasers

Consider the CO_2 molecule, it can take on various energy states depending upon some form of vibration and/or rotation. These states are quantised, that is, they can only exist at particular energy levels or not at all (consider by analogy a length of string held

Table 1.2 Performance characteristics of gas metalworking lasers [57]

Property		Types of CO_2 lasers				
		Slow flow or sealed	Axial flow	Transverse flow	Transverse excited atmospheric pressure	Waveguide
Wavelength (μm)		10.6	10.6	10.6	10.6	10.6
Standard beam operating mode	CW	•	•	•		•
	Pulsed				•	•
	Q-switched					
Average output power (W)		3–100	50–5,000	2,000–25,000		0.1–40
Output stability (percentage variation)			0.25–5	0.25–5	0.25–5	0.25–5
Pulse energy (J/pulse)					0.03–75	
Repetition rate (pulses/s)					1–300	
Beam diameter (mm)		3–4	5–70	5–70	5–100	1–2
Beam divergence (mrad)		1–2	1–3	1–3	0.5–10	10
Wall plug efficiency (%)		5–15	5–15	5–15	1–10	5
Most suitable applications	Cutting	•	•	•		
	Drilling				•	
	Part marking				•	•
	Welding		•	•		
	Cladding		•	•		
	Alloying		•	•		
	Microsoldering	•				•
	Heat treatment		•	•		
	Cleaning				•	

CW continuous wave

at each end, it will only resonantly vibrate at fixed frequencies as with a violin, *etc.*). The basic energy network possible with CO_2 is shown in Figure 1.16. The gas mixture in a CO_2 laser is subject to an electric discharge causing the low-pressure gas (usually around 35–50 Torr) to form a plasma. In the plasma the molecules take up various excited states as expected from the Boltzmann distribution [$n_i = C \exp(-E/kT)$, where n_i is the number of molecules in energy state i, E is the energy of state i, k is the Boltzmann constant (1.3805×10^{-23} J K^{-1}), T is the absolute temperature and C is a constant]. Some will be in the upper state (00^01), which represents an asymmetric oscillation mode. (The notation is 001 for asymmetric oscillation and the superscript 0 indicates no spin quanta; 100 represents the symmetric oscillation and 010 the bending. They can all have several quanta in each state, so two bending quanta would create the 020 state and 00^11 would be an asymmetric oscillation with a rotational spin). By chance this (00^01) molecule may lose its energy by collision with the walls of the cavity or by spontaneous emission. Through spontaneous emission the state falls to the symmetric oscillation mode (10^00) and a photon of light of wavelength 10.6 μm is emitted travelling in any direction dictated by chance. One of these photons, again by chance, will be travelling down the optic axis of the cavity and will start oscillating between the mirrors. During this time it can be absorbed by a molecule in the (10^00) state, it can be diffracted out of the system or it can strike a molecule which is already excited, in the

Table 1.3 Performance characteristics of solid-state metalworking lasers [57]

Property		Types of solid-state lasers					
		Ruby	Nd:glass	Nd:YAG (CW)	Nd:YAG (pulsed)	Yb:fibre	Er:fibre
Wavelength (μm)		0.694	1.06	1.06	1.06	1.03–1.1	1.5–1.6
Standard beam operating mode	CW			•			
	Pulsed	•	•		•	•	•
	Q-switched			•			
Average output power (W)		10–20		0.04–800	0.04–400	0.01–20,000	0.01–500
Output stability (percentage variation)		1–5	1–5	1–5	1–5	1–2	1–2
Pulse energy (J/pulse)		0.3–100	0.15–100		0.01–100		
Repetition rate (pulses/s)		0.01–4	0.1–1		0.05–300	10^{10}	
Beam diameter (mm)							
Beam divergence (mrad)		0.2–10	3–10	2–18	0.3–10		
Wall plug efficiency (%)		0.1–0.5	1–5	0.1–2	0.1–2	12.5	
Price range ($)		15,000–70,000	8,000–110,000	3,000–90,000	3,500–110,000		
Most suitable applications	Cutting			•			
	Drilling	•	•	•	•		
	Part marking				•		
	Welding	•		•	•	•	•
	Cladding						
	Alloying			•	•		
	Microsoldering					•	•
	Heat treating			•			
	Cleaning				•		

Table 1.4 Performance characteristics of excimer lasers

Property	Types of excimer lasers			
	ArF	KrCl	KrF	XeCl
Wavelength	193	222	248	308
Average output power (W)	0.3–100	2–4	2–100	0.02–150
Output stability (percentage variation)	5		5	5
Pulse energy (MJ/pulse)	0.2–500	0.1–250	20–15,000	1.2–5,000
Repetition rate (pulses/s)	1–250	10–500	1/60–1,000	1–1,000
Pulse width (ns)	1.2–65	1.2–6	1.5–1,200	1.2–80
Beam diameter (width × height, mm^2)	1 × 30 to 30 × 10	1 × 3 to 8 × 30	5 × 20 to 100 × 100	1 × 3 to 45 × 45

($00^{0}1$) state. At this moment it will stimulate that excited molecule to release its energy and fall to a lower energy state, thus emitting another photon of identical wavelength, travelling in exactly the same direction and with the same phase. One can imagine the incident photon shaking the energy free by some form of resonance. The two photons travelling in the same direction with the same phase now sweep back and forth within the cavity generating more photons from other excited molecules. The excited state be-

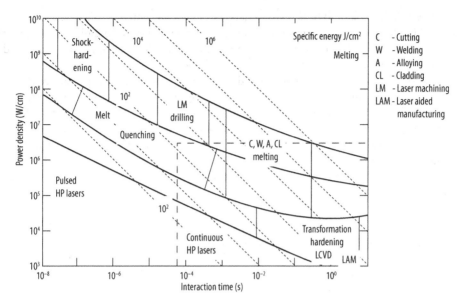

Figure 1.15 Operational regimes for various processing techniques. *HP* high-powered, *LCVD* laser chemical vapour deposition

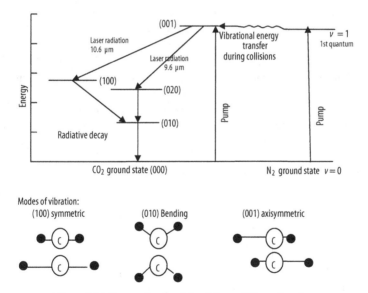

Figure 1.16 Energy levels of the CO_2 and N_2 molecules

comes depleted and so by the Boltzmann distribution more and more of the energy is passed into that state, giving a satisfactory conversion of electrical energy into the upper state. It is necessary for another condition to hold, and that is that the excited state which lases should be slow to undergo spontaneous emission and the lower state should be faster in losing its energy. This allows an inversion of the population of excited species to exist and thus makes a medium which is more available for the stimulated emission process (amplification) than for absorption.

In fact if that were the whole story, the CO_2 laser would not be special nor one of the most powerful lasers available today. The CO_2 laser is helped by a quirk of nature whereby nitrogen, which can only oscillate in one way (being made of two atoms), has an energy gap between the different quanta of oscillation which is within a few hertz of that required to take cold CO_2 and put it into the asymmetric oscillation mode – the upper laser level (00^01). Thus, by collision with excited nitrogen, cold CO_2 can be made excited. One of Kumar Patel's first experiments in 1964 used this phenomenon. He excited the N_2 in a discharge tube and passed it into the CO_2 gas, which made it lase even though the CO_2 was not in the plasma area. The only way the nitrogen can lose its energy is by collision with the tube walls or with a molecule which will absorb that energy or spontaneous emission. Its lifetime is long and the efficiency of the CO_2 laser is thus high (15–20 %). Since only cold CO_2 will undergo this reaction with nitrogen, the efficiency is a function of the gas temperature, since the lower energy levels must clear to ground quickly.

Thus, the design of a CO_2 laser, in common with that of all lasers, is built around the requirement of cooling; in this case to have cool CO_2 gas. Firstly, the gas mixture in the laser is around 78 % He for good conduction and stabilisation of the plasma, 13 % N_2 for this coupling effect and 10 % CO_2 to do the work. The efficiency of the laser is not a strong function of the gas composition except for certain impurities. Secondly, the gas is cooled by conduction through the walls for slow flow lasers or by convection in the fast axial flow and transverse flow lasers.

Slow flow lasers are cooled through the walls of the cavity. The general arrangement is shown in Figure 1.17. Typical operating figures are 20 l min^{-1} gas flow, 7 l min^{-1} coolant flow, 20 °C temperature and a gain of around 30–50 W m^{-1}.

Fast axial flow lasers are cooled by convection of fast-circulating low-pressure gas. The general arrangement is shown in Figure 1.13. Pumping can be by DC, AC or RF discharge. Two 5-kW lasers, the AF5 from Ferranti Photonics having a zigzag of tubes and the TRUMPF TLF5000 with a square cavity, are shown in Figure 1.18. The Wegmann Baasel Triagon laser (not shown) has a triangular cavity.

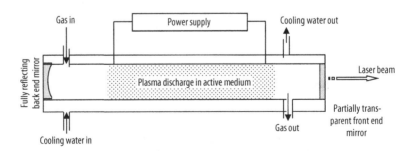

Figure 1.17   Basic construction of a slow flow laser

Figure 1.18 Examples of industrial fast axial flow lasers: **a** Ferranti Photonics AF5, and **b** TRUMPF TFL5000 Turbo

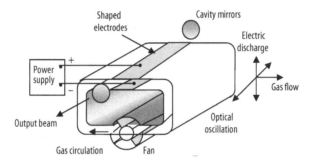

Figure 1.19 General construction of a transverse flow laser

Transverse flow lasers are also convectively cooled by fairly fast flowing low-pressure gas. The general arrangement is shown in Figure 1.19. The original AVCO 15-kW laser was of this type in 1971. The current lasers include the UTRC 25-kW laser (Figure 1.20), and the MLI laser (Figure 1.14).

The main radiation from the CO_2 laser is at 10.6-μm wavelength from the transitions of (001) → (100) (asymmetric to symmetric), or to a lesser extent at 9.6 μm from (001) → (020) (asymmetric to double bending). But owing to the presence of additional energy from rotation states (symbolised by the superscript 1), the emission can vary between 9 and 11 μm; which one is made to lase depends on whether the cavity has a selective mechanism within it, such as a grating (see Chapter 2).

The quantum efficiency of CO_2 lasers is 45 %, that is, the ratio of energy for (001–100)/001 (Figure 1.16). So far, the operating efficiency achieved is around 15–20 % for electric discharge to optical power but only around 12 % for wall plug efficiency (optical energy out divided by the total electrical energy into the system). Typical values for the quantum and wall plug efficiencies for the main types of industrial laser are given in Table 1.5.

Figure 1.20 United Technology UTRC 25-kW transverse flow laser

Table 1.5 Efficiency of main types of industrial lasers

Type	Wavelength (μm)	Quantum efficiency (%)	Wall plug efficiency (%)
CO_2	10.6	45	12
CO	5.4	100	19
Nd:YAG	1.06	40	4
Nd:glass	1.06	40	2
Diode-pumped YAG	1.06	40	8–12
Diode GaAs	0.75–0.87	≈ 80	50
Diode GaP	0.54	≈ 80	50
Excimer KrF	0.248	≈ 80	0.5–2

1.3.1.2 Carbon Monoxide Lasers

The CO laser is constructed in a similar way to the CO_2 laser – as are all gas lasers. The energy diagram is shown in Figure 1.21. It has the advantage of a quantum efficiency of near 100 % and thus promises to have a wall plug efficiency twice that of the CO_2 laser, although this is seriously reduced by the cooling requirements. For high-powered lasers, an improvement in efficiency could be significant since a 100-kW CO_2 laser would require a power supply of at least 0.83 MW. This is getting near to being a small power station; half that value would be more practical. However, the CO laser currently operates best at very low temperatures at around 150 K (liquid nitrogen temperatures) and requires extensive power for refrigeration, which may affect this potential efficiency advantage. Higher-temperature operation has been achieved by adding xenon to the gas mixture. Currently, operating efficiencies of around 19 % have been reported, but these are reduced to nearer 8 % when the power for cooling is considered.

These lasers emit at 5.4-μm wavelength, which is an interesting absorption area for water and therefore of medical interest. Small sealed CO lasers are sold for that area of application. High-powered CO lasers for material processing are not currently commercially available but the designs being considered are similar to those for a fast axial flow system with added cooling from liquid nitrogen or special refrigeration. Some

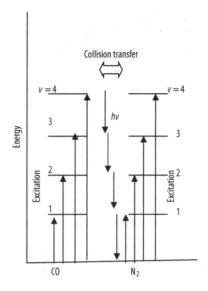

Figure 1.21 Vibrational energy levels for the CO and N_2 molecules

designs include substantial cooling by pressure expansion (Joule–Thompson cooling) but these tend to be noisy. The CO laser output power is very sensitive to temperature. Either way, there is an added expense compared with the CO_2 laser.

1.3.1.3 Excimer Lasers

The excimer laser has the energy diagram shown in Figure 1.22. The name derives from the excited dimer molecules (strictly excited complex molecules), which are the lasing species. There are several gas mixtures used in an excimer laser, usually noble gas halides; they are shown in Table 1.6. These lasers are slightly different in that the gain is so strong that they do not need an oscillator, although their performance is improved

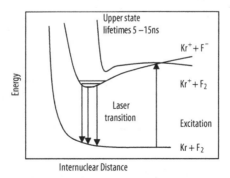

Figure 1.22 Energy levels for excimer laser emissions

Table 1.6 Range of wavelengths for different gas mixtures in an excimer laser

Gas mixture	Wavelength (nm)
F_2	158
ArF	193
KrCl	222
KrF	248
XeCl	308
XeF	354

with one. An electric discharge is generated, for example, in a gas mixture of Kr, F_2, Ne and He at around 4 atm. An excited dimmer Kr^+F^- is formed, with a lifetime of 5–15 ns, which undergoes the stimulated emission process. It generates ultraviolet photons in a brief pulse for each discharge of the condenser bank into the gas mixture. The photons are spread over a waveband of around ±0.4 nm, which is not as sharp as for other lasers. The pulses are usually very short, around 20 ns (a piece of light around 6 m long!), but very powerful, typically around 35 MW (the energy per pulse is thus 0.7 J). Owing to the lack of a resonant oscillation, the mode from these lasers is very poor. The process has more to do with amplified spontaneous emission or superradiance than with laser oscillation.

The construction of an industrial excimer laser is illustrated in Figure 1.23. Excitation is by a 50–100-ns duration 35–50-kV pulse across the electrodes with peak current densities of around 1 kA cm^{-2}. Preionisation to some 10^8 cm^{-3} is needed to avoid electron avalanching (sparking). This is achieved by flooding the cavity with ultraviolet light from small spark discharges in the cavity. The high photon energy (see Section 2.1) of ultraviolet light and the potential of very high peak power densities of around 10 MW cm^{-2} to 200 TW cm^{-2} (thermofusion experiments) makes a niche for ultraviolet laser processing which can be achieved by excimer or frequency-quadrupled Nd:YAG radiation.

The optics of the excimer laser are made of fused silica, crystalline CaF_2 or MgF_2. One of the cavity mirrors has a highly reflective aluminium coating on the rear surface

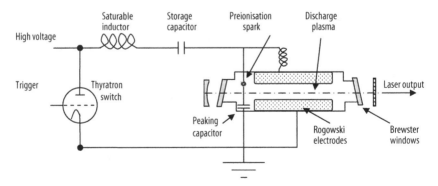

Figure 1.23 Basic construction of an excimer laser with a preionisation electric discharge and matched circuits

– to protect it from the corrosive atmosphere of the halogen gases. The output window is usually 8 % reflective. The window will degenerate with the formation of lattice defects due to the intense exposure. The extent of the deterioration can sometimes be monitored by viewing the build-up of fluorescence from the window when the laser is operating.

The output wavelength can be changed by changing the gases (see Table 1.6) and sometimes the optics, if coated optics have been used. The size of the output beam is determined by the aperture from the cavity, normally a rectangle (*e.g.*, $20-30 \times 10 \, \text{mm}^2$) with a high divergence of 2–10 mrad owing to the fact that the cavity has a high Fresnel number and the system is almost superradiant. To reduce this poor mode, special optics have to be fitted. One of the more successful techniques is to use the main laser as an amplifier of a seeding beam from a laser oscillator of a few microjoules.

The gas mixture is typically 4–5-mbar halogen gas, 30–500-mbar argon, krypton or xenon as required and the rest is 4–5-bar helium or neon. The halogen gas is usually supplied as a 5 % mixture in helium to reduce the danger. The gas slowly degrades through corrosive action, forming dust particles which are filtered out of the gas stream. This gives lifetimes of around 10^6 pulses per fill. The running costs are high, at \$10–30 per hour (2009 prices), due mainly to the maintenance and equipment costs. Fitting cryogenic traps can improve the lifetime and reduce the costs to as little as \$1–2 per hour [12]. For further reading, see Duley's book "UV Lasers" [14].

1.3.2 Solid-state Lasers

Solid-state lasers have the active medium held in an insulating dielectric crystal or amorphous glass. The lasing action comes from energy jumps between discrete electronic energy levels of the dopant such as rare earth ions or transition ions with unfilled outer shells or defect centres known as colour centres. The main industrial solid-state lasers include Nd^{3+}:YAG, Er^{3+}:YAG, Yb^{3+}:YAG, ruby (Cr^{3+}:Al_2O_3), titanium sapphire (Ti^{3+}:Al_2O_3) and alexandrite (Cr^{3+}:$BeAl_2O_4$). The host material for the neodymium or other rare earth element may be YAG ($Y_3Al_5O_{12}$), yttrium lithium fluoride (YLF), yttrium aluminium perovskite (YAP; $YAlO_3$), yttrium vanadate (YVO_4) or phosphate or silica glass. Table 0.1 and Table 1.3 list some of the wavelengths available from solid-state lasers.

Solid-state lasers have the advantage of relatively long lifetimes for the excited states, which allows higher energy storage than for gas lasers and hence allows them to be Q-switched to give very high peak powers in short pulses (10^{15}-W peak power is the potential output of the Vulcan laser at Rutherford Appleton Laboratory, Didcot, UK).

1.3.2.1 Neodymium–doped Yttrium Aluminium Garnet Lasers

Pure $Y_3Al_5O_{12}$ is a colourless optically isotropic crystal with the cubic structure of garnet. If around 1 % of the yttrium rare earth is substituted by the alternative rare earth neodymium, the lattice will then contain Nd^{3+} ions. These ions can undergo the transitions shown in Figure 1.24. The Nd^{3+} ions absorb at specified absorption bands,

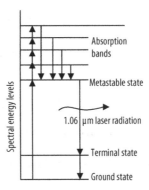

Figure 1.24 Neodymium energy levels

and decay to a metastable state from which lasing action can occur to a terminal state. This terminal state requires cooling to reach the ground state. The cooling is usually achieved with deionised water flowing around the YAG rod and the flash lamp. The rod and flash lamps are situated at different foci of a reflective elliptical cavity, which is either gold-plated or made of alumina. The quantum efficiency is 30–50 %. Using krypton flash-lamp pumping, the operating efficiency is low, approximately 2 % since the pumping is done with a broadband illumination of which only a proportion is able to excite the neodymium ions in the crystal (see Figure 1.25). It thus lacks the natural coupling between N_2 and CO_2 lasers. This means that considerable energy has to be pumped into the crystal rod, giving a serious cooling problem. For this reason the YAG laser is currently limited to around 400 W per 100-mm length of rod before serious beam distortion, due to thermal effects, occurs, or, worse still, the rod cracks. It is also the reason for the studies on different geometries, such as fibres, discs, slabs and tubes (see section 1.2.1.3). The total power from the system may be increased by the use of a master oscillator–power amplifier arrangement or by optically coupling multiple beams through bundles of fibres as illustrated in Figure 1.26. A more efficient pumping system is to use diode lasers of the appropriate frequency to pump the neodymium with greater precision. The operating efficiency then rises to 8–10 %. The preferred pumping frequency for Nd:YAG is 0.809 µm (see Figure 1.25). A further advantage of diode pumping is the lifetime of the diodes, which is some 10 times longer than that of lamps (currently 1,000 h for lamps but 10,000 h for diodes).

Other solid-state lasers for material processing include Nd:YVO$_4$ operating at 1.06 µm, Yb:YAG operating at 1.03 µm and Er:YAG operating in the "eye-safe" region of 1.54 µm ("eye-safe" means the radiation will be absorbed on the cornea and not penetrate to the retina with a 10^5 times amplification in intensity). These lasers have been developed since they appear more optimal than Nd:YAG for diode pumping.

The overall construction of a Nd:YAG laser is shown in Figure 1.27 for pumping by a lamp or diode. It consists of the standard cavity design with the active medium being neodymium, Nd^{3+} ions, in a YAG crystal rod mounted at one of the foci of an elliptical chamber made of gold-coated metal or plain ceramic. The ceramic chambers are thought to give more uniform illumination to the YAG rod. At the other focus is

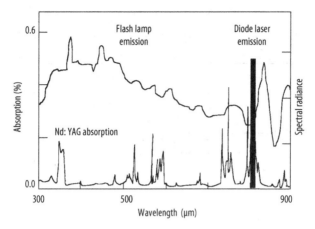

Figure 1.25 Nd:YAG absorption spectrum compared with the emission spectra of a Kr flash lamp and a GaAs diode laser

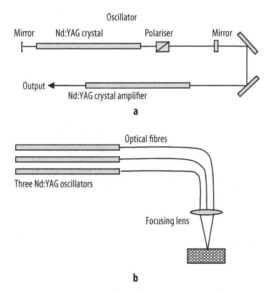

Figure 1.26 Possible arrangements for achieving powers of over 1 kW with a Nd:YAG laser: **a** master oscillator–power amplifier, and **b** coupled fibres

a krypton lamp, whose emission spectrum suits Nd^{3+}. Mounted in the optical cavity is an aperture for mode control and possibly a Q-switch for rapid shuttering of the cavity to generate fast, short pulses of power. A Q-switch ("Q" stands for "quality") is a device which spoils the lasing oscillation in a controlled way. It could be a mechanical chopper, a dye which can be bleached, an optoelectric shutter or an acousto-optic switch. Few of these devices can stand much power at present, so they are confined to resistance trimming with 10–20 W lasers. Modulation of these switches gives a wide range of pulse rates, typically 0–50 kHz. The speed of these shutters allows the energy to build up in

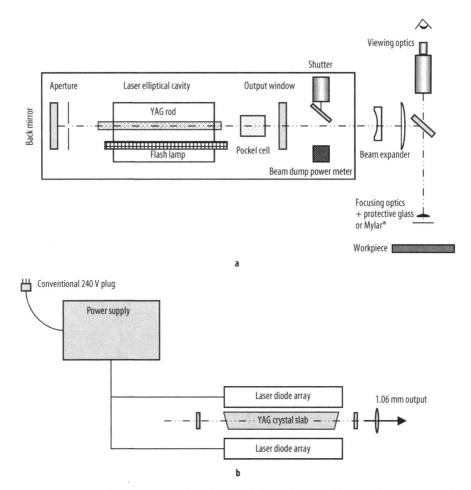

Figure 1.27 a General construction of a Nd:YAG solid-state laser, and **b** general arrangement for diode laser pumping of a YAG slab solid-state laser

the cavity while lasing action is inhibited and so when the shutter is open rapidly, very high peak powers can often be achieved. For example, a 20 W Nd:YAG laser Q-switched may produce a 6 ns pulse of 1 mJ per pulse, which is 166 kW. Individual pulse shaping is an aspect of Nd:YAG laser engineering which offers a new freedom in manipulating optical energy, a freedom that has yet to be exploited.

For lower power output from a Nd:YAG laser, the beam can be passed to a barium borate or a lithium niobate crystal. These are nonlinear optical devices which if swamped in photons will absorb two or more photons to rise to higher energy states. This energy can then be lost in one step, giving radiation of twice the photon energy, which will have half the wavelength. This means that by simply passing the 1.06-μm beam from the Nd:YAG through an aligned crystal, the emerging radiation will be 0.530-μm, or green, light, with around 30 % efficiency. This is known as "frequency doubling". Repeating this will give ultraviolet light – a serious competitor to the excimer laser, particularly when diode-pumped YAG lasers are considered (see Section 2.2.1.4).

A modification to the rod geometry is the introduction of slab or face-pumped YAG crystals. These are costly but may have the ability to generate 1 kW of power with sufficient cooling area to avoid distortion of the beam during operation. The first Face Pumped Laser (FPL), popularly known as Slab laser due to its geometry, was discovered by Martin and Chernock [U.S. Patent # 3,633,126] at General Electric Corporate R&D center in Schenectady, New York. This geometry reduces thermal lensing compared with the rod design, allowing stronger pumping.

1.3.2.2 Neodymium Glass Lasers

The Nd:glass lasers have the same energy diagram for Nd^{3+} as the YAG laser but the energy conversion is better in glass. However, the cooling problems are more severe owing to the poor conductivity of glass and so the Nd:glass lasers are confined to slow repetition rates, approximately 1 Hz. At higher repetition rates the beam divergence (or ease of focusing) becomes unacceptable for material processing. The beam from a glass laser is more spiked than that from a YAG laser as seen in Figure 1.28. It is more prone to burst mode operation.

1.3.2.3 Diode-pumped Solid-state Lasers

There is a problem with flash-lamp-pumped Nd:YAG lasers, as already noted, in that only a few percent of the flash lamp power is actually absorbed by the Nd^{3+} ions and so used in the lasing action; the waste energy heats up the YAG rod, causing distortion and variations in the refractive index. This leads to poor pulse-to-pulse consistency (approximately 10–15 % variation) and low beam quality (M^2 around 15–100). The lamps have a lifetime of a few hundred hours and require substantial power supplies to drive them. These problems can be eliminated by using diode lasers instead of flash lamps to excite the Nd^{3+}, as illustrated in Figure 1.27. The wall plug efficiency of diodes is around 30–40 % and all the light is emitted centred on a strong absorption line of Nd^{3+} at 808 nm. The power supply and cooling requirements are greatly reduced and M^2 values of as little as 1.1 have been reported [15]. The remaining problem is the cost of the high-powered diodes required to do the pumping. As the size of market for laser diodes increases, so the price is likely to fall. It is thus very likely that for reasons of cost and quality lamp pumping will be phased out within the next few years.

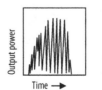

Figure 1.28 The spiky nature of Nd:glass pulses (*left*) compared with Nd:YAG pulses (*right*)

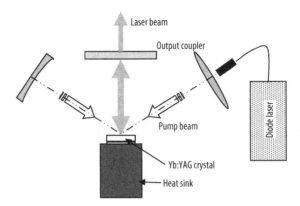

Figure 1.29 Basic construction of a disc laser

1.3.2.3.1 Disc Lasers

One version of a diode-pumped solid-state laser is a disc geometry in which a coin-shaped YAG crystal doped with ytterbium forms the lasing medium. This is irradiated with a multireflected pump beam from an InGaAs diode shining onto one side of the "coin"; the other side is fixed to a heat sink. A disc, 0.3 mm thick and 7 mm in diameter, doped with ytterbium up to 25 % (maximum dopant rate for neodymium is only 1.5 %) can produce over 500 W of high-quality beam from the top surface of the "coin" at 1.03-μm wavelength. The reason for the high power and quality is the superior cooling and higher dopant rates possible with this geometry, as well as the cavity design. Figure 1.29 shows one design for a disc laser.

1.3.2.3.2 Fibre Lasers

Meanwhile, diode-pumped fibre lasers are being developed [16]. These lasers are doped plastic or glass fibres that are end- or side-pumped by diode lasers. A diagram of the construction of a fibre laser is shown in Figure 1.30. IPG Photonics is marketing a 2-kW CW fibre laser based on ytterbium operating at 1,085 nm. It gives a spot size down to 50 μm, providing a power density of $100 \, \mathrm{MW \, cm^{-2}}$ from a unit measuring $110 \times 60 \times 118 \, \mathrm{cm^3}$, which includes the power supply and the air cooling system. It has a beam quality 10 times better than that of a standard Nd:YAG laser. The fibre can be very thin, 100 μm, and hence the only way oscillations can be contained is by waveguiding within the fibre as a monomode transverse excitation (*i.e.*, Gaussian). The wall plug efficiency is stated as 20 %, whereas the lifetime for the pumping diodes is reckoned to be 100,000 h, indicating several years of maintenance-free operation. The 700-W version of a fibre laser was able to cut through 50 mm of steel. The 2-kW version could weld steel from several metres distance, the mode being so good. With this sort of performance these lasers appear to have much to offer.

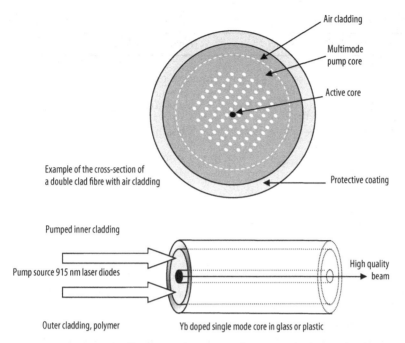

Air cladding

Multimode
pump core

Active core

Example of the cross-section of
a double clad fibre with air cladding

Protective coating

Pumped inner cladding

Pump source 915 nm laser diodes

High quality
beam

Outer cladding, polymer

Yb doped single mode core in glass or plastic

Figure 1.30 Basic structure of a fibre laser. The cavity reflectors are built into the fibre either as a loop feedback at one end and a cleaved fibre reflector at the other or as internally etched Bragg reflectors (thin layers of different refractive index). These lasers thus have minimal alignment problems. The core diameter may be as small as 100 μm

1.3.2.4 Semiconductor Lasers

These lasers are becoming the most important laser material processing tool both for pumping solid-state lasers and for direct application to surface heating and welding. They have the advantages of being compact, efficient, with a quick modulation response and reliability. The principal materials used are shown in Table 1.7.

Diode lasers [17] are currently the most efficient devices for converting electrical into optical energy. Their wall plug efficiency may reach up to 50 %. In a diode laser the excited state is that of the electrons in the conduction band. The two states, electrons and holes, come together in an active region set at a p–n junction in a semiconductor material. A current flow induces electrons to jump from the conduction band down to the valence band and give up the energy difference between these two Fermi levels as radiation ($h\upsilon$). This is illustrated in Figure 1.31 for a double heterojunction. There is a spread of energy within the Fermi levels that is a function of temperature and current density (1 nm in wavelength per degree Celsius); hence, laser diodes are tunable over a small range, but require careful temperature and current control. There is a critical carrier density required to start lasing action (approximately 10^{18} cm^{-3}). The early homojunction style between p–n GaAs required a current density of around 50,000 A cm^{-2} with consequent enormous heating problems. It was the invention in

Table 1.7 Some of the main semiconductor laser materials and their emission wavelength ranges

Material (active/cladding/substrate)	Emission wavelengths (μm)
Group III–V compounds	
AlGaAs/AlGaAs/GaAs	0.7–0.9
GaInPAs/InP/InP	1.2–1.6
GaInP/AlGaInP/GaAs	0.66–0.69
AlGaAsSb/AlGaAsSb/GaSb	1.1–1.7
GaInN/AlGaN/sapphire	0.38–0.45
Group IV–VI compounds	
PbSnTe/PbSnSeTe/PbTe	6–30
PbSSe/PbS/PbS	4–7
PbEuTe/PbEuTe/PbTe	3–6
Group II–VI compounds	
ZnCdSe/ZnSSe/GaAs	≈ 0.5

1970 of the double heterostructure which made the laser diode into a significant tool. The Nobel prize in physics in 2000 was awarded for this invention. Thus, laser diodes are usually made of multiple layers forming double heterojunctions to reduce the critical current density and allow operation at room temperature. The construction of a vertical cavity surface emitting diode laser is illustrated in Figure 1.31. It consists of a p–n junction (*e.g.*, AlGaAs) separated by an active layer of, for example, AlGaAs, with all layers having different doping levels. Bragg reflectors consisting of multilayers of GaAlAs of different refractive index are deposited on each side of the active region to make an optical oscillator. The epitaxial purity of the structure is crucial to its endurance.

Lasing action can also be taken from the edge of the thin layer of the active medium using the cleaved ends of the crystal as mirrors; they have a reflectivity to air of approximately 30 %. A light-emitting diode (LED) operates on a similar principle but is operated below the critical current and may not have the cleaved reflective ends. Some diodes have a built-in variation of refractive index which acts as a waveguide for the laser radiation.

The best developed diode laser materials are GaAs and GaAlAs emitting at 750–870 nm and InGaAs emitting in the range 900–1,000 nm. They emit over a bandwidth and hence can be tuned by fitting a grating as one of the cavity mirrors. The power is such that they can also, in some cases, be frequency-doubled. The total power conversion efficiency is a few percent for the low power units and up to 30 % for the commercial arrays. The power increases with the volume of the active layers. Lasers with stripes a few micrometres wide can generate CW powers up to about 100 mW, a 50-μm-wide stripe can generate 0.5 W and a 500-μm-wide stripe can generate 4 W. Further power is achieved by making linear or stacked arrays. A linear array 1 cm wide can generate up to 20 W CW or produce peak powers up to 100 W in quasi-CW operation. Stacking such arrays can give a few kilowatts of power from a laser head about the size of a human fist – which must give pause for thought! Unfortunately the divergence is around 30–40°, which makes them resemble the output from a torch rather than a laser, but does open up the possibility of using this efficient generator of light for general illumination. Some advances are expected through a master oscillator–power amplifier

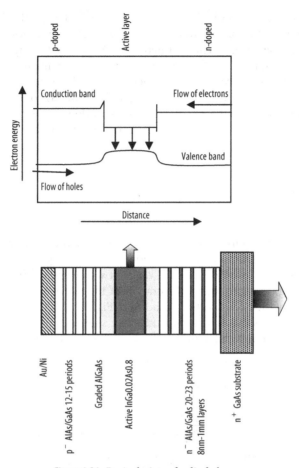

Figure 1.31 Basic design of a diode laser

approach. The master oscillator generates a single longitudinal mode beam, which is amplified in a wide-area power amplifier. The whole construction is monolithic. The output is a larger version of the input and therefore considerably easier to focus. Other approaches are to feed multiple beams into a fibre delivery system.

Currently the applications are predominantly for low-power systems, of a few milliwatts, as used in compact disc (CD) and DVD players, bar code readers, optical storage systems, laser printers, short-haul optical links, for voice, data and cable television, and lecturers' pointers. Material processing diode lasers which accounted for barely 0.3 % of the total diode laser market in 2001 are rapidly becoming reliable tools for surface treatments with powers of several kWs. Currently, their main application is in pumping other solid-state lasers (Section 1.3.2), but they are also being considered for welding, cutting, annealing, soldering, drilling, ceramic sealing, lithography, inspection, marking and welding plastics, metal and silicon, rapid-prototyping, desktop manufacturing, embossed holograms and grating manufacture. This is quite a list, but diodes have yet to establish a niche for themselves. The medical applications may prove to be a large market (see Chapter 11). However, the largest market of all is likely to be their use as LEDs for illumination in place of incandescent bulbs. For such applications they offer savings in electricity, savings in maintenance (lifetimes of 10–20,000 h compared with 1,000 h) and give better illumination. They are currently used on some traffic lights, with re- markable savings for the councils concerned. The development of the GaN diode on a $LiAlO_2$ or $LiGaO_2$ base producing blue light allows the use of fluorescence to give white light illumination and hence an entrée into the domestic lighting market, a huge market with implications for strange architectural designs based on the use of small intense lighting.

1.3.3 Dye Lasers

Dye lasers are one of the most readily tunable lasers [16]. They can also operate at high powers with pulse lengths from CW to femtoseconds. They are, however, extremely inefficient. They work through the absorption of a pumping laser (*e.g.,* a green cop- per vapour laser) and emit over a wide range of wavelengths, which can be selected by cavity tuning, changing the concentration of the dye and changing the pressure. An illustration of their construction is shown in Figure 1.32. The range of wavelengths and some of the dyes that have been studied are shown in Figure 1.33. The wide flu- orescence comes from the multitude of vibration and rotation states possible with the

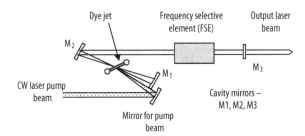

Figure 1.32 Basic design of a dye laser, showing the pump beam exciting the fast flowing dye in the dye cell. The generated laser beam passes through a frequency-selective element that determines which frequency will lase [56]

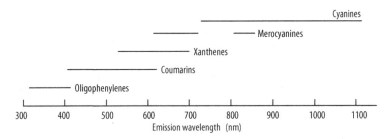

Figure 1.33 Approximate range of emission wavelengths for various dye chemical classes. The cyanines extend beyond the 1,100-nm range

large molecules of the dyes (molecular weight around 175–1,000). These lasers are used in material processing for isotope separation (see Section 1.4.15), and photodynamic therapy (see Section 11.3.2.9). They are also used in the Guide Star project (to excite the sodium atoms high in the stratosphere as a source for the adaptive optical control of telescope mirrors), LIDAR, laser cooling and optical trapping.

1.3.4 Free-electron Lasers

The free-electron laser does not depend on excited states but depends on synchrotron radiation. Synchrotron radiation is emitted when an electron changes direction; the energy involved appears as radiation. The laser consists of a circuit for relativistic electrons streaming around in a ring. As part of the circuit there is a magnetic wiggler, which is a short-wavelength magnetic field that causes the electrons to make a wiggly path, emitting photons on each turn in the same direction as the travelling electrons. These machines can generate radiation from deep infrared to X-radiation. Owing to the velocity distribution within the flow stream, there is considerable spectral spread in the output beam. It is sometimes known as a "rainbow laser".

The general arrangement is illustrated in Figure 1.34.

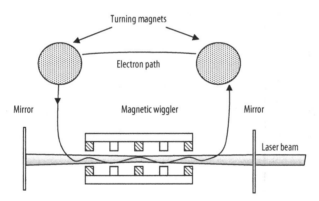

Figure 1.34 Basic construction of a free-electron laser, showing the electron recirculation path and the magnetic wiggler

1.4 Applications of Lasers

The laser was invented in 1960 and was soon dubbed "a solution looking for a problem". So new was the tool that our thinking had not caught up with the possibilities. Today the story is distinctly different. Table 1.8 lists most of the areas of application. They fall into three groups: optical uses, power uses, as in material processing, and ultrahigh power uses for atomic fusion. The range of applications is briefly discussed here only by way of background to material processing and to illustrate some of the possibilities with optical energy, many of which are not currently applied in material processing.

Table 1.8 General applications of lasers

Application	Property of beam most used						Laser
	Monochromatic	Low divergence	Coherence	High power	Single mode	Efficient	
Powerful light		***		**			He-Ne, Ar
Alignment	*	***					He-Ne
Measurement of length	***	**	***		*		He-Ne, ruby, Nd:glass
Velocity measurement	***		***				He-Ne, Nd:glass
Holography	***	**	***	**	***		All, mainly visible
Speckle interferometry	***		***				He-Ne
Inspection	***		**	**			He-Ne, ruby
Pollution detection	***		***	**			Dye, GaAs
Analytical techniques	***	**		*			Nd:YAG
Recording	***	***	**		*		GaAs, GaAsP
Communications	***	***	***				He-Ne, GaAs, I$_2$
Heat source	*	**		***	**	***	CO$_2$, Nd:YAG or Nd:glass, excimer
Medical	***	***					CO$_2$, ruby, Ar, excimer
Printing	***	***		**			He-Ne, Ar
Isotope separation	***			***			Dye, Ar, Cu
Atomic fusion	***	***		***			CO$_2$, Nd:glass

* useful property, ** important property, *** essential property

1.4.1 Powerful Light

The beam from a laser can have a low divergence and hence can be projected to make a bright spot, as with laser pointers for lecturing. When the spot is moved, a pattern is retained in the eye, generating a form of laser light show. Strange patterns can be made by rastering, the use of optical gratings, screwed-up transparent paper and many other optical components, leading to the mind-disorienting effects of a laser disco or laser light show.

The diode laser and LED (which has the same construction as a diode laser without the optical cavity and uses a lower current) are set to introduce a new approach to lighting in general. High-brightness LEDs are beginning to penetrate the world lighting market (estimated at $12 billion in 2001). One example is the use of LEDs in traffic lights, where they not only consume 85 % less energy than the incandescent equivalent, but also have only a third of the maintenance costs. It has been estimated [18] that if the 100 million traffic lights in the USA were all converted to LEDs, there would be a staggering $190 million saving in energy costs per year and an annual reduction in energy consumption of 3×10^9 kWh per year. The current growing applications are display back lighting (29 %), automotive uses (27 %), signs (26 %), electronic equipment (10 %), illumination (4 %) and traffic signals (4 %). The automotive uses have been boosted by the new legislation requiring a third brake light on cars, 90 % of which are fitted with LEDs in Europe since they are cheaper, faster and brighter. Research is currently ongoing to increase the brightness of diode lasers and LEDs for headlight applications (30 lm W^{-1} has so far been achieved commercially). The natural divergence of the beam will result in less dependence on reflectors for focusing the radiation. White-light LEDs are being developed by joining a blue LED and a yellow-emitting phospor, or coating the package surrounding an ultraviolet LED with a red–green–blue-emitting phosphor or combining the output of red, green and blue LEDs [19]. By 2010 low-level outdoor illumination for car parks, *etc.*, will probably be done by LEDs, and since LEDs require relatively small currents, proposals exist to have solar-powered, LED-illuminated Catseye® studs marking the edges and the centre of roadways. By 2015 indoor domestic lighting may be by LEDs and not incandescent bulbs. An LED may last 20,000 h, whereas an incandescent bulb may only last 1,000 h, and will also reduce the electricity bill for lighting by some 75 %. Currently, there is a way to go before this becomes reality since LEDs can have illuminations up to 120 lm, whereas incandescent bulbs have illuminations up to 1,700 lm and fluorescent tubes have illuminations up to 3,000 lm. An illumination of 1,000 lm for an LED is nearly available commercially.

1.4.2 Alignment

Laser theodolites with automatic level read-off are available. The assembling of autoparts and sequencing of wiring harnesses can be indicated by lasers. Tunnelling is now guided by lasers. The flexing of bridges, supertankers and the movement of glaciers are now recorded by laser alignment techniques (Figure 1.35a,b). As the laser beam is swept around, a line is marked out. If the sweeping mechanism is carefully levelled, then a rapid technique for ground levelling, used by farmers and road builders, is instantly available. The beam is usually passed through a pentaprism, which reduces the need to

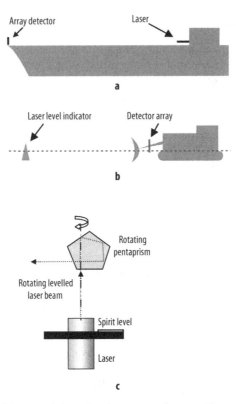

Figure 1.35 Examples of the use of a laser for alignment: **a** flexing of large structures such as ships and bridges, **b** automatic alignment of a bulldozer shovel for ground levelling, and **c** a laser level

align the prism carefully to the beam. This is then rotated about the beam axis to produce a level signal. The laser used is usually a visible, red, He–Ne laser (Figure 1.35c). This signal can be used directly by specially equipped bulldozers [20].

An optical lock has been postulated [21] that would be difficult to pick. The lock would consist of a cylinder with input fibres on one side and output fibres on the other side of the barrel. The key would connect these two sets of fibres by its own internal fibre system.

1.4.3 Measurement of Length

This can be done in several ways, four of which are interference, time of flight, occlusion times and triangulation.

1.4.3.1 Interference

Using the coherent nature of a laser beam, whereby the beam is a continuous wave CW stream, one can use the beam as a form of ruler in an interferometer, which can have

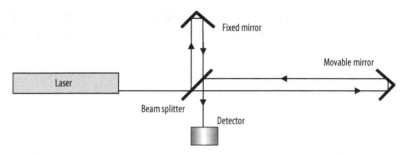

Figure 1.36 A Michelson-type interferometer for measuring length

very different path lengths for the two interfering beams. The basic design of the interferometer is shown in Figure 1.36. As the mirror is moved, the detector will sense a high and a low signal for the constructive and destructive interference between the two beams. Thus, the number of wavelengths or even quarter wavelengths (distance between a null in the signal to a peak) can be measured between one position of the movable mirror and the next. An accuracy of ±0.1 μm has been achieved this way for the positioning of machine tables using argon or He–Ne lasers with high levels of repeatability [22]; a CO_2 laser would have a quarter-wavelength accuracy of ±2.5 μm. Distances up to the coherence length of the laser beam can be measured in this way. This is often around 100 m or so. (The coherence length is the maximum path difference that two rays can travel and yet be able to interfere constructively. This is similar to the length of a CW stream from the laser. For a He–Ne laser it is around 70 cm and for a CO_2 laser a distance of several hundred metres is possible.)

Interference measurements of length can measure more than just length. When a monomode optical fibre is stretched, the stretching will affect the optical path length within the fibre. This change in length can be detected in an interferometer as shown in Figure 1.37. The sensor can house quite long lengths of fibre, which will increase the sensitivity. For the sensor to work, the beam must propagate as a single mode within the fibre, *i.e.*, there is only one path that it can follow. To make a measurement, the light from a laser diode is focused into a monomode fibre. A fibre coupler divides the beam into two fibres: one goes into the sensing head, where there is some form of fibre stretcher that responds to temperature, strain or (if a piezoelectric element) magnetic or electric fields; the other is the reference beam. With use of the reflection from the

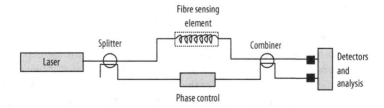

Figure 1.37 Measurement of temperature, strain or magnetic and electric fields

front and back surfaces of a monomode birefringent fibre (as in a Fizeau interferometer), accuracies of the order of 0.0005°C have been reported for a 1-cm-long sensing element [23].

1.4.3.2 Time of Flight

Distance can be measured by the flight time of a short pulse. This is how it was found that the moon wobbles and is moving away from us at a speed of 3.5 cm year^{-1}; it is also how rangefinders work. Any distance greater than 1 km can only be measured this way if a laser is used. The accuracy is approximately ±2 cm, depending on the pulse length. Currently, there is a hand-held device operating at 72 µW – and costing around $10,000 at 2003 prices – which is used by the customs to check the length of containers and lorries for finding false walls, *etc.* [24]. Laser radar [25] has been developed. The Firefly CO_2 imaging radar has been able to range and get Doppler images from targets over 800 km away. The use of coherent laser radar [26] has reduced the clutter on radar screens and can be used for range, direction and velocity measurements on hard targets. The laser is used to substitute trace rounds in military operations; a computer calculates the bullet trajectory. A more futuristic possibility is to use laser ranging for collision warning in cars, automatic speed control or even car guidance – in fact as a collision-avoidance system for any moving object [27].

The time of flight of a laser pulse is

$$\tau \pm \frac{\delta\tau}{k} = \frac{L}{c},$$

where $\delta\tau$ is the time for a half wavelength to pass, k is a constant, L is the target distance and c is the velocity of light.

For a target distance of 100 m

$$\tau \pm \frac{\delta\tau}{k} = \frac{0.1}{30,000} = 3 \times 10^{-7} \text{ s} .$$

For a meaningful measurement $\delta\tau/k < 3 \times 10^{-7}$ s, requiring pulses in the nanosecond range.

1.4.3.3 Occlusion Time

By measurement of the occlusion time of the scanning beam shown in Figure 1.38, the width of a wire can be measured while the wire is being made and travelling at speeds of a few kilometres per second. The technique can also be used on stationary objects as a form of micrometer.

The beam scans the object at a speed of $v = 2\omega f$, where ω is the angular velocity in radians per second of the mirror shown (the reflected beam rotates at twice the speed of the mirror) and f is the focal length of the lens in metres (see Figure 1.38). Thus, the occlusion time, Δt, gives the dimension of the component as $\Delta y = v\Delta t = 2\omega f \Delta t$. The accuracy of this system depends on the beam size at the plane of the object.

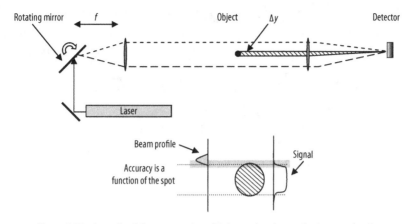

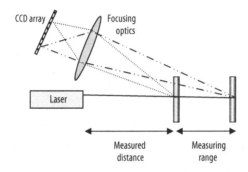

Figure 1.38 A method for measuring thickness by the occlusion method

Figure 1.39 The triangulation method for measuring distance

1.4.3.4 Triangulation Methods

The principle of this technique is shown in Figure 1.39. A beam of light from a He–Ne or red diode laser is aimed at the object to be measured, on which it produces a light spot. A high-resolution charged coupled device (CCD) camera which is built into the instrument views the spot and displays the image on an array of light-sensitive diodes. As the distance varies, the image on the array changes from one diode to the next. The active diodes indicate the distance. Typical measuring ranges are 300 mm to 10 m, with an accuracy of 0.05 % of the measured length. A reading takes approximately 50 ms.

1.4.4 Velocity Measurement

There are two types of laser velocity meters – the laser Doppler velocimeter and the laser Doppler anemometer.

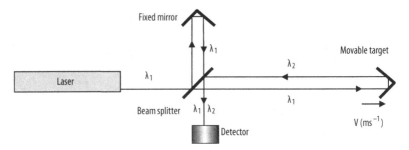

Figure 1.40 Layout for a laser Doppler velocimeter

1.4.4.1 Laser Doppler Velocimeter

The laser Doppler velocimeter measures the frequency shift in the radiation emitted and the radiation returning as illustrated in Figure 1.40. It works best with a cooperative target, but this is not strictly necessary. The outgoing radiation has a wavelength of λ_1, but while it is being reflected from the mirror moving with velocity v, this wavelength is extended or shrunk by $v(\lambda_1/c)$ – *i.e.*, the speed of mirror, v, times the time the wave is on the mirror, travelling at the speed of light, c, for a distance λ_1. Thus, the Doppler-shifted wavelength is

$$\lambda_2 = \lambda_1(1 + v/c).$$

These two waves will meet at the detector and will reinforce one another after a length of

$$[\lambda_1/(\lambda_1 - \lambda_2)]\lambda_2 = \lambda_{\text{beat}},$$

which is the beat wavelength (see Figure 1.41). The beat frequency is thus

$$[c(\lambda_1 - \lambda_2)]/(\lambda_1\lambda_2) = f_{\text{beat}}.$$

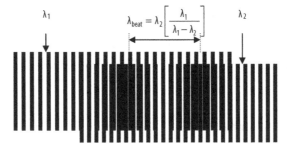

Figure 1.41 Beat frequency between two waveforms of slightly different wavelength arriving at the same location

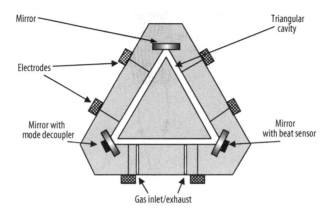

Mirror

Triangular cavity

Electrodes

Mirror with mode decoupler

Mirror with beat sensor

Gas inlet/exhaust

Figure 1.42 A laser gyroscope

Now,

$$\lambda_2 = \lambda_1 (1 + v/c) \, ;$$

therefore, by substituting for λ_2, the beat frequency is

$$f_{beat} = [c(\lambda_1 v/c)] / (\lambda_1 \lambda_2) = v/\lambda_2 \, .$$

The pulse flight time also records the distance, which is handy in a military context. These instruments can be used in measuring the high-speed movement of drop hammers and other machine movements or the length of moving material by integrating the speed and time.

A variation on this is to have a triangular laser as shown in Figure 1.42. As the triangle is rotated, the optical path length differs for the light rotating in one direction compared with that for the light rotating in the other direction. The result is a beat frequency which can be detected on the detector mirror. This device is thus a gyroscope which cannot be toppled. It is currently fitted in many airliners and missiles.

Instead of a triangular laser, longer path lengths and hence greater accuracy can be obtained with a fibre-optic gyroscope. Such a device is small, robust, lightweight and virtually maintenance-free. The general arrangement is illustrated in Figure 1.43. It operates on the principle of the Sagnac interferometer [28].

The difference in path lengths for the counterrotating rays, ΔL, caused by the rotation of $\underline{N}$ loops of fibre is given by

ΔL = the path length × (velocity difference/velocity of light) = $(2\pi r N)[(2r\omega)/c]$.

This would give a phase shift of $\Delta L (2\pi/\lambda)$. Phase shifts as small as 10^{-7} rad can be measured. The fibres have to maintain monomode polarisation to ensure that the two counterrotating beams follow exactly the same path.

The number of applications is growing rapidly. An automobile navigation system for driving around Tokyo has been fitted into some 500,000 Nissan and Toyota cars since

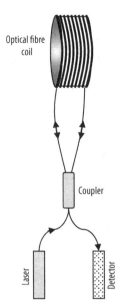

Figure 1.43 A fibre optical gyroscope

1987. This system uses CD-ROM maps, a computer information display and a dead-reckoning navigation system based on the laser gyroscope. This is being superseded by the Global Positioning System based on satellites. Optical gyroscopes are also used in commercial aircraft for an artificial horizon. A typical system would have four coils, one for each axis of rotation, plus one for cross-checking [29].

The laser Doppler velocimeter is ideal for heavy acceleration machines, such as missiles, autonomous vehicles, remote agricultural spraying and vehicles in hazardous zones, or underground regions where satellites are no help and magnetic rocks interfere with a compass.

1.4.4.2 Laser Doppler Anemometer

The other meter is the laser Doppler anemometer illustrated in Figure 1.44. This instrument is now standard equipment for those studying fluid flow phenomena. The split beam is reunited at the point in a flowing stream to be analysed. At this point the coherent beam will form Young's fringes, as illustrated in Figure 1.44, where the phases of the two beams are in step or out of step. As particles flow through the fringes they will reflect the beam (the system usually needs some smoke or dust addition for a good signal). This reflected signal will be detected by a photomultiplier as a series of flashes. The frequencies in this signal can be analysed by fast Fourier transformation and the velocity of the particles in the plane of the fringes is analysed, including the velocity variation and thus turbulence. The whole analysis is done without interfering with the flow in any way and can be done remotely; for example, to measure the flow within a diesel engine while it is operating. Since this method only measures one velocity vec-

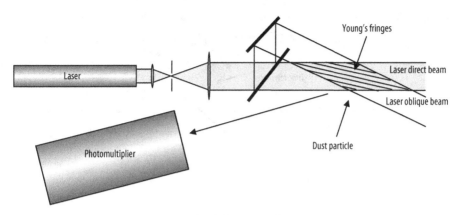

Figure 1.44 Principle of a laser Doppler anemometer

tor, a two-colour system may be used for measuring two dimensions simultaneously. By analysing the precise frequency of the return signal, the Doppler shift would indicate the velocity in the direction along the beam axis. Thus, this instrument is capable of measuring the velocity in all three dimensions simultaneously and swiftly.

1.4.5 Holography

This is a true three-dimensional form of photography which requires no lens in the camera! The arrangement for making a hologram (*holo* meaning "whole" and *gram* meaning "image") is shown in Figure 1.45a. The film is thus exposed to the direct beam, which can be considered as a time marker, and the reflected beam from the object, which gives data on the shape and illumination of the object as well as distance when its phase is compared with the phase of the time marker. The interference pattern which results on the photographic plate thus has information on the shape, illumination and time of arrival of the waves from the object all over the plate. If the light is shone through the developed hologram in the direction of the reference beam, as in Figure 1.45b, the wavefront is reconstructed as before. The definition is dependent on the grain size of the film. Special films have been developed for holography, for example, the Denysik holograms with less than 5-μm emulsion, and the subject has been advanced with the making of white light holograms. Holograms can be made using a light-sensitive plastic which causes a change in the refractive index on exposure. These holograms do not require developing. Moving holograms are being developed, although currently the framing speed is a little slow at two frames per second [30].

1.4.5.1 Holographic Interferometry

This is achieved by developing a hologram of, for example, a turbine blade and then mounting the image from the hologram in exact alignment with the blade itself. The blade viewed through the hologram will appear as usual; any misfit would show as

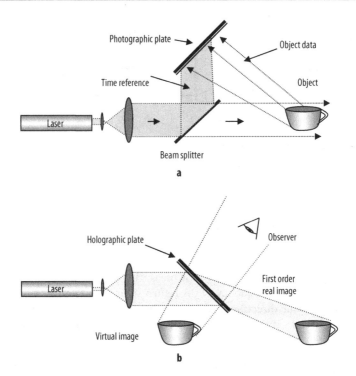

Figure 1.45 The arrangement for taking and observing a hologram: **a** taking a simple hologram with the object beam recording the shape of the object and the reference beam recording the phase time of arrival of the rays from the object, which is related to the distance from the photographic plate, and **b** viewing a hologram to reconstruct the wave front from the original object to give a virtual and a real image

a fringe pattern between the object itself and the image from the hologram. If the blade is now set in motion and the blade is illuminated by a stroboscope at the moment it is in line with the holographic image, any strain in the moving blade can be measured. This amazing tool allows flow visualisation, vibration analysis and many other unique measurements to be made, yet for some reason it is little used.

1.4.5.2 Commercial Holograms for Credit Cards, etc.

In 1969 the "rainbow hologram" or "Benton hologram" was invented [31]. Such a hologram is viewable in white light. It is made in two stages; firstly by taking a hologram of the object as just described and secondly by making another hologram from the image of this first hologram illuminated through a narrow slit, thus eliminating any parallax in the vertical plane. The second hologram when viewed with monochromatic light gives an image of the object and the slit. If it is viewed at a different wavelength, then the same image occurs, but the slit is in a different location. Thus, a spectrum is formed at the position of the slit. Viewing from the slit position will give a sharp image of the object of a colour depending on the position of the observer in the spectrum.

To make the inexpensive, mass-produced holograms for magazine covers and credit cards the second hologram is recorded on photoresist, which when developed leaves a surface relief pattern. This pattern is plated with nickel. On removal of the photoresist, a surface relief hologram made of nickel remains. This is mounted in a press and used to stamp out the pattern onto a heated plastic sheet. This can be done at high speeds of several thousand per hour.

1.4.6 Speckle Interferometry

When laser light is shone onto an object, the object appears speckled owing to the coherent nature of the laser light forming interference patterns on the retina of the eye. These patterns are a function of the roughness of the object being viewed. If this pattern is stored in a computer and played back on a video screen at the same time as the image is received a millisecond or so later, then any movement of the object during that time will be recorded as a fringe pattern. This is a tool similar to holographic interferometry but does not require a film. Figure 1.46 from the original work of Butters [32] shows the flow inside and outside a lit electric light bulb as visualised using speckle interferometry. The experimental arrangement is shown in Figure 1.47. In a more recent

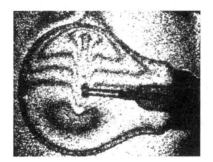

Figure 1.46 A speckle interferogram of an electric light bulb showing the convection currents inside and outside the bulb. (Photo courtesy of J.N. Butters)

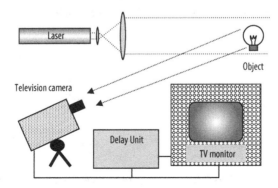

Figure 1.47 An arrangement for undertaking speckle interferometry

example [33] the blood flow in the brain has been monitored by looking at the change in shape of blood vessels by this technique. Magnetic resonance and positron imaging have also been used, but their spatial resolution is around 1 mm, whereas it is 25 μm for speckle interferometry, and their temporal resolution is minutes rather than real time, which can be achieved by this technique.

1.4.7 Measurement of Atmospheric Pollution and Dynamics

A burst of light of a particular frequency shone into the sky will interact with the molecules and aerosols in the air. The backscattered radiation can be used in pollution detection and for other purposes. The process is called LIDAR (light detection and ranging). The general arrangement for LIDAR is illustrated in Figure 1.48. The scattering is caused in various ways.

1.4.7.1 Rayleigh Scattering

In the upper atmosphere above 30 km the scattering is mainly from molecules, called Rayleigh scattering (see Section 2.4.1.1). The intensity of the backscattered signal is a function of the density of the air and hence the temperature. It will contain the incident frequency and the Doppler-shifted frequencies, giving further data on wind speeds.

1.4.7.2 Mie Scattering

Below 30 km scattering is mainly from aerosols and is hence Mie scattering (see Section 2.4.1.2). This region is called the troposphere and is where all our weather comes from. This scattered light again has signals on density and Doppler shifts, giving data on temperature and line-of-sight wind velocities.

1.4.7.3 Differential Absorption LIDAR

If the frequency of the light pulse is near the absorption band of an atomic or molecular species and it is accompanied by a pulse, either simultaneously or in sequence, at

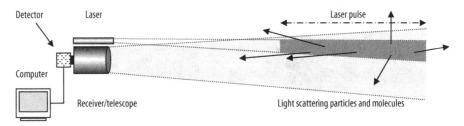

Figure 1.48 A simple LIDAR arrangement. (Copyright 2004, PennWell Corporation, *Laser Focus World*, used by permission [35])

a slightly different frequency, the differential absorption of the two backscattered frequencies gives data on the concentration of that particular species. This is known as differential absorption LIDAR. This process is extensively used for pollution detection of NO, O_3, SO_2, CH_4 and water vapour. It has been used to measure ozone holes and industrial emissions. It is now used from aircraft to patrol oil and gas pipelines. The sensitivity is around 50 to 100 ppm and 1,000 miles a day of pipeline can be checked [34].

1.4.7.4 Raman LIDAR

If the pulse of light is monochromatic or of very narrow spectral width, the spectrum of the scattered light from a molecular gas or liquid will contain lines at a wavelength different from that of the incident light owing to the interaction of the radiation with the quantised vibrational and spin states of the molecules affected. This extra energy from these inelastic interactions is either added (anti-Stokes) or subtracted (Stokes) from the backscattered light. The intensity of this Raman spectrum is around 1,000 times smaller than that of the main signal from the elastic Mie scattering. By comparing the values for nitrogen and water vapour, this Raman LIDAR technique allows the computation of the absolute water vapour concentration and by studying the rotational lines, one can determine the temperature.

1.4.7.5 Resonance LIDAR

If the pulse of light has energy equal to the energy of an allowed transition within the atom, then a fluorescence spectrum is generated at the same frequency by elastic absorption processes. The intensity gives data on the concentration of that species. The signal may also be altered by a Doppler shift, giving data on velocities. The constant ablation of meteors in the upper atmosphere has left a layer of alkali metals at an altitude of around 80–115 km. This atmospheric sodium layer is a popular target for LIDAR work. The spectral shape of the sodium D_{2a} line gives data on temperature and the Doppler shift gives velocities. The sodium fluorescence is also used by astronomers as a spot of light high in the atmosphere from which they can measure the wave front variations due to convection currents in the atmosphere. By using variable-shaped mirrors, they can correct the wave front when they view the stars. This adaptive optics technique could well have implications for material processing in the future.

LIDAR is used for atmospheric composition and dynamics, mapping, bathymetry, defence, oceanography and natural resource management [35]. It has recently been taken into space for global environmental monitoring and even mapping of Mars. It is anticipated that it will be used for three-dimensional imaging of real buildings for virtual reality environments.

One variation on the use of scattered and reflected radiation is an Australian idea for precision crop spraying. The crop spraying boom is fitted with multiple lasers which scan the ground and detect weeds by their reflectivity, which stimulates a squirt of weedkiller. The idea is in its infancy at present [36].

1.4.8 Inspection

1.4.8.1 Defect Inspection

If a laser beam is either focused as a line or scanned over the surface of an object, for example, a foil strip from a roll, any flaw will show as a variation in the reflectivity of the surface. A simplified arrangement is shown in Figure 1.49. In one example [37] three cameras pick up three different polarised images of the reflections from an incident polarised beam. Defects as small as 0.25 mm wide and 3 mm long can be picked up on a steel strip or hot galvanised strip and the location of the defect is marked. The uncut steel coil is then shipped to the user, usually an automobile manufacturer, whose press line will note the mark and discard the part made in that area. This saves time and material by using full-size coils each time.

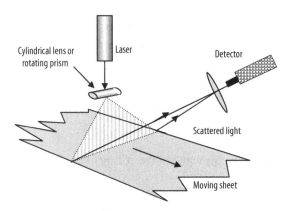

Figure 1.49 A line image or scan inspection system

1.4.8.2 Bar Code Readers

Scanning systems are now common in supermarket checkouts for reading bar codes. A bar code reader consists of a laser, which is made to scan by passing it through a rotating polygon or hologram of a set of lines. The laser is usually a diode or a He–Ne laser operating in the red waveband. The back-reflected light from the bar code is detected by a photodetector, which records the frequency pattern and compares this with a directory to identify the goods [38]. Two-dimensional matrix bar codes known as data matrix codes are now frequently used for storing from a few bytes to 2 kB. Up to 2,335 alphanumeric characters can be stored in a data matrix symbol. Lasers are also being used to read the code.

1.4.8.3 Fingerprint and Debris Detection

Fingerprint detection is enhanced by using a laser to create fluorescence (see also Section 1.4.9). Another use of fluorescence is to use a paint containing a gain medium and scattering particles which when illuminated with intense light from a green laser, for example, will shine red or yellow. If aircraft were painted in this way, then finding the debris after a crash would be greatly simplified. The detection system has been shown to work over distances of 1,000 m [39].

Scattered laser light has been used by semiconductor manufacturers to detect dust for some time. A recent variation is to use the deep ultraviolet to look for "nanodust". The scattered radiation has a characteristic depending on different types of dust and hence some indication of the source of the dust is also given [40].

1.4.9 Analytical Technique

A reasonably powerful (approximately 10 kW) pulse of focused radiation but of low energy (approximately millijoules) strikes the material to be analysed and evaporates a small part of the surface. This vapour is sucked into a mass spectrometer and analysed, a process known as laser-induced mass spectroscopy (LIMS).

An alternative technique is to create a laser-generated spark on the surface of the material and analyse the spectrum; this is also known as laser-induced-breakdown spectroscopy (LIBS).

Remote chemical sensing can be done through irradiating a gaseous, liquid or solid target and noting the Raman spectrum. This is the spectrum arising from the excited vibration states of molecules and hence can give a measure of the quantity of that chemical species. LIDAR is one version, noted in Section 1.4.7. Another example is to sort different plastic materials for recycling.

An alternative analytical route is by fluorescence, as just noted with fingerprints. In one example [41] a single laser pulse from a frequency-quadrupled Nd:YAG laser working in the ultraviolet is directed across the luggage conveyor in an airport. The pulse will break down any TNT vapour into NO, which will give a fluorescent signature. This technique should allow detection of nitrogen-based explosives from approximately 10-m distance. A further chemical analysis based on fluorescence is found in cell cytology (see Section 1.4.13, Chapter 11).

1.4.10 Recording

The CD player is now an established part of many homes. The information is stored on the underside of a disc in a digital form represented by pits of different length embedded in an aluminium-coated layer protected by a surface coating of plastic [42, 43]. The pits follow a helical pattern outward from the centre, similar to traditional grooved records only they are more compact. On the CD the pits carrying the information are sited on the flat top of tiny ridges, which serve to locate the signal. Unlike previous records with grooves or even magnetic tapes,the read head can be sited approximately 1 mm above the surface of the spinning disc. The read head is a low-powered GaAlAs diode laser tightly focused onto the surface (or GaN for blue light). As the CW beam is reflected from the pits, a digital signal is generated in the reflected beam which is read by a photodetector. The system appears simple but is actually a marvel of optical

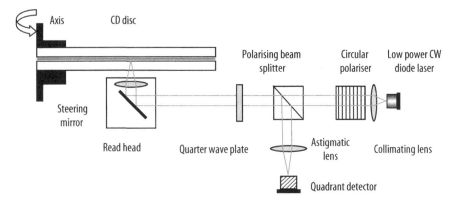

Figure 1.50 General arrangement for reading a compact disc (*CD*). A constant low-power laser beam scans a data track to read out data from the disc. The reflected light is modulated by the data mark pattern and is directed to a detector by a beam splitter. The quadrant detector gives signals for steering the beam on the track and the signal modulation is interpreted as the data on the disc

engineering. The components are shown in Figure 1.50. Firstly, the beam is collimated and circularly polarised; then via a polarising beam splitter and a quarter wave plate the outgoing beam and the reflected beam become polarised at right angles to each other, which allows good separation at the polarising beam splitter. The reflected signal passes to a quadrant photodetector. This detector not only reads the signal but also serves to sense the location relative to the ridges and the correctness of the focus, which are automatically corrected.

The home CD system may be less sophisticated and have three beams; two tracking the ridge and one reading the signal. The latest versions may have the beam splitter, objective and photodetector in a single miniaturised head using waveguide and holographic optical elements. These developments have made and will make CD players ever cheaper and smaller.

CD write systems are now on the market. Write once optical storage (WORM) devices are ideal for archival storage. They consist of an absorptive layer of tellurium sandwiched between two protective sheets of plastic or glass with an air gap to allow the drilling process. The recorder is similar to that shown in Figure 1.50, but with a more powerful laser which can burn holes in the tellurium. The storage capacity can exceed 10 GB on a double-sided 14-in. disc. Other WORM technologies include bleachable dyes or permanent phase changes.

Read–write CD systems are now on the market. One version is based on magneto-optical storage, illustrated in Figure 1.51. In this case the disc is made of a layer of ferromagnetic material such as TbFeCo or FeTbGd, which when locally heated above the Curie temperature becomes paramagnetic. This means the magnetic polarity can be set by an external field, which is supplied from a tiny coil. Reading these discs optically relies on the tiny difference in the polarisation of the reflected beam due to the Kerr effect. The variation is slight, approximately 0.5°. Data stored as magnetic signals are not good for archival work.

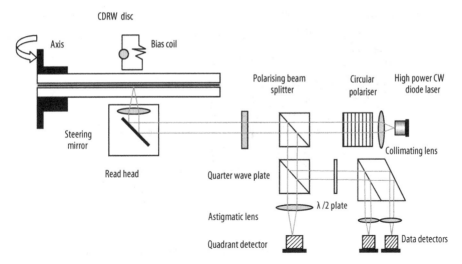

Figure 1.51 A magneto-optic read–write system. *CDRW* rewritable CD

The data density written to storage discs increases inversely as the square of the wavelength of the optical stylus, since it is area-dependent. So far, GaAlAs diode lasers have dominated the scene. They work at around 780 nm. Lasers based on GaAlAsP operating at 680 nm would raise the storage per centimetre from 80 MB cm^{-2} to 0.2 GB cm^{-2} and when frequency-doubled diodes are used at 428 nm or the latest blue diode lasers at 408 nm this would rise further to 0.4 GB cm^{-2}. Stacks of layers are being developed which would again increase the capacity of this amazingly compact storage system. The latest device using a blue GaN diode laser uses two recording layers separated by a phase-shift technique. It can be used as a read–write storage device storing 25 GB per layer and is capable of withstanding 10,000 repeated recordings with a lifetime of over 30 years [44]. The market for blue diode lasers is set to exceed \$2 billion before 2010.

However, this storage capacity is trivial compared with the potential from holographic storage. Via this route the whole *Encyclopaedia Britannica* could be stored on a device the size of a five pence piece or dime coin. In this case the page is encoded onto a laser beam by a spatial light modulator as a two-dimensional pattern of light and dark spots. This digital information is written as a single hologram in one go in the storage material by interaction of the laser beam with a reference beam as for normal holography. The arrangement is shown in Figure 1.52. Reading is by a less powerful laser fired from the direction of the reference beam and the output hologram is read by a CCD array. The data transfer speed of this system is approximately 1 Gb s^{-1}, with random access times of less than 1 µs. Changing the reference beam in wavelength, direction or phase allows many holograms to be stored within the same material. The beam angle can be changed quickly with acousto-optic devices (see Section 2.2.1.6). Photorefractive material is currently used as the storage medium. This has a change of refractive index where the beam power is sufficient. The development of storage media is the subject of much research today, since the goal at the end of the tunnel is of

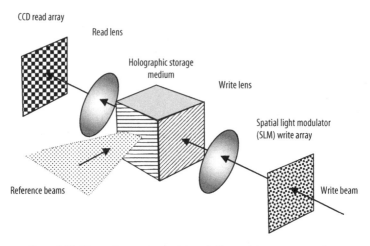

Figure 1.52 General arrangement for a holographic memory system

mind-boggling value [45]. One of these new media is Tapestry developed by InPhase for its holographic storage system. It consists of two independent but compatible polymers. The first polymer is thermally cured and acts as a scaffold, whereas the other can be altered by radiation to store the signal. Their arrangement is illustrated in Figure 1.53 [45].

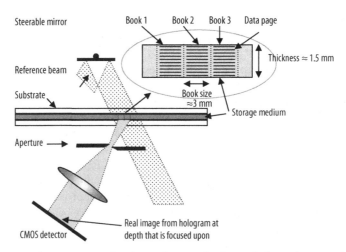

Figure 1.53 General arrangement for the InPhase Technologies holographic recording system. Each data page contains a single hologram of around 1 Mb and each book may contain 96 pages, with prospects of more in the future [42]

1.4.11 Communications

The beam from a CO_2 laser has a frequency of 10^{13} Hz. Most of us are content to listen to a radio station operating at around 1 MHz. Thus, there is potential for 10^7 1 MHz wavebands all operating simultaneously on one laser beam – if only we could read every wave of a light beam separately! Nevertheless, this accounts for the considerable interest in laser beams as a means of communication. If this level of technology could be achieved, then there would be an explosion in communication systems. The television telephone and high speed broadband would be commonplace. Currently only a hundred or so separate signals can be operated simultaneously in one optical fibre from one laser. Low-loss optical fibre technology with graded or stepped refractive index silica fibres is a new industry. Many new housing estates are being fitted with fibres as well as electric cables. The capacity is one attraction of optical communication; others are the lack of interference from outside sources, the security of transmission and the ability to transmit through space. Transatlantic fibre links have erbium-doped fibre laser amplifiers every 100–200 km. Today most transatlantic calls are made using fibre links as opposed to satellites [46].

1.4.12 Heat Source

This is the main subject of this book. The radiation from a 2-kW CO_2 laser can be typically focused to around 0.2-mm diameter. The power density is then $(2,000/\pi 0.1^2) = 0.6 \times 10^5$ W mm^{-2}. This is to be compared with 5–500 W mm^{-2} from an electric arc. For a nanosecond pulsed laser giving 0.3 J/pulse with the same 2 kW beam of 0.2 mm diameter, one can get a power density exceeding 10^{10}–10^{12} W cm^{-2}, which is approximately the total power generated by the power plants in the world (for this extremely short period of time). Only the electron beam can rival this value. Thus, the laser, even at quite modest power levels, offers the highest power density available to industry today. By defocusing, it also offers the lowest! This range of power intensity means the laser can evaporate any known material provided the beam can be absorbed or can give it any specified thermal experience. It is one of the most flexible and easily automated industrial energy sources. Today it is used in production processes in cutting, welding and surface treatments; it is also used in heat treatment, melting, alloying, cladding, direct metal deposition, machining, microlithography, stereolithography, bending, texturing, cleaning, engraving and marking. These processes are discussed in the following chapters. Not discussed is the tracking and shooting down of rockets [47].

The laser is also being considered for the ignition of howitzer shells. Currently, shells need a pyrotechnic primer – a cartridge loaded with energetic material that will expel hot gas and particles into the main charge – to fire the shell. These have to be replaced for each shot and hence they have to be supplied to the battlefield, which creates logistics and storage problems. The laser firing through an optical window directly into the charge would prevent all these problems from arising.

Lasers are also deployed for humanitarian services such as demining of antipersonnel mines. Nd:YAG lasers are used for detonation from a distance or for cutting them without disturbing the explosive inside [48, 49].

1.4.13 Medical Uses

The interaction of laser radiation with biological tissue has opened a whole new subject, that of phototherapy. The subject is discussed in Chapter 11, but in summary the laser can be used as a heat source to cut and weld tissue as well as encouraging fluid flow within a body. It can also photolytically break chemicals and be used as a tool for analysis by fluorescence, spectroscopy, scattering and absorption.

The medical application of lasers started as a uniquely useful tool in *eye surgery* for welding detached retinas, relieving the pressure from glaucoma by drilling holes, machining the eyeball to relieve myopia. In *surgery* the laser is a sterile cutting tool that cauterises as it cuts; this process is enhanced by adding it to the tools within an *endoscope*. The precision of surgery has developed such that surgical operations on individual chromosomes are now possible! The laser has been used as a heat source to clear blocked arteries or stones. In *dermatology* the laser can remove many cosmetic blemishes by selective photothermolysis. Procedures include removal of unwanted hair, pigment spots, tattoos, spider veins and wrinkles, and their have been some stunning results with removal of strawberry marks or birthmarks.

More recently, optically switchable drugs and dyes have been used in cancer treatment by a process called *photodynamic therapy*, in which the tumour is dyed with a porphyrin dye that will break down when irradiated at a certain frequency to create a free oxygen radical that will kill a nearby cell, hopefully the tumour. This is a new area of medicine with a massive future.

Cell cytology has surged ahead with the ability to count and separate cells by dye marking and measuring the fluorescence from the dye. The internal structure of a cell can be imaged by *optical coherence tomography*. *Diffuse optical tomography* measures tissue absorption and scattering, from which concentrations of haemoglobin and levels of inflammation can be deduced. *Raman spectral analysis* of breath or the liquid in one's eye has been used to detect numerous chemicals in the body, in particular sugar levels.

The laser will cause massive changes in our diagnostic and curing procedures. The future for lasers in medicine looks very exciting.

1.4.14 Printing

Laser printers are in many offices. They operate on the same principle as a Xerox[®][2] machine by forming an electrostatic field on a selenium drum, but with the capability of scanning 21,000 lines per minute from digital data. The pattern of the field is transmitted by a rastered laser beam driven from the computer data string. The quality from these machines is as for standard print, in fact the original edition of this book was produced this way. Desktop publishing has thus become a reality.

A similar rastering system can be used to make pits in photogravure rolls and computer-to-plate printing systems. This makes colour printing quick and simple and has led to a revolution in printing techniques for newspapers and magazines. Laser engraving of rubber rolls is now a standard technique for converting flat drawings to cylindrical rolls for the printing of wallpaper.

---

[2] Xerox[®] is a registered trademark of the Xerox Corporation. www.xerox.com

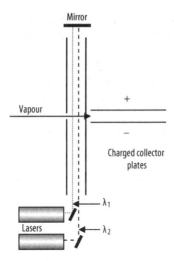

Figure 1.54 Arrangement for isotope separation by laser-selective ionisation

1.4.15 Isotope Separation

The energy levels for ionising an isotope, of, say, tritium or uranium required for atomic power uses, differ slightly from isotope to isotope. Passing the isotope vapour from one chamber to another at near vacuum and shining two frequencies of laser light from carefully tuned dye lasers through the passing gas will cause only one isotope to be ionised – not totally but with some level of efficiency. The ionised species is easily sorted by arranging an electric field in the second chamber as illustrated in Figure 1.54.

1.4.16 Atomic Fusion

The large AURORA excimer laser, the SHIVA CO_2 laser, NOVA glass lasers at Lawrence Livermore National Laboratory and Los Alamos National Laboratory in the USA and the Vulcan laser in the UK were built with the hope, amongst other things, of being able to squash deuterium to produce helium, with the release of the mass difference as energy according to Einstein's law of $E = mc^2$. The massive NOVA laser oscillator–amplifier system in the National Ignition Facility at Lawrence Livermore National Laboratory runs 192 beams into the target chamber, in which sits a 2–3-mm-diameter deuterium pellet conveniently mounted [50]. The beams are so timed that they all strike simultaneously. When all is working as expected, the laser could deliver 2-MJ pulses of around 8 ns duration or 0.25 PW (1 PW = 10^{15} W) [51]. At 23–40 TW of power some indication of success has been achieved (1 TW = 10^{12} W). The laser is currently used to generate X-rays which actually perform the implosion reaction similar to the reaction in a hydrogen bomb. Ultimately the idea is to drop the pellets into the beam to have a maintainable power supply. There is still a long way to go.

The latest Vulcan Nd:YAG oscillator–amplifier laser at Rutherford Appleton Laboratory is capable of delivering 100 TW and is to be upgraded to give petawatt pulses. This may enable physicists to observe reactions occurring in the interior of stars [52].

At Lawrence Livermore National Laboratory they are developing a more realistic approach to generating power from nuclear fusion. The approach is called laser inertial confinement fusion–fission (LIFF) energy, which combines a modest neutron-rich fusion source, such as just described, with a subcritical fission blanket of, for example, unenriched uranium to maintain the neutron flux. It would greatly reduce concerns about proliferation of nuclear fuel, minimise the amount of nuclear waste and eliminate concerns about meltdown and nuclear accidents. It is also considered likely to be able to consume most of the current nuclear waste in the form of spent nuclear fuel. It relies, however, on achieving the fusion reaction by laser ignition. If successful it would go a long way to solving the energy requirements of our crowded planet.

1.4.17 Stimulated Radioactive Decay?

Just as a laser works by a photon of the correct energy stimulating an excited state to give up its energy and move to a lower state, so it seems reasonable that radioactive decay can also be stimulated. Indeed that is the case and has resulted in atomic piles and reactors in which this occurs. Could this be developed to remove the problem of radioactive waste?

The Petawatt laser at Rutherford Appleton Laboratory has been used to fire at a gold target to generate γ-rays that in turn have transmuted iodine-129 (half life of 15.7 million years) into iodine-128 (half life of 25 min). The γ-rays knocked out a neutron to fulfil the reaction. Could that not be used in a stimulated reaction with some feedback mechanism as there is in the laser?

1.5 Market for Laser Applications

The overall sales of lasers into these many application areas is roughly shown in Figure 1.55 [53, 54]. The market for lasers in material processing has been a growth area for several years and is expected to continue to grow at 10–20 % per year for some time yet, though the market crash of 2008 may slow this up for a year or so.

It is hoped that this brief summary of the uses to which lasers have been put will let the reader see the immense new area of applications opened up by the laser and set in perspective the material processing applications discussed in the rest of this book. It may also stimulate the imagination to see some new potential applications and ideas for material processing. This is a new technology which still has a long way to go.

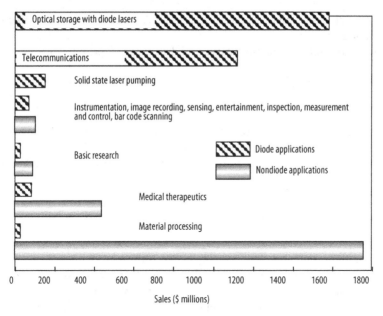

Figure 1.55 Worldwide diode and nondiode laser sales by application, illustrating the predominance of laser material processing in the market for nondiode lasers and the large market for optical storage (CDs, *etc.*) and telecommunications for diode lasers [48]

Questions

1. Dr. Manfred Muller of Munich is approached by the salesperson of Ludwick Laser Ltd claiming that it has improved the gain coefficient for its new YAG laser system by 10 times. Its cavity design includes a mirror with 98 % reflectivity and an output window with 65 % reflectivity. The length of the cavity is 30 cm.

   a. What is the power gain of the new system compared with the old one?
   b. What are the three basic components needed to make a laser?
   c. By how much can Dr. Muller reduce the length of the cavity to make a lighter laser with the same power as the old one to be mounted on a robot in view of the improved gain?

2. When designing a CW CO_2 laser, what are the variables under your control to vary the output of power?

3. If the gain coefficient for your 100 cm-long amplifier is 0.5 cm^{-1} what is the power gain?

4. In what way does the design of the laser cavity mirrors affect the beam structure?

5. Derive an equation for the power from a disc laser based solely on cooling considerations. Calculate what power would be expected from a disc laser if the thermal conductivity of the ytterbium-doped disc YAG material is 0.14 W cm^{-1} K^{-1}, the dimensions of the disc are 7-mm diameter and 0.3-mm thickness and the energy

conversion has a wall plug efficiency of 15 %? Assume the limiting temperature for laser action is 250°C.

6. Write short notes on the advantages and disadvantage the following types of lasers:

   a. disc laser;
   b. fibre laser; and
   c. diode laser.

   For each give a diagram of the construction: a brief outline of the lasing action and the materials used, together with approximate data on performance.

7. Write brief notes to describe the process and applications involved in the following:

   a. speckle interferometry;
   b. holography;
   c. LIDAR; and
   d. laser Doppler velocimetry.

8. Derive the expression $f_{beat} \approx 2v/\lambda_1$ between the beat frequency, f_{beat}, in a laser Doppler velocimetry instrument and the velocity, v, of the target in the direction of the instrument for a monitoring laser beam of wavelength λ_1.

9. List five characteristics that distinguish optical energy from other forms of industrial energy and give two examples of how those characteristics have been utilised.

10. How can a laser be used to measure the distance to the moon and also to calibrate precisely the distance movements of a machine tool?

11. If the moon is 384,400 km away from the earth and the distance is being measured with a nanosecond pulse, how long would the pulse take to make the return flight and what would be the accuracy of the measurement of the distance? The velocity of light is 300,000 km s^{-1}.

12. List the ways in which the laser can be used to store data.

13. What is holography?

14. Using a sketch of the experimental arrangement, describe how a hologram is made.

15. How could a laser be used to measure the strain on a rotating turbine blade?

References

[1] Maitland A, Dunn MH (1969) Laser physics. North-Holland, Amsterdam
[2] Verdeyen JT (1981) Laser electronics. Prentice-Hall, Upper Saddle River
[3] Duley WW (1976) CO_2 lasers: theory and applications. Academic, New York
[4] Thyagrajan K, Ghatak AK (1981) Lasers. Theory and applications. Plenum, New York
[5] Tarasov LV (1983) Laser physics (trans: Wadhwa RS). MIR, Moscow
[6] Schawlow AL, Townes C (1958) Infrared and optimal masers. Phys Rev 112:1940–1949
[7] Einstein A (1916) Strahlungs-emission und -absorption nach der Quantentheorie. Verh Dtsch Phys Ges 18:318–323
[8] Hänsch TA, Pernier M, Schawlow AL (1971) Laser action of dyes in gelatin. IEEE J Quantum Electron QE-7:47
[9] http://www.cord.org/cm/leot/course01_mod06
[10] Bass M (ed) (1995) Handbook of optics. McGraw-Hill, New York
[11] Kogelnik H, Li T (1996) Laser beams and resonators. Appl Phys 5(10):1550

[12] Crafer RC, Oakley PJ (1993) Laser processing in manufacturing. Chapman and Hall, New York

[13] Hall DR, Baker HJ (1994) Diffusion cooled large surface area CO_2/CO lasers. Proc SPIE 2505:12–19

[14] Duley WW (1996) UV lasers. Effects and applications in material science. Cambridge University Press, Cambridge

[15] Gitin M, Reingrube J (1995) Diode pumped solid state lasers show a bright future. Ind Laser Rev Dec 8–10

[16] Hill P (2009) Fibre laser hits 2 kW record mark. Opto & Laser Europe Jul/Aug 9

[17] Wintner E (1998) Semiconductor lasers. In: Schuöcker D (ed) Handbook of the Eurolaser Academy, vol 1. Chapman and Hall, New York, chap 6

[18] Hewett J (2002) Opto & Laser Europe May 41–43

[19] Opto & Laser Europe (2004) Photonics to revolutionise the world. Opto & Laser Europe Jan 24–25

[20] Kelly J (1995) Beam guides machines through complex curves. Opto & Laser Europe 24:21–22

[21] Hewitt J (2004) Unpickable optical lock aims to foil car criminals. Opto & Laser Europe Apr 11

[22] Steinmetz CR (1990) Laser interferometry operates at submicron level. Laser Focus World Jul 93–98

[23] Hariharan P, Creath K (2004) Interferometry. In: Brown TG et al. (eds) The optics encyclopedia. Wiley-VCH, Weinheim, p 954

[24] Morrison DC (1990) Laser technology enlists the anti-drug campaign. Lasers and Optronics May 31–32

[25] Waggoner J (1990) SDIO says laser radar works. Photonics Spectra Jul 18

[26] Nordstrom RJ, Berg LJ (1990) Coherent laser radar: techniques and applications. Lasers and Optronics Jun 51–56

[27] Arnt W (1990) Laser ranging keeps cars apart. Photonics Spectra Jul 133–134

[28] Reunert MK, Yoshiba B (1996) Fibre optic gyroscope: a new sensor for robotics. Sens Rev 16(1):32–34

[29] Martha H (1993) Fiber optic gyros help Tokyo drivers navigate. Photonics Spectra 27(1):18

[30] Carts YA (1990) Media lab develops holographic video. Laser Focus World May 95

[31] Benton SA (1969) Hologram reconstructions with extended incoherent sources. J Opt Soc Am 59:1545

[32] Butters JN (1983) Speckle interferometry and other technologies. In: Proceedings of the 1st international conference on lasers in manufacturing (LIM1), Brighton, November 1983, pp 149–160

[33] Opto & Laser Europe (2001) Speckle technique tracks blood flow inside brain. Opto & Laser Europe Aug 14

[34] Hogan H (2008) No longer walking the line. Photonics Spectra Jun 22–23

[35] Kinkade K (2004) Laser reveals critical changes in the earth's atmosphere. Laser Focus World May 154

[36] Anscombe N (2007) Crop spraying system targets weeds. Photonics Spectra Feb 36–37

[37] Cowdery E (2002) Keeping an eye out for bad marks. Materials World Jun 31–32

[38] Tsufura L (1995) Barcode scanning: on going evolution and development. Lasers and Optronics Jul 25–27

[39] Anonymous (1996) Laser paint assists in search and rescue. In: Photonics design and applications handbook. Laurin, Pittsfield, p H515

[40] Hirleman D (2002) Lasers – the non-destructive answer to dust investigation. Materials World Jun 30

[41] Hogan H (2002) Laser system detects explosives remotely. Photonics Spectra Apr 35

[42] Higgins TV (1995) Optical storage lights the multi-media future. Laser Focus World Sep 103–111

[43] Milster TD (2004) Data storage, optical. In: Brown TG et al (eds) The optics encyclopedia. Wiley-VCH, Weinheim, pp 227–274

[44] Lenth B (1994) Optical storage: a growth mass market for lasers. Laser Focus World Dec 87–91

[45] Hewett J (2004) Holographic drives set for long awaited debut. Opto & Laser Europe Jul/Aug 15–17

[46] Higgins TV (1995) Light speed communications. Laser Focus World Aug 67–74

[47] Anscombe N (2002) Defending the skies: the airborne laser. Optics & Laser Europe May 33–34

[48] Rothacher T, Lütthy W, Weber HP (2004) Demining with Nd:YAG laser. Rev Sci Instrum 75(4):1078–1080

[49] Heller A (2004) Laser burrows into the earth to destroy landmines. https://www.llnl.gov/str/October04/Rotter.html

[50] Bibeau C, Rhodes MA, Atherton J (2006) World's largest laser. Photonics Spectra Jun 50–60

[51] Wallace J (2003) Laser pulse delivers ignition-sized punch. Laser Focus World Aug 24–28

[52] Hatcher M (2002) Vulcan upgrade: power to the people. Opto & Laser Europe May 30–31

[53] Laser Focus World (2007) Laser Focus World Jan 82–100

[54] Laser Focus World (2007) Laser Focus World Feb 67–77

[55] Martin WS, Chernock, J.B., U.S. patent # 3,633,126, Multiple internal Reflection Face Pumped Laser

[56] Duarte FJ, Foster DR (2004) Lasers-dye. In: Brown TG et al (eds) The optics encyclopedia. Wiley, New York, p 1082

[57] Techtran Corporation, Lasers in metal working. Techtran Corporation (flyer)

"Stimulated emission is not a new phenomenon."

2 Basic Laser Optics

Boswell: Then, Sir, what is poetry? Johnson: Why, Sir, it is much easier
to say what it is not. We all know what light is; but it is not easy to tell
what it is

Boswell's Life of Johnson

Open the second shutter so that more light can come in

Attributed as the dying words of Johann Wolfgang von Goethe

(1749–1832)

In this chapter the basic nature of light and its interaction with matter is described and
the fundamentals of how such energy can be manipulated in direction and shape are
presented.

2.1 The Nature of Electromagnetic Radiation

Electromagnetic radiation has been a puzzle ever since man first realised it was there.
Pierre de Fermat (1608–1665) stated the principles of ray propagation: "The path taken
by a light ray in going from one point to another through any set of media is such as
to render its optical path equal, in the first approximation, to other paths closely adjacent to the actual one" (*i.e.*, the path will be the one with the minimum time: a concept
much in vogue at the time following in the traditions laid down by Euclid and Hero of
Alexandria). This is a rather complicated statement from which the laws of reflection
and refraction can be derived. Christian Huygens (1629–1695) [1] introduced the wave
concept of light to explain refraction and reflection. This he did through the "Huygens
principle" that *each point on a wavefront may be regarded as a new source of waves*. Sir
Isaac Newton (1642–1727) in 1704 unravelled the puzzle of colour and introduced the
concept of light consisting of a number of tiny particles moving through space and subject to mechanical forces, the "corpuscular theory" [2]. Albert Einstein (1879–1955) [3]
in 1905 invented the concept of the photon to explain the photoelectric effect and gave
birth to the quantum theory of radiation. In fact there is still some mystery left. For example, if light passes through two parallel slits and then falls on a screen, as in Thomas
Young's (1773–1829) famous double-slit experiment, a diffraction pattern is formed on

the screen. The phenomenon can be simply explained by assuming that the radiation passing through the slits expands as a wavefront from the slits and makes an interference pattern on the screen. It is difficult to explain the outcome of the experiment by assuming the light is a stream of particles. However, in the photoelectric effect light falling on a target will give off electrons of fixed energy, E, from the target regardless of the intensity of the incident light. E is given by

$$E = h\nu - p, \tag{2.1}$$

where h is Planck's constant (6.625×10^{-34} J s), ν is frequency (c/λ, where c is the velocity of light, i.e., 2.99×10^8 m s^{-1}, and λ is the wavelength of light in metres) and p is a constant characteristic of the material.

In the wave theory, the radiation would be spread over the surface and would not all be available for one electron.

This dichotomy between waves and particles varies in significance with the wavelength or energy of the "photons". Thus, at the long wavelengths from radio to blue light, the wave theory explains most phenomena observed for normal intensities. With X-rays and γ-rays, which are highly energetic photons of short wavelength, the particle theory explains most events.

The quantum theory, of which we are talking here, was initiated by Werner Karl Heisenberg (1901–1976) [4] and Erwin Schrödinger (1887–1961) [5] in 1926. It makes a link between these states through Neil Bohr's analysis of Planck's constant. He suggested that the constant is the product of two variables, one characteristic of the wave and the other of a particle. Thus, if the wave has a period, T, a wavelength, λ, particle energy, E, and momentum, p, Bohr suggested, on dimensional grounds amongst others, that $h = ET = p\lambda$. Thus, if the particle aspects are strong, then the wave aspects will be weak. It just happens that the size of Planck's constant is such that the electromagnetic spectrum takes us from strongly particle type radiation to strongly wave type radiation. Why Planck's constant is of such a size is unknown and must be left as an exercise for the readers and their heirs and successors! However, this concept that $\lambda = h/p$ suggests all matter with momentum has a wavelength. This was shown to be the case for electrons by Davisson and Germer in the USA and G.P. Thomson in the UK, but the size of h makes the wavelength very small. The wavelength of Earth, for example, would be calculated as follows. The mass of Earth $m = 5.976 \times 10^{24}$ kg and the velocity of Earth $v = 3 \times 10^4$ m s^{-1}; therefore, Earth's wavelength $\lambda = 6.625 \times 10^{-34}/(5.976 \times 10^{24} \times 3 \times 10^4) = 3.7 \times 10^{-63}$ m, which is a bit difficult to measure!

The momentum of a photon can be found from Planck's law $E = h\nu$ (justified from the photoelectric effect and other phenomena, where ν is the frequency) and Einstein's equivalence of mass and energy $E = mc^2$ (justified by experiments on nuclear disintegration).

Together these give

$$h\nu = hc/\lambda = mc^2, \tag{2.2}$$

and since the momentum $p = mc$ we have $p = h/\lambda$, which is the same as Bohr's relationship quoted earlier.

Table 2.1 Photon properties of different lasers

Device	Source of laser energy	Wavelength, λ (μm)	Frequency, ν (Hz)	Energy, E^a (eV)	(J $\times 10^{-20}$)
Cyclotron	Accelerator	0.1 (X-ray)	2.9×10^{15}	12.3	192
Free-electron laser	Magnetic wiggler	$1 \times 10^3 - 10^6$	$1 \times 10^8 - 10^{11}$	1×10^{-6}	$1 \times 10^{-2} - 10^{-5}$
Excimer laser	Atomic electron orbits	0.249 (UV)	1.2×10^{15}	4.9	79.4
Argon ion		0.488 (blue)	6.1×10^{14}	2.53	40.4
He–Ne laser		0.6328 (red)	4.7×10^{14}	1.95	31.1
Nd:YAG laser	Molecular vibration	1.06 (IR)	2.8×10^{14}	1.16	18.5
CO laser		5.4	5.5×10^{13}	0.23	3.64
CO$_2$ laser		10.6	2.8×10^{13}	0.12	1.85

^{a} Energy calculated from $E = h\nu$; 1 eV = 1.6×10^{-19} J.

Incidentally this suggests that the pressure, P, on a mirror from a photon from a CO_2 laser incident normally is (see Section 12.2.3.1) $2p = P = 2 \times 6.625 \times 10^{-28}/10.6 \times 10^{-6} = 1.25 \times 10^{-28}$ N s per photon, not of any great significance until one considers the avalanche of photons possible with the laser.

The energy of a photon from a CO_2 laser is given as 1.85×10^{-20} J in Table 2.1, where it is compared with the energy of photons from other optical generators. Thus, in a 1-kW CO_2 laser beam there will be a flux of $1,000/1.85 \times 10^{-20} = 5 \times 10^{22}$ photons per second and the overall force will be 6×10^{-6} N – still not very exciting, but possibly measurable. However, over the focused spot from this laser, of, say, 0.1 mm diameter, the pressure would be $(4 \times 6 \times 10^{-6})/[\pi(0.1 \times 10^{-3})^2] = 760$ N m^{-2}. This is equivalent to a depression in molten steel of approximately 1 cm! This is very close to what is observed. One wonders whether we have missed something in ignoring photon pressure.

It is assumed that the velocity of a photon is always c, the velocity of light in a vacuum or the limiting velocity of all objects with finite rest mass, and that this is a universal constant. Photons do not behave as normal particles, which can have a variable velocity. The early explanations of refraction, for example, in which the wave theory explains the process by suggesting that the velocity of light varies from one medium to another, has to be interpreted as follows: the photon travels at the speed c always, but in passing through a medium, the wavefront slows owing to the absorption/re-emission processes taking place as the photon interacts with the molecules of the medium through which it travels. The reason for this universal constant is related to the concept of time: it has an uncanny ring that we have more thinking to do to understand this subject.

2.2 Interaction of Electromagnetic Radiation with Matter

When electromagnetic radiation strikes a surface, the wave travels as shown in Figure 2.1. Some radiation is reflected, some absorbed and some transmitted. As it passes

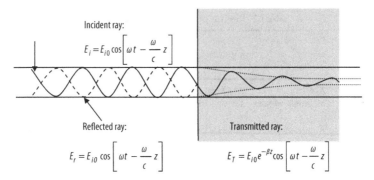

Incident ray:

$$E_i = E_{i0} \cos\left[\omega t - \frac{\omega}{c} z\right]$$

Reflected ray:

$$E_r = E_{i0} \cos\left[\omega t - \frac{\omega}{c} z\right]$$

Transmitted ray:

$$E_T = E_{i0} e^{-\beta z} \cos\left[\omega t - \frac{\omega}{c} z\right]$$

Figure 2.1 Phase and amplitude, E, of an electromagnetic ray of frequency ω travelling in the z direction striking an air–solid interface and undergoing reflection and transmission

through the new medium, it will be absorbed according to some law such as the Beer–Lambert law, $I = I_0 e^{-\beta z}$. The absorption coefficient, β, depends on the medium, the wavelength of the radiation and the intensity (see Section 2.2.1). The manner in which this radiation is absorbed, reflected or transmitted is considered to be as follows. Electromagnetic radiation can be represented as an electric vector field and a magnetic vector field as illustrated in Figure 2.2. When this passes over a small charged particle, the particle will be set in motion by the electric force from the electric field, E. Provided that the frequency of the radiation does not correspond to a natural resonance frequency of the particle, then fluorescence or absorption will not occur, but a forced vibration would be initiated. The force induced by the electric field, E, is very small and is incapable of vibrating an atomic nucleus. We are therefore discussing photons interacting with electrons which are either free or bound. This process of photons being absorbed by electrons is known as the "inverse bremsstrahlung effect". (The bremsstrahlung effect is the emission of photons from excited electrons.) As the electron vibrates so it

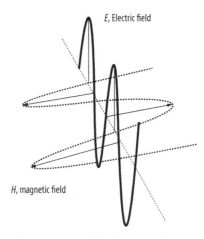

E, Electric field

H, magnetic field

Figure 2.2 The electric and magnetic field vectors of electromagnetic radiation

will either re-radiate in all directions (the reflected and transmitted radiation) or be re-strained by the lattice phonons (the bonding energy within a solid or liquid structure), in which case the energy would be considered absorbed, since it no longer radiates. In this latter case the phonons will cause the structure to vibrate and this vibration will be transmitted through the structure by the normal diffusion-type processes due to the linking of the molecules of the structure. We detect the vibrations in the structure as heat. The flow of heat is described by Fourier's laws on heat conduction – a flux equation $(q/A = -k\mathrm{d}T/\mathrm{d}x)$ (see Chapter 5). If sufficient energy is absorbed, then the vibration becomes so intense that the molecular bonding is stretched so far that it is no longer capable of exhibiting mechanical strength and the material is said to have melted. On further heating, the bonding is further loosened owing to the strong molecular vibrations and the material is said to have evaporated. The vapour is still capable of absorbing the radiation but only slightly since it will only have bound electrons; with sufficient absorption the electrons are shaken free and the gas is then said to be a plasma.

Plasmas can be strongly absorbing if their free-electron density is high enough. The electron density in a plasma is given by equations such as the Saha equation (2.3) [6], which assumes thermal equilibrium in the plasma so that standard free-energy changes can be calculated using conventional thermodynamic principles, which is not necessarily true with short laser pulses:

$$\ln\left(\frac{N_1}{N_0}\right)^2 = -5040\left(\frac{V_1}{T}\right) + 1.5\ln\left(T + 15.385\right), \qquad (2.3)$$

where N_1 is the ionisation density, N_0 is the density of atoms, V_1 is the ionisation potential (eV) and T is the absolute temperature (K).

This indicates that temperatures of the order of 10,000–30,000 °C are required for significant absorption (Figure 2.3) [7]. This sequence in the stages of absorption is illustrated in Figure 2.4.

It is interesting to note that the energy absorbed by an electron may be that of one or more photons; however, it will only be in extreme cases, such as the Vulcan laser operating at 1 PW or so that a sufficient number of photons would be simultaneously

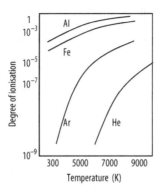

Figure 2.3 Degree of ionisation as a function of temperature

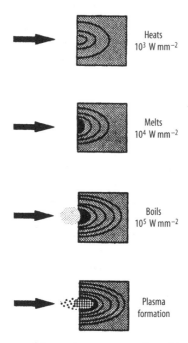

Figure 2.4 Sequence of absorption events varying with absorbed power

absorbed to allow the emission of X-rays during laser processing. This is a strategic advantage for the laser over electron beam processes, which require shielding against this hazard.

At these very high photon fluxes the electric field is sufficient to strip electrons from the atoms, which become charged and then repel each other. With femtosecond pulses (10^{-15} s) there is no time for conduction and so the material forms a solid-state plasma, similar no doubt to the interior of stars.

Incidentally, the mean free time between collisions of electrons in a conductor is calculated to be around 10^{-13} s. This means that only for extremely short laser pulses of around 1 ps (10^{-12} s per pulse) is it possible that the material would contain two temperatures not at equilibrium – the electron temperature and the atomic temperature. Also, for very short pulses non-Fourier conduction has been postulated [8], in which a compression or heat wave forms; this may be related to the acoustic signals noted in Section 12.2.3.1 or shock hardening mentioned in Section 6.19.

2.2.1 Nonlinear Effects

Ordinarily, the optical effects we experience are linear effects. When light interacts with matter, the matter responds in a proportionate way. Thus, we have the linear effects of reflection, refraction, scattering and absorption, all of which occur at the same frequency; the frequency of the light is not altered by the process. However, in 1961 Peter

Franken and others at the University of Michigan focused a high-powered ruby laser (red light) onto a quartz crystal and generated ultraviolet light mixed with the transmitted light. This was the birth of the new subject of nonlinear optics.

Today many electro-optic devices of practical importance depend upon nonlinear optical effects. These effects include second-harmonic generation as observed by Franken and his colleagues and optical rectification, the Pockel's electrooptic effect, sum and difference frequency mixing, the Kerr electro-optical effect, third-harmonic generation, general four-wave mixing, the optical Kerr effect, stimulated Brillouin scattering, stimulated Raman scattering, phase conjugation, self-focusing, self-phase modulation and two-photon absorption, ionisation and emission.

This exciting new area of physics has been opened up by the laser since the focused beam can generate huge electric and magnetic fields affecting the atomic dipoles (Lorentzian dipoles). At normal levels of radiation, several watts per square metre, the dipoles respond in one-to-one correspondence with the driving force, in fact linearly; however, at high levels of irradiation, several megawatts per square metre, the dipoles no longer respond linearly but more in the style of an overdriven pendulum and they exhibit a variety of harmonic oscillations. Via such effects it is possible to mix the frequencies of light waves. This is quite remarkable and against all the principles of the superpositioning of waves which were used to explain so much of earlier light theory, such as Young's experiment.

2.2.1.1 Fluorescence

If a solid or a liquid is strongly illuminated by a frequency of radiation that it is able to absorb, it will become excited. To lose this energy the structure may simply become hot, or re-radiate at the same frequency "resonance radiation" or at a lower frequency

Figure 2.5 Optical excitation causing fluorescence. *1* photoabsorption excites the molecule from the ground electronic state S_0 to a vibrationally excited state in the first singlet state S_1. *2* rapid radiationless decay occurs to a lower level of S_1 through intramolecular vibrational relaxation. *3* fluorescence decay occurs as the level falls back to the S_0 state at a higher vibrational level of that state. *4* radiationless decay to the ground state

"fluorescence". The lower frequency is predicted by Stokes law (Sir George G. Stokes, 1819–1903, Lucasian Professor of Mathematics at Cambridge University, who worked on spectroscopy, diffraction, viscosity – another Stokes law – and vector analysis). The reason is illustrated in Figure 2.5.

Fluorescence lifetime is an important diagnostic tool in medical studies to determine chemical groups such as amino acids and their environment – the subject is known as "*fluorimetry*" (see Chapter 11). Some materials emit very slowly and can be seen to glow after exposure as in the case of phosphors on watches and some TV screens – *phosphorescence*.

Fluorescent radiation is usually of a lower frequency than the stimulating radiation but it may be at a higher frequency (anti-Stokes radiation) if some extra energy is provided by the material being hot or a multiphoton event occurring.

Fluorescence of some materials can be stopped by irradiating them with infrared radiation. This has the effect of removing the excess energy in the structure of the material as heat. There are some commercial fluorescent screens on the market which will fluoresce in ultraviolet light from a lamp; the glowing screen can be used to image an infrared laser beam falling on it. On the other hand, a change in frequency can in some cases stimulate the fluorescence.

2.2.1.2 Stimulated Raman Scattering

If low-intensity light is transmitted through a transparent material, a small fraction is converted into light at longer wavelengths, with the frequency shift (Stokes shift) corresponding to the optical phonon frequency in the material. This process is called Raman scattering; see Figure 2.6. At higher intensities Raman scattering becomes stimulated and from the spontaneous scattering a new light beam can be built up. Under favourable

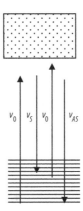

Figure 2.6 Raman scattering. Input radiation of v_o is inelastically scattered. In Stokes Raman scattering an overall transition to a higher vibrational state occurs, giving less energetic radiation of frequency v_S. In anti-Stokes Raman scattering the radiant shift is from a higher vibrational state to a lower one, giving more energetic radiation. Thus, Raman spectroscopy gives data on vibrational levels of a molecule, from which it can sometimes be identified

conditions, the new beam can become more intense than the remaining original beam. The amplification is equally high in the forward and the backward directions. This may lead to a situation where a large fraction of the radiation is redirected towards the light source rather than towards the target. This could be a problem with intense light being transmitted in fibres, but also forms the basis of certain detection techniques, such as LIDAR (see Section 1.4.7).

2.2.1.3 Stimulated Brillouin Scattering

The same process takes place with the acoustical phonons as opposed to the lattice vibrations. The corresponding frequency shift is much smaller. Acoustical phonons are sound waves and the frequency shift exists only for the wave in the backward direction. Again, at high intensities the Brillouin effect becomes a stimulated process and the Brillouin wave may become much more intense than the original beam. Almost the entire beam may be reflected towards the laser source.

2.2.1.4 Second-harmonic Generation

Light waves are not supposed to interact with one another, but in the case of nonlinear interactions the nonlinear radiation itself couples the energy from one beam to another. This would not be possible in a vacuum. One can imagine the overstimulated structure being distorted and so affecting the absorption of other beams. In second-harmonic generation the nonlinear polarisation wave moves through the structure at one velocity and the primary refracted wave moves at another. For them to interact constructively, the phase velocities of the two waves must match. This can be done by using birefringent crystals, such as lithium niobate ($LiNbO_3$), lithium borate (LiB_3O_5) and others as listed in Table 2.2, whose refractive index depends on the direction and polarisation of the propagating light. If a polarised light wave passes through a birefringent crystal at just the right angle, the phase velocities of the induced polarisation wave and the second-harmonic wave can be made equal. However, this does mean that the angle and the temperature of the crystal have to be very carefully maintained. Once done, though, the effect is near magic. Thus, for example, a beam from a Nd:YAG laser is shone into the $LiNbO_3$ crystal held in a temperature enclosure at the correct an-

Table 2.2 Common electro-optic materials

Quartz	CdS
$Ba_2NaNb_5O_{15}$	Ag_3AsS_3 (proustite)
$LiNbO_3$	$CdGeAs_2$
$BaTiO_3$	$AgGaSe_2$
$NH_4H_2PO_4$	$AgSbS_3$ (pyrargyrite)
KH_2PO_4	β-BaB_2O_4
$LiIO_3$	β-Barium borate
CdSE	$KTiOPO_4$
KD_2PO_4	LiB_3O_5

gle and the invisible infrared beam of 1.06 µm emerges, with some 30 % converted to green light at 0.53 µm. Frequency tripling can also be obtained from crystals of different structures.

2.2.1.5 The Kerr Effect

When light is reflected from a magnetised medium, its state of polarisation and even its amplitude are changed. This effect is known as the *Kerr effect* after John Kerr (1824–1907), a Scottish physicist who was one Lord Kelvin's first research students. When light is reflected from a surface, the surface electrons are moved by the incoming radiation electric field. If there is a magnetic field, then the direction of movement of the electrons will be affected as by the normal laws of electromagnetism and their angle of movement will be altered and hence the angle of polarisation with which they are emitted will be altered owing to their change in direction. The effect depends on the direction and strength of the magnetic field relative to the radiation.

There is an "optical Kerr effect", which is a third-order nonlinear polarisation effect which can cause a change in the refractive index of the material subject to high-intensity radiation (see Section 2.2.1.6).

One of the more bizarre effects using this optical Kerr effect is optical phase conjugation. In one form, called degenerate four-wave mixing, two beams converge in the material and set up a form of grating within the material; a third wave couples nonlinearly with the others to form a phase-conjugated wave. This principle is applied to phase-conjugated mirrors. Phase-conjugated mirrors return the light to the source; any distortions between the source and the phase-conjugated mirrors are automatically compensated because of the phase reversal. Phase-conjugated mirrors are finding their way into commercial lasers to mitigate beam distortions and applications in adaptive optics are under development [9]. Self-focusing fibres are also a possibility using this effect.

2.2.1.6 The Pockel Effect

When an electric field is applied to certain materials, the electrostatic forces can distort the locations of the molecules of the material and result in a redistribution of the internal charges, causing a change in refractive index for noncentrosymmetric crystals such as CdTe and GaAs and anisotropic materials such as $LiNbO_3$ and KDP. This is known as the *linear electro-optic effect*, or the *Pockel's effect*. The effect is used in a Pockel cell to spoil the lasing oscillations in some solid-state lasers by deflecting the beam. This is one form of Q switch known as an electro-optic Q switch (see Section 1.3.2.1).

For materials that have inversion symmetry, such as silicon, germanium, diamond and liquids and gases in general, the Pockel effect vanishes and the second-order electro-optic effect becomes noticeable, known as the optical Kerr effect (see the previous section).

2.3 Reflection or Absorption

The value of the absorption coefficient will vary with the same effects that affect the reflectivity. For opaque materials,

$$\text{Reflectivity} = 1 - \text{absorptivity}.$$

For transparent materials,

$$\text{Reflectivity} = 1 - (\text{transmissivity} + \text{absorptivity}).$$

In metals the radiation is predominantly absorbed by free electrons in an "electron gas". These free electrons are free to oscillate and re-radiate without disturbing the solid atomic structure. Thus, the reflectivity of metals is very high in the waveband from the visible to the DC, i.e., very long wavelengths; see Figure 2.7. As a wavefront arrives at a surface, then all the free electrons in the surface vibrate in phase, generating an electric field 180° out of phase with the incoming beam. The sum of this field will be a beam whose angle of reflection equals the angle of incidence. This "electron gas" within the metal structure means that the radiation is unable to penetrate metals to any significant depth, only one to two atomic diameters. Metals are thus opaque and they appear shiny.

The reflection coefficient for normal angles of incidence from a dielectric or metal surface in air ($n = 1$) may be calculated from the refractive index, n, and the extinction coefficient, k (or absorption coefficient as described above), for that material:

$$R = \left[(1-n)^2 + k^2\right] / \left[(1+n)^2 + k^2\right]. \tag{2.4}$$

For an opaque material such as a metal, the absorptivity, A, is

$$A = 1 - R,$$
$$A = 4n / \left[(n+1)^2 + k^2\right]. \tag{2.5}$$

Some values of these constants are given in Tables 2.3 and 2.4. The value of the reflectivity, R, shown in Table 2.3 is 1 for a perfectly flat clean surface – which is rarely the case.

The variation of the amplitude of the electric field, E, with depth, d, is given by the Beer–Lambert law for a wavelength λ in a vacuum as

$$E = E_0 \exp(-2\pi k d / \lambda).$$

The intensity is proportional to the square of the amplitude and hence the variation of intensity with depth is given by

$$I = I_0 \exp(-4\pi k d / \lambda). \tag{2.6}$$

For example, iron has a value of the extinction coefficient, k, of 4.49 (Table 2.4) for 1.06-µm radiation. Thus, the intensity would have fallen to $1/e^2$ (i.e., 0.13 times the incident value) after a depth of 0.038 µm; and for 10.6-µm radiation with $k = 32.2$ (Table 2.4), this depth becomes 0.052 µm.

Table 2.3 Complex refractive index and coefficient of reflection for some materials to 1.06-μm radiation [45]

Material	k	n	R
Al	8.50	1.75	0.91
Cu	6.93	0.15	0.99
Fe	4.44	3.81	0.64
Mo	3.55	3.83	0.57
Ni	5.26	2.62	0.74
Pb	5.40	1.41	0.84
Sn	1.60	4.70	0.46
Ti	4.0	3.8	0.63
W	3.52	3.04	0.58
Zn	3.48	2.88	0.58
Glass	0	1.5	0.04

Table 2.4 Refractive index and Brewster angles for various materials

Material	λ (μm)	Refractive index		Brewster angle
		k	n	
Al	1.06	8.5	1.75	60.2
	10.6	34.2	0.108	88.3
Fe	1.06	4.49	3.81	75.2
	10.6	32.2	5.97	88.2
Ti		3.48	2.88	70.8
Glass		–	1.5	56.3

2.3.1 Effect of Wavelength

At shorter wavelengths, the more energetic photons can be absorbed by a greater number of bound electrons and so the reflectivity falls and the absorptivity of the surface is increased (Figure 2.7).

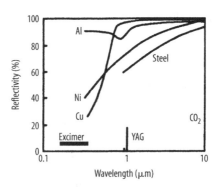

Figure 2.7 Reflectivity of a number of metals as a function of temperature

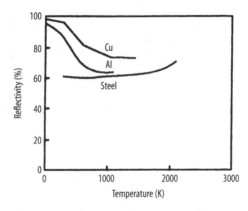

Figure 2.8 Reflectivity as a function of temperature for 1.06-μm radiation

2.3.2 Effect of Temperature

As the temperature of the structure rises, there will be an increase in the phonon population, causing more phonon–electron energy exchanges. Thus, the electrons are more likely to interact with the structure rather than oscillate and re-radiate. There is thus a fall in the reflectivity and an increase in the absorptivity with a rise in temperature for some metals, as seen in Figure 2.8 [10].

2.3.3 Effect of Surface Films

The reflectivity is essentially a surface phenomenon and so surface films may have a large effect. Figure 2.9 shows that for interference coupling the film must have a thickness of around $[(2n + 1)/4]\lambda$ to have any effect, where n is any integer. The absorption variation for CO_2 radiation by a surface oxide film is shown in Figure 2.10 [10,11]. One form of these surface films may be a plasma [12] provided that the plasma is in thermal contact with the surface.

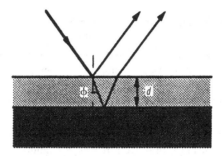

Figure 2.9 A surface film as an interference coupling, "antireflection" coating. If $2d/\cos\phi = [(2n + 1)/2]\lambda$, then there will be destructive interference of the reflected ray

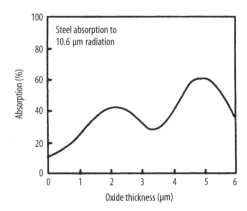

Figure 2.10 Absorption as a function of the thickness of an oxide film on steel for 1.06-μm radiation

2.3.4 Effect of Angle of Incidence

The full theoretical analysis of reflectivity was first done by Drude [13] from atomistic considerations of the electron flux in a radiant field, which he then applied to the Maxwell (1831–1879) equations. It is sometimes known as "Drude reflectivity". It showed a variation in reflectivity with both the angle of incidence and the plane of polarisation. If the plane of polarisation is in the plane of incidence, the ray is said to be a "p" ray (parallel); if the ray has its plane of polarisation at right angles to the plane of incidence, it is said to be an "s" ray (*Senkrecht* meaning "perpendicular"). The reflectivities for these two rays reflected from perfectly flat surfaces are given by:

$$R_p = \frac{[n - (1/\cos\phi)]^2 + \kappa^2}{[n + (1/\cos\phi)]^2 + \kappa^2} \tag{2.7}$$

$$R_s = \frac{[n - \cos\phi]^2 + \kappa^2}{[n + \cos\phi]^2 + \kappa^2}. \tag{2.8}$$

The variation of the reflectivity with angle of incidence is shown in Figure 2.11. At certain angles the surface electrons may be constrained from vibrating since to do so would involve leaving the surface. This they would be unable to do without disturbing the matrix, *i.e.*, absorbing the photon. Thus, if the electric vector is in the plane of incidence, the vibration of the electron is inclined to interfere with the surface at high angles of incidence and absorption is thus high; however, if the plane is at right angles to the plane of incidence, then the vibration can proceed without reference to the surface or angle of incidence and reflection is preferred. There is a particular angle – the "Brewster angle" – at which the angle of reflection is at right angles to the angle of refraction. When this occurs it is impossible for the electric vector in the plane of incidence to be reflected since there is no component at right angles to itself. Thus, the reflected ray

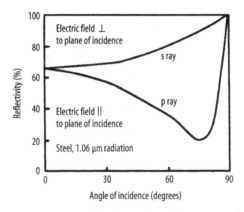

Figure 2.11 Reflectivity of steel to polarised 1.06-μm radiation

will have an electric vector mainly in the plane at right angles to the plane of incidence. This is the reason why Polaroid®[1] spectacles reduce the glare from puddles. At this angle the angle of refraction = (90° − angle of incidence) and hence by Snell's law (see Section 2.4) the refractive index, n = tan(Brewster angle). Any beam which has only one or principally one plane for the electric vector is called a "polarised" beam. Some values of the refractive index and the Brewster angles for different materials are given in Table 2.4.

Most lasers produce beams which are polarised owing to the nature of the amplifying process within the cavity which will favour one plane. Any plane will be favoured in a random manner, unless the cavity has folding mirrors, in which case the electric vector, which is at right angles to the plane of incidence on the folded mirrors, will be favoured because that is the one suffering the least loss.

2.3.5 Effect of Materials and Surface Roughness

Roughness has a large effect on absorption owing to the multiple reflections in the undulations (see Table 6.1, page 299). There may also be some "stimulated absorption" due to beam interference with sideways-reflected beams [14]. Provided the roughness is less than the beam wavelength, the radiation will not suffer these events and hence will perceive the surface as flat. The reflected phase front from a rough surface, formed from the Huygens wavelets, will no longer be the same as the incident beam and will spread in all directions as a *diffuse reflection*. It is interesting to note that it should not be possible to see the point of incidence of a red He–Ne beam on a mirror surface if the mirror is perfect.

---

[1] Polaroid® is a registered trademark of the Polaroid Corporation 4350 Baker Road Minnetonka, MN 55343-8684, USA. www.polaroid.com

2.4 Refraction

On transmission the ray undergoes refraction described by Snell's law (Willebrord Snell, 1591–1626, Professor of Mathematics at Leiden University, Holland): "The refracted ray lies in the plane of incidence, and the sine of the angle of refraction bears a constant ratio to the sine of the angle of incidence":

$$\sin \varphi / \sin \psi = n = v_1 / v_2 , \tag{2.9}$$

where n is the refractive index, φ is the angle of incidence, ψ is the angle of refraction, v_1 is the apparent speed of propagation in medium 1 and v_2 is the apparent speed of propagation in medium 2.

The apparent change in the velocity of light as it passes through a medium is the result of scattering by the individual molecules. The scattered rays interfere with the primary beam, causing a retardation in the phase. Consider a plane wave striking a very thin, transparent sheet whose thickness is less than the wavelength of the incident light [15], as shown in Figure 2.12. Let the electric vector have a unit amplitude and then it can be represented at a particular time as $E = \sin(2\pi x / \lambda)$. If the scattered intensity is small, then the intensity reaching some point, P, will be essentially the intensity of original wave plus a small contribution from all the light scattered from all the atoms of the sheet. Now the energy scattered by one atom will be proportional to its scattering cross-section, σ, which is that part of the area of the atom presented to the oncoming radiation. Thus, the scattered amplitude is proportional to $\sqrt{\sigma}$. If there are N atoms per cubic centimetre, the total scattered amplitude per square centimetre would be proportional to $Nt\sqrt{\sigma}$; where t is the thickness. Since it is assumed that $t \sim \lambda$, the waves leaving the sheet will all be in phase. At point P, however, their phases will differ by the different distances travelled, R. We can calculate the net effect by summing the scattered amplitudes of all the atoms over the surface, E_s – allowing for the amplitude being proportional to $1/R$:

$$E + E_s = \sin\left(\frac{2\pi x}{\lambda}\right) + \sqrt{\sigma} Nt \int_0^\infty \frac{2\pi r\, dr}{R} \sin\left(\frac{2\pi R}{\lambda}\right) .$$

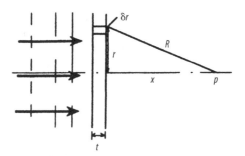

Figure 2.12 Radiation passing through a thin transparent layer

Since $x^2 + r^2 = R^2$ and x is constant, we have $r\,dr = R\,dR$, and the integral may be written as

$$\int_0^\infty \frac{2\pi}{R} \sin\left(\frac{2\pi R}{\lambda}\right) r\,dr = 2\pi \int_x^\infty \sin\left(\frac{2\pi R}{\lambda}\right) dR = \frac{2\pi\lambda}{2\pi}\left[-\cos\left(\frac{2\pi R}{\lambda}\right)\right]_{R=x}^{R=\infty}.$$

(The integral limits are 0 to ∞ for r and x to ∞ for R.)

At $R = \infty$, the quantity in brackets is equal to zero and so we have

$$E + E_s = \sin\left(\frac{2\pi x}{\lambda}\right) + \sqrt{\sigma} N t \lambda \cos\left(\frac{2\pi x}{\lambda}\right).$$

This is of the form $\sin A + B \cos A$, where B is assumed to be very small. Under these conditions we may write

$$\sin(A + B) = \sin A \cos B + \cos A \sin B \approx \sin A + B \cos A.$$

Therefore,

$$E + E_s = \sin\left(\frac{2\pi x}{\lambda} + \sqrt{\sigma} N t \lambda\right),$$

which shows that the phase of the wave at point P has been altered by the amount $N t \lambda \sqrt{\sigma}$. However, we know that the presence of a sheet of refractive index n and thickness t would have retarded the phase by

$$2\pi(n-1)t/\lambda;$$

hence

$$\sqrt{\sigma} N t \lambda = \frac{2\pi}{\lambda}(n-1)t$$

and so

$$n - 1 = \frac{1}{2\pi} N \lambda^2 \sqrt{\sigma}. \tag{2.10}$$

This derivation is not precise (it has not allowed for absorption) but it has shown the nature of the refraction process and how the material properties affect the refractive index. For example, introduce a strain and the value of N may vary, and so on. It does not show how n varies with λ since the scattered intensity does not just depend upon σ but also depends on $1/\lambda^4$ – the Rayleigh scattering law. The normal form of a dispersion curve (refractive index versus wavelength) is known as a Cauchy equation,

$$n = A + B/\lambda^3 + C/\lambda^4,$$

a semiempirical equation which is useful away from absorption bands.

2.4.1 Scattering

So far we have assumed that the medium through which the light is passing is uniform, but if it consists of numerous inhomogeneities acting as re-radiating centres the phenomenon of scattering is observed in which light may appear to no longer travel in straight lines: the back glare of car headlights in fog is an example. The extent of the scattering depends on the particle size and density. It comes in various forms.

2.4.1.1 Rayleigh Scattering

Particles much smaller than the wavelength of the incident light (for example, molecular clusters or imperfections in the silica lattice of a fibre) will scatter the radiation in the form of a spherical wave. The extent of this power loss depends on the number of particles and the wavelength. It has been found that this effect is proportional to $1/\lambda^4$. This is the reason the sky is blue, but it can also be a limiting factor in the design of fibres and some optics. For example, the attenuation of a laser beam, $E_{\text{attenuation}}$, passing through a plasma cloud, as in laser welding, could be described by the equation [16]

$$E_{\text{attenuation}} = P\left[1 - e^{-(Q_{\text{sca}}+Q_{\text{abs}})\pi r^2 Nz}\right],$$

where P is the laser beam power (W), r is the average radius of the particles (m), N is the number of particles per cubic metre and z is the beam path length (m).

The Rayleigh scattering efficiency is given by

$$Q_{\text{sca}} = \frac{8}{3}\left(\frac{2\pi r}{\lambda}\right)^4\left(\frac{m^2 - 1}{m^2 + 2}\right),$$

with the complex refractive index $m = (n + ik)$ and λ the wavelength.

2.4.1.2 Mie Scattering

When the diameter of the particles is approximately the size of the incident wavelength, the scattering is less dependent on the wavelength. This is known as Mie scattering [17]. It is possibly very relevant to laser material processing as Hansen and Duley [18] reported. Within the keyhole or interaction zone, when there is some form of boiling or ablation, there is almost certainly an aerosol which will cause scattering of the incident beam, thus affecting the focus and processing conditions. Some interesting results were recorded by Akhter [19] when laser welding with a powder feed in which the absorption was enhanced by the presence of the powder. The calculations of Hansen and Duley [18] showed, for particles of radius r, that for $2\pi r/\lambda \gg 1$ there was strong forward scattering, a form of refocusing of the beam. This is a subject area which will merit further study in the years to come (see also Sections 2.2.1.2, 2.2.1.3).

2.4.1.3 Bulk Scattering

For particles much greater than the wavelength of incident radiation the scattered intensity is almost independent of the wavelength. This is the reason why snow and fog are white. Some of this form of radiation transfer must be present in blown powder laser cladding processes.

2.5 Interference

Light waves are electromagnetic disturbances that travel through space. A vibrating electric charge sets up changing electric and magnetic fields around it which spread through space at the speed of light in the form of spherical waves oscillating transverse to the direction of travel. Enough "spherical" wavelets integrate to make a wavefront of any given shape. The description of the relationship between these electric and magnetic fields is given in Maxwell's famous set of four equations, from which all electromagnetic phenomena can be deduced – although that requires some effort! From them it is possible to show that the velocity of light $c = 1/\sqrt{(\mu_0 \varepsilon_0)}$, where μ_0 is the magnetic permeability of space and ε_0 is the electric permeability of space representing the storing of energy in inductive or capacitive form – which is the basis of the oscillation.

Since they have a transverse wave form, for normal energies these waves can be linearly superimposed (see Section 2.2.1). Thus, for two waves travelling in opposite directions a standing wave may form, as in the laser cavity. Two waves of similar frequency but of slightly different direction travelling in the same direction gives rise to a standing transverse wave form – an interference pattern used in the laser Doppler anemometer and Michelson interferometer (Section 1.4) or mode structures as observed coming from a laser. If several beams of slightly differing frequency are collinear, this could create almost any wave form. If they are all sinusoidal wave forms, they can be separated analytically into their constituent waves by Fourier analysis. Two waves travelling with the same frequency and direction but with different planes of polarisation will give rise to elliptical or circularly polarised beams. The addition or subtraction of waves is known as interference.

2.6 Diffraction

On striking a sharp edge, the electromagnetic waves will spread and not remain as a collimated stream. One can imagine waves on water striking an edge, such as a harbour wall, after which they will expand into the harbour. The divergence angle of the wave stream is a function of the wavelength, the longer ones spreading more than the shorter ones. Thus, the roar from a distant road will have a lower note than that from a road roar nearer the source. This diffraction phenomenon was first noted by Francesco Grimaldi (1618–1663) and was demonstrated elegantly in Young's double-slit experiment. Diffraction often leads to interference as two beams overlap. If the beams have a plane front (far field), then the phenomenon will be described as Fraunhofer diffraction (after Joseph von Fraunhofer, 1787–1826) and if they have curved front (near field), then it will be described as Fresnel diffraction (after Augustin-Jean Fresnel, 1788–1827, who did the first analytical analysis of diffraction). The calculation of diffraction from a slit is given in Section 2.8.

2.7 Laser Beam Characteristics

The energy from a laser is in the form of a beam of electromagnetic radiation. Apart from power, it has the properties of wavelength, coherence, power distribution or mode, diameter and polarisation. These are now discussed in the following sections.

2.7.1 Wavelength

Since the invention of the laser in 1960, many hundreds of lasing systems have been developed but only a few of commercial significance in material processing. Some of the wavelengths of the important material processing lasers are shown in Table 0.1.

The wavelength depends on the transitions taking place by stimulated emission. The wavelength may be broadened by Doppler effects due to the motion of the emitting molecules or by related transitions from higher quantised states as with the CO laser. On the whole, the radiation from a laser is amongst the purist spectral forms of radiation available. Very high spectral purity can be achieved by using a frequency-selecting grating as the rear mirror of the laser optical cavity, but this is rarely worth the effort for material processing. In consequence, if one wishes to achieve a very short pulse of light, for example, of 1 fs (a beam of light around 0.3 μm long!), it is not possible without first making a laser with a broader waveband, as is required by the Fourier series, which defines such a short pulse wavefront. But that is a problem for others who are not so involved in material processing.

2.7.2 Coherence

The stimulated emission phenomenon means that the radiation is generating itself and in consequence a continuous waveform is possible with low-order mode beams. The length of the continuous wavetrain may be many metres long. The comparison of laser light with standard random light is illustrated in Figure 2.13. This long coherence length allows some extraordinary interference effects with laser light, as noted in Chapter 1, such as length gauging, speckle interferometry, holography and Doppler velocity measurement. This property has not yet been used in material processing. In years to come it may be that someone will be able to use it as a penetration meter or to carry out subtle experiments with interference-banded heat sources.

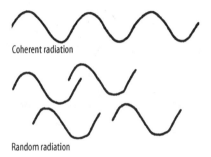

Coherent radiation

Random radiation

Figure 2.13 Comparison of the electric vector phase for coherent and random radiation

2.7.3 Mode and Beam Diameter

A laser cavity is an optical oscillator. When it is oscillating there will be standing electromagnetic waves set up within the cavity and defined by the cavity geometry. It is possible to calculate the wave pattern for such a situation and it is found that there are a number of longitudinal standing waves at slightly varying angles. The number of such off-axis standing waves is related to the Fresnel number $(a^2/\lambda L)$ (see Section 1.2.1.2). These standing waves interfere with each other giving a transverse standing wave which emerges from the cavity as the mode structure of the beam. For a nonamplifying, cylindrical cavity the amplitude of the transverse standing wave pattern, $E(r, \varphi)$, is given by a Laguerre–Gaussian distribution function of the form

$$E(r, \varphi) = E_0 \left(\frac{\sqrt{2}r}{w(z)} \right)^n L_p^n \left(\frac{2r^2}{w^2(z)} \right) \exp\left(-\frac{r^2}{w^2(z)} \right) \left(\begin{Bmatrix} \sin \\ \cos \end{Bmatrix} n\varphi \right),$$

where $E(r, \varphi)$ is the amplitude at point r, φ, $w(z)$ is the beam radius at point z along beam path, r is the radial position, φ is the angular position, n is an integer and

$$L_p^n(x) = e^x \frac{x^{-4}}{p!} \frac{d^p}{dx^p} \left(e^{-x} x^{p,n} \right),$$

which is the generalised Laguerre polynomial (Edmond Laguerre 1834–1886). Some low-order polynomials are

$$L_0^n(x) = 1,$$
$$L_1^n(x) = n + 1 - x,$$
$$L_2^n(x) = 1/2(n+1)(n+2) - (n+2)x + 1/2x^2.$$

The intensity distribution is found from the square of the amplitude:

$$P(r, \varphi) = E^2(r, \varphi).$$

These are the classical mode distributions for a circular beam. The distributions for a square beam are similar, but with Hermite polynomials. A plot of the amplitude and

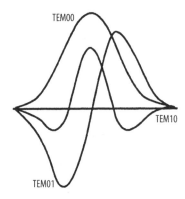

Figure 2.14 Amplitude variation for various modes

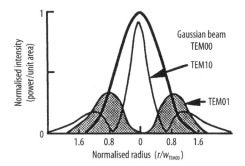

Figure 2.15 Intensity distribution for various modes

spatial intensity distributions which this expression represents for various orders of mode is shown in Figures 2.14 and 2.15. Typical mode patterns that would be made from such beams are shown in Figure 2.16.

The classification of these transverse electromagnetic mode patterns is by (TEM_{plq}) where p is the number of radial zero fields, l is the number of angular zero fields and q is the number of longitudinal zero fields.

Most slow flow lasers operate with a near perfect TEM00 or TEM01* mode. The TEM01* mode is made from an oscillation between two orthogonal TEM01 modes as illustrated in Figure 2.16.

Most fast axial flow lasers also give a beam with a low-order mode since they have long, narrow tubes – low Fresnel number $(a^2/\lambda L)$ – (see Section 1.2.1.2). The modes from these lasers may be slightly distorted owing to plasma density variations.

Transverse flow lasers usually have multimode beams of indeterminate ranking. They are either quasi-Gaussian – in that they are a single lump of power – or asymmetric owing to the transverse amplification being different across the cavity owing to the heating of the gas as it traverses. To reduce this effect some cavities are ring-shaped – see Section 1.2.1.3.3.

The higher the order of the mode, the more difficult it is to focus the beam to a fine spot, since the beam is no longer coming from a virtual point.

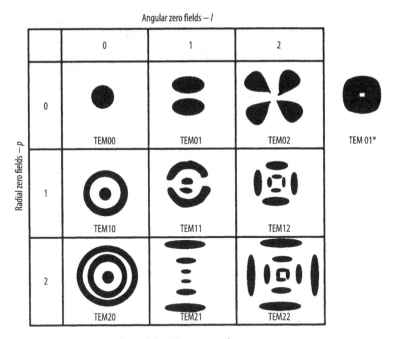

Figure 2.16 Various mode patterns

A question arises in material processing as to what is the beam diameter. For example, the data in Figures 2.14 and 2.15 were calculated with the mathematical radius, $w(z)$, the same. This is obviously not related to the diameter which affects heating processes. Sharp *et al.* [20] argue that the beam diameter should be defined as that distance within which $1/e^2$ of the total power exists. (See Chapter 12 for methods of measuring the beam diameter.)

2.7.4 Polarisation

The stimulated emission phenomenon not only produces long trains of waves but these waves will also have their electric vectors all lined up. The beam is thus polarised. Many of the early lasers and some of the more modern ones which do not have a fold in the cavity will produce randomly polarised beams. In this case the plane of polarisation of the beam changes with time – and the cut quality may show it! To avoid this it is necessary to introduce into the cavity a fold mirror of some form. Outside the cavity such a fold would make no noticeable difference. Inside the cavity it is a different matter since the cavity is an amplifier and hence the least-loss route is the one being amplified in preference to the others – in fact almost to their total exclusion. Polarised beams have a directional effect in certain processes, for example, cutting, owing to the reflectivity effects on the sloping cut front shown in Figure 2.11 and discussed in Section 2.3.4. Hence, material processing lasers are usually engineered to give a polarised beam which is then fitted with a circular polariser – see Section 2.9.2.

Polarisation plays a role in the reflection and scattering of all light. If the electric vector is all aligned in one direction, then the beam is "linearly polarised". If it has two vector directions at right angles to each other of equal intensity, it is said to be "circularly polarised" – if the field rotates clockwise to an observer looking into the beam then it is said to be "right-circular polarised" as opposed to "left-circular polarised". With one vector stronger than the other it is "elliptically polarised". The "extinction ratio" is the ratio between the maximum and minimum intensities of the beam after passing through a polarisation filter. Birefringent crystals have fast and slow indices of refraction for different states of polarisation. Certain molecules, notably quartz and sugars, can rotate the plane of polarisation of transmitted beams. Known forms of life are overwhelming composed of amino acids with left-handed optical activity and use sugars that are right-handed – unlike laboratory-prepared sugars and amino acids. A meteorite discovered in Australia in 1969 contained a surprising quantity of amino acids with this same bias towards left-handedness, thus posing some interesting questions. Bees are considered to navigate by the polarisation of the sunlight scattered from the atmosphere [21].

2.8 Focusing with a Single Lens

To manipulate the beam, to guide it to the workplace and shape it, there are many devices which have so far been invented. These devices are now discussed together with the basic theory of their design. In nearly all of them the simple laws of geometric optics listed in Table 2.5 are sufficient to understand how they work, but to calculate the precise spot size and depth of focus one needs to refer to Gaussian optics and diffraction theory.

2.8.1 Focused Spot Size

2.8.1.1 Diffraction-limited Spot Size

A beam of finite diameter is focused by a thin lens onto a plate as shown on Figure 2.17. The individual parts of the beam striking the lens can be imagined to be point radiators of a new wavefront. The lens will draw the rays together at the focal plane and constructive and destructive interference will take place there. When two rays arrive at the screen and they are half a wavelength out of phase, then they will destructively interfere and the light intensity will fall; the converse will occur when they arrive in phase. Thus, if ray AB (Figure 2.17) is $\lambda/2$ longer than ray CB, point B will represent the first dark ring of what is known as a "Fraunhofer diffraction pattern" (assuming the wavefronts are planar). The central maximum will contain approximately 86 % of all the power in the beam. The diameter of this central maximum will be the focused beam diameter, usually measured between the points where the intensity has fallen to $1/e^2$ of the central value.

Table 2.5 Gaussian optical properties

Terminology

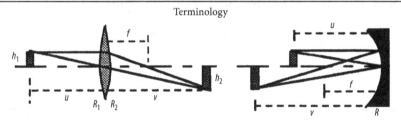

$m = h_2/h_1$ = magnification; R_1, R_2 = radii of curvature; n_1, n_2 = refractive index of two medium.

	Spherical surface	Plane surface
Reflection	$\frac{1}{u} + \frac{1}{v} = \frac{1}{f}$	
	$f = -\frac{R}{2}$	$f = -\infty$
	$m = -\frac{v}{u}$	$m = +1$
	Concave: $f > 0$, $R < 0$	
	Convex: $f < 0$, $R > 0$	
Refraction at single surface	$\frac{n_1}{u} + \frac{n_2}{v} = \frac{n_2-n_1}{R}$	$v = -\frac{n_2}{n_1}u$
	$m = -\frac{n_1 v}{n_2 u}$	$m = +1$
	Concave: $R < 0$	
	Convex: $R > 0$	
Refraction at a thin lens	$\frac{1}{f} = \frac{n_2-n_1}{n_1}\left(\frac{1}{R_1} - \frac{1}{R_2}\right)$	
	$\frac{1}{u} + \frac{1}{v} = \frac{1}{f}$	
	$m = -\frac{v}{u}$	
	Concave: $f < 0$	
	Convex: $f > 0$	

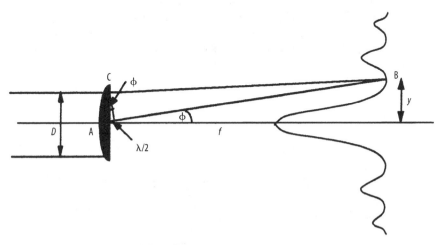

Figure 2.17 The diffraction-limited spot size

For a rectangular beam with a plane wavefront, the first dark fringe will occur when the beam path difference between the centre and the edge rays, d, is $\lambda/2$

$$d = \lambda/2 = (D/2)\sin\varphi.$$

That is, when $\lambda = D\sin\varphi$, or for other fringes when $m\lambda = D\sin\varphi$.

From geometry, $2y = 2f\tan\varphi$, and for small angles $\tan\varphi = \sin\varphi = \lambda/D$.

Therefore, $2y = d_{min} = 2f\lambda/D$.

For plane front circular beams there is a correction of 1.22 and so the equation becomes

$$d_{min} = 2.44f\lambda/D. \tag{2.11}$$

For Gaussian beams there is sometimes a further small correction. The focal spot size for a multimode beam will be larger because the beam is coming from a cavity having several off-axis modes of vibration and therefore not all coming from an apparent point source. This correction for a TEM_{plq} beam is

$$d_{min} = 2.44(f\lambda/D)(2p + l + 1). \tag{2.12}$$

Radial nulls, p, are more damaging to the focal spot size than angular nulls, l. For example, the expected spot size for a CO_2 laser beam 22 mm in diameter with a TEM01 mode focused by a 125 mm focal length lens would be expected to be $d_{min} = 2.44[(125 \times 10.6 \times 10^{-3})/22] \times 2 = 0.29$ mm, whereas a TEM10 beam would be expected to focus to 0.44 mm.

2.8.1.2 M^2 Concept of Beam Quality

An unmodified laser beam diverges by diffraction from its initial waist value of D_0 at an increasing rate as shown in Figure 2.18, and reaches a maximum value only at infinity. This maximum value is the far-field divergence, $\Theta_{0\infty}$. If a lens focuses the beam, it forms a new waist, D_1. The beam converges towards and diverges away from this new waist with a far-field divergence of $\Theta_{1\infty}$ where

$$D_0\Theta_{0\infty} = D_1\Theta_{1\infty} = \text{constant}.$$

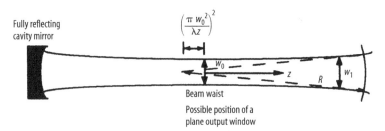

Figure 2.18 Variation of radius of curvature of the phase field with distance. A small value of R is known as the "near field", whereas a large value is known as the "far field"

This constancy of $D\Theta$ values through the system with aberration-free optics allows the calculation of spot size, depth of focus, Rayleigh length and curvature of phase fronts.

To be able to use this property, we need to define a quality factor comparing the actual beam divergence, Θ_{act}, with the divergence from a Gaussian laser beam with the same initial waist size, Θ_r. Consider a laser cavity giving an actual beam divergence of Θ_{act} and having a beam waist radius W_0. A Gaussian beam originating from the same virtual origin as the actual beam would have a divergence Θ_{Gauss} and a beam waist radius w_0 defined by the Gaussian beam propagation equation for a diffraction-limited Gaussian beam (TEM00) (see Figure 2.18):

$$w^2(z) = w_0^2 \left[1 + \left(\frac{\lambda z}{\pi w_0^2} \right)^2 \right] \qquad (2.13)$$

where $w(z)$ is the beam radius at a distance z from the waist position of radius w_0 for a beam of wavelength λ.

In the far field, z becomes large; hence,

$$\left(\frac{\lambda z}{\pi w_0^2} \right)^2 \gg 1$$

and hence Θ_{Gauss}, which is equal to $w(z)/z = \lambda/\pi w_0$ from Equation 2.13.

It can be seen that $\Theta_{Gauss} w_0 = \lambda/\pi = $ constant for all Gaussian beams as noted above.

Using the same propagation equation, the divergence, Θ_r, of a Gaussian beam with the same waist radius as the actual beam, W_0, is

$$\Theta_r = \lambda/\pi W_0 .$$

If we define the ratio $M = \Theta_{act}/\Theta_{Gauss}$, this equals W_0/w_0 since the Gaussian comparator beam and the actual beams have the same virtual origin at a point at a distance l from the waist. Thus, $\Theta_{act} = W_0/l$ and $\Theta_{Gauss} = w_0/l$, making

$$\Theta_{act}/\Theta_{Gauss} = W_0/w_0 = M$$

and

$$w_0 = W_0/M .$$

Then

$$\Theta_{Gauss} = \lambda/\pi \left(W_0/M \right) ;$$

therefore,

$$\Theta_{act} = M \left(\lambda M/(\pi W_0) \right) .$$

But

$$\Theta_r = \lambda/(\pi W_0)$$

and thus

$$M^2 = \Theta_{act}/\Theta_r . \tag{2.14}$$

This is the comparator which we sought. It is sometimes expressed as $Q = M^2$, which avoids the rather tedious argument just presented [22]. In a recent International Organization for Standardization (ISO) standard it is also described as $1/K$, where K is yet another measure of quality. All are based on the same comparison with Gaussian beams.

Applying this to a lens, we have

$$\Theta_{act} = \frac{D_L}{2f} \text{ and } \Theta_r = \frac{2\lambda}{\pi d_{min}} ;$$

therefore,

$$d_{min} = \frac{4M^2 f \lambda}{\pi D_L} .$$

It can be seen that this quality factor, M^2 or Q, allows real beams of higher-order mode than the basic Gaussian TEM00 to be treated as Gaussian by using a modified wavelength, $M^2 \lambda$.

Thus, knowing M^2, one can calculate various beam characteristics:

1. The beam diameter, D, at any distance along the beam path, z, from the beam waist is given from the basic propagation equation:

$$D_z = D_0 \left[1 + \left(\frac{4M^2 \lambda z}{\pi D_0^2} \right)^2 \right]^{\frac{1}{2}} \tag{2.15}$$

2. The wavefront radius, R_z, at any distance, z, from the beam waist is given by

$$R_z = z \left[1 + \left(\frac{\pi D_0^2}{4M^2 \lambda z} \right)^2 \right] . \tag{2.16}$$

3. The Rayleigh range, R, which is the distance from the beam waist of diameter D_0 to the position where it is $\sqrt{2}D_0$, is

$$R = \left(\frac{\pi D_0^2}{4M^2 \lambda} \right) . \tag{2.17}$$

The Rayleigh range is the multiplier in the equations for D_z and R_z:

$$D_z = D_0\left[1 + \left(\frac{z}{R}\right)^2\right]^{\frac{1}{2}} \text{ and } R_z = z\left[1 + \left(\frac{R}{z}\right)^2\right].$$

4. The depth of focus is the distance either side of the beam waist, D_0, over which the beam diameter grows by 5 % (see also Section 2.8.2):

$$DOF = \pm 0.08\pi\frac{D_0^2}{M^2\lambda}. \tag{2.18}$$

5. Focused spot size. Since $D_0\Theta_{\infty 0} = D_1\Theta_{1\infty}$ for all aberration-free optical systems, then $D_1 = D_0\Theta_{0\infty}/\Theta_{1\infty}$ for a focusing lens placed at the beam waist, the preferred place since the wavefront is plane at that location.

$$\Theta_{1\infty} = D_0/2f$$
$$\Theta_{1\infty} = 2M^2\lambda/(\pi D_0). \tag{2.19}$$

Therefore,

$$d_{min} = f\Theta_{0\infty} = 4fM^2\lambda/(\pi D_0).$$

For a focusing lens placed z millimetres from the beam waist,

$$d_{min} = f\Theta_{0\infty}\left(\frac{D_0}{D_z}\right) = \frac{4fM^2\lambda}{\pi D_z}.$$

At this point it is interesting to note that $d_{min} = f\Theta_{act\infty}$ is independent of wavelength if Θ_{act} is mainly decided by the cavity optics. This is a result of M^2 being inversely proportional to λ. Thus, there is no particular focusing advantage in using shorter-wavelength lasers for a given cavity, for example, either CO or CO_2 lasers using the same cavity.

The usefulness of M^2 is apparent from the above equations. However, like all good things in life there is a snag – how to measure $\Theta_{act\infty}$?

The beam expands as described in Equation 2.13 and shown in Figure 2.18. However, unless one measures the beam expansion at infinity, one is likely to measure something other than $\Theta_{act\infty}$ such as the trigonometric divergence, Θ_T, or the local divergence, Θ_L [22]. Figure 2.19 shows these three values. They can be calculated approximately from the wave propagation equation (2.13):

$$D_z = D_0\left[1 + \left(\frac{z}{R}\right)^2\right]^{\frac{1}{2}} \tag{2.20}$$

where the Rayleigh range $R = \frac{\pi D_0^2}{4M^2\lambda}$.

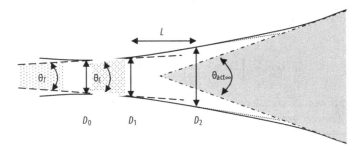

Figure 2.19 The various angles of divergence discussed in establishing the value of the beam quality factor M^2

Now $dD_z/dz \to \Theta_\infty$ as $z \to \infty$; thus, by differentiating, we get

$$\frac{dD_z}{dz} = \frac{D_0}{R^2}\left[\frac{z}{\left[1+\left(\frac{z^2}{R^2}\right)\right]^{\frac{1}{2}}}\right].$$

As $z \to \infty$

$$\left(1+\frac{z^2}{R^2}\right)^{\frac{1}{2}} \to \left(\frac{z^2}{R^2}\right)^{\frac{1}{2}} = \frac{z}{R};$$

hence,

$$\Theta_\infty = \frac{dD_z}{dz}\bigg|_{z=\infty} = \frac{D_0}{R^2}\left[\frac{z}{z/R}\right] = \frac{D_0}{R}.$$

Also

$$\Theta_T = \left(\frac{D_z - D_0}{z}\right) = \frac{D_0\left\{\left[1+(z/R)^2\right]^{\frac{1}{2}}-1\right\}}{z},$$

and

$$\Theta_L = \frac{D_1 - D_2}{L}.$$

This divergence can be corrected to infinity if Θ_T is multiplied by $[(v+1)/(v-1)]^{\frac{1}{2}}$, where $v = D_z/D_0$, and noting from Equation 2.20 that

$$v = \left[1+\left(\frac{z}{R}\right)^2\right]^{\frac{1}{2}},$$

then

$$\Theta_{\text{T corrected}} = \frac{D_0}{z} \left[\frac{(v-1)(v+1)^{\frac{1}{2}}}{(v-1)^{\frac{1}{2}}} \right] = \frac{D_0}{z} \left(v^2 - 1 \right)^{\frac{1}{2}}.$$

But

$$v = \left[1 + \left(\frac{z}{R} \right)^2 \right]^{\frac{1}{2}}.$$

Therefore,

$$\Theta_{\text{T corrected}} = \frac{D_0}{R} = \Theta_\infty.$$

Similarly, if Θ_L is multiplied by $\left[v/(v^2 - 1)^{\frac{1}{2}} \right]$, the value is corrected.

   Thus, to calculate M^2 for a given beam:

(a)   find the beam waist from the cavity optics, *e.g.*, output diameter for a flat output window is D_0;
(b)   find the beam diameter, D_z, at a known distance from the beam waist, z (two or three readings at different distances would help to confirm each other);
(c)   calculate $\Theta_T = (D_z - D_0)/z$;
(d)   multiply by the correction factor with $v = D_z/D_0$ to obtain

$$\Theta_\infty = \Theta_T \left[\frac{(v+1)}{(v-1)} \right]^{\frac{1}{2}}; \text{ and}$$

(e)   from Equation 2.19 derive

$$M^2 = \left(\frac{D_0 \Theta_\infty \pi}{4\lambda} \right).$$

All other beam calculations follow.

   However, notice in Figure 2.19 the significant understatement of Θ_∞ defining M^2 which is often used in laser specifications.

   For example, consider a CO_2 laser with a 15-mm beam diameter from the flat output window, whose beam has expanded to 30 mm after an 8 m beam path:

$$\Theta_T = (30 - 15)/8000 = 1.87 \, \text{mrad},$$
$$v = 30/15 = 2.$$

The corrected value

$$\Theta_\infty = \Theta_T \left(\frac{2+1}{2-1} \right)^{\frac{1}{2}} = 1.73 \Theta_T = 3.23 \, \text{mrad}.$$

There are two values of M^2, one based on Θ_T and one on the correct value of Θ_∞:

$$M_{\Theta_T}^2 = \frac{15 \times 1.87 \times 10^{-3} \times \pi}{4 \times 10.6 \times 10^{-3}} = 2.07\,,$$

and

$$M_{\Theta_\infty}^2 = \frac{15 \times 3.23 \times 10^{-3} \times \pi}{4 \times 10.6 \times 10^{-3}} = 3.57\,.$$

The Rayleigh range

$$R = \frac{\pi D_0^2}{4M^2\lambda} = \frac{\pi \times 15^2}{4 \times M^2 \times 10.6 \times 10^{-3}}\,;$$

therefore,

$$R = 16.7 \times 10^3/M^2\,,$$
$$R_{\Theta_T} = 8.05\,\text{m}\,,$$
$$R_{\Theta_\infty} = 5.17\,\text{m}\,.$$

The focal spot size for this laser is $f\Theta_\infty$ so for a 125-mm focal length lens

$$d_{\text{min}\ \Theta_T} = 125 \times 1.87 \times 10^{-3} = 0.234\,\text{mm}$$
$$d_{\text{min}\ \Theta_\infty} = 125 \times 3.23 \times 10^{-3} = 0.403\,\text{mm}$$

which gives a 72 % error based on $d_{\text{min}\ \Theta_T}$! Thus, corrected values of M^2 must be used in optical calculations.

2.8.1.3 Spherical Aberration

There are two reasons why a lens will not focus to a theoretical point; one is the diffraction-limited problem discussed earlier and the other is the fact that a spherical lens does not have a perfect shape. Most lenses are made with a spherical shape since this can be accurately manufactured without too much cost and the alignment of the beam is not so critical as with a perfect aspherical shape. The net result is that the outer ray entering the lens is brought to a shorter axial focal point than the rays nearer the centre of the lens, as shown in Figure 2.20. This leaves a blur in the focal point location. The plane of best geometric focus (the minimum spot size) is a little short of the plane of the planar wavefront – the paraxial point. The minimum spot size, d_a, is given by

$$d_a = K(n; q; p)\left[\frac{D_L}{f}\right]^3 S_2 = 2\Theta_a S_2\,,$$

where Θ_a is the angular fault (half-angle), S_2 is the distance from the lens, D_L is the diameter of top hat beam mode on the lens, f is the focal length of the lens and $K(n; q; p)$

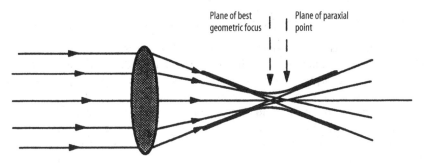

Figure 2.20 Spherical aberration of a single lens focusing a parallel beam

is a a factor dependent on the refractive index, n, the lens shape, q, and the lens position, p:

$$K(n; q; p) = \pm \frac{1}{128n(n-1)} \left[\frac{n+2}{n-1} q^2 + 4(n+1)pq + (3n+2)(n-1)p^2 + \frac{n^3}{n-1} \right],$$

where q is the lens shape factor $(r_2 + r_1)/(r_2 - r_1)$, r_2 and r_1 are radii of curvature of the two faces of the lens and p is the position factor $1 - 2f/S_2$.

Figure 2.21 shows the variation of spherical aberration with lens shape. The optimum shape is when the refraction angles at both faces of the lens are approximately equal. Note that there is a huge difference between a planoconvex lens mounted one way rather than the opposite way around.

Other lens faults are:

1. mechanical and optical axis are not correctly aligned – leading to coma effects; and
2. lens surface is not correctly spherical – leading to astigmatism if it has a cylindrical element.

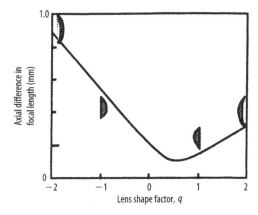

Figure 2.21 Spherical aberration of a ray 1 cm off the optic axis passing through a lens of focal length 10 cm, diameter 2 cm and refractive index 1.517. (After Jenkins and White [15])

2.8.1.4 Thermal Lensing Effects

In optical elements which transmit or reflect high-power radiation there will be some heating of the component which will alter its refractive index and shape. As the power changes or the absorption changes, so will the focal point and the spot size. The two main elements usually concerned are the output coupler on the laser and the focusing lens, although the beam guidance mirrors could also be involved if adequate water cooling is not supplied. Transmissive optics can only be cooled from the edge or by blowing filtered, dry air onto the lens surface. Transmissive optics have a thickness chosen according to the pressure differential across them. Rarely is much thought given to thermal lensing, yet there is an optimum thickness to balance cooling with distortion [23].

Thermal lensing is mainly due to the rise in temperature of the optic causing variations in the refractive index (dn/dT) and only slightly in the expansion of the optics (dl/dT). The physical and optical constants for the principal infrared materials are given in Table 2.6. The focal length shift for a thin lens is given approximately by the quasi-statistical formula [24]

$$-\Delta f = \left(\frac{2APf^2}{\pi k D_L^2} \right) \frac{dn}{dT} \times 100,$$

where Δf is the change in focal length (%), A is the absorptivity of the lens material (m^{-1}), P is the power incident on the lens (W), n is the refractive index of the lens material (dimensionless), k is the thermal conductivity of the lens (W m^{-1} K^{-1}), D_L is the incident beam diameter on the lens (m) and T is the temperature (K).

Using this equation with $f/D_L = 10$ (*i.e.*, an F10 optic) and $P = 2$ kW with a thin uncooled lens, the change in focal length due to thermal distortion for various materials is 0.02 % for ZnSe to 2.6 % for germanium.

By comparison, because of the geometric effects of thermal distortion on a 4-mm-thick lens, 38 mm in diameter made of ZnSe, which has a temperature difference of 14 °C between the centre and the edge owing to the passing of a laser beam of 1,500 W [25], a change of focal length of approximately 4 μm would be expected owing to the change in the shape of the lens. This can be calculated from geometrical considerations and the simple lens formula for the focal length:

$$f = R/(n-1),$$

where R is the radius of curvature of a planoconvex lens.

The expansion of the middle of the lens is expected to be

$$\beta l \Delta T = 7.57 \times 10^{-6} \times 0.04 \times 14 = 4.2 \,\mu\text{m}.$$

The effect this has on the lens focal length is

$$\Delta f = \delta R/(n-1) \sim -4.2/1.04 = 4 \,\mu\text{m}.$$

Table 2.6 Thermal and optical constants for principle infrared materials for 10.6 µm radiation

Material	Absorptivity (m^{-1} × 10^{-6})	Refractive index (n)	dn/dT (×10^{-6} °C^{-1})	Thermal conductivity (W m^{-1} K^{-1})	Specific heat (J kg^{-1} °C^{-1})	Thermal coefficient expansion (×10^{-6} °C^{-1})	Density (kg m^{-3})	Thermal diffusivity (m^2 s^{-1} × 10^{-6})
ZnSe	0.05	2.403	64	18	356	7.57	5270	9.6
CdTe	0.18	2.674	107	6.2	210	5.9	5850	5.05
GaAs	1	3.275	149	48	325	5.7	5370	27.5
Ge	3	4.003	408	59	310	5.7	5320	35.7
Si	150	3.418	160	156	716	2.56	2330	9.3
KCl	0.014	1.455	−33	6.5	683	36	1980	4.8
Quartz IR grade		1.45	10	1.4	745	0.55	2200	0.85

Data from II–VI handbook 1991.

There is a further problem since much of the absorption is on the surface. This creates a temperature gradient in the depth direction. The thicker the optic, the more bowed will be the internal isotherms. Such aberrations will affect the M^2 value of the beam.

In considering these issues, it is best to choose the material that will absorb less heat and show the least affect from being heated, *e.g.*, ZnSe. The lens for high-power work should be cooled on the edge and by surface blowing if possible. It should also be as thin as the pressure differential will allow.

2.8.1.5 Beam Flight Tubes

An unexpected aspect of thermal lensing is to be found in the design of beam flight tubes which are used to pass the beam safely in open air if it cannot be transferred by a fibre, as with CO_2 radiation. For long flight tubes a mirage effect may be set up within the tube owing to thermal gradients caused by heating of the tube from sunshine or radiators, *etc.* Self heating of the gas in the tube by the absorption of the beam may distort or bend the beam. In both cases this would upset the alignment of a large gantry system.

To overcome this problem, flight tubes are often purged with dry nitrogen or helium gases which do not show self-heating problems for reasonable levels of power transmission; for ultrahigh powers a vacuum is recommended. A 10 % increase in divergence has been found when the tube is filled with air as opposed to nitrogen or helium.

2.8.2 Depth of Focus

The depth of focus is the distance over which the focused beam has approximately the same intensity. It is defined as the distance over which the focal spot size changes by $\pm 5\%$.

Considering the focusing beam to converge with an angle whose tangent is $D/(2f)$, by similar triangles we get

$$f/z_\mathrm{f} = D/1.05 d_\mathrm{min} = D/1.05(2.44 f\lambda/D)$$
$$z_\mathrm{f} = \pm 2.56 F^2 \lambda\,,$$

where the F number equals f/D. Allowing for multimode beams,

$$z_\mathrm{f} = \pm 2.56 F^2 M^2 \lambda\,. \tag{2.21}$$

Table 2.7 shows some figures for the focal spot size and the depth of focus given by different lenses with beams of different mode structures.

Table 2.7 Effects of F number on the focal length and depth of focus for different mode structures and wavelengths

Wavelength (μm)	F number Mode (f/D)		Diffraction, d_{min} (mm)	Depth of focus, z_f (mm)	Spherical aberration, d_{min}^a	
					Bi-convex (mm)	Plano-convex (mm)
10.6	2	Top hat	0.26	0.08	0.5	0.36
	5	TEM00	0.13	0.5	0.08	0.06
	5	TEM01*	0.26	1.0		
	5	TEM20	0.65	2.5		
	10	TEM00	0.26	2.0	0.02	0.015
1.06	2	Top hat	0.026	0.008	0.5	0.36
	5	TEM00[b]	0.013	0.05	0.08	0.06
	5	TEM01[b]	0.026	0.1		
	5	TEM20[b]	0.065	0.25		
	5	Multi	−0.4	−2.0		
	10	TEM00	0.026	0.2	0.02	0.015

[a] For a beam diameter of 20 mm and bi-convex $q = 0$, $K = 0.1$ and plano-convex $q = 1$, $K = 0.073$. (Where the value is smaller than the diffraction-limited spot size, it means that spherical aberration is less significant. The two values should be added to obtain the approximate expected spot size.)
[b] Current industrial high-powered YAG lasers cannot achieve this level of mode purity, but it can be seen that the incentive to do so is high the new fibre laser can do so using monomode fibres.

2.9 Optical Components

2.9.1 Lens Doublets

We have so far discussed the single simple lens. A doublet is an alternative to an aspherical lens for overcoming the effects of spherical aberration. We have just noted that spherical aberration becomes the main issue for short focal length lenses of less than $F/5$. If such a short focus is needed, then the doublet is a cheaper option than an aspherical lens. Table 2.8 shows a comparison of lens types.

The effect of a doublet compared to a singlet is illustrated in Figure 2.22 [26]. This figure illustrates the advantages to be found for doublets at low F numbers.

Table 2.8 Comparison of basic lens types [15]

Type	Advantage	Disadvantage
Singlet	Low cost	High SA at low F number, no colour correction
Air space aplanat (doublet)	Excellent TWD	Cost, no colour correction
Cemented achromatic doublet	Better SA than singlet	Low power only fair TWD for low F number
Air space achromat (triplet)	Colour correction OK, low SA at low F number	Cost

The colour correction is not relevant for single-frequency lasers.
SA spherical aberration, TWD transmitted wavefront distortion

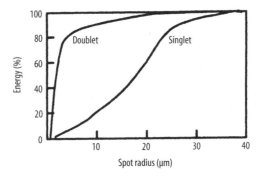

Figure 2.22 Encircled energy at different radii for singlet and doublet F/5 lenses focusing 1.06-μm radiation from a Nd:YAG laser

2.9.2 Depolarisers

When a polarised beam strikes a mirror surface at 45° to the plane of polarisation, the beam takes up two new planes: the p plane, parallel to the plane of incidence, and the s plane, perpendicular to the plane of incidence. When this reflected beam strikes a mirror having a surface coating which is $\lambda/4$ thick in the direction of propagation of the beam, then the p-polarised beam (parallel to the plane of incidence) will penetrate the film, whereas the s-polarised beam (perpendicular to the plane of incidence) will be reflected. The p-polarised beam will be reflected from beneath the film at the metal surface and so rejoin the main beam but then it will be phase-shifted by $2(\lambda/4)$. Thus, the final beam will be one in which the plane of polarisation alternates between two states at right angles with every beat of the wave form. This gives the impression to a viewer from the end of the beam that the plane of polarisation is rotating. The beam is said to be "circularly" polarised. Some care has to be taken with these carefully designed coatings, which are usually of MgF_2, because they are slightly hygroscopic and cannot be safely wiped clean. Nevertheless depolarisers are now fitted to nearly all commercial cutting machines.

An alternative to circularly polarised beams is radial polarisation (see Figure 2.23a). An intracavity conical prism has recently been introduced to make the beam from a fibre laser radially polarised; that is, when the polarisation axis is always radial from the centre of the beam. One such device is a double cone where the faces of the cone meet the beam at the Brewster angle, thus ensuring that only the radially polarised component of the incident beam enters the second collimating cone [27]. An alternative is a subwavelength circular grating in which normally reflected radiation will be polarised perpendicular to the grating rulings, i.e., radially [28]. The system is of potential use for optical tweezers and high-resolution microscopy, but it also shows considerable advantages for material processing, where it has been reported to increase the cutting speed by a factor of 10–50 % compared with a linearly polarised beam [29]. The circular grating radial polariser can be mounted as the fully reflecting mirror inside the laser cavity, in which case it would have a convex GaAs lens on its surface to reduce diffraction losses (Figure 2.23b). The radius of curvature of the convex surface should

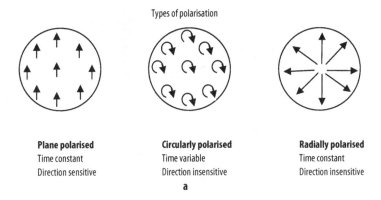

Types of polarisation

Plane polarised   **Circularly polarised**   **Radially polarised**
Time constant   Time variable   Time constant
Direction sensitive   Direction insensitive   Direction insensitive

a

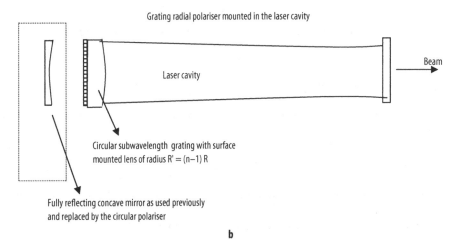

Grating radial polariser mounted in the laser cavity

Beam

Laser cavity

Circular subwavelength grating with surface
mounted lens of radius R' = (n−1) R

Fully reflecting concave mirror as used previously
and replaced by the circular polariser

b

Figure 2.23 a Different types of polarisation, and **b** an arrangement for mounting a radial polariser within the laser cavity to give a radially polarised beam [28]

be $r' = (n-1)R$, where R is the original radius of curvature of the fully reflecting mirror prior to installing the radial element in its place.

2.9.3 Collimators

A collimator or beam expander is often used in installations where the beam path is long or the laser produces such a small beam diameter that it is difficult to focus without having the lens very close to the work piece and therefore vulnerable to spatter. A transmissive beam expander is illustrated in Figure 2.24 for the Galilean and Keplerian designs. The general principle is that the new beam size will be $D_2 = D_1 f_2 / f_1$. For long beam path work the beam divergence is one of the main criteria [30].

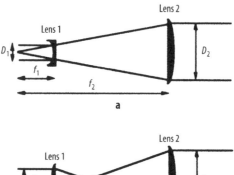

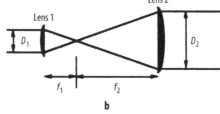

Figure 2.24 Examples of beam expanders and collimators: **a** Galilean, and **b** Keplerian

Most CO_2 laser cutting tables, gantries or robot systems have long beam paths and therefore need an optic to make the beam as parallel as possible. Failure to do so would mean the wavefront curvature and beam size would vary with position on the table and hence the focus would vary over the processing area. The objective of a beam expander or collimator is to locate the beam waist in the middle of the range of movement of the focusing optic to minimise this variation.

Beam expanders are usually marked on their barrels with the magnification, m, and the focus setting, G. G is not the beam expander focal length but is equal to m times the focal length.

For a simple lens where there is no change in beam size either side of the lens

$$\frac{1}{R'} = \frac{1}{f} - \frac{1}{R},$$

where R is the wavefront radius of curvature before the beam expander, R' is the wavefront radius of curvature after the beam expander and f is the focal length of the lens.

For a beam expander this becomes

$$\frac{1}{R'} = \frac{1}{G} - \frac{1}{(m^2 R)},$$

where in this case $D_2 = mD_1$ and the beam waist, for a beam of quality M^2, is located at [31]

$$z_{\text{waist}} = \left(\frac{m^2 uG}{m^2 u + G} \right) \left\{ 1 + \left[\frac{4M^2 \lambda \left(\frac{uG}{(m^2 u + G)} \right)}{\pi D^2} \right] \right\}^{-1},$$

where u is the distance of the object from the first lens of the collimator.

The important aspect to note here is that there is a limiting value of G beyond which normal diffraction dominates. When $G = \infty$, the beam waist is at the beam expander optic $z = 0$.

2.9.4 Metal Optics

2.9.4.1 Plane Mirrors

The reflectivity of a mirror is a function of the material; therefore, most mirrors are made of a good conductor (good reflector) coated with gold for infrared radiation. The gold may be further coated with rhodium to allow gentle cleaning. New optics based on coated silicon are also used. In the case of some lasers they may be sufficiently thin to allow gentle flexing, giving some control over beam mode structure, if mounted within the cavity. The reason for having good conductivity mirror substrates, apart from reflectivity, is the need for good cooling. This is usually achieved by water but may be by air blast. Above 1 kW, mirrors must be cooled to avoid distortion. Mirror materials can be ranked against a figure of merit:

$$FOM = k/A\beta, \tag{2.22}$$

where k is the thermal conductivity (W m^{-1} K^{-1}), A is the absorptivity (1 – reflectivity) and β is the linear coefficient of expansion (K^{-1}).

The figure of merit for a number of mirror materials is shown in Table 2.9.

The flatness of mirrors is achieved by careful machining. The most popular technique is single-point diamond machining. The flatness is measured on an interferometer and recorded as $\lambda/*$, for example, $\lambda/5$ means that there is a variation in flatness of one fifth of a wavelength over the mirror surface. The mirror must be mounted very carefully to avoid any mechanical distortion of this order. It must also be hard and tough, take a good polish and be cleanable.

Cleaning mirrors is done by placing a soft lens tissue on the mirror and allowing a drop of methanol or isopropyl alcohol to fall on it. The tissue is then drawn over the face of the mirror until it is dry. This will prevent scratching and also drying stains.

Table 2.9 Properties of metal optic materials

Property	Cu	Mo	Si
Thermal conductivity, k (W m^{-1} K^{-1})	390	133	156
Coefficient of expansion β ($\times 10^{-6}$ °C^{-1})	16.7	5.4	2.6
Density (kg m^{-3})	8,960	1,020	240
Hardness (Mohs)	3.0	6.0	7.0
Young's modulus ($\times 10^6$ MPa)	5.6	15.6	9.6
Specific heat (J kg^{-1} °C^{-1})	385	272	716
Reflectivity 0° AOI uncoated	0.99	0.98	0.98
Figure of merit, $k/A\beta$ ($\times 10^9$ W m^{-1})	2.33	1.23	3.0

AOI angle of incidence

Never rub a mirror surface. If a mirror becomes tarnished or damaged in any way, it is usually best to regrind and recoat it. If possible, mirrors should always be mounted so that they avoid dust falling on them.

2.9.4.2 Metal Focusing Optics (Parabolic Mirrors)

With the growing use of very high powered lasers with average powers over 5 kW, transmissive optics are near the limit of their thermal stress resistance. Most operators of such equipment prefer to use metal optics for focusing, collimating and guiding. One focusing element which uses the least number of mirrors is an off-axis parabolic mirror. They are very good if they are properly aligned, but they are very sensitive to alignment. Various arrangements are illustrated in Figure 2.25.

2.9.5 Diffractive Optical Elements – Holographic Lenses

2.9.5.1 Diffractive Optical Elements

Reflecting or transmissive plates finely etched or micromachined to two, three or 16 levels can be made in the form of a hologram and can thus reflect an image of any required shape. The early versions, known as "kinoforms" [32], had a reflectivity of

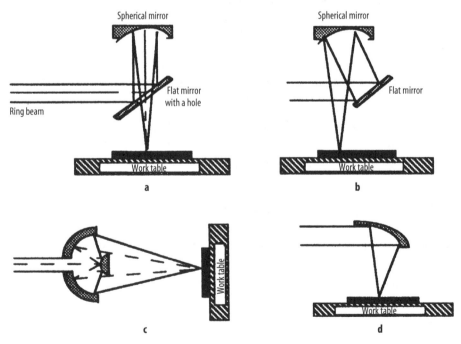

Figure 2.25 Various ways of focusing using mirrors: **a** beam on mirror axis, **b** beam off axis, **c** Cassegranian lens, and **d** parabolic mirror

around 30 % and there was some noise on the image at the edges. Modern versions are made of reflective material and are considerably more efficient [33].

The number of applications of diffractive optical elements is growing. At present they are used in processing for:

- multipoint soldering;
- beam shaping for uniform heating; and
- marking.

The optical applications include:

- Fresnel lenses;
- antireflection structures;
- achromatic lenses (the dispersion of a glass prism is in the opposite sense to a grating structure);
- coherent laser addition systems; and
- polarisation beam splitters.

2.9.5.2 Phase Plates

A variation on the etched diffractive optical elements is to insert thin surfaces into the beam path to change the phase from of a multimode beam [34].

2.9.6 Laser Scanning Systems

There are many occasions when a line beam is required. This can be achieved by a cylindrical lens or a scanning system [35]. These scanning systems can be based on oscillating aluminium mirrors as shown in Figure 2.26a. These systems have the weakness of giving a nonuniform power distribution owing to the turn point at the end of each oscillation. To avoid this, a rotating polygon is often used as shown in Figure 2.26b and c. This device has the problem of a varying velocity over the scan owing to the varying angle of incidence. Zheng [36] developed a double-polygon system which overcame that problem. There is some considerable geometry involved in designing these systems [35]. Computer control of the mirror oscillation allows the scanning of any pattern and hence laser marking, engraving, *etc.*

2.9.7 Fibre Delivery Systems

There are a variety of fibres being considered for delivering power beams for material processing [37]. The advantages appear obvious by analogy with electricity. There are, however, some difficulties which need to be faced when delivering power down a fibre. The first is the problem with the insertion into the fibre. The fibres are often a fraction of a millimetre in diameter, and thus when the focused beam is directed at the entry point into the fibre any dirt will cause catastrophic absorption. Once in the fibre the intensities are, of course, very high – for example, a 2-kW beam in a 0.5 mm-diameter

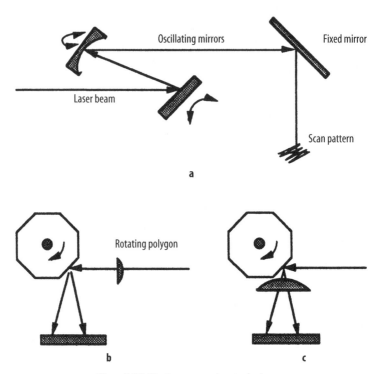

Figure 2.26 Various scanning techniques

fibre would have an intensity of around 10^6 W cm^{-2}. Compare this value with the published values of damage thresholds for fibres shown in Table 2.10 [37] and the problem becomes apparent. If the fibre is made larger to reduce this value, then the focusability is reduced and a major property of the laser beam is lost. The finest focus of a beam from a multimode fibre is an image of the end of the fibre. This will never be anything like the fineness possible with a straight laser beam. The usual limit is a magnification of a half. A further problem is that of high loss due to nonlinear events such as Raman, Brillouin and Rayleigh scattering noted earlier. There is thus a limit on the power transmission of high-quality, high-powered beams in fibres; however, this limit is very high. Currently 5 kW is being routinely delivered down 0.4-mm-diameter fibres. One alternative is a multiplicity of fibres as shown in Figure 1.26 of three beams being focused through one lens.

There is a growing market for fibre optic delivery systems for Nd:YAG lasers [38]. Many high-powered Nd:YAG lasers of greater than 1 kW are now sold with only a fibre optic delivery option. This is partly due to an appreciation that the multimode output from such lasers is not seriously affected by passing the beam down a 400-µm fibre and partly by the greater freedom which fibre delivery gives the operator. For example, the laser can be in its own room some distance away and can be used to serve several workstations all in separate enclosures, which could be separated by up to 1 km or so.

Table 2.10 Published values of the damage thresholds in various fibres [37]

Pulse duration	Wavelength (μm)	Material	Transmission loss (dB km^{-1})	Core diameter (μm)	Breakdown Power (W)	Intensity (W cm^{-2})
40 fs	0.620	SiO_2	~ 7	3	1.5×10^5	$> 2 \times 10^{12}$
5 ps	0.615	SiO_2	20	3	1.5×10^3	2×10^{10}
18 ns	0.248	SiO_2	2,000	1,000	1.6×10^7	$> 2 \times 10^9$
100 μs	1.06	SiO_2	1	10	~ 500	$> 5 \times 10^6$
CW	1.06	SiO_2	1	~ 10	100	5×10^6
500 ns	2.94	ZrF_4	12	100	~ 800	$> 1 \times 10^7$
CW	5.2	As_2S_3	900	700	100	2.6×10^4
CW	10.6	KRS-5	200–1,000	250	20	$> 4 \times 10^4$
CW	10.6	Hollow	1,000	3,000	800	

The fibres are made from extremely pure silica, often prepared from silane gas to avoid any impurities due to transition metal ions such as copper, iron and cobalt and hydroxyl ions. Such impurities are kept in the range of 1 ppb. The structure of the fibre consists of a core (the inner part of the fibre, Figure 2.27), the surrounding cladding of lower refractive index and an outer plastic protective coating; beyond that there is usually some form of metal sheathing. This metal sheathing may have within it thermal detectors to warn of damage. The light is confined to the core by total internal reflection at the core–cladding interface which occurs owing to the lower refractive index of the cladding. There are two main types of fibre: step-index fibre and graded-index fibre, as illustrated in Figure 2.27.

In step-index fibres the light rays take a zigzag path down the fibre until the rays homogeneously fill the core. The output beam has the diameter of the fibre core with an intensity pattern which is essentially flat-topped, although this will vary as the fibre

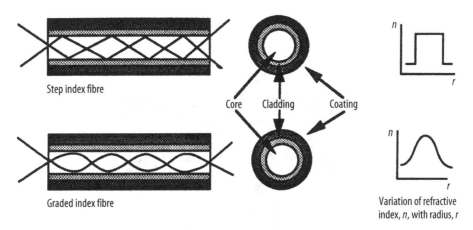

Step index fibre

Graded index fibre

Core   Cladding   Coating

Variation of refractive index, n, with radius, r

Figure 2.27 The structure of step-index and graded-index fibres

is bent. In the graded-index fibre the quantity of dopants affecting the refractive index varies across the fibre diameter, usually having a parabolical variation in refractive index. The rays propagate in an undulating manner. With a parabolic refractive index profile the path lengths of all rays are nearly equal for every angle of propagation. This is the condition for conserving the beam–parameter product: waist diameter $d_{waist} \times$ angle of divergence, Θ_∞ (see Section 2.8.1.2 on the M^2 concept of beam quality). The output power profile from a homogeneously filled graded-index fibre yields an intensity distribution similar to the refractive index profile. In real fibres the beam parameters increase because of the finite size of the fibre, imperfect fibre geometry, inhomogeneities or impurities of the silica material, bending and imperfect fibre coupling.

2.9.7.1 Fibre Coupling

The laser beam will propagate along the fibre with low loss when it is coupled into the fibre within the maximum angle of acceptance (Θ_{max}) determined by the numerical aperture of the fibre [39]:

$$NA = \sin(\Theta_{max}/2),$$

where NA is the numerical aperture and equals the square root of the differences between the refractive indices of the core axis, n_{core}^0, and the cladding, n_{clad},

$$NA = \left[\left(n_{core}^0\right)^2 - \left(n_{clad}\right)^2 \right]^{\frac{1}{2}}.$$

Typical values of the numerical aperture for fused silica range from 0.17 to 0.25 (*i.e.*, acceptance angles up to 28°). Any higher values require increased dopant concentration, running the risk of disturbances in the refractive index. The basic coupling requirements are (Figure 2.28)

$$d_{in} < d_{core} \text{ and } \sin(\Theta_{in}/2) < NA.$$

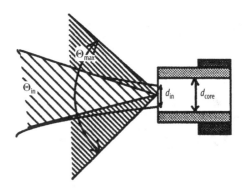

Figure 2.28 Fibre coupling geometry

There are more complex problems when the beam is elliptical or suffers from astigmatism. There is also the problem of thermal lensing discussed in Section 2.8.1.4 which could alter the value of d_{in} as a function of time. In practice, the core diameter and the numerical aperture are chosen so that they exceed the beam parameters d_{in} and Θ_{in} by factors of 1.5–3 to guarantee safe operation. Fibre coupling optics are usually based on a telescope system.

Preparation of the end face of the fibre which is to receive the focused laser beam is critical. It is usually prepared by cleaving or very careful polishing.

2.9.7.2 Fibres for Lasers Other than the Neodymium-doped Yttrium Aluminium Garnet Laser

Table 2.10 lists some of the fibre material available for other wavelengths. In addition to these, CO radiation at 5.4 μm wavelength can be passed down metal halide fibres such as fibres of CaF_2 and zinc halides. CO_2 radiation at 10.6 μm can be transmitted down special thallium-based fibres with heavy loss, giving a power limit currently of approximately 100 W. An alternative is hollow waveguides, such as thin-bored sapphire tubes (0.5–1-mm internal diameter) coated internally with a dielectric coating of lower refractive index than that of air. These devices have passed several kilowatts of 10.6-μm radiation over distances of several metres. The losses, α, of such waveguides have been calculated by Miyagi and Karasawa [40] to be

$$\alpha \propto I/D_{core}^3 \text{ and } \alpha \propto 1/R,$$

where D_{core} is the tube internal diameter and R is the bend radius. Thus, very fine waveguides would suffer serious loss, but large-diameter waveguides would be difficult to focus.

In general, engineers still think like electricians and would prefer fibre delivery of power, regardless of the fact that by so doing they are discarding one of the significant characteristics of optical energy – that it is one of the few forms of energy which can be transmitted through air or space without the need for a conductor.

2.9.8 Liquid Lenses

Liquid optics have a certain appeal in that they should be unbreakable, highly flexible, easily cleaned, easily cooled and possibly cheap. Various versions of liquid optics have been attempted or are being developed; but remember that you are currently reading this with a form of liquid optics – your eye!

1. *Gas jets of different refractive index such as cold nitrogen.* This works as a weak cylindrical lens to a beam passing at right angles to the jet.
2. *Water stream as a waveguide.* The water waveguide has been commercialised by Synova [41]. The laser beam from a Nd:YAG laser at 1,064 nm or a Yb:YAG laser at 1,070 nm or a frequency-doubled Nd:YAG laser at 532 nm is focused through a clean, deionised water pool into a fine orifice (25–150-μm diameter) from which

it emerges with the water jet flowing under a pressure of 2–50 MPa at approximately
1 l/h. The beam is homogenised and waveguided down the water jet by internal reflections. The working distance can be anywhere up to 1,000 times the nozzle diameter. Some very fine cutting has been demonstrated, producing parallel-sided cuts
with a greatly reduced heat-affected zone (HAZ). The cut edges are clean and ablation products are removed in the water. There are no noxious gases, low mechanical pressures and no focal position problems. The applications to date have been in
dicing SiC chips, cutting organic LED (OLED) masks, cutting medical stents and
cutting hard materials such as diamond and cubic boron nitride.

3. *Enclosed liquid surface whose shape can be controlled.* A liquid optic is now being
used in mobile phones developed by Varioptic [42]. In these devices a small droplet
of oil is held in a tiny water chamber – it has to be small so that surface tension
forces overcome waves and gravity effects. The design is illustrated in Figure 2.29.
By applying an electric charge, one can make the oil surface curve in a controlled
manner, since the electric charge affects the wettability of the oil on the walls of the
chamber. The advantages are significant: no moving parts, little power to drive it
(less than 15 mW), small size (8 mm diameter × 2 mm thickness) and focal range
from F5 to infinity. The applications expected are as autofocus units for mobile

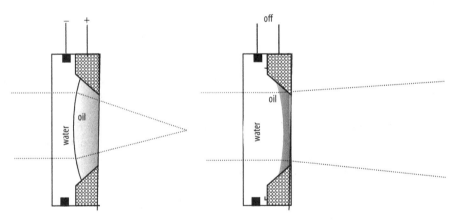

Figure 2.29 The Varioptic liquid lens based on electrowettability. On application of a voltage, the
meniscus shape changes owing to surface tension effects. This changes the power of the lens. The
changes are both rapid and reversible

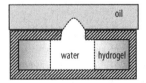

Figure 2.30 Example of a liquid lens: an expanding hydrogel ring creates a meniscus at a water–oil
boundary

phones, cameras, webcams, bar code readers, biometric readers for face, iris and fingerprint recognition and medical endoscopes, fibre scopes and dental cameras. An alternative is a small pool of liquid whose volume can be changed by pressure either pushing more fluid into the chamber or by the chamber changing size. The chamber changing size is an interesting concept illustrated in Figure 2.27. It has been developed by Jiang and Dong [43] of University of Wisconsin. The chamber holding water is made of a hydrogel that can respond to a stimulus. The chamber is covered with a water-repellent sheet with a hole in it. As the hydrogel expands, the water is pushed upwards but is pinned at the edges of the hole, thus forming a well-defined meniscus. This design is getting close to the way our eyes work.

4. *Enclosed liquid or polymer whose refractive index can be controlled* [43]. A nematic liquid crystal (a substance in which the molecules are oriented in parallel but not arranged in well-defined planes) can change refractive index depending on the orientation of the molecules within it. If a layer of such material is placed between two transparent electrodes, one possibly being hemispherical, when a voltage is applied, the electric field varies symmetrically about the centre of the electrodes – being highest at the edge and lowest at the centre. The orientation of the molecules varies with the electric field, causing a corresponding change in the refractive index; *i.e.*, a lens is formed of variable focal length depending on the electric field. There are problems with this type of lens, namely, astigmatism, distortion, light scattering and from the laser material processing point of view an inability to transmit large powers. Developments are taking place on the types of liquid crystals (they must not rotate in the electric field) and the mixture in which they are contained.

2.9.9 Graded-index Lenses

Graded-index (GRIN) lenses may be radially or axially graded. They are usually shaped optics with the index gradient acting as an aspherical corrector.

Radial graded-index lenses are used in the input scanner section of photocopiers or fax machines. Axial graded-index lenses are often used as objective lenses in CD players and laser diode collimators. They have a potential future in camcorders and military optical systems, since they allow colour-corrected low F numbers and a reduced number of optical components in a zoom system [44].

2.10 Conclusions

Radiant energy is one of the most adaptable forms of energy available today. Not only can it be shaped by the devices just discussed, it can also have properties in polarisation, wavelength and power. It can interact with itself to give interference and diffraction effects or even change to double its frequency. This makes the subject of optical engineering one of the stronger subjects in the future of engineering.

Questions

1. a. Derive a formula for the relationship between the focal length of a lens and the resulting minimum spot size.
   b. Using this formula, show how the focal length is related to the depth of focus.
   c. List the occasions on which this formula may not work.
   d. A 2 kW CO_2 laser beam of 19 mm diameter and $M^2 = 2$ is to be expanded to form a wider parallel beam. This beam is then used to cut cloth using a galvanometer-driven focusing concave mirror. The concave mirror is mounted 3 m above the cutting table. The required spot size for successful cutting has to be less than 400 µm. How big should the beam be when it is incident in the concave mirror? What is the radius of curvature of the mirror?
2. a. A spot size of 150 microns or µm is required from a CO_2 laser whose raw beam from the laser cavity is 19 mm in diameter and $M^2 = 2.5$. What lens is required to achieve this?
   b. Using this lens from (a), how could one change the spot size to 180 or 120 µm?
   c. For F numbers less than 5, spherical aberration may be a problem. What spot size could be achieved with an F number of 5?
   d. What could be done to achieve a 100 µm spot size?

References

[1] Huygen C (1690) Traite de la lumière, 1678. Leiden
[2] Newton I (1704) Opticks, 1st edn
[3] Einstein A (1905) Über einen die Erzeugung und Verwandlung des Lichtes Betreffenden Heuristischen Gesichtspunkt. Ann Phys 17:132
[4] Heisenberg W (1932) The development of quantum mechanics. Nobel lecture, 11 December 1932
[5] Schrödinger E (1933) The fundamental idea of wave mechanics. Nobel lecture, 12 December 1933
[6] Cobine JD (1941) Gaseous conductors. McGraw-Hill, New York
[7] Nonhof CJ (10988) Material processing with Nd-YAG lasers. Electro Chemical Publications, Ayr
[8] Hector LG, Kim WS, Ozisiki (1990) Propagation and reflection of thermal waves in finite mediums due to axisymmetric surface waves. In: Proceedings of the XXII ICHMT international symposium on manufacturing and material processing, Dubrovnik, August 1990
[9] Gray EG (ed) (1972) American Institute of Physics handbook, 3rd edn. McGraw-Hill, New York
[10] Juptner W, Rohte W, Sepold G, Teske K (1980) Cutting with high capacity CO_2 laser beams. DVS Ber 63:222
[11] Patel RS, Brewster MQ (1988) Effects of oxidation on low power Nd-YAG laser metal interactions. In: ICALEO '88 proceedings, Santa Clara, October–November 1988. Springer, Berlin/IFS, Kempston, pp 313–323
[12] O'Neill W (1990) Mixed wavelength laser processing. PhD thesis, University of London
[13] Drude P (1922) Theory of optics (English edn). Longmans, Green, New York
[14] Kielman F (1985) Stimulated absorption of CO_2 laser light on metals. In: Proceedings of the NATO Advanced Study Institute on laser surface treatment, San Miniato, Italy, September 1985, pp 17–22
[15] Jenkins FA, White HE (1983) Fundamentals of optics, 2nd edn. McGraw-Hill, London
[16] Greses J, Hilton P, Barlow CY, Steen WM (2002) Plume attenuation under high power Nd:Yag laser welding. In: ICALEO 2002 proceedings, Phoenix, October 2002, LIA, Orlando, paper 808
[17] Bohren CF, Huffman DR (1983) Absorption and scattering of light by small particles. Wiley, New York

[18] Hansen F, Duley WW (1994) Attenuation of laser radiation by particles during laser material processing. J Laser Appl 6(3):137–143

[19] Akhter R (1990) Laser welding of zinc coated steel. PhD thesis, University of London

[20] Sharp M, Henry P, Steen WM, Lim GC (1983) An analysis of the effects of mode structure on laser material processing. In: Waidelich W (ed) Proceedings of Laser'83 optoelectronic conference Munich, June 1983, pp 243–246

[21] Matthews SJ (2002) Back to basic – polarisation; an eye on polarity. Laser Focus World Nov 115–119

[22] Greening D (1994) Quality factor reveals beam divergence problem. Opt Laser Eng Apr 25–28

[23] Langhorn C, Kanzler K (1994) Thermal focusing in CO_2 lenses. Industrial Laser Review Dec 15–17

[24] Miyamoto I, Nanba H, Maruo H (1990) Analysis of induced optical distortion in lens during focussing high power CO_2 laser beam. Proc SPIE 1276:112–121

[25] Barik S, Giesen A (1991) Finite element analysis of the transient behaviour of optical components under irradiation. Proc SPIE 1441:420–429

[26] Lowrey WH, Swantner WH (1989) Pick a laser lens that does what you want it to. Laser Focus World May 121–130

[27] Kozawa Y, Sato S (2005) Generation of a radially polarised laser beam by the use of a conical Brewster prism. Opt Lett 30(22):3063–3065

[28] Lambda Research Optics (2009) Radial polarizer for CO_2 laser systems. http://www.lambda.cc/1800.pdf

[29] Niziev VG, Nesterov AV (1999) Influence of beam polarisation on laser cutting efficiency. J Phys D Appl Phys 32:1455–1461

[30] Zoske U, Giesen A (1999) Optimisation of beam parameters of focussing optics. In: Proceedings of the 5th international conference on lasers in manufacturing (LIM5), Stuttgart, September 1988. IFS, Kempston, pp 267–278

[31] Ellis N (2000) Understanding beam expanders. Industrial Laser User (19):19–21

[32] Patt PJ (1990) Binary phase gratings for material processing. J Laser Appl 2(2):11–17

[33] Taghizadeh MR, Blair P, Layet B, Barton IM, Wddie AJ, Ross N (1997) Design and fabrication of diffraction optical elements. Microelectron Eng 34(3–4):219–242

[34] Casperson LW (1994) How phase plates transform and control laser beams. Laser Focus World May 223–228

[35] Stutz GE (1990) Laser scanning systems. Photonics Spectra Jun 113–116

[36] Zheng HU (1990) In process quality analysis of laser cutting. PhD thesis, University of London

[37] Weber HP, Hodel W (1987) High power transmission through optical fibres for material processing. In: Industrial laser annual handbook. Laser Institute of America, Orlando, pp 33–39

[38] Walker R (1990) Fibreoptic beam delivery leads to versatile systems. Industrial Laser Review Jul 5–6

[39] Beck T, Reng N, Richter K (1993) Fibre type and quality dictate beam delivery characteristics. Laser Focus World Oct 111–115

[40] Miyagi M, Karasawa S (1990) Waveguide losses in sharply bent circular hollow waveguides. Appl Opt 29(3):367–370

[41] Hewett J (2007) Laser water jet cools and cuts in the material world. Optics and Lasers Europe Mar 17–19

[42] Ruffin P (2007) Autofocus liquid lenses target new applications. Optics and Lasers Europe Oct 17–18

[43] Jiang H, Dong D (2006) Liquid lenses shape up. Optics and Lasers Europe Nov 24–26

[44] Atkinson LG, Kindred DS (1996) An old technology, gradient index lenses, finds new applications. In: Photonics design and applications handbook, book 3, Laurin, Pittsfield, pp H-362–H-367

[45] Higgins TV (1994) Non-linear optical effects are revolutionising electro optics. Laser Focus World Aug 67–74

"Now you know the difference between a moon beam and a laser beam!"

3 Laser Cutting, Drilling and Piercing

> Measure a thousand times and cut once
>
> *Old Turkish proverb*

3.1 Introduction

The idea of cutting with light has appealed to many from the first time they burnt paper on a sunny day with the help of a magnifying glass. Cutting centimetre-thick steel (Figure 3.1) with a laser beam is even more fascinating!

Laser cutting is today the most common industrial application of the laser; in Japan around 80% of industrial lasers are used in this way. Apart from the fascination, which is rarely a driving force for investment in hard industry, the reason is most probably that in cutting there is a direct process substitution into an established market and the laser, in many cases, happens to be able to cut faster and with a higher quality than the competing processes. The comparison with alternative techniques is listed in Table 3.1.

Figure 3.1 Metal cutting of 5-mm-thick stainless steel with a CO_2 slab laser of 2 kW. (Courtesy of AILU through http://www.designforlasermanufacture.com and Rofin-Baasel UK)

Table 3.1 Comparison of different cutting processes [102]

Quality	Laser	Punch	Plasma	Nibbling	Abrasive fluid jet	Wire EDM	NC milling	Sawing	Oxy flame	Ultrasonic
Rate	✓	✓	✓	×	×	×	×		×	×
Edge quality	✓	✓	×	×	✓	✓	✓	×	×	✓
Kerf width	✓	✓	×	×	✓	✓		×		
Scrap and swarf	✓	✓		×	✓			×		✓
Distortion	✓		×		✓		✓		×	
Noise	✓	×	×		×				×	
Metal and nonmetal	✓	×	×	×	✓	×	✓			✓
Complex shapes	✓	×	✓							
Part nesting	✓	×			✓				✓	
Multiple layers	×	✓			×				✓	
Equipment cost	×				×	×		✓	✓	×
Operating cost						✓		✓	✓	×
High volume	✓	✓							✓	
Flexibility	✓	×	✓	✓	✓	×		✓	✓	
Tool wear	✓	×	✓	×	✓	×	×	×	✓	×
Automation	✓	✓	✓	×	✓	× ✓	× ✓		×	
HAZ	✓	×	×		✓	× ✓	× ×	×	×	
Clamping	✓	✓	✓	✓	✓	×	×	×	✓	
Blind cuts	✓	×	✓	✓		✓	×			×
Weldable edge	✓	✓	×	✓	✓	✓	✓	✓	✓	×
Tool changes	✓	×	✓		✓	✓		×	×	✓

Further comparisons can be found in Powell and Wykes [102].
EDM electric discharge machining, *NC* numerical control, *HAZ* heat-affected zone, ✓ point of particular merit, × point of particular disadvantage

The significant advantages of the laser are seen in the table but possibly these need some further explanation. Thus, the advantages can be divided into two categories – cut quality and process characteristics:

Cut quality characteristics:

1. The cut can have a very narrow kerf width giving a substantial saving in material. (Kerf is the width of the cut opening.)
2. The cut edges can be square and not rounded as occurs with most hot jet processes or other thermal cutting techniques.
3. The cut edge can be smooth and clean. The cut is reckoned to be a finished cut, requiring no further cleaning or treatment.
4. The cut edge is sufficiently clean that it can be directly rewelded.
5. There is no edge burr as with mechanical cutting techniques. Dross adhesion can usually be avoided.
6. There is a very narrow HAZ, particularly on dross-free cuts. Usually there is a very thin resolidified layer of micron dimensions. Thus, there is negligible distortion.
7. Blind cuts can be made in some materials, particularly those which volatilise, such as wood or acrylic.
8. Cut depth is limited and depends on the laser power. The current range for high-quality cuts with 2–5-kW laser power is 10–20 mm.

Process characteristics:

1. It is one of the faster cutting processes.
2. The workpiece does not need clamping, although it is usually advisable to do so to avoid the workpiece shifting with the table acceleration and for locating when using a computer numerical control (CNC) program.
3. There is no tool wear since the process is a noncontact cutting process, but the lens must be kept clean.
4. Cuts can be made in any direction; but see Section 3.6.1.3 on polarisation.
5. The noise level is low.
6. The process can easily be automated with good prospects for adaptive control in the future.
7. Tool changes are mainly "soft", that is, they are only programming changes. Thus, the process is highly flexible.
8. Some materials can be stack-cut, but there may be a problem with welding between layers.
9. Nearly all materials can be cut. They can be friable, brittle, electric conductors or nonconductors, hard or soft. Only highly reflective materials such as aluminium, copper and gold can pose a problem, but with proper beam control these can be cut satisfactorily.

3.2 The Process – How It Is Done

The general arrangement for cutting with a laser is shown in Figure 3.2. The principle components are the laser itself with some shutter control, beam guidance train, focus-

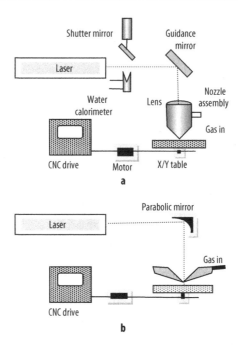

Figure 3.2 General arrangement for laser cutting: **a** using transmissive optics, and **b** using reflective optics. *CNC* computer numerical control

ing optics and a means of moving the beam or workpiece relative to each other. The shutter is usually a retractable mirror which blocks the beam path and diverts the beam into a beam dump which doubles as a calorimeter. When the beam is required, the mirror is rapidly removed by a solenoid or pneumatic piston. The beam then passes to the beam guidance train, which directs the beam to centre on a focusing optic. The focusing optic can be either transmissive or reflective; the transmissive optics are made of ZnSe, GaAs or CdTe for CO_2 lasers or quartz for YAG or excimer lasers; the reflective optics consist of parabolic off-axis mirrors or on-axis spherical mirrors (see Section 2.9.4). The focused beam then passes through a nozzle from which a coaxial gas jet flows. The gas jet is needed both to aid the cutting operation and also to protect the optics from spatter. In the case of the metal optics, an "air knife" is often used which blows sideways across the exit from the optic train, thus deflecting any smoke and spatter. For cutting processes which rely on melt removal by the gas jet there is a problem for the metal optics systems. The main reason for using metal optics is either that one runs a research school with clumsy students or that the power of the laser is such that the transmissive optic is near the limit of its thermal stress tolerance. In fact, if the transmissive optics are likely to break, then one uses metal optics. To achieve a gas jet suitable for cutting (more than 20 m s^{-1} and reasonably well focused) without interposing a transmissive element, a set of centrally directed nozzles [1] or a ring jet can be used; see Figure 3.2b. It is very important to the process that the beam, optic and jet are all lined up.

In cutting, the laser evaporates a hole through the material and then the "hole" is traversed to make a cut. The drilling and piercing processes are different from cutting in a line because they do not have an open-sided hole. Since every cut must start with piercing, if it does not start at an edge, our first topic here will be drilling and piercing.

3.3 Laser Drilling and Piercing

3.3.1 Introduction

Some of the first experiments with high-powered lasers used a ruby laser and its power was measured in the number of Gillette razor blades that it could pierce in a single shot. In those days that was an extraordinary thing to be able to do – a single pulse of light to knock a hole in sheets of steel. Historically, laser drilling was the first industrial application by Western Electric, using a ruby laser in 1965 to drill holes in diamond dies for wire extrusion, followed in the early 1970s by cutting processes.

Today, precision laser-drilled holes are used for a variety of applications. One of the main applications is boundary layer film cooling in jet engine components such as turbine blades and combustion chambers. State-of-the-art military engines involve over 1.2 million cooling holes, most of which are laser-drilled. Numerous high-density electronic packages employ laser-drilled vias for interconnecting layers; automobile injection nozzles, baby's teats for milk bottles and irrigation pipes all contain laser-drilled holes; specialist holes in surgical tooling; inkjet nozzles; CD discs and many more combine to form a huge and growing market for fine precise holes all made at high speed by laser.

The laser has an advantage in this market since:

- it can drill holes fast;
- burr and spatter can be controlled to some extent;
- it can drill any material that will absorb the radiation regardless of hardness;
- the diameter and shape of the hole can be controlled by trepanning methods; and
- it can pierce at almost any angle.

The processing challenges that are being addressed concern:

- increased speed;
- reduction or control of taper;
- elimination of spatter;
- reduction or elimination of the resolidified layer on the hole wall;
- precise cross-section shape – round, square or star-shaped, *etc.*;
- repeatability;
- high-aspect-ratio holes; and
- drilling through coated material.

3.3.2 Drilling Process Variations

The importance of laser drilling as an industrial process has led to many variations on how to achieve quick, high-quality holes with good repeatability. They include (see Figure 3.3):

- *Single-shot drilling*: One pulse makes and finishes the hole.
- *Double-pulse drilling*: The energy is divided between two pulses which follow each other in very rapid succession to interact with the plasma more efficiently.
- *Percussion drilling*: Single or multiple shot with no movement of the workpiece or the beam.
- *Trepanning*: Rotating the beam around the perimeter of the hole, a form of cutting.
- *Helical trepanning*: Starting near the middle of the hole and rotating around the perimeter, gradually deepening the hole with each rotation spirally machining into the workpiece, sometimes also with a change in the focal position following the hole base downward.

Using these methods holes can be made:

- *Normal to the surface* – the usual method
- *At an angle to the surface* – particularly relevant for turbine blades. No particular problem has been found with making angled holes except that at oblique angles the hole depth would usually have to be greater. At the start of drilling at an angle, the plume is emitted at right angles to the surface, which is an advantage, but once the keyhole has formed, it aligns with the hole axis [2]
- *As blind holes* – for electronic vias.

These process variations will now be discussed in turn. The physical mechanism describing how the beam penetrates a material is roughly the same for all of these processes. Percussion drilling is chosen as the example to discuss the physical mechanisms.

3.3.3 Percussion and Single- or Double-shot Drilling

Laser percussion drilling techniques are the quickest methods for achieving a hole by comparison with almost any other process, but they are not as precise as trepanning techniques. Percussion drilling is the preferred method in many aerospace applications

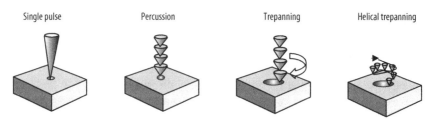

Figure 3.3 Different styles of drilling. (After Dausinger [31])

for cooling holes owing to its speed, sufficient accuracy and repeatability. It requires pulses of between 10^5 and 10^7 W cm^{-2} [3].

3.3.3.1 How Can Light Drill a Hole? The Physical Picture

In drilling, which relies on vaporisation, the focused beam first heats up the surface to boiling point and so generates a keyhole. The keyhole causes a sudden increase in the absorptivity owing to multiple reflections causing the hole to deepen quickly. Figure 3.4 [4] shows the calculated change in drilling speed as the hole deepens. As it deepens, so vapour is generated and escapes. This evaporation exerts a reaction force on the melt surface as the vapour accelerates away; also the temperature gradient across the surface of the melt exerts forces through variations in surface tension. Both of these forces drive the melt to the side of the forming hole. At the side the melt is driven by the high pressure at the base of the hole and the drag forces of the escaping vapours to flow up the walls and out as spray with the vapour [5]. When the laser pulse finishes, this flow will cease and the melt will fall back as splatter around the lip of the hole or remain as a recast layer within the hole. In general, the more melt there is, the poorer the quality of the resulting hole.

This is the usual method of drilling for pulsed lasers or in the cutting or drilling of materials which sublime and do not melt, such as wood, carbon and some plastics. Rykalin *et al.* [6] showed that at approximately 3×10^6 W cm^{-2} metals start to evaporate and at 10^8 W cm^{-2} a plasma is formed that will stop or prevent further drilling by blocking the beam through absorption within the electron cloud in the plasma. This will occur when the plasma frequency approaches the laser frequency.

This simplified view of laser drilling contains the overall picture but added to this are the detailed interactions of the incoming radiation within the keyhole and the fluid flow of the melt that is to be ejected. The incoming radiation passes into the forming hole through the exiting hot gas and dust, which will have a scattering effect on the beam. The beam will be absorbed by Fresnel absorption on the walls of the keyhole as it is waveguided to the base of the hole or reflected out of the hole, but it will also be absorbed by the electrons in the hot gas, which in turn will radiate to the walls of the forming hole. If the power is sufficient (more than 10^8 W cm^{-2}), a plasma whose

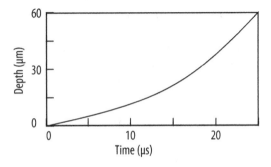

Figure 3.4 Time variation of keyhole depth. (After Noguchi *et al.* [4])

frequency is near that of the laser beam will form and effectively block the beam from reaching the substrate. The pressure will still be there, which is the basis for the process of shock peening (see Section 6.19).

The melt generated will flow up the walls of the hole owing to the high pressure at the base and the drag forces from the exiting gases. The speed of ejection of the gases will in most cases be supersonic and shock waves will form in the area of the hole. Added to this picture is the coaxial assist gas blowing down onto the hole entrance. This will initially create a pressure countering the exiting gases, but, if it is oxygen, it may react with the molten metal and generate more heat. On breakthrough, the assist gas pressure will help to drive the vapour through the hole and help scour the sides.

Much of this was witnessed by French *et al.* [7] using filming at 42,000 frames per second in which they identified three stages in the development of a hole: stage 1 – wide conical ejection of melt; stage 2 – melt ejection from the hole in a ribbonlike form; stage 3 – when the hole is deeper and the assist gas is oxygen, there is spark droplet ejection in all directions. They found that the assist gas affected the velocity of the ejected material: for an increase in assist gas pressure from 1 to 6 bar the velocity of the melt ejection was reduced by 50 % from the observed value of 17–27 m s^{-1}.

3.3.3.2 Rate of Penetration.

3.3.3.2.1 Lumped Heat Capacity Model

The rate of penetration of the beam into the workpiece can be estimated from a lumped heat capacity calculation, assuming the heat flow is one-dimensional and all of it is used in the vaporisation process – that is, that the heat conduction is zero. This fairly gross assumption is not ridiculous if the penetration rate is similar to or faster than the rate of conduction.

Thus, the volume removed per second per unit area equals the penetration velocity, V (m s^{-1}). The volume removed per seconds equals the power divided by the heat capacity of a unit volume, and we have

$$V = F_0 / \left\{ \rho \left[L + C_p \left(T_v - T_0 \right) \right] \right\} \quad (\text{m s}^{-1}),$$

where F_0 is the absorbed power density (W m^{-2}), ρ is the density of the solid (kg m^{-3}), L is the latent heat of fusion and vaporisation (J kg^{-1}), C_p is the heat capacity of the solid (J kg^{-1} °C^{-1}), T_v is the vaporisation temperature (°C) and T_0 is the temperature of the material at the start (°C).

If we substitute values into this equation, we can derive the approximate maximum penetration rate possible for different materials.

Assuming we have a 2 kW laser focused to 0.2 mm beam diameter, the power density will be

$$F_0 = P / \pi r^2 = 2,000 \times 10^6 / \pi 0.1^2$$
$$= 6.3 \times 10^{10} \text{ W m}^{-2}, \text{ or } 6.3 \times 10^6 \text{ W cm}^{-2}$$

The rate of penetration of such a beam into various materials is shown in Table 3.2. These penetration figures are of the same order as those found experimentally [8]. If the penetration rate is around $1 \, \mathrm{m \, s^{-1}}$, then the vapour velocity from a cylindrical hole would be defined by the ratio of the densities of vapour and solid to be of the order of $V \rho_v / \rho_s = 1{,}000 \, \mathrm{m \, s^{-1}}$. At these sonic speeds compression effects and variations in the hole shape will mean the actual velocity of exit of the vapour is much less, but nevertheless sonic flow and shock waves will occur (there is usually a distinct bang with each laser pulse). Such high-velocity flow will be capable of considerable drag in eroding the walls of the forming hole.

Thus, in this form of drilling or piercing the material is removed partly as vapour and partly as ejecta. Gagliano and Peak [9] estimated from their experiments that around 60% of the material was removed as ejecta. This figure was partially confirmed in Voisey et al.'s [10] detailed analysis of ejecta during drilling. They found using 0.5-ms pulses that the melt ejection fraction varied between 35 and 60% for most metals except the really heavy metals, such as tungsten, for which the fraction fell to nearer 10%. They also showed that the size of the ejected particles grew with a reduction in the pulse energy or an increase in the length of the pulse; larger ejected particles coming from long, lower-powered pulses. The average particle size for a 2.5 J, 0.5 ms pulse was found to be around $10 \, \mu\mathrm{m}$. Using high-speed photography, they found the velocity of ejection averaged approximately $8 \, \mathrm{m \, s^{-1}}$ for a 1.4 J, 0.5 ms pulse and $13 \, \mathrm{m \, s^{-1}}$ for a 2.4 J, 0.5-ms pulse – a figure in agreement with that of Ng and Li [11]. There was, of course, a range of velocities, presumably dependent on particle size and direction.

There are a number of side effects from this almost explosive evaporation. One is the recoil pressure required to accelerate the vapour away. Bernoulli's equation is able to give a rough estimate of the value of this pressure for an exit velocity of $1{,}000 \, \mathrm{m \, s^{-1}}$, even though it assumes incompressible flow, as

$$\Delta P = \rho_v V^2 / 2 = 4 \times 10^6 \, \mathrm{N \, m^{-2}} .$$

One atmosphere is $10^5 \, \mathrm{N \, m^{-2}}$. A pressure rise of this order (40 atm) will cause a rise in the vaporisation temperature (given by the Clapeyron–Clausius equation $\frac{dp}{dT} = \frac{\Delta H}{T \Delta V}$). This pressure causes stress in the surface, which is amplified by the thermal stresses generated in the heated material. Together they represent quite a considerable stress. If this can be applied very quickly, in a few nanoseconds (10^{-9} s), then the effect is similar to being hit, as in shot peening, as noted earlier.

3.3.3.2.2 Melt Fraction During Drilling

The quality of the hole or cut is determined to a large extent by the quantity of melt which may build up and cause debris on the surface or erosion marks on the wall. Thus, it is interesting to calculate how quickly the boiling point is reached and how much of the laser pulse energy is used in evaporation or melting.

For one-dimensional heat flow with constant energy input it can be shown (see Chapter 5) that the surface temperature at any time, t, after the start of irradiation is

Table 3.2 Material properties, penetration speeds, V, and time to evaporate, t_v, for a beam of power density 6.3×10^{10} W m^{-2} [103]

Material	Material properties								Process properties	
	ρ (kg m^{-3})	L_f (kJ kg^{-1})	L_v (kJ kg^{-1})	C_p (J kg^{-1} °C^{-1})	T_m (°C)	T_v (°C)	K (W m^{-1} K^{-1})	α (×10^{-5} m^2 s^{-1})	V (m s^{-1})	t_v (µs)
Tungsten	19,300	185	4,020	140	3,410	5,930	164	6.07	0.64	3
Aluminium	2,700	397	9,492	900	660	2,450	226	9.30	1.9	0.6
Iron	7,870	275	6,362	460	1,536	3,000	50	1.38	1.0	0.3
Titanium	4,510	437	9,000	519	1,668	3,260	19	0.81	1.2	0.09
304 stainless steel	8,030	≈ 300	6,500	500	1,450	3,000	20	0.5	0.97	0.4

given by

$$T(0, t) = (2F_0/K) [\alpha t)/\pi]^{1/2}, \qquad (3.1)$$

where α is the thermal diffusivity $(K/\rho C_p)$ $(\mathrm{m^2\ s^{-1}})$ and K is the thermal conductivity $(\mathrm{W\ m^{-1}\ K^{-1}})$.

Therefore, the time required to reach the boiling point on the surface, t_v, is

$$t_v = (\pi/\alpha) [(T_B K)/(2F_0)]^2.$$

The estimated time for a 2-kW laser beam to cause vaporisation is shown in Table 3.2. The thermal gradient at that time would have penetrated [assuming a Fourier number $(x^2/\alpha t) = 1)$] to around 2 μm for iron and hence it can be seen that the HAZ is expected to be small in this case. The previous calculation, based upon ignoring the heat conduction, is thus not so wide of the mark.

The energy and peak power of a pulse are also critical in determining how much of the energy will be used to evaporate as opposed to melt. If the pulse is sufficiently short, the thermal penetration distance is limited, being proportional to $\sqrt{(\alpha t_{\mathrm{pulse}})}$. It is thus possible to do a heat balance on this small volume and estimate how much of the energy is available for evaporating material.

Such an energy balance on the volume $\left[\sqrt{\propto t_{\mathrm{pulse}}} \left(\frac{\pi d^2}{4} \right) \right]$ gives

$$E = A\rho \sqrt{\propto t_{\mathrm{pulse}}} \frac{\pi d^2}{4} \left[C_p (T_m - T) + L_f + m' \{ C_p (T_v - T_m) + L_v \} \right],$$

where A is a constant and m' is a multiple of the sensible heat required to evaporate a unit mass of material.

From this equation we get

$$E/d^2 \sqrt{t_{\mathrm{pulse}}} = A\rho \sqrt{\alpha \pi}/4 \left[C_p (T_m - T) + L_f \right] + m' \{ A\rho \sqrt{\alpha \pi}/4 C_p (T_v - T_m) + L_v \},$$

which can be simplified as

$$\frac{E}{\left(d^2 \sqrt{t} \right)} = A' + Bm',$$

where A' and B are constants dependent on the material being drilled.

If $m' > 1$, then there is sufficient energy in the pulse to evaporate all the heat-affected volume. Using the values for iron and aluminium (see Table 3.2), we find that:

Material	A' $(\mathrm{J\,m^{-2}\,s^{-1/2}})$	B $(\mathrm{J\,m^{-2}\,s^{-1/2}})$
Iron	7.1×10^9	5.1×10^{10}
Aluminium	6.4×10^9	7.0×10^{10}

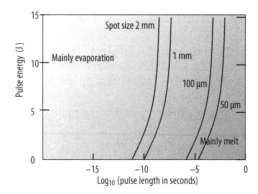

Figure 3.5 The pulse energy and duration required to evaporate all the material in the heat-affected zone, assuming no thermal conduction. The *lines* represent just sufficient energy for evaporation

Figure 3.5 shows how the pulse energy varies with pulse duration to supply just sufficient energy to evaporate the affected volume. In this simple calculation remember that the depth of the evaporated volume is equal to $\sqrt{(\alpha t_{\text{pulse}})}$. So the shorter the pulse, the smaller the thickness of the treated zone. Hence, a larger spot means a thinner depth of penetration is required if the whole volume is to be evaporated with the energy supplied. If the pulse energy increases, then the depth of penetration must increase to maintain the equation for evaporating all the affected material. This is counterintuitive owing to the condition that all the material has to be evaporated. A more intuitive equation would result if we had considered pulse power instead of energy. Since the power is proportional to $1/\sqrt{t}$ if all the material were to be evaporated. Shorter pulse times (hence more intense pulses) by a factor of $(58/76) = 0.76$ (the ratio of A' and B for iron and aluminium) are required if melting aluminium is to be avoided.

The figure shows clearly that longer pulses of low energy generate more melt than short sharp pulses and that pulses longer than 0.3 ms even with a very fine 50-µm spot size will generate significant melt that has to be removed. A nanosecond pulse is a good pulse length to use to give minimum melt. Many papers have shown that shorter pulses leave less spatter and recast material.

3.3.3.2.3 One-dimensional Heat Flow Model

The simple view of the drilling mechanism just discussed can be greatly elaborated. In the paper by Chan and Mazumder [12], they assumed that the laser radiation was absorbed by the surface of the liquid melt at the bottom of the hole. They then modelled the heat flow from the solid to the liquid–solid interface and through the liquid melt layer to the liquid–vapour interface. At the liquid–vapour interface the power is absorbed and was modelled for a moving boundary by the Stefan energy balance. The vapour is driven off with considerable force owing to the rate of evaporation and thermal expansion to near sonic speeds within the distance of a molecular free path or so. In this layer, known as a *Knudsen layer*, there will be a discontinuity in the temperature,

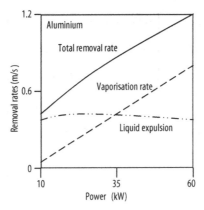

Figure 3.6 Vaporisation, liquid expulsion and total removal rates for aluminium versus laser beam power. (After Chan and Mazumder [12])

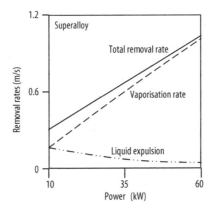

Figure 3.7 Vaporisation, liquid expulsion and total removal rates versus laser power for superalloy. (After Chan and Mazumder [12])

density and pressure. It is a form of shock wave. The final part of the model calculates the gas dynamics required to drive the vapour away to maintain the mass balance. The recoil pressure from the evaporation is substantial, being typically several atmospheres as we saw earlier in the simple Bernoulli analysis. The rate of material removal as melt or vapour was calculated and is shown in Figures 3.6–3.8. It can be seen that the overall rate of drilling is of the order of 1 m s^{-1}, in line with the earlier model. In Figure 3.7, for the example of superalloy, the higher the power, the less will be the melt. The calculated Mach number of the exiting gases for a one-dimensional model is shown in Figure 3.9.

3.3.3.3 Double-pulse Drilling Format

When using very short nanosecond pulses, Forsman *et al.* [13] showed that if the pulses were very close together, an enhancement of around 10 times in the removal rate could

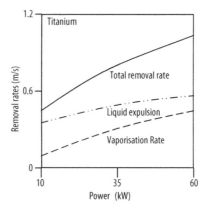

Figure 3.8 Vaporisation, liquid expulsion and total removal rates versus laser beam power for titanium. (After Chan and Mazumder [12])

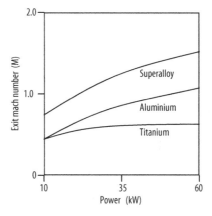

Figure 3.9 Mach number of vapour ejection versus beam power for three different materials. (After Chan and Mazumder [12])

be achieved. Using picosecond pulses, Campbell *et al.* [14] showed a 100% improvement in drilling speed. In this technique the total pulse energy is the same, *i.e.*, one large pulse or two half pulses. The effect is thought to be an interaction of the beam with the laser-generated plasma. The first pulse generates plasma, since 25 μJ per pulse for 10.6 ps is a power density of 1.2×10^{11} W cm^{-2}, which is quite sufficient to generate plasma. The plasma frequency is defined by $\omega_p = \sqrt{\frac{n_e e^2}{\varepsilon_0 m_e}}$ and the laser frequency by $\omega_l = 2\pi \frac{c}{\lambda}$, where n_e is the electron number density, e is the electric charge of the electron, ε_0 is the permittivity of free space, m_e is the mass of an electron, c is the velocity of light and λ is the laser wavelength.

If the electron number density is such that the plasma frequency is smaller than the laser frequency, then the plasma will be transparent to the beam, but the beam will suffer refraction effects in the steep thermal and density gradients. The refractive index of the plasma is approximately given by the equation

$$N \sim \sqrt{1 - \frac{\omega_p^2}{\omega_i^2}} = \sqrt{1 - \frac{n_e e^2}{\epsilon_0 m_e \omega_i^2}}.$$

Hence, the variation of the refractive index with the electron number density is given by

$$\frac{dN}{dr} \approx -\frac{1}{2} \frac{e^2}{\epsilon_0 m_e \omega_i^2} \frac{dn_e}{dr}.$$

Focusing will occur if the electron density gradient in the radial direction is positive and defocusing will occur if it is negative. The initial plasma plume will have a very high electron concentration at the core and be defocussing but after a very short time the electrons will have moved to the outside by mutual repulsion and the gradient will be reversed; if, at that moment, a very short interval after the first pulse, the second one arrives, then it will be focused. This is a possible explanation for the observed increased performance with the double-pulse technique. As an alternative Forsman et al. [13] suggested that the second pulse may interact with the cooling ablation products to give a second plasma wave that clears the debris more efficiently.

It is a feature of this process that if the pulse separation is too long or too short the effect is not noticed.

3.3.3.4 The Spatter and Recast Layer Problems

Laser-drilled holes are inherently associated with spatter deposition due to the incomplete expulsion of the ejected melt. This was most elegantly witnessed by Rohde and Dausinger [15] using techniques from high-speed filming at 42,000 frames per second to shining light through the hole while it was forming. The results showed the intermittent nature of droplets clearing the hole and debris being mainly ejected from the edges of the hole. This molten material will resolidify at the hole exit and within the hole as a recast layer. Spatter is unacceptable in most applications requiring the surface to be abrasive blasted or even reground. It is likely to cause an unquantifiable effect on fluid flow if used for cooling holes and it may interfere with another surface to which it has to fit. The recast layer is usually very thin but may have properties different from those of the background material and hence may in some cases be undesirable or even become a stress raiser if cracks form in it. The best way to remove it is to arrange for the processing parameters to give the minimum amount of melt and the maximum amount of vapour, that is, as we have seen, short intense pulses. This works, but is often unproductively slow, in which case the alternative is to coat the surface of the workpiece with a removable but enduring layer, e.g., CCl_4 [16], Na_2CO_3 solution or washing-up detergent [17, 18]. Otstat et al. [19] used paraffin wax or silicone grease. Low and Li [20] recommended using an antispatter composite coating containing a mixture of ceramic particles embedded in an elastomer base, which they showed worked well [20, 21].

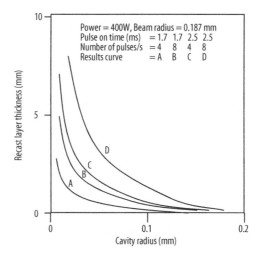

Figure 3.10 Recast layer thickness for 400 W and 0.187-mm spot radius for variation in pulse length and frequency. (After Kar and Mazumder [22])

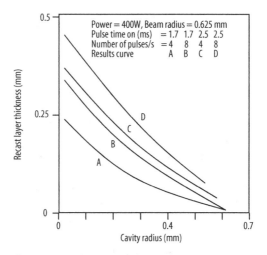

Figure 3.11 Variation of recast layer thickness for 400 W and 0.625 mm spot size with radial position, pulse length and repetition frequency. (After Kar and Mazumder [22])

The thickness of the recast layer was estimated by Kar and Mazumder [22]. They showed that the recast layer thickness, calculated as the melt layer thickness at the end of the pulse, was thinner at the mouth of a hole than at the bottom and was thinner for shorter, more intense pulses This is illustrated in Figures 3.10 and 3.11. Their model also showed that the taper angle decreases with an increase in pulse intensity, as expected. It also decreases with a greater number of pulses; showing that there is a clean-up effect (Figure 3.12).

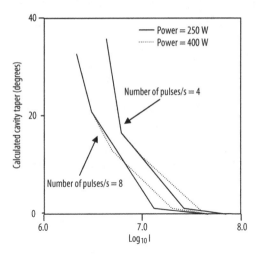

Figure 3.12 Calculated taper angle versus laser intensity at t = 10 s. (After Kar and Mazumder [22])

3.3.3.5 The Taper Problem

Laser-drilled holes can be made to have conical, inverted conical or egg timer shapes as well as cylindrical ones. The shape is partly dependent on the beam shape and how it expands within the hole. This is particularly the case for single-shot drill holes; with multishot percussion drilled holes it is possible to vary the power and focus between shots and so adjust the resultant hole shape. A parametric study of the causes of taper was made by Ghoreisha *et al.* [23]. Their results are summarized in Figure 3.13, which

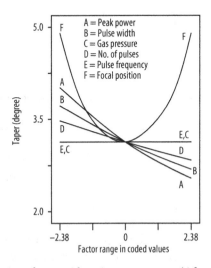

Figure 3.13 Statistical variation of taper with various parameters. (After Goreisha *et al.* [23], copyright 2001, Laser Institute of America, Orlando, Florida, all rights reserved)

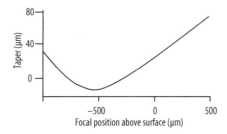

Figure 3.14 Influence of focal position on the taper of a hole. (After Witte *et al.* [24], copyright 2007, Laser Institute of America, Orlando, Florida, all rights reserved)

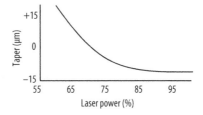

Figure 3.15 The influence of laser power on the degree of taper. (After Witte *et al.* [24], copyright 2007, Laser Institute of America, Orlando, Florida, all rights reserved)

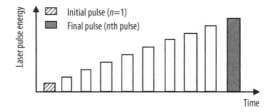

Figure 3.16 The rising pulse energy used in sequential pulse delivery pattern control. (Low *et al.* [20])

shows that assist gas pressure and pulse frequency did not affect taper. The main parameters were focal position, followed by pulse power, pulse duration and number of pulses. The variation with focal position is clearly seen in Figure 3.14 [24] the variation with laser power is shown in Figure 3.15.

One successful technique to reduce taper is sequential pulse delivery pattern control, in which each successive pulse is an increment more powerful than the pulse before as illustrated in Figure 3.16 [25]. It seems that by this process the first lower-powered pulses remove material quickly as melt over a smaller diameter and achieve breakthrough, after which the succeeding more powerful pulses clear out the partially formed hole mainly by evaporation and melt ejection downwards supported by the assist gas. The reported results showed a reduction in the upward-directed melt from 80 to 60% before breakthrough, and after breakthrough the material was ejected downwards. The reduced upward-ejected material resulted in a reduction in the spatter area diameter from 6.7 to 2.7 mm and downward removal rates from 20 to 28% using constant pulse power from 34 to 39% using sequential pulse delivery pattern control [25]. This result observed with the ramping up of the power in a sequence of pulses was also observed by French *et al.* [26].

The preferred method for avoiding taper is to use a trepanning method, described later.

3.3.4 Drilling Ceramic-coated Material

Modern turbine blades operate in an environment which is above the melting point of the blade material, in order that the engine efficiency can be as high as possible. To avoid the blade melting, it is cooled by a flow of gas (usually gas at near 1,000 °C) through the cooling holes just discussed but also by cladding the blade with a ceramic coating (thermal barrier coating), usually made of yttria (approximately 4–8%)-stabilised zirconia ceramic sprayed onto a bond coating of NiCrAlY. Voisey *et al.* [27] found that drilling through such a multiple layer at normal incidence led to no decrease but a possible increase in interfacial toughness. There was some microstructural damage at the NiCrAlY–superalloy bond interface, but not at the NiCrAlY–ceramic interface. Inclining the hole angle did not lead to a major reduction in toughness of the interface. In fact there was some evidence that the dense resolidified layer of ceramic in the interfacial region around the drilled holes can act to deflect and arrest a propagating crack.

3.3.5 Trepanning

Trepanning is a process in which the beam initially pierces and then moves around the perimeter of the proposed hole essentially to cut out the shape of the hole. Almost any shaped hole can be cut this way – round, square or star-shaped. In the case of a circular hole there are various proprietary optics that will move the beam in very small circles to give fine holes with straight sides.

There are basically three approaches to trepanning. The first involves acceleration, translation, deceleration and "settling" of either the part or the laser beam for each laser pulse. This approach takes additional time, but has the benefit of enabling so-called *spiral trepanning*, where the initial and final laser penetration can occur at a location within the perimeter of the hole, thus avoiding any melt flow problems at the start and end of the process. The trepanning orbit may be done several times, for example, if it is done twice, this would involve one orbit to trepan and a second to "clean up" the hole.

The second method is to keep the beam on continuously and cut out a circle. This has the disadvantage of starting and stopping on the hole perimeter. It is however the easiest to engineer with proprietary optics [28] such as the TGSW working head in which the beam passes through a series of movable prisms illustrated in Figure 3.17 or the ILT working head using a Dove prism as illustrated in Figure 3.18 or the spinning beam optics of Harris and Brandt [29] (Figure 3.19).

The third method is to have a programmable beam circuit described as "programmable continuous motion automated spiral trepanning" or "PC-MAST" for brevity.

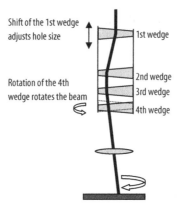

Figure 3.17 The optical arrangement in the trepanning head of Forschungsgesellschaft für Strahlwerkzeuge. (After Witte *et al.* 2007 [24], copyright 2007, Laser Institute of America, Orlando, Florida, all rights reserved)

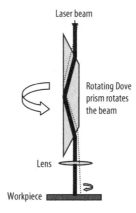

Figure 3.18 The beam rotation achieved with a "Dove prism" in the ILT trepanning head

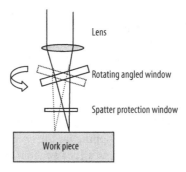

Figure 3.19 A spinning beam apparatus [28]

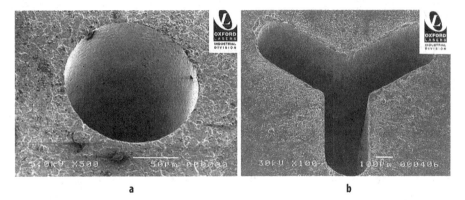

Figure 3.20 Examples of the quality of trepanned holes made by laser: **a** an example of a laser-trepanned hole used in diesel fuel injectors (the typical diameter is 40–150 μm through 1-mm steel), and **b** a spinneret hole showing the versatility of trepanned laser holes, which can be of almost any required shape. (Courtesy of AILU through http://www.designforlasermanufacture. com and Oxford Lasers)

This has the advantage of saving time through not stopping for each pulse and of being able to start and stop in the interior of the hole.

In all these processes it is necessary to have a pulse frequency such that the material does not resolidify between pulses. Also, the pulse overlap needs to be sufficient to give a smooth edge. The overlap needs to be higher for small holes owing to the greater curvature; hence, the pulse frequency is usually higher for smaller holes. In fact, Jacob and Smithfield [30] showed that the optimum number of pulses to drill a cylindrical hole is independent of the diameter! This also means the drill time is independent of the hole diameter. There is an advantage in trepanning with a high-brightness laser, such as a fibre or disc laser, over percussion drilling since such lasers can be focused very finely. The penetration is very swift with the intense beam but if that focus is smaller than the required hole size, then the beam would have to be defocused for percussion drilling, which would slow up the penetration and create more melt. With trepanning the high penetration speed of an intense spot is always used. The hole quality can be exceptionally good (Figure 3.20).

3.3.6 Helical Trepanning

This process is the same as for trepanning except that full penetration is not achieved with the first pulses but a hole is created by machining down in a spiral manner to the breakthrough point. It achieves a high level of precision but is slower than percussion drilling.

3.3.7 Applications of Laser Drilling

3.3.7.1 Cooling Holes in Turbine Blades, Vanes and Combustion Chambers

Modern "effusion-cooled" combustion liners contain more than 60,000 holes compared with the previous traditional designs of only 30,000. The main method of production had been electric discharge machining (EDM) or electrochemical machining (ECM). They made multiple holes all at the same angle with high levels of roundness, negligible taper and with little HAZ. Laser-trepanned holes just managed to compete, with some holes needing two orbits [31]; the second orbit was needed to clean out the final sizing and remove the small amount of debris. However, laser percussion drilling, even with a bell mouth and some taper, was found to be "fit for purpose" in providing the correct air flow characteristics [32]. This gave a surge in productivity. The laser manufacturers responded with improved beam quality, stability, high peak powers and full automation of focus and spot size with control allowing variations within a programme. They also developed techniques for drilling on the fly while parts are rotating and various other refinements. When holes are drilled into cooling channels, back wall protection is required, which is usually some plastic insert. Blades and vanes are often coated in a thermal barrier coating (usually yttria-stabilised zirconia), which makes EDM or ECM processing impossible. The work of Voisey *et al.* [27] already mentioned has shown that the laser can drill such material and improve the adhesion of the thermal barrier coating layer. The laser is replacing the multihead EDM and ECM approach owing to its flexibility in drilling at different angles and to various sizes. However, the competition is still on, with the EDM process also being developed. One such development is the "fast hole" technique using hollow tooling and high-pressure flushing liquids. None the less, laser tooling will always be simpler and cheaper.

3.3.7.2 Inkjet Nozzles

Inkjet technology is constantly finding new markets which have grown from computer printers to printing textiles, marking, depositing compounds for medical diagnosis or even depositing customer-formulated aromas and small-scale rapid prototyping. Inkjet printers work by ejecting minute droplets of ink through carefully drilled precision holes. The driving force for the ink is either piezoelectric or thermally generated bubbles. There are currently three main methods for making the holes: electroforming, silicon etching or excimer micromachining. In the range of 300–1,200 dots per inch, the excimer route is considered the best all-round process [33]. The holes, which will be below 20 μm in diameter for nanolitre or even picolitre droplets, can be made round, square or elliptical. Ultraviolet ablation is preferred since it is a cold cutting route with little collateral damage. The polyimide material for the nozzle plate is coated with a few micrometres of a water-soluble matter so that the recondensed carbon debris from the laser-generated plasma can be subsequently washed off. Drilling is also made through the polyimide into a backing material of low-tack dicing tape to ensure clean holes of similar shape along the nozzle plate.

3.3.7.3 Via Drilling

Our age is being dominated by electronic gadgets which are becoming progressively smaller and smarter, be they telephones, calculators, BlackBerry®[1] devices, televisions, cameras, notepads, control systems for cars, aeroplanes, and machines, security systems, bar code readers, CD players – the list is becoming endless. Within these devices is the electronic circuitry mounted on small boards and made of finely etched circuits (see Chapter 8 on ablation processes). The density of such circuits is greatly enhanced if they are stacked in a single block with interconnectors between the layers. Drilling these via connectors is becoming the domain of the laser. A microvia is defined as a blind hole with a diameter of less than 100 μm. It is usually drilled from the top and/or bottom layer(s) to the first or second adjacent internal layers. The layered material is often layers of copper and polymer. The techniques currently used are laser drilling, mechanical drilling, punching, plasma etching and photo-defined chemical etching. Mechanical drilling is suitable for microvias larger than 150 μm. Below 200 μm the cost of mechanical drilling increases dramatically. Laser drilling of microvias now occupies around 94% of the market for diameters less than 200 μm. The CO_2 laser and ultraviolet lasers (either excimer or frequency-quadrupled Nd:YAG at 266 or 355 nm) are the tools of choice. The CO_2 laser can produce microvias with a high throughput because of its high absorption coefficient and high pulse rate. It also has the advantage of the radiation being reflected from the copper layers and therefore the lasing process can be made self-limiting. However, the longer wavelength limits the focusing precision to microvias greater than 40 μm. Ultraviolet lasers can produce clean holes in most polymers with better resolution, but they suffer from higher costs, especially if an excimer laser is used. Exitech uses a dual laser approach to partially drill the three-layer, two-material substrate. First, holes in the uppermost copper layer are trepanned to 50–100 μm in diameter using an ultraviolet laser. The exposed dielectric underneath this is then drilled with a CO_2 laser using the copper as a mask. This part of the process self-terminates when the beam reaches another copper layer beneath the dielectric. The Exitech system can drill 10,000 complete holes per second by this process [34]. Frequency-doubled Nd:YAG green light at 532 nm has also been used [35]. The polymer of a circuit board is partially transparent to this radiation and so as the hole develops through the top copper layer the rounded base of the hole acts as a concave lens defocusing the beam a little as it enters the polymer layer; which is a new angle on drilling.

GaN is expected to replace GaAs and silicon in the next generation of high-power-density microwave monolithic integrated circuits thanks to its excellent qualities. One of the most promising materials for the substrate for such circuits is sapphire, which is strong, a good conductor, chemically stable and relatively cheap. Interconnects for these devices require holes through the sapphire, which is typically more than 100 μm thick. Wet or dry etching of this chemically stable material is difficult and mechanical drilling is almost impossible for such fine holes, less than 100 μm in diameter. An ultrafast laser can do it with precision, reliability and speed. To remove sapphire material, an energy of

---

[1] BlackBerry® is a registered trademark owned by Research In Motion Ltd and is registered and/or used in the USA and countries around the world. http://www.blackberry.com

some 9.9 eV is required to overcome the band gap to break down the structure. This can be achieved either by an extreme ultraviolet beam of around 100 nm, which requires working in a vacuum and is very costly, or by an ultrafast laser using picosecond, high-intensity pulses, typically 10 ps, 1–5 kHz, 7.5 µJ per pulse of green light at 527 nm from a YLF laser focused to a 10 µm spot and trepanned to make 70 µm diameter vias [36]. The material is thought to be removed by multiphoton ionisation, which for 527 nm radiation would require five photons acting together to overcome the band gap. The quality of the holes is remarkably fine.

3.3.7.4 Aspiration Holes in Miscellaneous Objects

Plastic irrigation pipes and cigarette paper are difficult to make holes in by simply punching them since any flap left over will probably reseal the hole. The laser makes a clean hole with no "flap". It can also do it very rapidly, particularly if diffractive optics are used to make multiple beams all drilling simultaneously [37]. Cigarette paper can be perforated with multiple beams at paper speeds of 0.8 m s^{-1}. A typical installation for perforating irrigation pipes would use a 500 W CO_2 laser making four holes per second of 0.5 mm diameter. Aerosol valves, babies' teats in feeding bottles and pump spray nozzles (0.04-mm holes) are further examples.

3.3.7.5 Engineering Holes

Fine lock pin holes in MONEL®[2] metal bolts is an example of a multitude of small but essential holes that the laser can drill well and fast. Some materials are particularly difficult to drill, for example, HASTELLOY®[3] [38], which is "gummy" to drill, making mechanical drilling very slow at around 60 s per hole and causes extrusions at both ends of the hole which have to be cleaned. Mechanical punching is fast but is limited to holes greater than 3 mm in diameter. ECM (electrochemical machining) is too slow at 180 s per hole but does give a neat hole; EDM (electrodischarge machining) is expensive and slow at 58 s per hole. Electron beam drilling is fast at 0.125 s per hole but needs a vacuum chamber and is more expensive than a Nd:YAG laser; a Nd:YAG laser took 4 s per hole. The holes were made by trepanning the required size over a range of sizes.

3.3.7.6 Rock Drilling

This is not an application but it is a dream. How good it would be to drill deep into the earth with only a laser beam! Or how convenient to be able to clear the slag from the mouth of a steel-carrying torpedo ladle, which is currently done with pneumatic drills.

---

[2] MONEL® is a registered trademark of Special Metals Corporation, New Hartford, New York, USA. http://www.specialmetals.com

[3] HASTELLOY® is a registered trademark name of Haynes International Inc. http://www.haynesintl.com

A small hole can be drilled in rock as for metal but there is less evaporation and a considerable quantity of melt. Granite, like most rocks or slag, forms small glassy beads which can be mechanically removed. Then there is the problem of water and mud in real earth drilling which makes the job look hopeless. There is a glimmer of hope in the work of Kobayashi *et al.* [39] in which they found that the laser will make its own waveguided hole through the water to start a drilling process and that more material can be removed by thermal spallation owing to stress cracking than could ever be removed by melting/evaporation. Leong *et al.* [40] trepanned samples of limestone, sandstone and shale. Percussion drilling of sandstone with a 4 kW CO_2 laser and high-pressure nitrogen assist gas achieved a drilling rate of 45 m h^{-1}, which compared well with the expected 15 m h^{-1} using rotary drills. The removal mechanism was mainly by spallation. So all is not dead, but there is a way to go yet.

3.3.8 Monitoring the Drilling Process

In-process sensing and control is discussed in more detail in Chapter 12.

Several methods for breakthrough detection in drilling have been developed based on acoustic and optical signals.

3.3.8.1 Acoustic Monitoring of Laser Drilling

Experienced operators of laser drilling machines can hear when breakthrough has been achieved and whether the process is behaving normally. This signal can also be detected if the acoustic pressure and the spectrum are analysed [41]. These crude overall signals are adequate for control purposes, but it is thought-provoking to consider that if the drilling process is a heavy disturbance within a tubelike hole, then some resonant frequency would be likely to be present telling us how deep the hole is as well as whether there is breakthrough. This has not been found yet, but by analogy with organ pipes it seems very likely that such a signal should be there.

3.3.8.2 Optical Techniques

Observing the plasma plume and thus finding the focus by chromatic aberration (see Section 12.2.2.1) also gives a reading on the light intensity, which fades rapidly when penetration is achieved [42].

3.3.8.3 Summary of Laser Drilling

- The main methods are single shot, percussion, trepanning and helical trepanning.
- Material is removed as melt and vapour. The longer and less powerful the pulse, the more will be the quantity of melt and *vice versa*. The shorter and more intense the pulse, the more material will be removed as vapour, usually giving a cleaner hole.
- Molten material can cause spatter and a recast layer.

"… and this piece has been perforated all over with such fine holes, only x-rays can get through!"

3.4 Methods of Cutting

The general arrangement for cutting shown in Figure 3.2 can be used to cut in seven different ways, as shown in Table 3.3.

3.4.1 Vaporisation Cutting/Drilling

This process was discussed in Section 3.3.

3.4.2 Fusion Cutting – Melt and Blow

Once a penetration hole has been made or the cut has been started from the edge, then it is possible with a sufficiently strong gas jet to blow the molten material out of the cut kerf and so avoid having to raise the temperature to the boiling point or much beyond the melting point. It is thus not surprising to find that cutting in this manner requires approximately half the power needed for vaporisation cutting. Note the ratio of the latent heat of fusion for melting and boiling in Table 3.2 which is approximately a factor of 20.

Table 3.3 Different ways in which the laser can be used to cut

Method	Concept	Relative energy
Vaporisation		40
Melt and blow		20
Melt burn and blow		10
Thermal stress cracking		1
Scribing	Perforation	1
"Cold cutting"	hv high energy photons or multi photons	100
Burning stabilised laser cutting		5

The process can be semiquantitatively modelled by assuming all the energy enters the melt and is removed as such before significant conduction occurs. This assumption is not so daft since the HAZ for good cuts by this method rarely exceeds a few micrometres.

We thus have a simple lumped heat capacity equation based on the heat balance on the material removed, similar to Equation 3.1, as shown in Figure 3.21.

The balance is

$$\eta P = w t V \rho \left(C_p \Delta T + L_f + m' L_v \right), \qquad (3.2)$$

where P is the incident power (W), w is the average kerf width (m), t is the thickness (m), V is the cutting speed (m s^{-1}), m' is fraction of melt vaporised, L_f is the latent heat of fusion (J kg^{-1}), L_v is the latent heat of vaporisation (J kg^{-1}), ΔT is temperature rise to cause melting (K), η is the coupling coefficient and ρ is the density (kg m^{-3}).

Rearranging this equation, we get

$$(P/tV) = (w\rho/\eta)(C_p \Delta T + L_f + m' L_v) = f(\text{material}). \quad (\text{J m}^{-2}). \qquad (3.3)$$

Apart from the values of the coupling efficiency, η, and the kerf width, w, which is a function of spot diameter and to some extent speed, the other variables are all material constants. Thus, it is reasonable to expect that the group (P/tV) is reasonably constant for the cutting of a given material with a given beam. A collation of the data from the

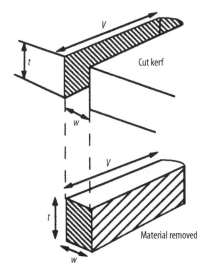

Figure 3.21 Volume melted and removed during cutting

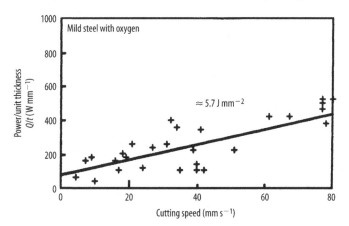

Figure 3.22 P/t versus V for mild steel

literature is presented in Figures 3.22–3.24. The straight line correlation is significant considering all the unspecified different cutting methods used by the various authors – particularly in regard to the effects of polarisation and gas jets as discussed later. So it is possible to draw up a chart of the severance energy (P/tv) (J mm^{-2}) required for unit area severed for different materials, as in Table 3.4 [43, 44].

This at least establishes the main operating parameters. What is actually happening at the cutting front is of considerable complexity. Figure 3.25 shows the cut front in section. The beam arrives at the surface and most of it passes into the hole or kerf, some may be reflected off the unmelted surface and some may pass straight through. At slow speeds the melt starts at the leading edge of the beam and much of the beam passes clean through the kerf without touching the material if it is sufficiently thin [45].

Table 3.4 Average severance energies for CW CO_2 laser cutting found experimentally from a variety of sources. Note, these figures do not apply to Nd:YAG pulse cutting, where the mechanism is different: for example, for mild steel Nd:YAG values are between 15 and 200 J mm^{-2} (Principally from Powell *et al.* [43, 44])

Material	Lower value of P/Vt (J mm^{-2})	Higher value of P/Vt (J mm^{-2})	Average of P/Vt (J mm^{-2})
Mild steel + O_2	4	13	5.7
Mild steel + N_2	7	22	10
Stainless steel + O_2	3	10	5
Stainless steel + Ar	8	20	13
Titanium + O_2	1	5	3
Titanium + Ar	11	18	14
Aluminium + O_2			14
Copper + O_2			30
Brass + O_2			22
Zirconium + O_2			1.7
Acrylic sheet	1	3	1.2
Polythene	2.7	8	5
Polypropylene	1.7	6.2	3
Polystyrene	1.6	3.5	2.5
Nylon	1.5	5	2.5
ABS	1.4	4	2.3
Polycarbonate	1.4	4	2.3
PVC	1	2.5	2
Formica	51	85	71
Phenolic resin			2.7
Fibre glass (epoxy)			3.2
Wood: pine (yellow)			23
Oak			26
Mahogany			24
Chipboard	45	76	59
Fibreboard			50
Hardboard			23
Plyboard	20	65	31
Glass			20
Alumina	15	25	20
Silica			120
Ceramic tile			19
Leather			2.5
Cardboard	0.2	1.7	0.5
Carpet (auto)			0.5
Asbestos cement			5.0

ABS acrylonitrile–butadiene–styrene

The absorption takes place on the steeply sloped cut front (θ is approximately 14° to the vertical [46]) by two mechanisms: mainly by Fresnel absorption – that is, direct interaction of the beam with the material – and secondly by plasma absorption and re-radiation. The plasma build-up in cutting is not very significant owing to the gas blowing it away. Thus, the power density on the cut front is $F_0 \sin \theta \approx F_0 \times 0.24$. This causes melting and the melt is then blown away by the drag forces from the fast-flowing gas stream and the pressure drop along the kerf depth. At the bottom of the kerf the

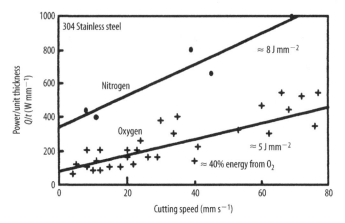

Figure 3.23 P/t versus V for stainless steel

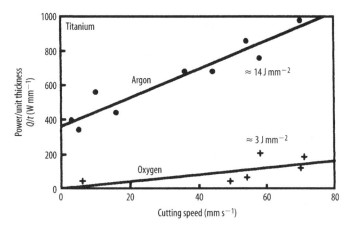

Figure 3.24 P/t versus V for titanium

melt is thicker owing to flow from above, deceleration of the film and surface tension retarding the melt from leaving. The gas stream ejects the molten droplets at the base of the cut into the atmosphere. In blowing through the kerf, the gas would entrain the surrounding gas in the kerf and generate a low-pressure region further up the cut length. This can have a detrimental effect by sucking the dross back into the cut. In fact the problem of the removal of the dross from the bottom edge is further complicated by the wettability of the workpiece to the melt and the flow direction of the gas jet. Thus, cutting thin tin plate is difficult owing to the dross clinging to the molten tin plate and the poorly directed gas jet which is emitted from a slot in thin material. The gas stream not only drags the melt away but will also cool it. In fact both momentum and heat transfer will occur. The extent of the cooling can be calculated. The heat removal by convection is described by

$$Q = hA\Delta T.$$

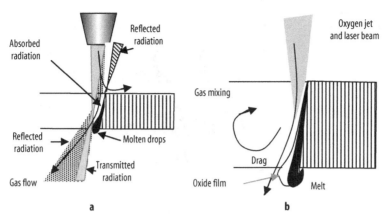

Figure 3.25 Interactions at the cutting front: **a** optical energy transfer, and **b** mass and momentum transfer

This is an equation which defines the heat transfer coefficient, h. The value of h has been determined for many geometries [47]. It is usually quoted as Nu = $f(\text{Re}, \text{Pr})$. (The Nusselt number is a function of the Reynolds number and the Prandtl number.). An approximate, and high, estimate of h can be derived, such as $h < 100\ \text{W m}^{-2}\ \text{K}^{-1}$. The heat loss at the cut front now becomes

$$Q = 100 \times t \times w \times \Delta T.$$

For a thickness t = 2 mm, a kerf width w = 1 mm and ΔT = 3,000 K – all high values – Q = 0.6 W. Thus, the cooling effect of the gas is negligible compared to the few thousand watts being delivered by the beam. This is mainly due to the small area involved in cooling during laser cutting.

In fusion cutting the action of the gas is to drag the melt away and little else. The design of the nozzle and the alignment of the nozzle with the laser-generated kerf are important areas of concern inasmuch as they affect the drag of the gas on the melt.

As the cut rate is increased, the beam automatically couples to the workpiece more efficiently by less of the beam being lost through the kerf [45]. Also the beam tends to ride ahead onto the unmelted material. When this occurs, the power density increases since the surface is not sloped and so melting proceeds faster and is swept down into the kerf as a step. As the step is swept down, it leaves behind a mark on the cut edge called a striation (see Figure 3.26). The cause of striations is a subject of some dispute, and there are many theories: the step theory just outlined; the critical droplet size causing the melt thickness to pulsate in size before it can be blown free [48]; and the sideways burning theory (Section 3.4.3). There are conditions under which reduced striations occur. These are governed by gas flow or by pulsing at higher frequencies to the natural striation frequency [43]. A further feature of the cut face is that there is often, but not always, a break in the flow lines (Figure 3.27) [49]. This may be due to the start of the first reflection of the beam on the cut face, the end of the burning reaction (see the next

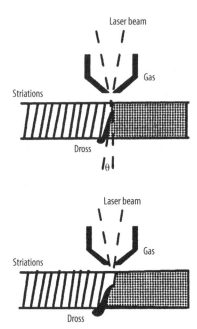

Figure 3.26 The stepwise formation of striations

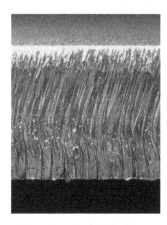

Figure 3.27 Striations on the cut face of a piece of stainless steel cut with 3-kW laser power at $1.5\,\mathrm{m\,min^{-1}}$. Note the three zones: (1) initial penetration; (2) first reflection and flow; (3) washout. (Purtonen and Salminen [49])

section), a laminar /turbulent flow transition or it may be a shock wave phenomenon. Currently this is not well understood.

It appears that the cut face has three parts. The first is the direct penetration of the beam, which shows as fine vertical striations, then a shelf zone, presumably where most of the beam absorption occurs, and finally a washout region where the superheated melt flows out of the kerf, melting more material on the way [50].

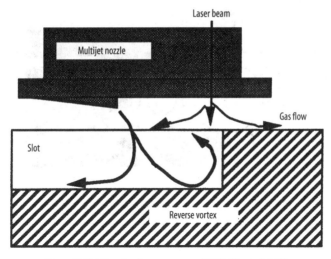

Figure 3.28 The slotting process of O'Neill *et al.* [51]

Blind cuts can be achieved by introducing a strong side jet as well as the central jet. The process was invented by O'Neill *et al.* [51]. In this process the two jets can be combined to give a reverse vortex within the kerf driving the dross upwards and out. Figure 3.28 illustrates the arrangement. Blind slots 3–4 mm deep are possible, but the flow structure fails if slots are overlapped, as in machining. Laser caving [52] is a slower version of this concept.

3.4.3 Reactive Fusion Cutting

If the gas in the previous method is also capable of reacting exothermically with the workpiece, then another heat source is added to the process. Thus, the cut front becomes an area of many activities. Figure 3.25b illustrates this. The gas passing through the kerf is not only dragging the melt away, as just seen, but is also reacting with the melt. Usually the reactive gas is oxygen or some mixture containing oxygen. The burning reaction starts, usually at the top, when the temperature reaches the ignition temperature. The oxide is formed and is blown into the kerf and will cover the melt lower down. This blanketing will slow the reaction, and may even be the cause of the break in the striation lines just noted. It can be seen from Figures 3.22–3.24 that the amount of energy supplied by the burning reaction varies with the material; with mild steel it is 60%, with stainless steel it is also 60% and with a reactive metal such as titanium it is around 90%. Thus, cutting speeds are usually at least doubled compared with those in melt and blow cutting using this technique. As a general rule, the faster the cut, the less heat penetration and the better the cut quality. However, since there is a cutting reaction taking place, some chemical change in the workpiece may be expected. With titanium this can be critical since the edge will have some oxygen in it and will be harder and more liable to cracking. With mild steel there is no noticeable effect except a very thin

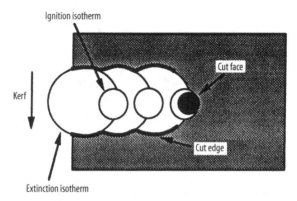

Figure 3.29 Striation formation due to sideways burning

resolidified layer of oxide on the surface of the cut. An advantage is that the dross is no longer a metal but is usually an oxide, which for mild steel flows well and does not adhere to the base metal as strongly as if it were metal. With stainless steel the oxide is made up of high melting point components such as Cr_2O_3 (melting point of approximately 2,180 °C) and hence this freezes more quickly, causing a dross problem. It is the same with aluminium.

Owing to the burning reaction a further cause of striations is introduced. In slow cutting, at speeds less than the speed of the burning reaction, the ignition temperature will be reached and then burning will occur proceeding outwards in all directions from the ignition point as illustrated in Figure 3.29. This mechanism is mainly apparent as a cause for striations if the cut is slow. In this case very coarse striations are revealed as illustrated in the slow cut in Figure 3.30.

3.4.4 Controlled Fracture

Brittle material which is vulnerable to thermal fracture can be quickly and neatly severed by guiding a crack with a fine spot heated by a laser.

The laser heats a small volume of the surface, causing it to expand and hence to cause tensile stresses all around it. If there is a crack in this space, it will act as a stress raiser and the cracking will continue in the direction towards the hot spot. The speed at which a crack can be guided is swift, of the order of 1 m s^{-1}. This is as hoped for until the crack approaches an edge, when the stress fields become more complex and difficult to forecast. As a cutting method for glass it is superb. The speed, edge quality and precision are very good. The only problem is that for straight cuts snapping is quicker and for profiled cuts one usually needs a closed shape, as for the manufacture of car wing mirrors. If someone could solve the control problem on completing the closed form, then a significant process would have been developed. This process requires that the surface is not melted since that might damage the edge. It thus requires very little power. Typical figures are shown in Table 3.5.

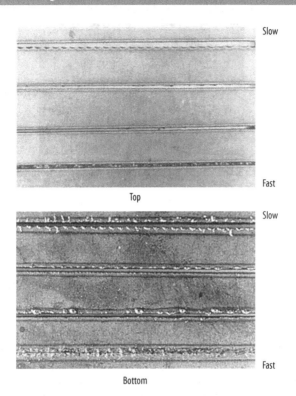

Figure 3.30 Top and underside views of cuts in mild steel made at various speeds. The coarse striations formed at lower speeds are clearly seen

Table 3.5 Controlled fracture cutting rates

Material	Thickness (mm)	Spot diameter (mm)	Incident power (W)	Rate of separation (m s^{-1})
99% Al$_2$O$_3$	0.7	0.38	7	0.3
	1.0	0.38	16	0.08
Soda glass	1.0	0.5 × 12.7	10	0.3
Sapphire	1.2	0.38	12	0.08
Quartz (crystal)	0.8	0.38	3	0.61

3.4.5 Scribing

This is a process for making a groove or line of holes either fully penetrating or not, but sufficient to weaken the structure so that it can be mechanically broken. The quality, particularly for silicon chips and alumina substrates, is measured by the lack of debris and little HAZ. Thus, low-energy, high-power-density pulses are used to remove the material principally as vapour.

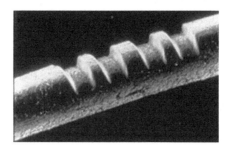

Figure 3.31 A human hair carved using an excimer laser

3.4.6 Cold Cutting

This is a new technique only recently observed with the introduction of high-powered excimer lasers working in the ultraviolet and with ultrashort very high power pulses of the order of picoseconds. The energy of the ultraviolet photon is 4.9 eV (Table 2.1). This is similar to the bond energy for many organic materials. Thus, if a bond is struck by such a photon, then it may break. On the whole, if it did, it would reform and no one would be any the wiser. However, once the die has been rolled, the bond could reform in another way, as with sunbathing and the generation of a tan (or carcinogens!). Ultraviolet light is just at the beginning of the biologically hostile radiation range which goes on to X-rays and γ-rays. When this radiation is shone onto plastic with a sufficient flux of photons so that there is at least one per bond [53, 54], then the material just disappears without heating, leaving a hole with little to no debris or even edge damage. Figure 3.31 illustrates how a human hair can be machined this way. With ultrashort powerful pulses multiphoton events will occur in which the sum of two or more photons is sufficiently energetic to break bonds in the same way as with ultraviolet light (see Section 8.4 on ablation bond breaking).

This exciting new technique for encraving seems to be a dream come true for the electronics manufacturer and certainly the electronics industry has not been slow to take it up [55]. Marking is another attractive application.

The potential medical applications include a dazzling array of possibilities in microsurgery and engineering with single cells, as well as more conventional tumour ablation (see Chapter 11).

3.4.7 Laser-assisted Oxygen Cutting – the LASOX Process

By using the laser as a "match" to ignite the metal in an oxygen stream, one can cut very thick sections with relatively little laser power [56, 57]. For example, a 1 kW laser is able to cut 50 mm steel at 200 mm min^{-1} reliably. Thicknesses up to 80 mm have been cut in the laboratory this way. The process, illustrated in Figure 3.32a, is essentially oxygen cutting and has the same wide kerfs, of approximately 4 mm. There are, however,

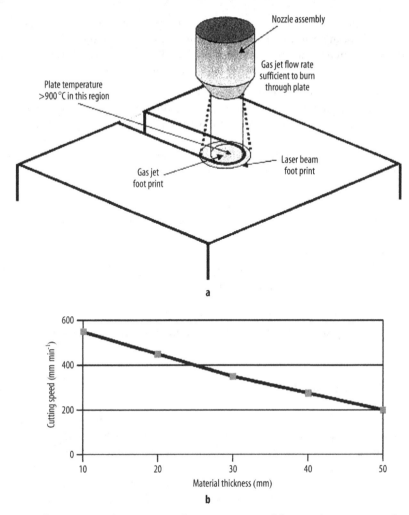

Figure 3.32 The LASOX process: **a** general arrangement, and **b** typical cutting rates for mild steel [56, 57]

significant advantages for the "LASOX" process over plasma and oxy/fuel systems. The advantages so far identified are:

1. square cut top edges as opposed to rounded ones;
2. reduced taper, generally less than 2°;
3. faster piercing;
4. typical kerf widths of 2.5 mm when cutting in the thickness range of 20–50 mm;
5. little variation in roughness over the smooth cut edge;
6. the process is omnidirectional and independent of laser polarisation;
7. the required laser power is approximately 1 kW, with no advantage in having more power; in fact extra power may stimulate edge damage by sideways burning;

8. cutting speeds, shown in Figure 3.32b, are comparable to those of plasma and oxy/fuel cutting, varying from 500 mm min^{-1} for a 15 mm thick plate in mild steel to 200 mm min^{-1} for a 50-mm-thick plate; and
9. there is less distortion owing to the absence of hot surface jets.

The process arrangement is to have a short-focus lens focusing the beam within the nozzle chamber, from where it expands to heat a portion of the surface which is larger than the area of the impinging coaxial oxygen jet. The crucial aspect of the process is that the full width of the oxygen stream should strike plate material that has been heated to above the ignition temperature of 900–1,000 °C for mild steel. If this LASOX condition is met, then a stable and controlled reaction occurs, giving a good quality cut.

The process can also achieve deep penetration piercing in a fraction of the time taken by standard processes (less than 5 s compared with 60 s). Trepanned holes, less than 10 mm in diameter, can be cut with tolerances of ±0.25 mm, with almost no taper. This process opens a new range of thickness that can be cut by laser since the conversion of a LASOX system back to standard oxygen-assisted laser cutting involves only a straightforward change of optics. In standard oxygen-assisted laser cutting it is difficult to cut thicknesses greater than 25 mm, regardless of the laser power, owing to the problems with fluid flow in a narrow kerf. LASOX was invented by O'Neill and developed by a consortium under the National Shipbuilding Research Program [57].

3.5 Theoretical Models of Cutting

The simple model presented in Section 3.4.2 covers a surprising amount of detail in describing the laser cutting process. For a more detailed analysis care has to be taken in describing the heat flow into the cut face as a line source [48] or as a cylindrical source [58], and the fluid flow of the gas stream [51]. Analytical models, however, are limited in their ability to model detail in real-world problems. Thus, numerical models have been attempted and some useful semiquantitative models, such as Olsen's model of melt thickness on a cut front (see Section 5.9), have been developed. These models are discussed in more detail in Chapter 5.

3.6 Practical Performance

Laser cutting is a multiparameter problem and hence sometimes difficult to understand regarding the interrelationship between all the parameters. The parameters can be grouped as follows:

- Beam properties: spot size and mode; power, pulsed or CW; polarisation; wavelength;
- Transport properties: speed; focal position;
- Gas properties: jet velocity; nozzle position, shape, alignment; gas composition; and
- Material properties: optical; thermal.

3.6.1 Beam Properties

3.6.1.1 Effect of Spot Size

The principal parameters in laser cutting are laser power, traverse speed, spot size and material thickness as seen in the simple model (Section 3.4.2). One of the most important of these is the spot size. This acts in two ways; a decrease in spot size will, firstly, increase the power density, which affects the absorption, and, secondly, decrease the cut width. Lasers with stable power and low-order modes – usually true TEM00 modes, as opposed to irregular mountain modes! – cut considerably better than other lasers. Figure 3.33 [59] shows the effect of mode on cutting performance and Figure 3.34 shows the results of Sharp [60] in cutting mirrors, which is not possible with any other form of beam. Notice with Sharp's work that he was using only relatively little power to cut 0.5 cm-thick gold-plated copper mirrors! In the hole drilling work of Shaw and Cox [61], they obtained holes with an aspect ratio of 100 using only 100 W of power. They attribute this amazing performance to a very low-order-mode YAG laser with low power to avoid explosive erosion effects. Poor mode structures tend to produce cuts which are comparable to those obtained with a good plasma torch. The spot size is controlled by the laser design, which establishes the mode, and the optics, which decides how fine the focus will be – see Section 2.8. Usually a lens F number of around 5 is selected since that is as low as one can get before spherical aberration becomes potentially significant.

High-brightness laser beams may be able to penetrate deeply but they do so with a narrow kerf. For cutting thick plate a narrow kerf may be a disadvantage owing to the difficulty of blowing the dross clear and oxygen starvation deeper in the slot. Wider cut widths, as in the LASOX process (see Section 3.4.7), can result in greater cut depths. Another way of enlarging the kerf is to spin the beam. This process, first tested at

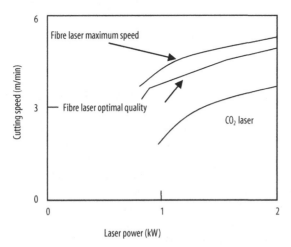

Figure 3.33 The effect of mode on cutting performance as illustrated between a high-brightness fibre laser and a standard CO_2 laser [59]

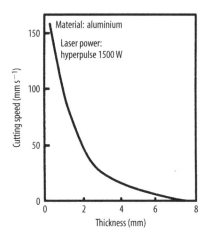

Figure 3.34 Cutting mirrors [60]

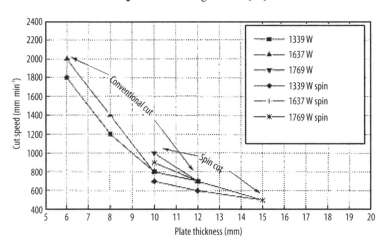

Figure 3.35 Cut speed versus laser power for conventional and spun laser beams [30]

The Welding Institute (now TWI), Cambridge, in 1985, has been developed further by Harris and Brandt [29] in Australia. The apparatus is shown in Figure 3.19 and the cutting results are shown in Figure 3.35. The spin speed was found to be optimum at 3,000 rpm for a 10-mm sheet and 2,000 rpm for a 12–15-mm sheet when using a 1.5 kW CO_2 beam. The kerf width is almost double that of conventional laser cutting (2.5 mm instead of 1.2 mm) and is presumably a function of the optical spin width. The nozzle diameter is larger and the oxygen pressure lower. The alignment of the beam in the nozzle is, of course, eccentric when the beam is stationary, so that in operation it rotates around the nozzle. This rotation means the beam is swiftly passing across the cut front and periodically heating deep behind the front onto the trailing edge of the flowing dross. The result is possibly more turbulence within the kerf, producing dross particles which are less spherical than is conventional.

3.6.1.2 Effect of Power

The overall effect of increasing the power is to allow cutting at faster speeds and/or greater depths, as shown in Figures 3.22–3.24. Equation 3.3 shows the general relationship. The potential disadvantage of increasing the power is that the cut width increases, side burning spoils the edge finish and sharp corners become rounded. A further aspect of high power, particularly relevant to high-brightness fibre lasers, is that the temperature of the melt may be sufficiently high to terminate the oxygen burning reaction and turn it into a reduction process with the consequent loss of the additional energy noted under reactive fusion cutting [62].

These problems can be controlled to a certain extent by pulsing. There are some laser cutting systems which are able to change from CW to pulsed on slowing into a corner to control side burning or corner rounding. The pulse keeps the power intensity constant, and therefore the depth of penetration, and the pulse rate can then be adjusted in line with the speed to control the overall heat input. Other systems that are not capable of switching during processing are often used in the pulsed mode when cutting fine shapes. The Nd:YAG laser is usually used in the pulsed mode to cut.

There are several styles of pulsing: simple power switching, which turns the beam on and off; switching with excess current giving a "superpulse" – this pulse may have an energy 2–3 times the CW value; Q-switching using a very high speed switch in the laser cavity – such as a Pockel cell – which can give several thousand times the CW power rating for a very short time such as a few nanoseconds; and spiking on the CW beam with short power surges generating a "hyperpulse", available on certain lasers such as PRC lasers (see Figure 3.36).

In superpulsing, with a square power versus time profile, it has been found [63] that the enhanced peak power aids in cutting high-reflectivity, high-conductivity materials

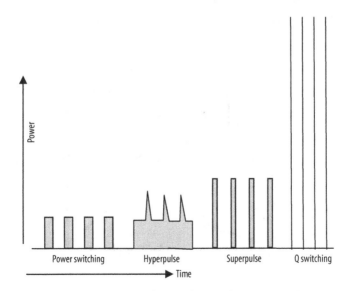

Figure 3.36 Different pulsing techniques

such as aluminium, copper and gold. In cutting aluminium, the thickness capability is doubled compared with CW processing for the same average power; the speed is also improved for the same thickness by some 20%. The reduced heat input allows angled cutting up to 40° without burnout of the thin edge. Also, such pulsing is found to reduce the quantity of clinging dross or burr. The Q-switching, superpulse and hyperpulse systems give enhanced penetration, whereas the pulse rate and speed determine the edge finish and striation pattern.

3.6.1.3 Effect of Beam Polarisation

Figure 3.37 illustrates the problem. The maximum cutting speed is doubled, cutting in one direction as opposed to one at right angles when cutting with a plane-polarised laser beam. Nearly all high-powered lasers have folded cavities, which favours the amplification of radiation whose electric vector is at right angles to the plane of incidence on the fold mirrors. That is, a horizontal folding will produce a beam polarised vertically. If the cavity is not folded or the folding has near-normal reflections, then the beam will still be plane-polarised but the plane of polarisation may move unpredictably with time. This is serious in view of Figure 3.37, so even these lasers are now equipped with a fold at the total reflecting mirror to stabilise the plane of polarisation from the cavity. The cause of the phenomenon shown in Figure 3.37 is that at the cutting face there is a glancing angle of incidence and, as observed in Chapter 2, there is a distinct difference in the reflection of a beam at such angles depending upon whether the electric vector is at right angles to the plane of incidence (s-polarisation) or in the plane of incidence (p-polarisation). If it is s-polarised, then it will suffer a high reflectivity as shown in Figure 2.11. If it is p-polarised, it will be preferentially absorbed. This can be imagined as due to the form of oscillation expected of the interacting electron, as dis-

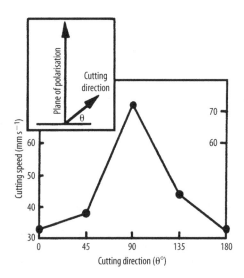

Figure 3.37 The effect of polarisation on the cutting performance with direction of cut [101]

cussed in Section 2.7.4. This polarisation phenomenon was first noted by Olsen [64]. Since then nearly all production cutting machines have been fitted with circular polarisers as described in Section 2.9.2. Such beams cut equally well in all directions and with a performance between that of the two plane-polarised beams (s and p). There is growing interest in radially polarised beams.

3.6.1.4 Effect of Wavelength

The shorter the wavelength, the higher the absorptivity (see Section 2.3.1), and the finer the focus for a given mode structure and optics train (see Section 2.8). Thus, YAG radiation is preferable to CO_2 radiation as a general rule, although, owing to the poor mode structure of most YAG lasers of any significant power, the spot sizes for both CO_2 and YAG are similar, with an advantage for the true TEM00 laser. Fibre lasers with monomode fibres give TEM00 beams at 1.03 or 1.55 µm wavelength whose penetration powers are significantly better than those of other lasers.

The results from Culham Laboratory on cutting plastics [65] using a CO_2 laser with 10.6 µm radiation and a CO laser with 5.4 µm radiation were obtained with the same resonator and optics and so represent a fairly good comparison between CO_2 10.6-µm radiation and that at 5.4 µm from a CO laser, with an advantage for the shorter wavelength. It is felt that the difference lies in the absorption of the beam on the cut face.

Some materials are transparent to certain wavelengths and this leads to some interesting problems in cutting. One such material is human tissue (see Chapter 11).

Very short wavelengths, in the ultraviolet, have energetic photons of energy similar to the bond energy in organic material (*e.g.*, approximately 4.6 eV). They are thus able to directly break bonds. If a sufficient flux of photons can simultaneously sever several adjacent bonds, then the material will effectively fall apart, resulting in a cut but without the need for heating. This is known as "cold cutting". Whether the mechanism just described is correct or not is questioned since some heating has been noticed, but that may be a secondary effect as opposed to the main event.

A similar effect can be achieved with multiphoton interactions using lower-energy photons [66]. Infrared radiation from an ultrafast mode-locked laser (see Chapter 2) can produce very short pulses of tremendous power, for example, a 1 mJ pulse of 200 fs duration has a power of 5 GW. This power is sufficient to ionise the solid material in a time during which heat can only flow some 0.1 nm, even travelling at $1,000 \, \text{m s}^{-1}$. Such a laser is more user-friendly than an excimer laser and the application areas for cold cutting and ablation are growing fast in the electronics industry.

3.6.2 Transport Properties

3.6.2.1 Effect of Speed

The faster the cutting, the less time there is for the heat to diffuse sideways and the narrower the HAZ. The kerf is also reduced owing to the need to deposit a certain

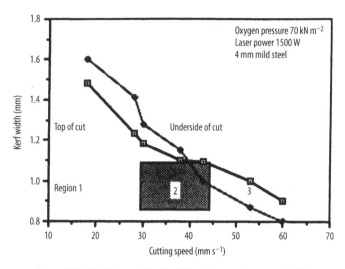

Figure 3.38 Variation of kerf width with cutting speed [68]

amount of energy to cause melting. Hence, with a Gaussian beam there is a "sharpened pencil" effect in that as the speed rises so there is only sufficient energy at the tip of the Gaussian curve and not at the root to cause melting and hence cutting. The kerf width varies with speed as shown in Figure 3.38 [67]. The three regions shown in the figure are due to side burning at slow speeds, stable cutting at medium speeds and failure of the dross to clear in the higher-speed region. The faster the speed, the better the cut finish until this last region is reached.

3.6.2.2 Effect of Focal Position

The surface spot size determines the surface power intensity and whether penetration will occur, but optimum cutting may be obtained by having the minimum spot size below the surface. The problem is related to absorption on the cut face and how to keep the energy together (see the discussion in Section 4.4.6). Very deep cuts are rarely achieved with any great quality since the beam spreads out and suffers multiple reflections. There are exceptions; consider Figure 3.39, a 5-cm cut with parallel walls in block board. How could the beam do that? It must have been waveguided down a slot whose walls are made of graphite – not a normal material to consider for reflections!

A new type of lens first described by the FORCE Institute in Denmark is the Dual Focus™[4] lens. This lens, developed by V&S Scientific, has a central area with a longer focal length than the surrounding area as shown in Figure 3.40 [50, 68]. This has the effect of increasing the depth of focus of the beam. The manufacturer recommends that

---

[4] Dual Focus™ is a registered trademark of V&S Scientific Ltd., Unit 2 Caxton Pl, Caxton Way, Stevenage, Herts, SG1 2UG, UK

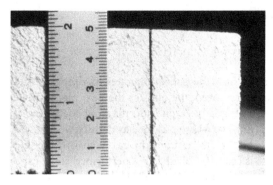

Figure 3.39 A laser cut through -cm-thick block board

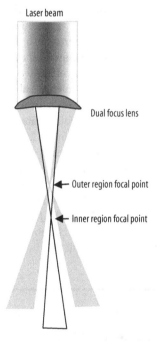

Figure 3.40 Beam structure with a dual-focus lens

the upper focus is on the upper surface to be cut and the lower focus is on the lower surface. The resulting cut has a narrower kerf than found with a single lens cut (*e.g.*, 0.5 mm top and 0.38 mm bottom, compared with 0.86 mm top with a single lens and 0.48 mm bottom); the cutting speed is increased (approximately 20%) probably owing to the increased intensity within the kerf; there is less dross attached to the bottom; and piercing is faster with reduced dross or material eruption. In cutting stainless steel, this latter point can save complete operations, such as the hand removal of eruptions to prevent collisions with the nozzle [69].

3.6.3 Gas Properties

3.6.3.1 Effect of Gas Jet Velocity

It has been described how the gas jet operates by dragging the melt out of the cut. The quicker it can be removed, the quicker the next piece can be melted. Thus, since the drag force per unit area is $\frac{1}{2}c_f\rho u^2$ and the drag coefficient c_f is a function of the Reynolds number ($\rho u d/\mu$ (where ρ is the gas density, u is the gas velocity, d is a typical dimension, *e.g.*, kerf width, and μ is the gas viscosity), the velocity of the gas in the slot on the cut face is critical. Gabzdyl [70] performed some experiments by directing the jet at various angles into the cut front with little effect. Increasing the gas jet velocity increased the cutting rate up to a point as seen in Figure 3.41. It was a puzzle as to why there should be this fall off in cutting speed with nozzle pressure. Some early workers suggested cooling was the problem, but the calculation in Section 3.4.2 showed this to be incorrect or at least very unlikely. Kamalu and Steen [71] performed some Schlieren experiments and showed that there was a density gradient field adjacent to the cut surface which could be affecting the focus at the cut front; however, the density gradient field was like a lens and hence the effect was difficult to justify to the extent shown in Figure 3.41. Fieret *et al.* [72] at Culham Laboratory took some surface pressure measurements and showed that there was a series of shock phenomena associated with the high-pressure jets. The structure is illustrated in Figure 3.42. The results were plotted as pressure fields (Figure 3.43). The first Mach shock disc is expected to influence cutting when the nozzle distance is around 2 mm – the very distance at which most people were working! Using their pitot system, they worked through a number of nozzle shapes in the hope of finding the best shape to avoid this problem. It was shown that a nozzle having an orifice with an odd number of lobes, such as "⋆", avoided the shock disc problem. However, it introduced the further problem of how to keep such a shape when there is the chance of beam clipping or simple back reflection, which might damage the nozzle. Undeterred by all this fine science, the job shop users were

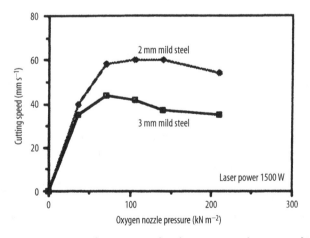

Figure 3.41 Variation of cutting speed with oxygen nozzle pressure [71]

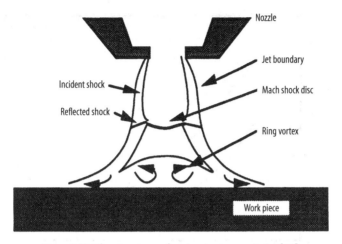

Figure 3.42 The structure of an impinging sonic jet [72]

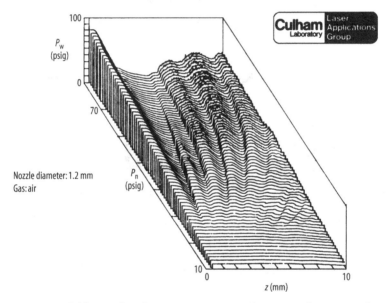

Figure 3.43 Pressure field on a plate from an impinging jet for various distances and pressures. P_n is the nozzle pressure on the workpiece and z is the nozzle to workpiece distance [72]

happily building nozzles to go to even higher pressures and cutting with Laval super-sonic nozzles at pressures of 14 bar or so using specially designed optics to withstand the pressure (Figure 3.44) [73]. This has been found to have beneficial effects. There are basically two cutting regions: low pressure, 1–6 bar, with oxygen and high pressure, 10–20 bar, for inert gases such as nitrogen, as shown in Figure 3.45 [74]. Too high a gas velocity with a reactive gas causes excessive side burning, whereas the inert gases need all the drag they can get.

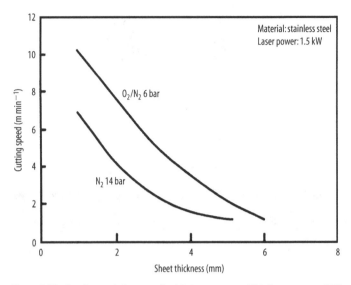

Figure 3.44 Cutting stainless steel with inert gas and high pressures [73]

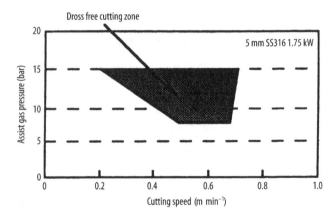

Figure 3.45 Effects of N_2 assist gas pressure on the optimum cutting window [74]

Multiple-nozzle systems are being used in several areas. The so-called clean-cut nozzle of Amada shown in Figure 3.46 [75] operates at around 1-atm pressure on the inner jet for lens protection and at around 5 atm in the outer ring jet. The effect is to produce burrless, striation-free cuts. For example, Mitsubishi claims to have cut 4-mm aluminium (A5052 Al 2.4% Mg alloy) with only 1.8-kW CO_2 power by this method. Amada and Prima Industrie have also pioneered a ring nozzle using a water spray on the outer nozzle which reduces depth of hardening (HAZ), dross, roughness, fume and smoke and cutting errors due to expansion while cutting [76].

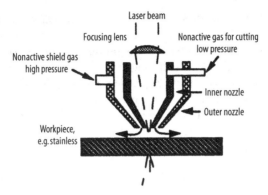

Figure 3.46 A high-pressure ring nozzle used for the "clean cut" technique [75]

3.6.3.2 Effect of Nozzle Alignment

The quality of the cut is affected by the alignment of the nozzle with the laser beam. An exhaustive set of experiments was undertaken by Gabzdyl [77], who methodically misaligned the beam and the jet. The alignment affects both the roughness of the cut and the way the dross clears the kerf; for example, it is possible to deliberately misalign the beam to make the specimen clear of dross but with all the dross clinging to the waste material.

3.6.3.3 Effect of Gas Composition

In some results from BOC, Zheng [78] and Chen [79] showed that the gas composition has an effect on the cut quality. There is an advantage if pure oxygen is used; a mere 1% impurity will reduce the cutting performance seriously, as shown in Figure 3.47 [74]. This point was analysed by O'Neill and Steen [80], who considered the natural mixing occurring within the kerf during cutting. They showed that cutting at depths greater

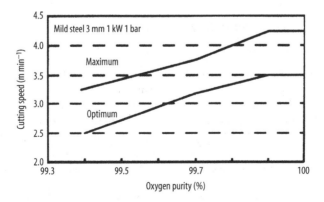

Figure 3.47 Variation in cutting performance with oxygen purity [74]

than 15 mm would be difficult owing to this mixing reducing the effectiveness of the reactive gas.

Reactive cutting has a greater tendency to produce striations. Also, oxygen can form oxide layers on the cut edge. Hence, cutting with inert gas is often favoured, particularly with stainless steel, to avoid the dross problem resulting from the formation of high melting point oxides of chromium. Inert gas cutting is usually performed at very high pressures (14 bar or so) with Laval supersonic nozzles to enhance the drag forces. The cut quality can be superb.

3.6.3.4 The Water-jet-guided Laser

As a variation on using gas to remove the molten product, a water jet has been used as a waveguide for the laser beam [81]. By inserting a pulsed 100–500 W Nd:YAG laser beam into a laminar water jet, whose diameter may be only 50 μm, one can use it to cut, drill, ablate and surface-structure metals, ceramics, plastics, composites and semiconductor material with an outstanding degree of quality and with a large operating stand-off distance of up to 100 mm. The effect of the water jet is to greatly reduce any thermal effects. The scouring action is also sufficiently strong that a high level of parallelism is obtained. It is also suited to cutting sandwich structures owing to the long working distance; depths of 50 mm are reported [82]. The energy for cutting comes entirely from the waveguided laser, the water jet serves solely to scour and cool. The cut surface is similar to a sandblasted surface. Cut depths are between 0 and 3 mm, with a minimum kerf width of 0.05 mm, compared with 0.15 mm for a laser alone. There is little fume or debris since it is all absorbed in the water stream. The main disadvantage is the presence of the water, which needs mopping up in some way. There is, however, not much of it. The Synova MicroJet®[5] [83] has a water jet of 1 l/h at a pressure of 2–50 MPa and a nozzle diameter of 25–150 μm. The working distance is up to 1,000 times the nozzle diameter. It requires a water stream of deionised water to avoid too much absorption.

3.6.4 Material Properties

3.6.4.1 Effect of Optical Properties – Reflectivity

For an opaque material the absorptivity = (1 – reflectivity); therefore, one might expect that high-reflectivity materials would be more difficult to cut. This is the case, but not quite as dramatically as the above argument suggests because the reflectivity is not only a function of the material but is also a function of the surface shape and the presence of surface films (such as oxides), surface plasmas and nonlinear events with multiphoton interactions from high-intensity beams. Owing to the important effect of thin films such as oxides, the absorption can be strongly time dependent [84]. Also, owing to the coupling effect with plasmas and the known decrease in reflectivity with temperature,

---

[5] MicroJet® is a registered trademark of Synova SA, Ch. de la Dent d'Oche, 1024 Ecublens, Switzerland. http://www.synova.ch

Table 3.6 Effects of surface treatment on cutting speeds [87]

Material	Power (W)	Polished velocity (mm s^{-1})	Untreated velocity (mm s^{-1})	Shot-blasted velocity (mm s^{-1})
C263 Ni alloy	600	12.7	12.7	21.1
N80 Ni alloy	400	12.7	16.9	21.1
L2% Cr steel	200	12.7	25.4	25.4

there is a further cause of a time dependency in the absorption as the material heats up. There is also a significant difference in the cutting rate depending on the surface finish as given in Table 3.6 [85]. Presumably wave formation on the melt front film would have a noticeable effect on the absorption but this has yet to be shown. Keilmann [86] introduced the concept of "stimulated absorption" based upon the standing wave pattern on the cut front. The electromagnetic standing waves arise from re-radiation from surface protuberances, a hard theory to visualise, but one that is known with radio waves circling the earth.

3.6.4.2 Effect of Thermal Properties

The ease with which a material can be successfully cut depends upon the absorptivity, the melting point of the material or oxide formed, char tendency and brittleness associated with the coefficient of thermal expansion. In fact the questions are:

1. Can sufficient power be absorbed?
2. Will this power cut successfully or damage the material?

Materials can be ranked by these properties, as in Table 3.7. See also the list of the cuttability of many materials on pages 3–6 in the *Industrial Laser Annual Handbook 1990*, by PennWell Books (Tulsa, OK, USA).

Table 3.7 Behaviour of different materials to laser cutting

Property	Material
High reflectivity (need for fine focus)	Au, Ag, Cu, Al, brass
Medium/high reflectivity	Most metals
High melting point	W, Mo, Cr, Ta, Ti, Zr
Low melting point	Fe, Ni, Sn, Pb
High oxide melting point (dross problems)	Cr, Al, Zr
Low reflectivity	Most nonmetals
Organics	
Tendency to char	PVC, epoxy, leather, wood, rubber, wool, cotton
Less tendency to char	Acrylics, polythene, polypropylene, polycarbonate
Inorganics	
Tendency to crack	Glass, natural stones
Less tendency to crack	Quartz, alumina, china, asbestos, mica

3.6.5 Practical Tips

3.6.5.1 Cornering and Edge Burning

There is a danger of burnout when cutting corners, thin slices or wherever the workpiece temperature approaches the ignition temperature. This can be avoided to some extent by:

1. using pulse power;
2. ramping the power in line with the speed (some systems have this facility within their software);
3. overshoot and return, creating a "Mickey Mouse ear"; and
4. using a circular water spray around the oxygen jet [76].

3.6.5.2 Workpiece Moves During Cutting

Vibration must be eliminated from the workstation-support structure. This is a basic design requirement for laser cutting systems. The workpiece must not be allowed to move owing to the acceleration forces on the table nor must it be able to tip as pieces are cut free. Thus, it must be lightly clamped and properly supported underneath. X/Y tables have to be rigid, and many are in consequence very heavy.

3.6.5.3 End Discontinuity

There is usually a blemish on the lower side at the end of a cut, where the assist gas jet can flow either side of the last half-thickness. This is not a problem with thin materials of less than 3 mm. For thicker materials pulsed power should be used for the last millimetre. Alternatively this last part will have to be separately finished.

3.6.5.4 Damage at the Initial Pierce Hole

Piercing the sheet from anywhere except the edge will create spatter. This splatter is a danger to the optics and the hole quality. There are several strategies for piercing, such as firing the laser as it approaches the workpiece; lowering the nozzle to the workpiece and then backing off to a safe height; and simply lowering to the focus position and firing. The first is to be preferred, if the software allows the laser to be controlled during a workstation movement. The last has dangers, particularly if a capacitance height sensor is used, since the plasma may affect the height reading and then the machine will take evasive action as though it has hit an obstacle! Some of the problems from piercing can be avoided by:

1. starting on scrap metal and then moving into the cut line; and
2. piercing by careful pulsing.

3.7 Examples of Applications of Laser Cutting

The main industrial application of lasers, at present, is in cutting. This work is increasingly done by laser "job shops". The costed example in Section 3.8 shows that the cost-effectiveness of the laser is due to its speed and the high-quality cut produced which reduces or eliminates after-treatment and hence makes significant manufacturing cost savings. However, these gains are only real if the laser can be kept working, owing to the high capital investment involved in a laser facility (in 2008 around £ 100,000–500,000 in equipment alone). Thus, it makes sense to bring the work to the laser in the form of job shops. Some 90% of the present job shops offer a specialist service in cutting. There has been a considerable growth of job shops in recent years. Around 30% offer a service in laser engraving or marking using YAG or CO_2 lasers in addition to cutting. The job shops are now being taken as a part of the manufacturing process and increasingly the job shop is being involved in the design stage of a component. This is partly as a result of new management techniques such as just in time (JIT) and materials requirement planning II (MRPII) and others. They are also increasingly being expected to take responsibility for the manufacture of complete components with a design team using the mutual expertise of the contractor and the job shop. This is a significant shift in manufacturing practice and has resulted in some remarkable cost-effective developments. The applications span the manufacturing industry from aerospace to food processing and toy manufacture.

The applications for lasers in cutting are too numerous and therefore hard to list. Laser cutting is one of the neatest and fastest profile cutting process available today. A typical job shop could have a turnaround time from the drawing to the article of a few hours (if pushed!), since the set-up time is only that required to program the cutting table. This is hard to compete with unless the requirement is for more than 10,000 or so pieces, when some hard automation, as with a stamping process, might be cheaper.

The ease of profile cutting has opened up several interesting novelties in design. For example, cutting tabs and slots could make assembly easier. It would also allow self-jigging for low-cost fixtures; it could have built-in error proofing by making the tabs or slots of different sizes. Another idea is to cut slots on bends that are made near previously drilled holes to avoid distorting the hole when bending is done. A variation on that is to cut a series of perforation slots along a bend line so that bending could be more accurate with cheaper tooling or could even done by hand [87].

3.7.1 Die Board Cutting

One of the first industrial applications of the laser was for die board cutting [88]. It used a BOC Falcon laser at 200 W installed by William Thyne in the UK in 1971. The die boards which are cut by this fully automatic machine are used in the manufacture of cartons. The laser replaced a process of sticking blockboard pieces together to make slots in which knives for cutting or creasing could be mounted. In the laser process the slots are simply cut in the blockboard and the knives are mounted in the laser-made groove. The process takes around one tenth of the previous time. Nearly all cartons are now made this way with full computer aided design (CAD)/computer aided manufacture software to drive the laser and design the carton.

3.7.2 Cutting of Quartz Tubes

Quartz tubes are used for car halogen lamps. Thorn EMI uses 500 W CO_2 lasers operating on a twin-position cutting arrangement. The process was installed because there was a saving of material (approximately 1 mm per cut at 4,000 cuts per hour, equivalent to 4 m of tube per hour) and a significant reduction in fume and dust, giving a saving in fume extraction and a better working environment.

3.7.3 Profile Cutting

This is mainly a job shop activity for the display industry, typewriter parts, gun parts, medical components, valve plates, gaskets, stained glass and many other applications. The accuracy of cutting is around a few micrometres, with a very fine finish for certain materials. One Australian firm specialises in making filter meshes, another in chainsaw parts.

3.7.4 Cloth Cutting

Garment cutting by laser is on the whole too slow since the competing processes stack-cut with a saw. Stack-cutting cloth by laser is not easy owing to welding, charring or smoke damage. Single-thickness cutting of thick material by laser is, however, excellent. Thus, it is used for cutting car floor carpets and seat covers. General Systems in Canada has a fully automatic laser machine using four lasers simultaneously to cut car fabrics. Cutting sailcloth, material for car air bags, lace fabrics and edging of embroideries are all now processed by laser.

3.7.5 Aerospace Materials

Hard and brittle ceramics such as SiN can be cut 10 times faster by laser than by diamond saw. Titanium alloys cut in an inert atmosphere are used in airframe manufacture. The laser saves around 17.6 man-hours per plane for Grumman Corporation in the manufacture of one stabiliser component compared with chemical milling.

Aluminium alloys are similarly advantaged by using the laser, which has to be well tuned and of high power, > 1 kW. Cost savings of 60–70% compared with routing or blanking have been recorded.

Boron-epoxy and titanium-coated aluminium honeycomb plates can also be cut by laser. Stainless steel pressed parts are three-dimensional profile cut by several aircraft manufacturers with a view to subsequent welding of the cut edge.

3.7.6 Cutting Fibre Glass

The advantages of cutting fibre glass by laser are the reduction of dust, no cracking of the edges and no tool wear, all of which are problems with drilling or sawing. Water jets tend to fray the edges.

3.7.7 Cutting Kevlar®[6]

A nylon-based epoxy armour plate used for a variety of reasons where strength and lightness are required has been one of the marvel materials of today. It is also a gift for the laser user since there are very few alternative techniques which can cut it (*e.g.*, abrasive water jets, which tend to fray the edges). However, see Chapter 13 on safety, because the fumes can be poisonous. Some of the best results have been obtained with supersonic assist gas.

3.7.8 Prototype Car Production

The ease with which profiling can be done allows prototype car production [89,90] to be much faster compared with the use of nibblers; around 10 times as many components can be made in a given time. The cutting of sunroofs in cars as an assembly option is now done by robotically guided lasers. Also the cutting of the holes for left-hand-drive or right-hand-drive vehicles is done by laser on the assembled car [91,92].

A list of the numerous applications for the laser within the car industry would fill up several pages. Only a sample list is given here. The first application in the automobile industry was by General Motors in 1972 at its Delco Remy plant at Anderson, Indiana, USA, using four Coherent General 300 W CO_2 lasers to cut ignition coils. Since then cuts have been made in three-dimensional parts, power-steering pump wipers, door lock holes, dashboard appliques, three-dimensional axle carriers, PVC auto roof liners, parts for punch die sets, seat covers, exhaust systems, air-conditioning ports, wiper blades, air bags, hydroformed tubes,the parts for tailored blanks and many more items.

3.7.9 Cutting Alumina and Dielectric Boards

This is done by both through cutting and scribing. It is a common application.

3.7.10 Furniture Industry

Cutting timber of any hardness up to depths of 4 cm is possible. Since there is no mechanical stress, very tight nesting of parts can be arranged, giving a significant saving

---

[6] Kevlar® is a registered trademark of E.I. du Pont de Nemours and Company. http://www2.dupont.com

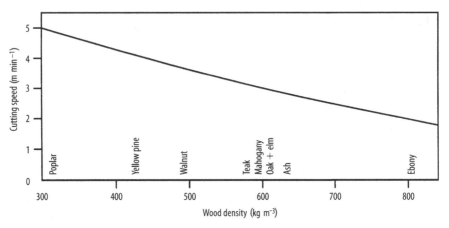

Figure 3.48 Cutting speed as a function of wood density for 10-mm-thick timber having 12% moisture. The laser power was 350 W and the assist gas was at 3–6 bar (Data from Powell [93])

Figure 3.49 An example of laser engraving

of material. The cutting rates are, however, similar to those for a band saw. The rate varies with the density of the wood as shown in Figure 3.48 [93]. This is as expected from Equation 3.3, in which the cutting speed is expected to be inversely proportional to the density. Charred surfaces, if not too badly charred, can be glued. Another application is laser engraving by machining very detailed patterns into wood (Figure 3.49). This is done by rapidly scanning a focused laser beam over a reflective mask, for example, copper. Masks for dropout patterns are made by chemically etching thin copper sheets on a Mylar®[7] backing. A development of this is to inlay the engraved area with metal or other wood as in marquetry. Some fine artistic work has been done this way, particularly for the Arab market.

---

[7] Mylar® is a registered trademark of DuPont Teijin Films. http://www.dupontteijinfilms.com

3.7.11 Cutting Paper

Cutting paper by laser profile cutting has led to a new fashion in stationery and led to a near craze in 1989 in patterned suntans achieved while sunbathing under an appropriate cut pattern!

3.7.12 Flexographic Print Rolls

Laser-engraved rubber rolls are a precise and fast way of transferring a flat picture to a cylindrical roll for the printing of wallpaper and other articles.

3.7.13 Cutting Radioactive Materials

Work on radioactive materials is considerably easier with optical energy than with other forms of energy since the generator is outside the hot zone and the only material which may become contaminated is the workpiece, fixtures and a few minor optics. It is also possible to transmit optical energy over long distances and so the work may not have to be confined to special areas. One of the advantages of the laser in cutting is the lack of fume. There is some fume but there is relatively little compared with any alternative. It is thus an attractive concept for the dismantling and repair of nuclear power stations.

3.7.14 Electronics Applications

Cutting of circuit boards has been mentioned. Resistance trimming of circuits, functional trimming of circuits and microlithography are new manufacturing processes introduced by the laser. The growing use of the excimer laser is of current interest. Hole drilling through circuit boards to join circuits mounted on both sides has advantages as already mentioned. The excimer laser can do this without risk of some form of conductive charring.

Optical lithography [94] to engrave structures in the manufacture of integrated circuits has been keeping pace with Moore's law (*the number of devices per unit area doubles every 18 months*) by changing to shorter and shorter wavelengths. Devices are now entering the subwavelength region with the introduction of phase-shift technology which is showing the possibility of features down to 64 nm using 193-nm radiation.

3.7.15 Scrap Recovery

Careful cutting of old telephone switches allows the recovery of the considerable precious metals content [95].

3.7.16 Laser Machining

This is similar to laser engraving on wood (Section 3.7.10; it has recently been achieved on steel in a process called "lasercaving" [52]. The rate of removal of material is slow, being around 35 mm^3 min^{-1} when using a 300 W finely focused beam and a carefully designed nozzle operated at only 1-bar pressure.

3.7.17 Shipbuilding

The laser is now powerful enough to cut the 15 mm plate required for ships and by use of the LASOX process (see Section 3.3.7) thicknesses up to 50 mm are now possible. It also cuts with little HAZ and therefore low distortion. To the plasma cutting engineers in shipbuilding this is amazing. The laser has virtually removed the need for expensive hammer flattening or after-machining processes. It can also be used to mark the pieces with part numbers, datum points, stiffener positions and flanging lines. The accuracy of the cutting process is such that it gives savings in the welding operation which follows. Vosper Thorneycroft [96] achieves an accuracy of 0.3 mm over a 10-m cut length when cutting 15-mm mild steel or 22-mm plywood. This process alone allows a reduction in the production time from contract to boat of approximately 1 year!

3.7.18 The Laser Punch Press

Currently the productivity of some laser cutting systems is close to that achievable with a mechanical punch press [97]. This is something of a "Holy Grail" for laser processing. Once some form of parity can be attained with punch press productivity, then a vast market will open up for the laser. The laser can cut any shaped hole, as well as the outline of the piece. Thus, the punch press, requiring special tools for each size and shape of hole and being incapable of profile cutting, would not compete. The current progress has been possible mainly through the more efficient use of the CNC movements. Firstly, the use of linear motors with accelerations up to 7.6 G (although only 2 G is used for stability reasons) reduces positioning time; secondly, the piercing is done on the fly; thirdly, the CNC does not stop between blocks of information. The laser beams are now of a higher quality and hence capable of greater penetration and speed, but this is secondary compared with the gains from proper table management. A 6 kW CO$_2$ laser can cut 1-mm mild steel at 30 m min^{-1}, ample speed to compete with a turret punch, which can make 200 holes per minute. Standard laser cutting with rapid transfer, pierce and cut gives only 50–60 holes per minute. However, using high-acceleration tables and a fully integrated CNC system, Fanuc has achieved 450 holes per minute [98], whereas Finn Power International in 2002 claimed 700 round 1.5 mm diameter holes per minute with 2.3 mm pitch in 1 mm mild steel or 600 holes per minute of 2 mm diameter with a 3.5 kW laser. The laser is bound to have a heavy impact on machine tools in the very near future.

3.7.19 Manufacture of Bikes and Tubular Structures

The profiling ability of the laser is shown to advantage when fitting tubular pieces together as for bikes, banisters and tables [99].

3.7.20 Cutting and Welding of Railcars

Another application requiring accuracy over large areas with minimal distortion is the cutting and welding of railcars.

3.8 Costed Example

The case studies in Figure 3.50 come from data given at a meeting at Ferranti Photonics in Dundee in 1987. The numbers are still relevant but they mainly illustrate that the laser is likely to be cost effective for small runs of between 3–10,000 pieces depending on the exact shape to be cut. To make one or two pieces is cheaper by hand and to make many thousands would probably be more economically done with hard automation.

3.9 Process Variations

3.9.1 Arc-augmented Laser Cutting

It has been found [100] that if an electric arc is located near the laser-generated event, then the arc will automatically root at the high-temperature zone and the magnetic pinch caused by this hot zone will constrict the arc to near the size of the laser beam for low-current arcs up to 80 A. Above this current the cathode jet from the arc is too

Case 1:   Shape to be cut in 1mm mild steel

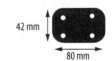

	Press Tool	Laser
Capital cost in tooling – design and manufacture of compound die	£1800	–
Time per piece (300 mm cut length at 2.0 m/min with a 500 W laser)	0.5s	9s
Cost/piece (at £100 per hour for laser hire)	£1800/n	£0.25
Optimal production range – breakeven	>7200	<7200
Delivery time – approximate	6 weeks	1 hour

Case 2:   Shape to be cut in 4mm stainless steel

	Press tool	Laser
Capital cost in tooling	£5500	–
Time per piece (600 mm cut length at 0.8 m/min for 500 W laser)	1s	45s
Cost per piece (£100 per hour for laser hire)	£5500/n	£1.25
Optimal production range – breakeven	>4400	<4400
Delivery time – approximate	6–8 weeks	3 hours

Figure 3.50 Costing example

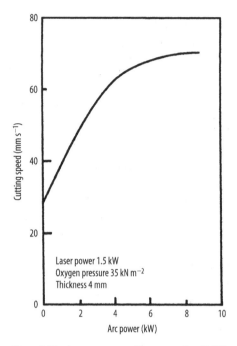

Figure 3.51 Arc-augmented laser cutting [100]

strong and the two energy sources may not be located in the same place nor would the arc be constricted. When the arc is on the same side of the workpiece as the laser, then a damaged cut top edge will probably result as the arc will root on one side or the other. However, when the arc is on the underside of the workpiece, then the cutting process can be speeded up by a factor of around 2, as shown in Figure 3.51, which shows a plot of the cut speed against power. Only where the curve starts to turn is the quality of the cut different from that of a pure laser cut. The weakness of this process is the need to have the arc on the side opposite the laser. Nevertheless, the process is there for anyone wishing to cut twice as fast for the same size of laser.

3.9.2 Hot Machining

The use of the laser to heat the material and thus soften it just prior to mechanical machining was the subject of considerable research in the USA during the early 1980s. The process works by reducing the cutting force by about 50%, provided that the heating is correctly located, and does not allow for quenching from the bulk material in-between heating and machining, thus resulting in transformation hardening instead of softening! The weakness with this process is the capital cost of the laser compared with that of a plasma torch, which could be used to do the same function.

3.10 Future Developments

The areas for development of laser cutting [101] must include:

1. Increasing the energy input to the cut region by having:

   a. higher-powered lasers;
   b. additional energy sources;
   c. improved coupling; and
   d. smaller spot size.

2. Increasing the ease of removal of the cut products – usually molten – by:

   a. increasing the drag; and
   b. increasing the fluidity.

3.10.1 Higher-powered Lasers

Some significant results have been obtained with superpulsing to obtain instantaneous high powers. Too high a power may create explosive conditions which spoil the quality. High-powered lasers are being vigorously sought by various research programmes, not only that, but they found them with high-brightness fibre lasers, only to find the problem moves to those discussed below in Section 3.10.4 on kerf width and thermodynamics.

3.10.2 Additional Energy Sources

Traditional cutting uses oxygen as an additional energy source and the results of that have been seen. Arc augmentation has been demonstrated (see Figure 3.51) but needs some invention to locate the arc successfully on the same side as the laser and to overcome the arc initiation problems. Some areas for study are in the gas composition. One of the more original ideas is to blow iron powder into a cut in aluminium and thus start a small thermite process [93].

3.10.3 Improved Coupling

Polarisation and the coupling improvements with high power density (by increasing the plasma coupling route) are currently practised. Radial polarisation is showing some promise. Little is understood of the action of thin films on the melt front. There may be a certain gas mixture which would generate just the correct film thickness for a given material.

3.10.4 Smaller Spot Size

This would increase the power density and locate the energy more efficiently at the melt zone. It can be achieved by better designed laser cavities giving Gaussian mode beams and using shorter focal length lenses corrected for spherical aberration. Fibre lasers using monomode fibre achieve this and the results have been disappointing. This is due to the very narrow kerf widths restricting the outflow of the melt, but also possibly because the temperatures achieved at these very high power densities are so high that the oxidation reaction is terminated and becomes reducing as predicted by the thermodynamics of the reaction.

Smaller spot sizes can also be achieved by using shorter-wavelength lasers. The processing capability of a Gaussian high-powered ultraviolet beam has yet to be seen, but it should be awesome. However, there is a safety aspect when using shorter wavelengths which will need to be addressed.

3.10.5 Increased Drag

This is the area of greatest promise. There are so many novel flow conditions which have not been examined yet. The whole subject of jet/slot fluid flow is little understood at the time of writing. The combination of a laser with a water jet (Section 3.6.3.4) may have a future.

3.10.6 Increased Fluidity

The melt could be made to flow better if the workpiece were vibrated ultrasonically, the process were done under supergravity conditions or the melt were more fluid because its chemistry is changed. For example, to cut with chlorine instead of oxygen would give either a vapour halide and no dross or a very fluid dross. Chlorine gas is denser than oxygen and therefore would have more drag. There is also an exothermic reaction involved. Chlorine should therefore give improved performance – pity that the operator might get killed! In fact all these options need thinking about but seem difficult to engineer. Is it not true that the enjoyment of engineering is part dreaming, part doing?

3.11 Worked Example of Power Requirement

Question: It is hoped to cut 10 mm thick mild steel by laser. What laser power is required and how fast will it cut?

1. From Table 3.4 we find that mild steel requires $5.7\,\mathrm{J\,mm^{-2}}$ to sever using O_2 assist gas. That is $P/Vt = 5.7\,\mathrm{J\,mm^{-2}}$. If $t = 10\,\mathrm{mm}$, then $P/V = 57\,\mathrm{J\,mm^{-1}}$.
2. Penetration of the laser into mild steel is a function of the focusability of the beam. One kilowatt of a low-order-mode beam would penetrate this thickness, but a higher-order-mode beam would find it difficult.

Thus, a qualitative judgement is required to determine the laser power for the required penetration. Let us say 1.2 kW. In that case the expected speed would be 21 mm s^{-1}.

The burning rate in oxygen is around 10–15 mm s^{-1} (see Figure 3.38). This speed is a little too close to 21 mm s^{-1}. If the burning rate were to be the faster, then a very poor cut quality would result. The burning rate can be controlled by the gas composition. Thus – depending upon economics – a 2-kW laser would be preferred and the expected cutting rate would be 35 mm s^{-1} using pure oxygen. Process economic evaluation can now proceed from there.

Questions

1. What are the main ways in which a laser can be used to drill plastics?
2. With what approximate velocity will vapour emerge from the forming hole during laser drilling?
3. How does spatter form?
4. Why is there a recast layer?
5. How fast would you expect a 3-kW laser focused to 200-µm spot diameter to drill through 5-mm-thick stainless steel?
6. How could you make a square inverted pyramid shaped hole with a 200-µm side length?
7. What measures could you take to reduce spatter?
8. Can taper be controlled?
9. Write brief notes and draw a sketch of six ways in which a laser can sever material. List against each method the relative amount of energy required to achieve a cut.
10. List the laser beam properties that affect the cutting performance.
11. Write a detailed description on why polarisation matters in laser cutting.
12. What is meant by a "circularly polarised beam" and how is a circularly polarised beam achieved?

References

[1] Gabzdyl J (1989) Effect of laser mode and coaxial jet on laser cutting. PhD thesis, University of London
[2] Schneider M, Berthe L, Muller M, Fabbro R (2007) Influence of incident angle on laser drilling. In: ICALEO 2007 proceedings. LIA, Orlando, paper 1205, pp 641–646
[3] Low DKY, Li L (2002) Hydrodynamic physical modeling of laser drilling. Trans ASME 124:852–862
[4] Noguchi S, Ohmura E, Harp WR,Tu J, Hirata Y (2008) Modelling of laser drilling considering multiple reflection of the laser evaporation and melt flow. In: ICALEO 2008 proceedings, Temecula, October 2008. LIA, Orlando, pp 674–683
[5] Andrews JG, Athey DR (1976) Hydrodynamic limit to penetration of material by a high-power beam. J Phys D Appl Phys 9:2181
[6] Rykalin N, Uglov A, Zuev I, Kokora A (1995) Laser and electron-beam treatment of metals In: Rykalin W (ed) Laser drilling. Mir, Moscow, chap 9
[7] French PW, Hand DP, Peters C, Shannon GJ, Byrd P, Steen WM (1998) Investigation of the Nd:YAG laser percussion drilling process using high speed filming. In: ICALEO'98 proceedings. LIA, Orlando, part 1, sect B, pp 1–10

[8] Yilbas BS (1986) The absorption of incident beams during laser drilling. Opt Laser Technol 8:27–32

[9] Gagliano FP, Peak UC (1971) The influence of the time development of the temperature during surface heating. IEEE J Quantum Electron QE-7(6):277

[10] Voisey KT, Rodden W, Hand D, Clyne TW (2001) Melt ejection characteristics during laser drilling of materials. In: ICALEO 2001 proceedings, Jacksonville, October 2001. LIA, Orlando, paper P505

[11] Ng GKL, Li L (2001) An investigation into the role of melt ejection in repeatability of entrance and exit hole diameters in laser percussion drilling. In: ICALEO 2001 proceedings, Jacksonville, October 2001. LIA, Orlando, paper 1807

[12] Chan CL, Mazumder J (1987) One dimensional steady state model for damage by vaporisation and liquid expulsion due to laser material interaction. J Appl Phys 62(11):4579–4586

[13] Forsman AC, Banks PS, Perry MD, Campbell EM, Dodell AL, Armas MS (2005) Double pulse machining as a technique for the enhancement of material removal rates in laser machining of metals. J Appl Phys 98:033302

[14] Campbell BR, Lehecka TM, Semak VV, Thomas JG (2007) Effect of the double pulse format for picosecond pulse laser drilling in metals. In: ICALEO 2007 proceedings. LIA, Orlando, paper M906, pp 45–51

[15] Rohde H, Dausinger F (1995) The forming process of a through hole drilled with a single laser pulse. In: ICALEO'95 proceedings, San Diego, November 1995. Springer, Berlin/IFS, Kempston, pp 331–340

[16] Hamilton DC, James DJ (1976) Hole drilling with repetitively pulsed TEA CO2 laser. J Appl Phys D Appl Phys 9:L41–L43

[17] Sharp CM, Mueller ME, Murthy J, McCay MH, Cutcher J (1997) A novel anti-spatter technique for laser drilling applications to surface texturing. In: ICALEO'97 proceedings, San Diego, November 1997. Springer, Berlin/IFS, Kempston, pp 41–50

[18] Kamalu J, Byrd P (1997) Laser drilling of mild steel using an anti-spatter coating. In: ICALEO'97 proceedings, San Diego, November 1997. Springer, Berlin/IFS, Kempston, pp 156–164

[19] Otstat RS et al (1969) Method for machining with laser beam. US Patent 3,440,388

[20] Low DKY, Li L, Byrd PJ (2001) The influence of temporal pulse train modulation during laser percussion drilling. Opt Laser Eng 35:149–164

[21] Low DKY, Li L, Byrd PJ (2003) Spatter prevention during the laser drilling of selected aerospace materials. J Mater Process Technol 139:71–76

[22] Kar A, Mazumder J (1990) Two-dimensional model for material damage due to melting and vaporisation during laser irradiation. J Appl Phys 68(8):3884–3891

[23] Goreisha M, Low DKY, Li L (2001) Hole taper control in laser percussion drilling using statistical modeling. In: ICALEO 2001 proceedings, Jacksonville, October 2001. LIA, Orlando, paper P528

[24] Witte R, Liebers R, Moser T (2007) Micro drilling with Nd:YAG lasers in the nanosecond regime. In: ICALEO 2007 proceedings. LIA, Orlando, paper M1004, pp 260–268

[25] Low DKY, Li L, Corfe AG, Byrd PJ (2001) Spatter free laser percussion drilling of closely spaced array of holes. Int J Mach Tool Manuf 41:361–377

[26] French PW, Naeem M, Sharp M, Watkins KG (2006) Investigation into the influence of pulse shaping on drilling efficiency. In: ICALEO 2006 proceedings, Phoenix. LIA, Orlando, paper 310

[27] Voisey KT, Thompson JA, Clyne TW (2001) Damage caused during laser drilling of thermal spray TBC on superalloy substrates. In: ICALEO 2001 proceedings, Jacksonville, October 2001. LIA, Orlando, paper B1802

[28] Fohl C, Wartenberg S, Dausinger F (2005) Trepanning optic for high precision laser drilling of metals. DOPS NYT 20(2):27–32

[29] Harris J, Brandt M (2002) The cutting of thick steel plate using a spinning Nd:YAG laser beam. Industrial Laser User 26:24–25

[30] Jacob P, Smithfield RI (2008) Precision trepanning with a fibre laser. In: ICALEO 2008 proceedings, Temecula, October 2008. LIA, Orlando, paper 303, pp 130–138

[31] Dausinger F (2001)Drilling high quality micro-holes. Industrial Laser User Mar 25–27

[32] French P, Naeem M, Watson KG (2003) Laser percussion drilling of aerospace material using a 10 kW peak power laser using a 400 μm optical fibre delivery system. In: ICALEO 2003 proceedings, Jacksonville. LIA, Orlando, paper 503

[33] Lizotte T, Ohar O, Waters SC (2002) Excimer lasers drill ink jet nozzles. Laser Focus World Jun 165–170

[34] Spedding V (2002) Exitech makes its mark in material processing. Opto & Laser Europe Jun 27

[35] Zhang C, Quick NR, Kar A (2007) Microvia drilling with a green laser. In: ICALEO 2007 proceedings. LIA, Orlando, paper M1006, pp 274–281

[36] Batta N, Nagai S, Li M (2007) Via hole machining in sapphire using ultrafast laser. In: ICALEO 2007 proceedings. LIA, Orlando, paper M102, pp 9–11

[37] Migliore L, Ozkan A (2001) Multiplex CO_2 drilling with diffractive optics. ICALEO 2001 proceedings, Jacksonville, October 2001. LIA, Orlando, paper A103

[38] Bright R, Jacobs P, Aindow M, Marcus H (2007) The influence of pulse parameters on the laser drilling of Hastelloy X. In: ICALEO 2007 proceedings. LIA, Orlando. paper 1201, pp 613–620

[39] Kobayashi T, Kiyonobu O, Kazuyoshi T, Umezu S, Okatsu K (2007) Underwater rock drilling by $CO2$ laser. In: ICALEO 2007 proceedings. LIA, Orlando, paper 905

[40] Leong K, Xu Z, Reed C (2001) Drilling rock. Industrial Laser Solutions article 172576

[41] Jacobs P, Hayman M, Marisco T, Denney P, Ilumoka A, Bright R (2007) Acoustic phenomena during laser drilling. In: ICALEO 2007 proceedings. LIA, Orlando, paper 108, pp 64–70

[42] Fox MDT, French P, Hand DP, Jones JDC (1990) Optical focus control for laser percussion drilling. In: ICALEO'99 proceedings, San Diego. LIA, Orlando, sect C, pp 1–10

[43] Powell J (1993) $CO2$ laser cutting. Springer, London

[44] Powell J, Kaplan AFH, Petring D, Kumar RV, Al-Mashikhi SO, Voisey KT (2008) The energy generated by oxidation reaction during laser-oxygen cutting of mild steel. In: ICALEO 2008 proceedings, Temecula, October 2008. LIA Orlando, paper 1504

[45] Duley WW, Gonzalves JN (1972) Can J Phys 50:215

[46] Olsen FO, Emmel A, Bergamnn HW (1989) Contribution to oxygen assisted $CO2$ laser cutting. In: Steen WM (ed) Proceedings of the 6th international conference on lasers in manufacturing (LIM6), Birmingham, UK, May 1989. IFS, Kempston, pp 67–79

[47] RH Perry (1984) Perry's chemical engineers' handbook, 6th edn. McGraw-Hill, London, sect 10.12

[48] Schouker D (1990) Physical mechanism and theory of laser cutting. In: Industrial laser annual handbook. PennWell Books, Tulsa, pp 65–79

[49] Purtonen T, Salminen A (2008) Fibre laser cutting of stainless steel in non-vertical positions. In: ICALEO 2008 proceedings. Temecula, October 2008. LIA, Orlando, paper 1507

[50] Powell J, Tan WK, Maclennan P, RuddD, Wykes C, Engstrom H (2001) Cutting stainless steel with a dual focusTM lens. Industrial Laser User 22:11–13

[51] O'Neill W, Elboughey A, Steen WM, Volgsanger M (1995). Selective removal of steel by a slotting process. In: ICALEO'95 proceedings, San Diego. LIA, Orlando, pp 158–167

[52] Ebert G, Sutor U (1990) Laser caving offers new machining method. Industrial Laser Review Aug 23–26

[53] Duley WW (1987) Excimer laser etching of organic polymers. In: Laser advanced manufacturing processes (LAMP'87), Osaka, Japan. High Temperature Society Japan, paper 7A02, pp 585–594

[54] Duley WW (1996) UV laser effects and applications in material science. Cambridge University Press, Cambridge

[55] Tonshoff HK, Butje R (1988) Material processing with excimer lasers. In: Hugel H (ed) Proceedings of the 5th international conference on lasers in manufacturing (LIM5), Stuttgart. IFS, Kempston, pp 35–47

[56] O'Neill W, Gabzdyl JT (2000) New developments in laser-assisted oxygen cutting. Opt Lasers Eng 34:355–367

[57] Gabzdyl JT, Penn W, Cahill P, Koch J (2002) Cutting 50mm shipbuilding parts with less than 2 kW of CO_2 laser power. In: ICALEO 2002 proceedings, Phoenix, October 2002 LIA, Orlando, sect D, paper 910

[58] Bunting KA, Cornfield G (1975) Trans ASME J Heat Transf Feb 116

[59] Woods S (2007) Fibre lasers for cutting: a review. Laser Focus World Ma

[60] Sharp CM (1988) $CO2$ laser cutting of highly reflective materials. In: ICALEO'87 proceedings, San Diego. Springer, Berlin/IFS, Kempston, pp 149–153

[61] Shaw LH, Cox MJ (1989) High aspect ratio Nd:YAG laser welding. In: Proceedings of advances in joining and cutting processes '89. Harrogate, UK. TWI, Great Abington, paper 53

[62] Powell J, Kaplan AFH, Petring D, Kumar RV, Al-Mashikhi SO, Voisey KT (2008) The energy generated by oxidation reaction during laser-oxygen cutting of mild steel. In: ICALEO 2008 proceedings, Temecula, October 2008. LIA, Orlando, paper 1504

[63] Kugler T, Naeem M (2002) Welding and cutting improvements with super modulated, high BQ Nd:YAG lasers. In: ICALEO 2002 proceedings, Phoenix, October 2002. LIA, Orlando, paper 506

[64] Olsen F (1981) Studies of sheet metal cutting with plane polarised CO2 laser. In: Proceedings of Laser'81 optoelectronics conference, Munich. Springer, Berlin, pp 227–231

[65] Spalding IP (1990) EUREKA 119 report. Culham Laboratory, Abingdon

[66] Girkin J, Ness K (12) Material processing using ultra fast lasers; from the research laboratory to the production line. Industrial Laser User 12:30–32

[67] Steen WM, Kamalu JN (1983) Laser cutting. In: Bass M (ed) Laser material processing. North Holland, Amsterdam, chap 2

[68] Powell J, Tan WK, Maclennan P, Rudd D, Wykes C, Engstrom H (2000) Laser cutting of stainless steel with a dual focus lenses. J Laser Appl 12(6):224–231

[69] V&S Scientific (1999) Laser cutting with dual focus™ lenses. Industrial Laser User 16:27–28

[70] Gabzdyl J (1989) Effect of laser mode and coaxial gas jet on laser cutting. PhD thesis, University of London

[71] Kamalu JN, Steen WM (1981) TMS paper A81-38. AIME, Warrendale

[72] Fieret J, Terry MJ, Ward BA (1986) Aerodynamic interactions during laser cutting. Proc SPIE 668:53–62

[73] Weick JM, Bartel W (1989) Laser cutting with oxygen and its benefits for cutting stainless steel. In: Steen WM (ed) Proceedings of the 6th international conference on lasers in manufacturing (LIM6), Birmingham, UK, May 1989. IFS, Kempston, pp 81–89

[74] Gabzdyl J (1996) Process assist gas for cutting of steels. Industrial Laser User Aug 23–24

[75] Kawasumi H (1990) Laser processing in Japan. In: Industrial laser annual handbook. PennWell Books, Tulsa, pp 141–143

[76] Banchi C (1996) Laser cutting with water cooling. Industrial Laser Review Mar 11–13

[77] Gabzdyl JT, Steen WM, Cantello M (1987) Nozzle beam alignment for laser cutting. In: ICALEO'87 proceedings, San Diego. Springer, Berlin/IFS Kempston, pp 143–148

[78] Zheng HU (1990) In process quality analysis of laser cutting. PhD thesis, University of London

[79] Chen SL (1992) Effects of gas composition and ripple power on laser gas cutting. PhD thesis, University of London

[80] O'Neill W, Steen WM (1995) A theoretical analysis of gas entrainment operating during the laser cutting process. J Phys D Appl Phys 28:12–15

[81] Richerzhagen B (2001) The best of both worlds – laser and water jet combined in a new process: the water jet guided laser. In: ICALEO 2001 proceedings, Jacksonville, October 2001. LIA, Orlando, paper M901

[82] Richerzhagen B (2002) Industrial applications of the water-jet guided laser. Industrial Laser User 28:28–30

[83] Hewett J (2007) Laser water jet cools and cuts in the material world. Optics and Lasers Europe Mar 17–19

[84] Duley WW, Semple DJ, Morency JP, Gravel M (1979) Coupling coefficient for cw CO_2 laser radiation on stainless steel. Opt Laser Technol 11(6):313–316

[85] Forbes N (1976) Designing for laser cutting. Fabricator 6:5

[86] Keilmann F (1985) Stimulated absorption of CO_2 laser light on metals. In: Proceedings of NATO ASI on laser surface treatment, San Miniato, Italy. NATO, pp 17–22

[87] Main N (2000) Designing for laser cutting. Industrial Laser User (18):14–15

[88] Dawson P (1996) The use of lasers in the die board industry. Industrial Laser User (3):21–22

[89] Roessler DM (1990) New laser processing developments in the automobile industry. In: Industrial laser annual handbook. PennWell Books, Tulsa, pp 109–127

[90] Ainsworth SJ (1997) Lasers in car body manufacture. Industrial Laser User (6):19–21

[91] Hanike L (1988) Laser technology within the Volvo car group. In: Hugel H (ed) Proceedings of the 5th international conference on lasers in manufacturing (LIM5). Stuttgart. IFS, Kempston, pp 97–118

[92] Koons JN, Roessler DM (1996) Laser cutting right hand drive vehicles. Industrial Laser Review Sep 7–12

[93] Powell J (1983) CO2 laser cutting. Springer, London

[94] McCarthy DC (2001) Optical lithography reaches a cross road. Photonics Spectra Dec 75–104

[95] Peak UC, Gagliano FP (1972) Thermal analysis of laser drilling processes. IEEE J Quantum Electron QE-8, pp 112–119

[96] Perryman I (1997) Fabrication time of first-of-class vessels, from contract signing to structural steel completion, has just been cut by a year. Industrial Laser User May 17

[97] Karube N et al (1997) Fast contour cutting using linear motors. Industrial Laser Review Jun 7–10

[98] Tulloch MH (1996) Nd:YAG laser cuts tough chrome moly bike tubes. Photonics Spectra Feb 18

[99] Clarke J, Steen WM (1978) Arc augmented laser cutting. In: Proceedings of Laser'78 Conference, London

[100] Steen WM (1987) Future developments. In: Soares ODD, Perez Amo M (eds) Applied laser tooling. Nijhoff, Dordrecht, chap 5

[101] Kamalu JN (1983) Laser cutting of mild steel: gas flow, arc augmentation and polarisation effects. PhD thesis, University of London

[102] Powell J, Wykes C (1989) A comparison between CO2 laser cutting and competitive techniques. In: Steen WM (ed) Proceedings of the 6th international conference on lasers in manufacturing (LIM6), Birmingham, UK, May 1989. IFS, Kempston, pp 135–153

[103] American Society for Metals (1985) Metals handbook (desk edition). American Society for Metals, Metals Park

"Darling I think you went through a red light!"

"Never mind the power, its the frequency tells us where we are."

4 Laser Welding

> If you can't beat them join them
>
> *Anon*
>
> By uniting we stand, by dividing we fall
>
> *John Dickinson (1732–1808),*
>
> *The Liberty Song Memoirs of the Historical Society of Pennsylvania*
> *volume xiv*

4.1 Introduction

The focused laser beam is one of the highest power density sources available to industry today. It is similar in power density to an electron beam. Together these two processes represent part of the new technology of high-energy-density processing. Table 4.1 compares the power density of various welding processes.

At these high power densities all materials will evaporate if the energy can be absorbed. Thus, when one welds in this way a hole is usually formed by evaporation. This

Table 4.1 Relative power densities of different welding processes

Process	Heat source intensity $(\mathrm{W\,m^{-2}})$	Fusion zone profile
Flux-shielded arc welding	$5 \times 10^6 - 10^8$	
Gas-shielded arc welding	$5 \times 10^6 - 10^8$	low / high
Plasma	$5 \times 10^6 - 10^{10}$	low / high
Laser or electron beam	$10^{10} - 10^{12}$	defocus / focus

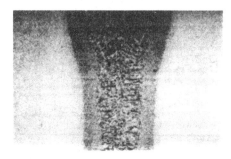

Figure 4.1 Micrograph of the transverse section through a laser weld showing the fusion and heat-affected zones

Table 4.2 Relative joining efficiencies of different welding processes

Process	Approximate joining efficiency ($mm^2\ kJ^{-1}$)
Oxyacetylene flame	0.2–0.5
Manual metal arc	2–3
TIG	0.8–2
Submerged arc welding	4–10
High-frequency resistance welding	65–100
Electron beam	20–30
Laser	15–25

TIG tungsten–inert gas

"hole" is then traversed through the material, with the molten walls sealing up behind it. The result is what is known as a "keyhole" weld. This is characterised by its parallel-sided fusion zone and narrow width (Figure 4.1). Since the weld is rarely wide compared with the penetration, it can be seen that the energy is being used where it is needed in melting the interface to be joined and not most of the surrounding area as well. A term to define this concept of efficiency is known as the "joining efficiency". The joining efficiency is not a true efficiency in that it has units of square millimetres joined per kilojoule supplied. It is defined as Vt/P, the reciprocal of the specific energy, referred to in Chapter 3, where V is the traverse speed ($mm\ s^{-1}$), t is the thickness welded (mm) and P is the incident power (kW). Table 4.2 gives some typical values of the joining efficiency of various welding processes.

The higher the value of the joining efficiency, the less energy is spent in unnecessary heating – that is, generating a HAZ or distortion. Resistance welding is by far the best in this respect because the energy is mainly generated at the high-resistance interface to be welded. However, it can be seen that the electron beam and laser are again in a class by themselves. So how do they compare with other processes in their performance characteristics and can they be distinguished from each other? In fact what sort of market expectation can one foresee for laser welding? Is it a gimmick or a gift? The main characteristics of the laser to bear in mind are listed in Table 4.3. The ways in which these characteristics compare for alternative processes are listed in Table 4.4.

Table 4.3 Main characteristics of laser welding

Characteristic	Comment
High energy density – "keyhole" type weld	Less distortion
High processing speed	Cost-effective (if fully employed)
Rapid start/stop	Unlike arc processes
Welds at atmospheric pressure	Unlike electron beam welding
No X-rays generated	Unlike electron beam
No filler required (autogenous weld)	No flux cleaning
Narrow weld	Less distortion
Relatively little HAZ	Can weld near heat-sensitive materials
Very accurate welding possible	Can weld thin to thick materials
Good weld bead profile	No clean-up necessary
No beam wander in magnetic field	Unlike electron beam
Little or no contamination	Depends only on gas shrouding
Relatively little evaporation loss of volatile components	Advantages with Mg and Li alloys
Difficult materials can sometimes be welded	General advantage
Relatively easy to automate	General feature of laser processing
Laser can be time-shared	General feature of laser processing

Table 4.4 Comparison of welding processes

Quality	Laser	Electron beam	TIG	Resistance	Ultrasonic
Rate	✓	✓	×	✓	×
Low heat input	✓	✓	×	✓	✓
Narrow HAZ	✓	✓	×		✓
Weld bead appearance	✓	✓	×		✓
Simple fixturing	✓	×	×		
Equipment reliability	✓		✓	✓	
Deep penetration	×	✓		×	
Welding in air	✓	×		✓	
Weld magnetic materials	✓	×	✓	✓	✓
Weld reflective material	×	✓	✓	✓	✓
Weld heat-sensitive material	✓	✓	×	×	✓
Joint access	✓			×	×
Environment, noise, fume	✓	✓	×	×	×
Equipment costs	×	×	✓		
Operating costs	–	–	–	–	–

✓ point of merit, × point of disadvantage

It can be seen from these tables that the laser has something special to offer as a high-speed, high-quality welding tool. One of the larger welding applications at present is that of "tailored blank" welding for the car industry (see Section 4.6); a process conceived in the 1980s and now accepted as the way to handle pressed products. The laser also has advantages in areas requiring the welding of heat-sensitive components such as heart pacemakers, pistons assembled with washers in place and thin diaphragms on larger frames. In 1995 a Ministry of International Trade and Industry study in Japan suggested that the laser was capable of doing some 25% of the then current industrial welding applications, whereas only 0.5% were actually being done by laser. Some adjustment has taken place since then, but there is still plenty of slack to be taken up. Thus, the potential for expansion was and is vast. As will be seen in this chapter, developments of laser welding have opened up the applications zone to thicker-section welding, as well as the thinner microwelding and new materials such as plastics, alloys and ceramics (see Table 4.8). Hand-held welding equipment is making laser welding accessible to the artistic community as well as the industrial community. The process has many superior qualities and hence expansion of the number of applications is almost certain; growth is currently held back by the need to train engineers in the use of optical energy.

4.2 Process Arrangement

As with laser cutting, welding relies on a finely focused beam to achieve the penetration. The only exception would be if the seam to be welded is difficult to track or of variable gap, in which case a wider beam would be easier and more reliable to use. However, in this case, once the beam has been defocused, the competition from plasma processes should then be considered. The general arrangement for laser welding is illustrated in Figures 4.2 and 4.3. Figure 4.3 illustrates the flexibility of the use of optical energy. It is in this area that laser users need to gain maturity. The advantages in welding, for example, a tube from the inside outwards is that inspection becomes straightforward, and thus considerable quality control costs might be saved. The optical arrangements possible for focusing a laser beam were discussed in Sections 2.8 and 2.9. Shrouding is a feature of all welding and the laser is no exception. However, shrouding is not difficult and coincides with the need to protect the optics from spatter. When welding high-reflectivity material, it is customary to tilt the workpiece by 5° or so, to avoid back reflections from entering the optics train and damaging O-rings or from being reflected

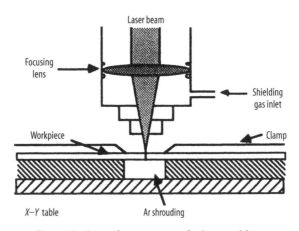

Figure 4.2 General arrangement for laser welding

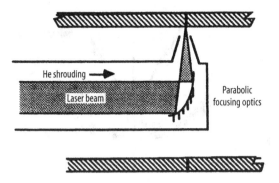

Figure 4.3 Arrangement for welding a pipe from the inside using metal optics

right back into the laser cavity and thus affecting the beam the instant it is to be used. Such feedback has an air of lack of control and is a threat to the output window of the laser. It might, however, be a good thing if properly controlled in that one might expect greater power when the reflectivity is high.

4.3 Process Mechanisms – Keyholes and Plasmas

There are two modes of welding with the laser illustrated in Figure 4.4. Conduction-limited welding occurs when the power density at a given welding speed is insufficient to cause boiling and therefore to generate a keyhole. The weld pool has strong stirring forces driven by Marangoni-type forces resulting from the variation in surface tension with temperature. This is discussed in more detail in Chapter 6. Most surface treatments in which melting occurs employ an out-of-focus beam, which results in conduction-limited weld beads. The alternative mode is "keyhole" welding, in which there is sufficient energy per unit length to cause evaporation and hence a hole in the melt pool. This hole is stabilised by the pressure from the vapour being generated. In some high-powered plasma welds there is an apparent hole, but this is mainly due to gas pressures from the plasma or cathode jet rather than from evaporation. The "keyhole" behaves like an optical black body in that the radiation enters the hole and is subject to multiple reflections before being able to escape (Figure 4.5). In consequence, nearly all the beam is absorbed. This can be both a blessing and a nuisance when welding high-reflectivity materials, since much power is needed to start the "keyhole", but as soon as it has started then the absorptivity jumps from 3 to 98% with possible damage to the weld structure.

There are two principal areas of interest in the mechanism of keyhole welding. The first is the flow structure since this directly affects the wave formation on the weld pool and hence the final frozen weld bead geometry. This geometry is a measure of weld quality. The second is the mechanism for absorption within the keyhole which may affect both this flow stability and entrapped porosity. The absorption of the beam is by Fresnel absorption (absorption during reflection from a surface) and inverse

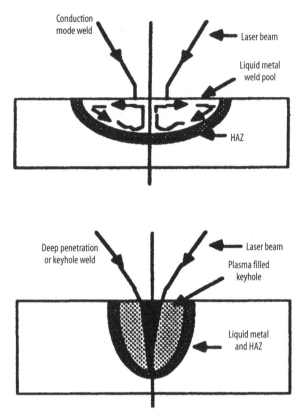

Figure 4.4 Conduction-limited and "keyhole"-type welds. *HAZ* heat-affected zone

bremsstrahlung leading to plasma re-radiation. The Fresnel absorption can be calculated for a given shape of the leading edge of the keyhole to be nonuniform [1]. The calculation must allow for the slope of the face, the mode structure of the original incident beam, polarisation effects and focal position. The plasma effects vary with polarisation and speed (see Section 4.4.3).

Some ingenious experiments have been done to visualise the keyhole. The most notable of these has come from the laser group in Osaka led by Arata and Matsunawa [2, 3]. Through these experiments high-speed videos of the keyhole entrance have been made – carefully illuminated by an argon laser and viewed through narrow band filters to avoid the glare from the plasma. These pictures show a roughly circular hole, with the dimensions of the focused beam diameter, very rapidly fluctuating in shape, pulsing in size and flapping from side to side. They also show that the plasma coming from the keyhole has two components. One is the metallic plasma from the boiling material, which fluctuates as directed by the shape of the keyhole, but tends to be directed backwards at slow speeds and more vertically at higher speeds. The other is the shroud gas plasma, which forms by interaction with the metal plasma. It is almost

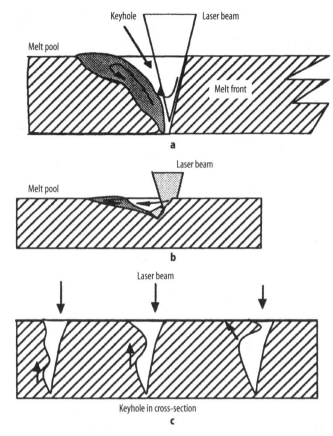

Figure 4.5 Approximate shape and flow pattern in laser welds

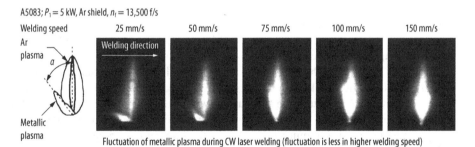

Figure 4.6 The variation of the metal and shroud gas plasmas from a laser keyhole weld. (Copyright 1996 and 2004, Laser Institute of America, Orlando, Florida, all rights reserved [70])

stationary relative to the laser beam but varies in intensity with the laser power and welding speed (see Figure 4.6).

Inside the keyhole, films have also been taken by a special microfocused X-ray system. The high-speed images were obtained using high-resolution fluorescent image in-

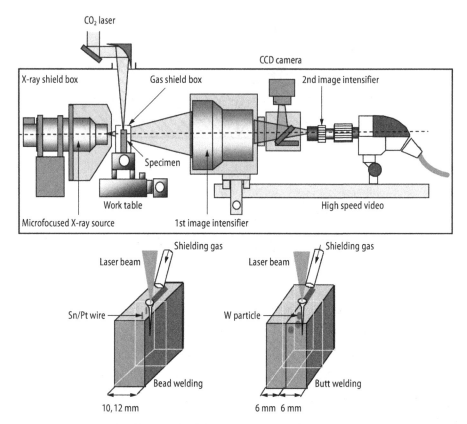

Figure 4.7 Equipment for filming a keyhole by X-rays as used by Osaka University. (Copyright 2003, Laser Institute of America, Orlando, Florida, all rights reserved [71])

tensifiers backed up with visible–visible image intensifiers (Figure 4.7). The Osaka team obtained images at the rate of 5,000 frames per second. They showed a hole, roughly illustrated in Figure 4.5. The hole walls are all in a high state of fluctuation with flow velocities up to 0.4 m s^{-1}. The thin melt on the leading edge appears to flow downwards with surface waves. Any hump on this surface would receive a higher intensity of the incident beam, causing it to evaporate, almost explosively, sending a jet of vapour into the rear melt pool, causing much oscillation of the pool and the potential for bubbles to be trapped as it solidifies. Flow in the pool itself has a vortex structure of one or two vortices which are rotating with considerable energy. This rotation was observed by introducing tungsten particles into the melt and observing their movement. Also a tin wire was placed inside the material and as soon as the weld struck it the tin flowed almost entirely over the pool, indicating intense stirring. This description is illustrated by the diagrams from Matsunawa's paper [3] (Figures 4.8, 4.9) and Arata's earlier work [2]. (Figure 4.10).

The vapour in the keyhole consists of very hot vapour from the material being welded together with shroud gas that has been sucked in owing to the pulsation of the

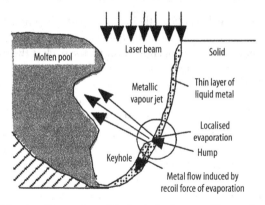

Figure 4.8 Side-view illustration of the keyhole shape and beam absorption [3]

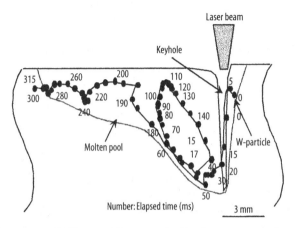

Figure 4.9 Liquid motion in CW laser welding as marked by the movement of a tungsten pellet [3]

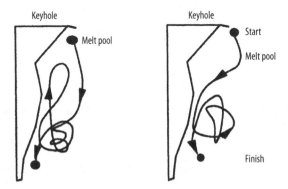

Figure 4.10 Flow in a "keyhole" weld mapped by a tungsten pellet [3]

keyhole. This vapour may be sufficiently hot to be partially ionised, forming a charged plasma. The flow of vapour out of the keyhole is fast, approaching sonic speeds, and hence it makes a snarling noise. This flow through the neck at the throat of the keyhole would be expected to show Helmholtz instabilities (similar to a flag flapping), with the higher velocities in the throat creating a low-pressure zone that would tend to close the throat. This may be one of the causes of the fluctuation noted, the other being the rapid fluctuation in fluid flow around the keyhole, driven by surface tension variations and sporadic boiling within the keyhole. The boiling reaction is very vigorous and causes a spray to form. This emerges as particles and dust. The temperature of the emerging vapour has been measured by Greses *et al.* [4] using a spectrographic technique. They showed that for Nd:YAG laser welding the vapour is not ionised and is only at a temperature of the order of 2,000 °C, whereas for CO_2 laser welding the vapour is ionised and is much hotter, at around 6,000–10,000 °C. The implication of these temperatures is that the composition of the shroud gas does not matter when welding with a Nd:YAG laser as much as it does with a CO_2 laser. This implication was confirmed experimentally by Greses *et al.* [4].

The charge in the CO_2 plasma is confirmed by the standing electric field around the plasma (see Section 12.2.3.3). However, the plasma or hot vapour in both cases interferes with the beam delivery. Figure 4.11 illustrates what can happen if there is no shroud gas to blow the plasma away when welding with 10 kW of CO_2 laser power. This blocking action can be caused either by absorbing the beam by inverse bremsstrahlung with the free electrons and re-radiating or by scattering in the particulate material being ejected or defocusing the beam in the steep refractive index gradients resulting from the steep thermal gradients in the gas.

Greses *et al.* [5] measured the particle sizes coming from the keyhole for both CO_2 and Nd:YAG laser welds made under the same power of 3.5 kW. They found that the CO_2 laser plume had the finer particles, with 93% below 100 nm and only 2% above 1 μm, whereas the Nd:YAG laser vapour had only 78% below 100 nm but 14% above 1 μm. Scattering by both Rayleigh and Mie scattering was calculated to be significant for Nd:YAG laser processing, whereas inverse bremsstrahlung absorption was higher with the CO_2 laser. Certainly Nd:YAG laser welding generates considerable dust compared with CO_2 laser welding.

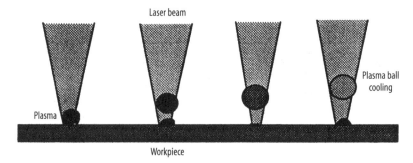

Figure 4.11 The blocking effect of the plasma if there is no side jet removing it

4.4 Operating Characteristics

The main process parameters are illustrated in Figure 4.12. They are as follows:

- Beam properties; power, pulsed or continuous; spot size and mode; polarisation; wavelength.
- Transport properties: speed; focal position; joint geometries; gap tolerance.
- Shroud gas properties: composition; shroud design; pressure/velocity.
- Material properties: composition; surface condition.

4.4.1 Power

4.4.1.1 Effect of Continuous Power

There are two main problems in welding: lack of penetration and the inverse "dropout". These are the boundaries for a good weld made with a given power as illustrated in Figure 4.13 [6]. The maximum welding speed for a given thickness rises with an increase in power. The fall-off shown at the higher power level of 2 kW is almost certainly due to the poorer mode structure given by most lasers when working at their peak power. However, from the results in Figure 4.14 [6] for higher power levels up to 5 kW, the fall-off may now be due to the same cause but may also be from plasma effects. The

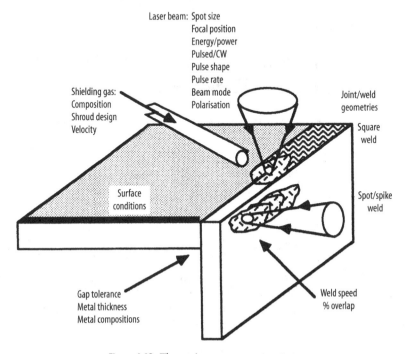

Figure 4.12 The main process parameters

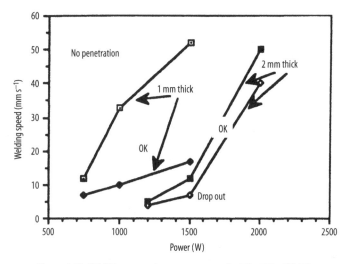

Figure 4.13 Welding speed versus power for Ti–6Al–4V [6]

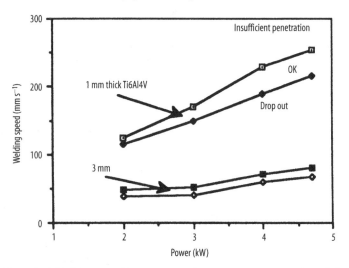

Figure 4.14 Welding speed versus power for a Laser Ecosse CL5 laser [6]

main point to note from these two graphs is that for more power the operating window is larger.

For high speeds the effects of sideways conduction during melting are slight and hence the Bessel functions discussed in the Swifthook and Gick model (Section 5.7) become soluble and an equation similar to that derived for cutting results is obtained. That is,

$$Y = 0.483X,\qquad(4.1)$$

in which,

$$Y = \frac{2vR}{\alpha} \quad \text{and} \quad X = \frac{P}{kgT},$$

where $2R = w$ is the weld width (m), α is the thermal diffusivity, *i.e.*, $k/\rho C_p$ (m^2 s^{-1}), g is the thickness (m), T is the temperature (K), P is the absorbed power, which is equal to $P(1 - r_f)$ (W), T_m is the melting point (K), r_f is the reflectivity and v is the welding speed (m s^{-1}).

Thus, we have

$$0.483P(1 - r_f) = vwg\rho C_p T_m . \tag{4.2}$$

This is a form of the lumped heat capacity model seen previously for cutting. This simple model for the maximum welding speed has neglected latent heat. What it has calculated is that 51.7% (*i.e.*, 1 – 0.483) of the delivered energy is lost to conduction outside the melting isotherm. A further 25% approximately will be used for latent heat effects and the resulting melting efficiency will be of the order of 23%, as observed. It has also been assumed that the power is distributed as a line source along the beam axis, the ultimate in fine focusing. In spite of these assumptions, the parametric relationships are enshrined in this equation. Incidentally, this formula would act as a useful rule of thumb to find out what welding speed should be possible for a given laser power if the beam is very finely focused. It is more accurate if the constant is taken as 0.25–0.3 as opposed to 0.483.

Penetration is inversely proportional to the speed for a given mode, focal spot size and power, as shown in Figure 4.15 [7] and expected from Equation 4.2.

The extent of the HAZ is a function of the welding speed. As an estimate it could be assumed that the edge of the heat wave causing the HAZ is when the Fourier number is 1 (see Section 5.9).

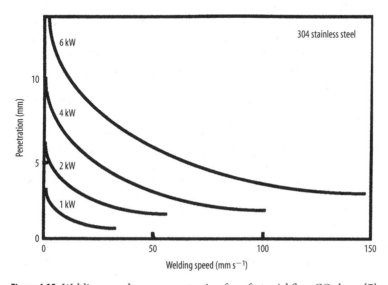

Figure 4.15 Welding speed versus penetration for a fast axial flow CO$_2$ laser [7]

Thus,

$$F = \frac{\alpha t}{x^2} = 1,$$

and hence

$$x = \sqrt{\alpha t} = \sqrt{\frac{\alpha D}{v}},$$

where x is the extent of the HAZ (m), α is the thermal diffusivity of material being welded (m^2 s^{-1}), t is the interaction time for conduction (s), D is the effective beam diameter on the weld pool (m) and v is the welding speed (m s^{-1}).

For example, if $D = 4$ mm and the thermal diffusivity of steel is 1.4×10^{-5} m^2 s^{-1}, then the HAZ is $(5.5/v)^{1/2}$: for a speed of 10 cm s^{-1} this would give a HAZ of 235 μm; for aluminium the value would be 600 μm.

4.4.1.2 Pulsed Power

The use of pulsed power introduces two more variables to be considered: pulse repetition frequency and percentage overlap. The welding speed is decided by the spot size × pulse repetition frequency ×(1 − overlap fraction). In fact, speed is independent of power. Penetration is a function of power and hence if the peak power is raised by pulsing or modulating the beam, as in superpulsing or hyperpulsing (see Figure 3.36), there can be greater penetration for a given average power. This effect can be more marked than expected. For example, for pulsing with a square pulse form at 100–500 Hz and a peak power of twice the average power, an improvement of 30% penetration has been reported when welding 304 stainless steel [8]. The increased peak power also means better welding of reflective material, since the keyhole is initiated more quickly. With use of a 1 kW average power Nd:YAG laser at 500 Hz and 2 kW peak power, 6181 aluminium was welded at nearly 3 times the speed of CW welding at 1 kW or with an improvement of 60% in penetration. Higher peak power also means greater tolerance to focal position (Figure 4.16) and pulsing means less energy is deposited in the workpiece, leading to reduced distortion.

These advantages for pulsing have been noted by others, with the further advantage that pulsing allows better control over the flow in the weld pool and can under the correct conditions reduce the formation of pores. Holtz [9] reported that a fast ripple (25 MHz) in the power during a shaped laser pulse leads to less spatter and smoother bead shapes. Pulsing was able at the correct speed and pulse rate to allow porosity-free lap welding of zinc-coated steel, as discussed in Section 4.4.7 [10]. It was also able to reduce the porosity in thick-section welding with a 20 kW laser beam by pulsing at the oscillation frequency of the weld pool, leading to the resonance effect illustrated in Figure 4.17 [11].

For welding the pulse is usually longer than for drilling and is shaped to have a smaller initial peak. The general operating areas are shown in Figure 4.18 [12].

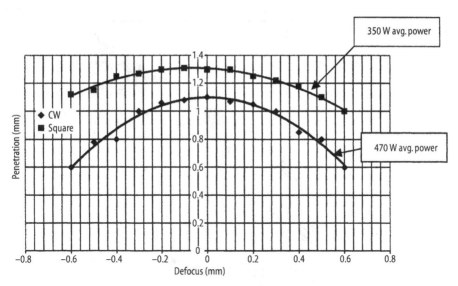

Figure 4.16 Variation of the penetration and depth of focus achievable with pulsing using a spot size of 0.3 mm on 304 stainless steel [8]

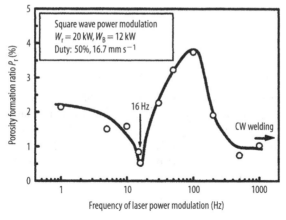

Figure 4.17 Effect of power modulation on porosity in 20 mm deep welding with 20 kW peak pulses from a CO_2 laser [11]

4.4.2 Spot Size and Mode

Spot size together with the power determines the power density and hence the penetration and ability to treat reflective materials. A low-order-mode structure, such as TEM00, can be more finely focused than higher-order modes. The fibre laser has a mode structure that can be more finely focused than the more traditional lasers and hence the penetration for a given power is better and the welding speed can be 50–100% faster with a fibre laser than with a Nd:YAG laser [13]. However, penetration and speed are not all the story. A 100 W Nd:YAG pulsed laser can weld reflective materials such

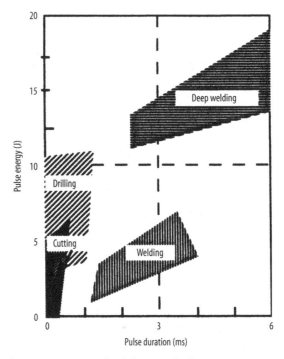

Figure 4.18 General operating regimes for different pulsed processes using a Nd:YAG laser [9]

as aluminium and copper alloys, whereas a 100 W fibre laser, which does not have the power enhancement of pulsing, is unable to do so [14]. There is another consideration; if the spot is very fine, the weld seam following has to be very accurate. Hence, many processes require the beam to be slightly broader than the minimum possible. This balance between penetration and weld width can be satisfied, at the expense of some speed, by dithering the beam, which can be done in various ways, such as spinning or moving in jerks as in walking.

4.4.3 Polarisation

At first sight one might think that polarisation will have no effect on laser welding since the beam is absorbed inside a keyhole and hence it will be absorbed regardless of the plane of polarisation. Note this is quite unlike cutting, where all the absorption has to take place on a steeply sloped cut front. This thought would be in essence correct, but some second-order events have been noted by Beyer *et al.* [1]. Figure 4.19 shows the slight variation in penetration thought to be due to polarisation effects. The resulting weld fusion zones are also wider for the case of s-polarisation (perpendicular to the plane of incidence) as expected since in this case the main absorption would be at the sides. The argument suggested for this phenomenon is that there are two absorption mechanisms. At slow speeds the plasma absorption dominates and the beam is

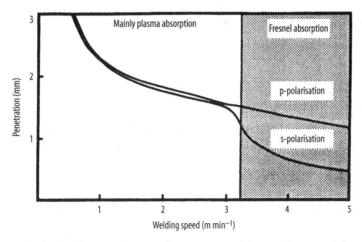

Figure 4.19 Influence of beam polarisation on welding performance [1]

absorbed by inverse bremsstrahlung effects in the keyhole generating a plasma which appears blue in argon-shrouded systems. As the speed increases, the Fresnel absorption (absorption by reflection on front face) gains in importance owing to the cooler plasma being less absorbing. However, no polarisation effects were noted with aluminium. This is still a puzzle and throws some questions on the whole theory.

4.4.4 Wavelength

Owing to the high absorptivity within the "keyhole" there is little operational difference when welding with long or short wavelengths. When welding with a conduction-limited weld, the surface reflectivity becomes paramount and the lower reflectivity with the shorter wavelengths gives a distinct advantage to excimer, Nd:YAG or CO lasers over the CO_2 laser.

However, there is another factor affecting absorption and that is the plasma formed owing to the very hot gases coming from the keyhole, as noted in Section 4.3. The plasma will have three effects:

1. Firstly, the electrons in the plasma are free to absorb photons and hence the plasma is going to block the beam depending upon the electron density. *i.e.*, the temperature.
2. Secondly, the hot plasma will cause density changes and hence changes in the refractive index, which will disperse the beam – a hot ball will behave like a concave lens.
3. Thirdly, there will be condensate and particles caught up in the fast-moving plasma, which will have associated scattering effects.

All of these events depend to some extent on the absorption which sustains the plasma. This is dominated by the inverse bremsstrahlung process [15, 16]. The absorption coefficient, α_{bremm}, due to inverse bremsstrahlung, was deduced by Raizer [17] to be

$$\alpha_{\text{bremm}} \approx n^2 \lambda^2 T^{3/2} \, ,$$

where n is the gas density, λ is the wavelength and T is the absolute temperature of the plasma.

Thus, for shorter wavelengths there will be less absorption and hence cooler and less absorbing plasma as found by Greses *et al.* [5]. This gives a significant advantage to shorter-wavelength lasers for welding and other plasma-generating processes. The penetration depth possible with a CO laser working at 5.4 μm is greater than that for a CO_2 laser simply because the CO laser can weld more slowly without suffering plasma-blocking problems [18]. The welding of volatile materials such as certain magnesium alloys and aluminium alloys appears to have reduced porosity when welded with Nd:YAG radiation compared with CO_2 radiation, and there is a reduced "nail head" to the weld ingot. Since shorter wavelengths are not so sensitive to the ionisation potential of the shroud gas, they can use cheaper gases.

4.4.5 Speed

The effect of speed on the welding process is principally described by the overall heat balance equation noted in Equation 4.2. However, in addition to these main effects there are some others. Firstly, there is the effect of speed on the weld bead and, secondly, there is the problem of shrouding high-speed welds.

4.4.5.1 Effect of Speed on the Weld Pool and Weld Bead Shape

As the speed increases so will the pool flow pattern and size change. In general, the flow in a laser keyhole weld pool is shown in Figures 4.4, 4.5, 4.8 and 4.9. At slow speeds the pool is large and wide and may result in dropout (Figure 4.20d). In this case the ferrostatic head is too large for the surface tension to keep the pool in place and so it drops out of the weld, leaving a hole or depression. This was described in detail by Matsunawa [19–21]. At higher speeds, the strong flow towards the centre of the weld in the wake of the keyhole has no time to redistribute and is hence frozen as a central ridge and an undercut at the sides of the weld, shown diagrammatically in Figure 4.20b. Also if the power is high enough and the pool large enough, then the same undercut occurs but the thread of the pool in the centre has a pressure which is a function of its surface tension and curvature [19–21]. This leads to pressure instability along the length of

a   b   c   d

Figure 4.20 Range of weld shapes varying usually with speed: **a** normal/good, **b** undercut, **c** humping (longitudinal section), and **d** dropout

the pool, causing the "pinch" effect in which those regions of high curvature flow to regions of lower curvature, resulting in large humps (Figure 4.20c). The pressure, p, in these regions would vary by

$$p = \gamma/r^2,\qquad(4.3)$$

where γ is the surface tension and r is the radius of curvature.

There is an intermediate region in which there is a partial undercutting and central string. All this has been mapped for certain alloys by Albright and Chiang [22] as shown in Figure 4.21.

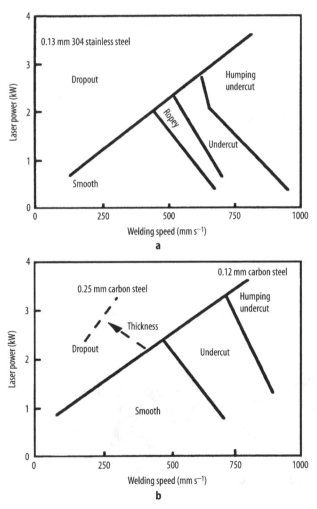

Figure 4.21 Map of weld bead profiles as functions of welding speed and laser power: **a** 0.12-mm-thick stainless steel, and **b** 0.12- and 0.25-mm-thick mild steel [22]

4.4.5.2 Effect of Speed on Shroud Arrangements

The faster the welding process, the shorter the weld pool. However, with increased speed the hot metal extends further beyond the welding point. Thus, trailing shrouds are usually needed to avoid atmospheric contamination.

4.4.6 Focal Position

There are suggestions [6, 23, 24] that the focal point should be located within the workpiece to a depth of around 1 mm for maximum penetration. What one should consider here is the need to have sufficient power density to generate a "keyhole" and then for that power to stay together within the keyhole to increase the penetration. Thus, the main parameters to consider would be the depth of focus and the minimum spot size. It was shown in Section 2.8.2 that the depth of focus, z_f, is given by

$$z_f = \pm 2.56 F^2 M^2 \lambda$$
$$\approx 50 F^2 \ \mu m \text{ for } M^2 = 2 \text{ and } \lambda = 10.6 \ \mu m \,. \tag{4.4}$$

And the minimum spot size, d_{min}, for a multimode CO_2 beam is given by

$$d_{min} = 2.4 F M^2 \lambda \,; \tag{4.5}$$

therefore,

$$d_{min} = 50 F \ \mu m \text{ for } M^2 = 2 \text{ and } \lambda = 10.6 \ \mu m \,. \tag{4.6}$$

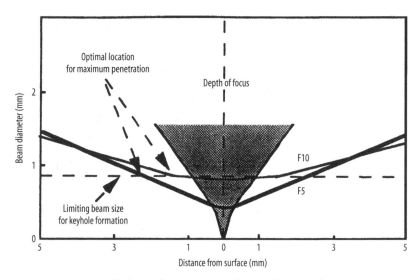

Figure 4.22 Beam diameter versus distance from the focus

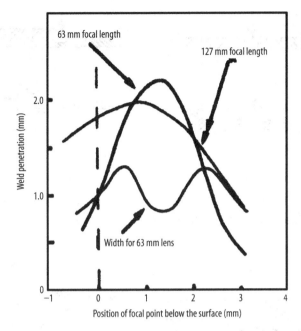

Figure 4.23 Effect of focal position on weld penetration for 1018 steel [24]

Figure 4.22 shows the beam diameter versus distance from the lens focus for two F numbers. The shaded area shows the parabolic relationship between the depth of focus, z_f, and F number and the minimum beam diameter, d_{min}. A certain power density, P/d_{min} or P/d_{min}^2, is required to form a "keyhole" for a given traverse speed. This is marked in Figure 4.22 by the horizontal line by way of illustration. From this analysis it can be seen that the optimal position of the focus for maximum penetration varies with F number as shown in Figures 4.22 and 4.23.

4.4.7 Joint Geometries

4.4.7.1 Joint Arrangements

Laser beams causing keyhole-type welds prefer a joint which helps the absorption and hence the formation of the keyhole. High-intensity welding processes are not sensitive to different thicknesses of the pieces to be joined. This allows some new types of joint to be considered. Figure 4.24 shows some of the variations which can be considered.

The flare weld was used by Sepold *et al.* [25] for very high speed welding of two strips, at speeds up to $4\,\mathrm{m\,s^{-1}}$, and is currently used for making seam welds in thick-section pipe. The plane of polarisation must be correct in this mode of welding or the beam will be absorbed before being reflected down to the point of the joint. It is, of course a very efficient joining technique. The "T" weld geometry has a surprise attached to it, in that as the keyhole penetrates at an angle into the workpiece it tends to turn upwards

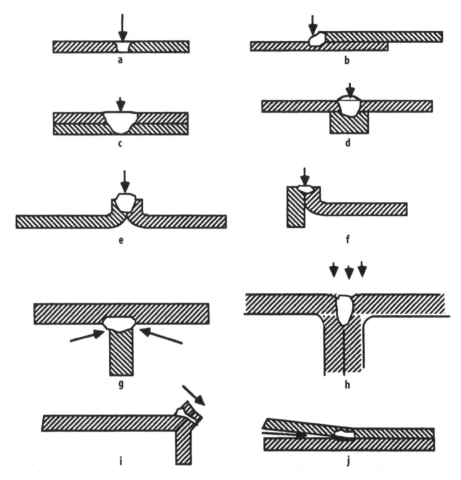

Figure 4.24 Various welding joint arrangements: **a** butt joint, **b** fillet or lap joint, **c** spot or lap weld, **d** spike or spot weld, **e** flange joint, **f** edge joint, **g** T-joint, **h** flare weld, **i** corner, and **j** kissing or flare weld

to allow full penetration around the base of the T. This very convenient event is the result of the reduced thermal load on the T side of the keyhole, which encourages the melting isotherm that way.

For a fully restrained weld, as in welding a cylinder into a hole, there is a tendency for cracking to occur if the weld is not fully penetrating owing to the stress raiser from the unwelded root. This can be overcome by using more power or welding more slowly, or redesigning the joint so that it is not so deep by flanging the underpart [26].

4.4.7.2 Effect of Gap

In butt joints the gap must be small enough that the beam cannot pass straight through the joint; that is, the gap should be smaller than half the beam diameter (less than

200 µm). For welds where there is a large gap the beam is sometimes rotated by rotating the lens off-axis from the beam. However, in these cases there is a chance of
some dropout or an underfill in the weld. This can be corrected by adding filler material as a wire [27] or as a powder [28, 29]. On the whole, laser welds do not require
filler material, they are "autogenous". One might question how this is possible when the
conservation of mass suggests that if there is a gap, then there will be a fall in the level
of the weld. In practice there is usually a rise in the level! This is due to the stresses in
the cooling weldment drawing the workpieces together and so squashing the melt pool.
Thus, a small gap can be tolerated without any filler.

The extent of the squeeze is proportional to the forces, which are in turn proportional to the contraction of the cooling weld. Thus, the gap which can be tolerated, g,
is approximately given by the following relationships [29]:

For butt welds $A\beta\Delta Twt_p = gt_p$. So

$$g = A\beta\Delta Tw, \tag{4.7}$$

where β is the coefficient of thermal expansion $(°C^{-1})$, ΔT is the temperature change,
the approximate melting point $(°C)$, w is the weld width (m), t_p is the sheet thickness
(m), g is the gap width (m) and A is a constant.

For lap welds (overlapping plates with the gap between the plates) $B\beta\Delta Tw2t_p = gw$.
So

$$g = B\beta\Delta T2t_p, \tag{4.8}$$

where B is a constant.

Welding with a gap in lap welding is essential if one is welding zinc-coated steel
or other volatile material situated between the lapped sheets. The only exception is to
weld with a carefully controlled pulse rate [30] such that the pores generated in one
pulse are removed by the next. The problem with volatile mid layers is the need to vent
the high-pressure vapours that will be generated. Zinc boils at 906 °C and steel melts
at 1,500 °C; the keyhole is even hotter. So as the keyhole enters the interlayer of zinc
there is a sudden evolution of vapour, which will destroy the weld continuity unless it
is vented. Akhter and Steen [31] have calculated the required size of the gap to allow
this venting. The situation, which is modelled, is shown in Figure 4.24a and b.

The volume of vapour generated per second at the interface, φ_{gen}, is

$$\varphi_{gen} = \frac{(w + 2b)Vt_{zn}\rho_s}{\rho_v}, \tag{4.9}$$

where V is the welding speed $(m\,s^{-1})$ and ρ_s and ρ_v are the densities $(kg\,m^{-3})$ of the
solid and the vapour.

The vapour escapes as it forms around the melt pool at velocity v_2. The volume escaping per second through the gap between the plates, φ_{esc}, is

$$\varphi_{esc} = \frac{v_2\pi(w + 2b)g}{2} \quad m^3\,s^{-1}. \tag{4.10}$$

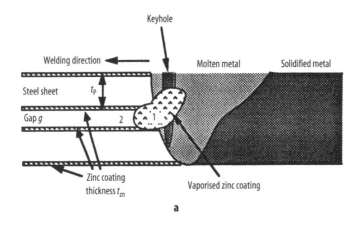

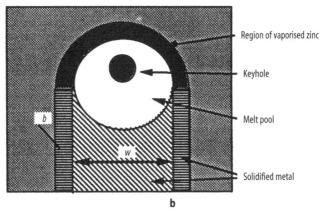

Figure 4.25 The welding of zinc-coated steel with a small gap between the sheets for the exhausting of the high-pressure zinc vapour generated during welding [31]: **a** side view, and **b** plan view [29]

The maximum velocity that can be exhausted through this gap is thus found by equating Equations 4.9 and 4.10:

$$v_2 = \frac{2Vt_{zn}\rho_s}{\pi g \rho_v} \quad \text{m s}^{-1}. \tag{4.11}$$

This escape velocity can only be achieved with an acceleration pressure. This pressure must not exceed the pressure of the ferrostatic head of the weld pool – $\rho_L g_c t_p$ (where g_c is the gravitational acceleration) – or the vapour would be expelled through the pool and destroy the weld quality. This velocity is given approximately by the Bernoulli relationship:

$$v_2 = \sqrt{\frac{2\Delta P_{12}}{\rho_v}} = \sqrt{\frac{2\rho_L g_c t_p}{\rho_v}}. \tag{4.12}$$

Thus, by eliminating v_2 between Equations 4.11 and 4.12, we have a relationship for the minimum value of the gap required to exhaust the vapour for sound welding of zinc-coated steel:

$$g_{min} \geq \frac{2t_{zn}V\rho_s}{\pi\sqrt{2\rho_v\rho_L g_c t_p}} . \qquad (4.13)$$

If the gap is smaller than this value, then some blowout is to be expected. A method for controlling the gap in production was suggested by Akhter and Steen [31] by dimpling the weld line.

However, the maximum value of the gap, if dropout is to be avoided, was given by Equation 4.8. Therefore, we have the range of gap values

$$B\beta\Delta T 2t_p \geq g \geq \frac{2t_{zn}V\rho_s}{\pi\sqrt{2\rho_v\rho_L g_c t_p}} .$$

Dividing throughout by t_p gives the relationship

$$2B\beta\Delta T \geq \frac{g}{t_p} \geq \frac{2t_{zn}V\rho_s}{\pi\sqrt{2\rho_v\rho_L g_c t_p^2}} .$$

This relationship is plotted on an operational diagram of the process in Figure 4.26, in which the constant B is

$$B = \frac{2\rho_s}{\pi\sqrt{2\rho_v\rho_L g_c}} = 1 .$$

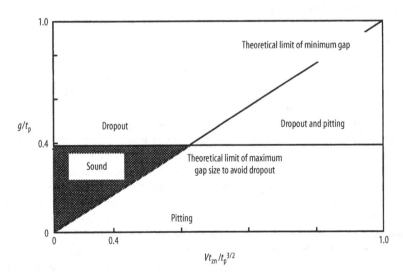

Figure 4.26 Operational diagram for the welding of zinc-coated mild steel with a gap for zinc vapour exhaust [29]

From Equation 4.2 we find that

$$\frac{P}{Vt_p w} = f(\text{material});$$

hence,

$$V = \frac{CP}{t_p w}.$$

From Figure 4.26 there is a maximum value to the function $Vt_{zn}/t^{3/2}$ and hence a maximum value to $Pt_{zn}/wt_p^{5/2}$. This suggests that there is a maximum value of the laser power above which it is impossible to weld zinc-coated steel. Putting in approximate values, this power is of the order of 5 kW. This means that above this value there is no gap large enough to exhaust the rate of vapour generation that will also be small enough to avoid dropout.

This problem with lap welding zinc-coated steels, which is of great interest to the car industry, can also be solved by careful control of the pulse rate and speed when welding with a Nd:YAG laser. Tzeng [32] and Katayama *et al.* [30] were able to lap-weld tightly clamped zinc-coated steel with porosity-free welds this way. Tzeng considered his process to be due to the overlap of pulses moving the porosity in a form of zone refining. Similar results have been reported using multiple Nd:YAG laser beams focused through a single lens [33]. Zero Gap Welding with alloying elements in between the layers of a lap configuration also provided defect and porosity free welding (Dasgupta *et al.* U.S. Patent # 6,479,168).

4.4.8 Gas Shroud and Gas Pressure

4.4.8.1 Shroud Composition

The gas shroud, which is necessary for the protection of the molten and cooling weld, can affect the formation of plasma, which may block or distort the beam and thus the absorption of the beam into the workpiece.

The formation of plasma is thought to occur through the reaction of the hot metal vapours from the keyhole with the shroud gas. It seems unlikely, in view of the fast emission of vapour from the keyhole, that the shroud gas could enter the keyhole; nevertheless it does so owing to the pulsating collapse and formation of the keyhole. The plasma formed above the keyhole with the shroud gas will be absorbing to an extent determined by the temperature and the ionisation potential of the gases involved. Table 4.5 lists the ionisation potential of the gases often encountered in laser processing. The reduced absorption of shorter-wavelength radiation was noted in Section 4.4.4.

Welding steel with both a CO_2 laser and a Nd:YAG laser, each at 3.5 kW, Greses *et al.* [4] showed that the welding plume for the Nd:YAG laser was simply a hot, dusty gas discharge at around 2,000 °C, whereas the plume for CO_2 laser welding was a plasma at around 6,000–10,000 °C. Thus, the ionisation potential of the shroud gas

Table 4.5 Ionisation potential of common gases and metals [72]

Material	First ionisation potential (eV)	Material	First ionisation potential (eV)
Helium	24.46	Aluminium	5.96
Argon	15.68	Chromium	6.74
Neon	15.54	Nickel	7.61
Carbon dioxide	14.41	Iron	7.83
Water vapour	12.56	Magnesium	7.61
Oxygen	12.50	Manganese	7.41

was not relevant to Nd:YAG laser welding at these powers. This is a very interesting piece of work since it means that Nd:YAG laser welding can use considerably cheaper shroud gas than can be used for CO_2 laser welding.

In welding with CO_2 lasers, the plasma-blocking effect will be less for those gases having a high ionisation potential; thus, helium is favoured, in spite of its price, as the top shroud gas in laser welding. The shroud underneath the weld could be of a cheaper gas, *e.g.*, argon, N_2 or CO_2. The difference in penetration can be significant. The variation in penetration with shroud gas composition and laser power for 10.6 μm radiation is shown in the results from RTM, Italy (Figure 4.27). The plasma blocking is greater with higher powers. The results of Alexander and Steen [34] (Figure 4.28) give these data a new slant. At slow speeds there is an advantage for helium but at high speeds there is an advantage for argon. The explanation is that the plasma is both good and bad in aiding absorption. If the plasma is near the workpiece surface or in the keyhole,

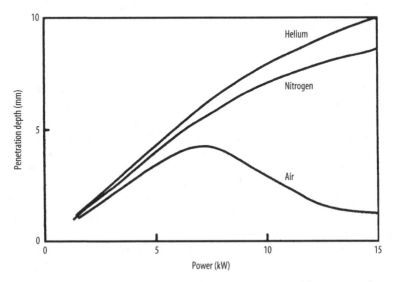

Figure 4.27 Variation in penetration with shroud gas composition and laser power for 10.6 μm radiation [34]

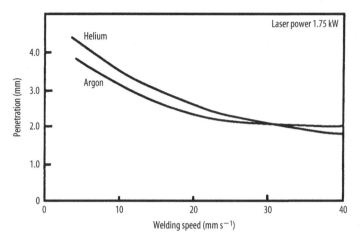

Figure 4.28 Penetration versus speed for helium and argon shroud gases [34]

it is beneficial (as in high-speed results of Alexander and Steen [34]). If, however, it is allowed to become thick or leave the surface, its effect is to block or disperse the beam. This effect of speed on the preferred gas composition was also noted by Seaman [24]. Because of this plasma effect it is usual to weld with a side-blown jet to help blow the plasma away.

If the shroud gas is slightly reactive with the weld metal then a thin film of, say, oxide may form which will enhance the optical coupling. The work of Jørgensen [35] shows there was greater penetration when the shroud gas contained 10% oxygen (Figure 4.29). This may, of course, be unacceptable for some welds, but it is worth noting.

4.4.8.2 Effect of Shroud Design

The shroud design must give total coverage of the melt and the reactive hot region of the weld. It must do so without having flow rates which may cause waves on the weld pool. As just noted, in welding a side jet is often added to blow the plasma away. In the case of welding zinc-coated steel, if the side jet blows backward along the new weld-ment, the zinc vapour will condense on the weldment and so enhance the corrosion protection [36]. The side jet can also be used to feed powder filler into the weld. For high-speed welds the shroud will need to have a trailing section. An interesting design invented by The Welding Institute [37] was a plasma disruption jet. It is illustrated in Figure 4.30. The concept is that if the fine 45° jet is correctly located, it will blow the plasma back into the keyhole and hence enhance the absorption. The welding performance is shown in Figure 4.31. The main benefits are for thicker-section welding. The weld fusion zone is altered to be more nearly parallel and the "nail head" can be eliminated by this process.

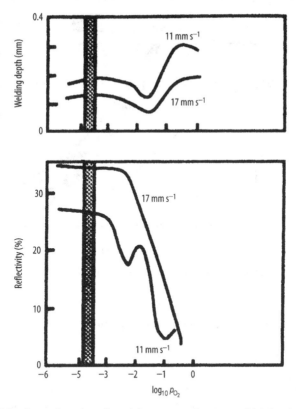

Figure 4.29 Weld depth as a function of partial pressure of oxygen, which is related to the reflectivity. (After Jørgensen [35])

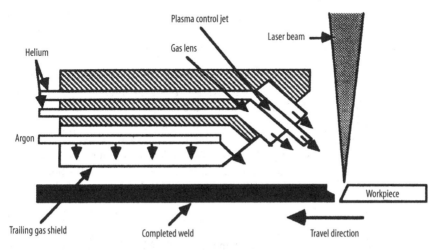

Figure 4.30 The design of a plasma disruption jet with trailing shroud [37]

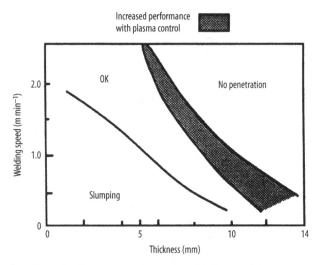

Figure 4.31 The effect of a plasma control jet for a 6-kW laser welding 18.8 stainless steel

4.4.9 Effect of Gas Pressure – Due to Velocity and Environment

4.4.9.1 Velocity Effects

The nozzle pressure affects the gas flow rate and hence the ability of the gas to either blow the plasma away or correctly protect the weld. There is a minimum flow rate for adequate protection and also one for the removal of the plasma. There is a maximum rate above which the weld pool flow is affected and the melt pool is ruffled, causing a poor bead.

4.4.9.2 Low-pressure Environment

Variation in the environment pressure has a dramatic effect on penetration, particularly at very low pressures (the results are shown schematically in Figure 4.32). This means that the penetration of the laser and that of the electron beam are not too dissimilar. The electron beam is by necessity working in a high vacuum and hence enjoys high penetration. There are two theories as to why this increase in penetration occurs. The first is that the lower pressure reduces the plasma density and hence the plasma is no longer blocking the beam. The second is that at the lower pressures the boiling point is reduced in a manner predicted by the Clapeyron–Clausius equation [38]:

$$\mathrm{d}p/\mathrm{d}T = \Delta H/T\Delta V. \tag{4.14}$$

Since the change in volume with the change in phase, ΔV, is negligible with melting, as opposed to boiling, there is not a similar effect on the melting point. Thus, the melting point and the boiling point become closer together at the lower pressures and hence the wall thickness of the keyhole will be thinner. A thinner liquid wall is easier to maintain

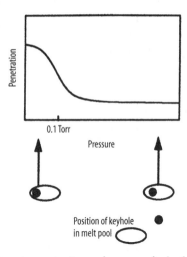

Figure 4.32 Relationship between penetration and pressure for both electron beams and lasers

and hence the keyhole is more easily stabilised. It is this stability which allows greater penetration. Arata [2] made a film illustrating the reduced plasma; it also shows the keyhole moving forward as illustrated in Figure 4.32. To separate these two theories is very difficult and will test people's imagination for some time yet.

4.4.9.3 Hyperbaric Pressures

Laser welding under water can be done up to depths of 500 m or so. The higher pressure intensifies the plasma formed and a side jet is needed. The penetration depth is reduced considerably by the increased pressure, of the order of 50% for higher pressures. There is also a greater tendency to form pores within the weld. There is little difference in the performance between 10 and 50 bar pressure.

4.4.10 Effect of Material Properties

The main material problems with laser welding, in common with most welding methods, are crack sensitivity, porosity, HAZ embrittlement and poor absorption of the radiation. For welds of dissimilar metals there is the additional problem of the possible formation of brittle intermetallics. Crack sensitivity refers to centreline cracking, hot cracking or liquation cracking. It is due to the shrinkage stress building up before the weld has fully solidified and is strong enough to take the stress. It is thus most likely in metal alloys having a wide temperature range over which solidification occurs, *e.g.*, those with high carbon, sulphur and phosphorous contents. Some alloys listed in order of crack sensitivity are given in Table 4.6. Cracking can be reduced or eliminated by using a high pulse rate, adding a filler or using preheat.

Porosity often results when welding material subject to volatilisation, such as brass, zinc-coated steel, Al/Li alloys or magnesium alloys. It may also be caused by a chemical

Table 4.6 Crack sensitivity rating of certain metals [7]

Material	Crack tendency	Composition								
		C	Si	Mn	Cu	Fe	Ni	Cr	Mo	Other
HASTELLOY® B2	High	0.12	1.0	1.0		4–6	Rem	1.0	26	V, Co
HASTELLOY® C4		0.12	1.0	1.0		4.5–7	Rem	15	16	V, Co
INCONEL®[1] 600		0.08	0.25	0.5	0.25	8.0	Rem	15.5	–	Al
INCONEL® 718		<0.08	–	–	0.15	18.5	52.5	19	3	Nb, Ti, Al
316 stainless steel		0.08	1.5–3	2.0		Rem	19–22	23–26		
310 stainless steel		0.25	1.5	2.0		Rem	19–22	24–26		
HASTELLOY® X		0.15				15.8	49	22	9	Co, W, Al
330 stainless steel		0.08	0.7–1.5	2.0	1.0	Rem	34–37	17–20		
Aluminium						Mg	Al			
2024	Low		0.6		4.4	1.5	Rem			

[1] INCONEL® is a registered trademark of Special Metals Corporation, New Hartford, New York, USA. http://www.specialmetals.com

Rem remainder

reaction in the melt pool, as with welding rimming steel, or melting with inadequate shrouding of metals such as spheroidal graphite cast iron. It may also be present in metals having a high dissolved gas content such as some aluminium alloys. Control may be achieved with proper attention to the shrouding system, adding a "killing" agent such as aluminium to rimming steel or controlling the pulse rate or spot size. The problems with root porosity were in Section 4.3.

Table 4.7 Laser welding characteristics for different alloy systems

Alloy	Notes
Aluminium alloys	Problems with: 1. reflectivity – requires at least 1 kW; 2. porosity; and 3. excessive fluidity – leading to dropout.
Steels	OK
Heat-resistant alloys, *e.g.*, INCONEL® 718, Jetehet M152, HASTELLOY®	OK but: 1. weld is liable to brittle; 2. segregation problems; and 3. cracking.
Titanium alloys	Better than slower processes owing to grain growth
Iridium alloys	Problem with hot cracking

Table 4.8 Laser weldability of dissimilar metal combinations [7]

	W	Ta	Mo	Cr	Co	Ti	Be	Fe	Pt	Ni	Pd	Cu	Au	Ag	Mg	Al	Zn	Cd	Pb	Sn
W																				
Ta	■																			
Mo		■																		
Cr		P	■																	
Co	F	P	F	G	■															
Ti	F			G	F	■														
Be	P	P	P	P	F	P	■													
Fe	F	F	G			F	F	■												
Pt	G	F	G	G		F	P	G	■											
Ni	F	G	F	G		F	F	G		■										
Pd	F	G	G	G		F	F	G			■									
Cu	P	P	P	P	F	F	F	F				■								
Au	–	–	P	F	P	F	F	F					■							
Ag	P	P	P	P	P	F	P	P	F	P		F		■						
Mg	P	–	P	P	P	P	P	P	P	P	P	F	F	F	■					
Al	P	P	P	P	F	F	P	F	P	F	P	F	F	F	F	■				
Zn	P	–	P	P	F	P	P	F	P	F	F	G	F	G	P	F	■			
Cd	–	–	–	P	P	P	–	P	F	F	F	P	F	G	■	P	P	■		
Pb	P	–	P	P	P	P	–	P	P	P	P	P	P	P	P	P	P	P	■	
Sn	P	P	P	P	P	P	P	P	F	P	F	P	F	F	P	P	P	P	F	■

■	Excellent
G	Good
F	Fair
P	Poor

The main advantages of laser welding are that it is a process having a very low hydrogen potential (which in other processes may cause hydrogen embrittlement), it gives less tendency to liquation cracking owing to the reduced time for segregation and it causes less distortion owing to the smaller pool size. Table 4.7 provides a summary of some of the laser welding characteristics for the main alloy systems.

The welding of dissimilar metals is only possible for certain combinations as shown in Table 4.8. Owing to the small fusion zone and relatively rapid solidification of the weld there is a greater range of welds possible with the laser than with slower processes. There is also a greater tendency to form metastable solid solutions, as discussed in Chapter 6.

4.4.11 Gravity

The experiments of Duley and Mueller [39] in the NASA KC-135 microgravity aircraft using a 25 W CW CO_2 laser to "weld" poly(methyl methacrylate) (PMMA) (acrylic), polypropylene and polythene showed that for hypogravity and hypergravity there was no change in the penetration depth, but there may be a reduction in the sheer strength with reduced gravity. In these experiments there was a notable change in the wave structure on the trailing edge of the keyhole. There were larger and faster waves with higher gravity.

4.5 Process Variations

4.5.1 Arc-augmented Laser Welding

It has been found that the arc from a tungsten–inert gas (TIG) torch mounted close to the laser beam interaction point will automatically lock onto the laser-generated hot spot. Steen and Eboo [40] found that the temperature only had to be around 300 °C above the surrounding temperature for this to happen. The effect is either to stabilise an arc which is unstable owing to its traverse speed or to reduce the resistance of an arc which is stable (Figure 4.33). The locking only happens for arcs with a low current and therefore a slow cathode jet: that is, for currents less than 80 A. The beauty of this process is that the arc is on the same side of the workpiece as the laser. The process allows a doubling of the welding speed for a modest increase in the capital cost. It does not enhance the penetration to any great extent [34]. At very high speeds there may be some problems with the weld bead profile since the weld pool is larger than with the laser alone. The increased pool size is, however, not as much as expected. Steen and Eboo [40, 41] showed by mathematical modelling of the heat flow from the laser and the arc separately that the combined effect of the two was not the expected addition of two effects. The only way these results fitted the data was if the arc radius was decreased by the hot core from the laser event (Figure 4.34). Thus, arc-augmented laser welding results in the arc rooting in the same location as the laser and it does so with a finer

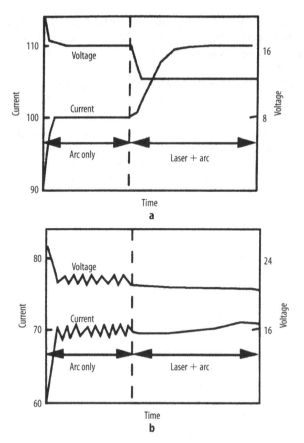

Figure 4.33 The coupling of an arc and a laser beam results in **a** the reduced resistance of the arc or **b** the stabilisation of the arc for high-speed welding [40, 41]

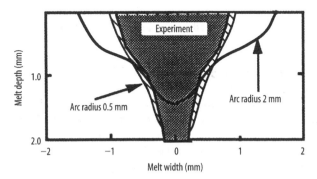

Figure 4.34 Comparison of experiment with theory for various theoretical arc radii during arc-augmented laser welding. (From Steen and Eboo [40])

radius than usual – a true form of adding energy to the laser event. If augmenting the laser is not appealing, then the process has another advantage in that it is a method for guiding an arc. The process is finding new applications for faster welding in tailored blank lines, thicker-section welding for shipping applications [42, 43] and the welding of zinc-coated steel. Gu and Mueller [44] lap-welded clamped galvanised steel at 5.5 m min^{-1} with the arc leading the laser. It appeared that this stabilised the keyhole and made a sufficiently large hole such that the zinc could be vented safely. In the case of thick-section welding, a 60% increase in speed was noted when using 12 kW of laser power and 6 kW of arc power on thicknesses up to 10 mm on material that had been laser-, plasma- or shear-cut but without any further treatment. For 12 mm thicknesses the oxide on the weld faces had to be removed or there would be porosity.

4.5.2 Twin-beam Laser Welding

If two laser beams are used simultaneously, then there is the possibility of controlling the weld pool geometry and hence the weld bead shape. Arata [2] using two electron beams demonstrated that the keyhole could be stabilised, causing fewer waves on the weld pool and giving a better penetration and bead shape. The recently devised Twistlas$^{®1}$ process of TRUMPF aims at the same effect. The weld is, however, wider and shallower but may have uses in controlling dropout when welding aluminium [45].

O'Neill and Steen [46], using both excimer and CO_2 beams simultaneously, showed that improved coupling for the welding of high-reflectivity materials, such as aluminium and copper, could be attained this way. The enhanced coupling was considered principally due to altering the reflectivity by surface rippling caused by the excimer blast (35 MW for 20 ns at 100 Hz) with a secondary effect coming from coupling through the excimer-generated plasma.

There are several references to twin-beam or mixed-beam processing: welding with a CO_2 laser and a diode laser [47] and welding with two pulsing Nd:YAG lasers [48]. The general effect is that the more energy there is, the greater the penetration, and if one beam is less focused, then the wider will be the weldment. There is some collaborative enhancement through reduced reflectivity and preheat effects.

4.5.3 Walking and Spinning Beams

Arata [2] suggested a method for avoiding the plasma by allowing the beam to dwell on a spot just long enough for the plasma to start forming and then to kick the beam forward to dwell on the next spot. He showed an improved penetration capability. The process is more efficient than pulsed welding.

The Welding Institute, UK, has experimented with a rotating beam which mimics arc welding, creating a larger pool and less problems with plasma. The performance is less than with a single stationary beam but offers some control over the weld bead and fit-up requirements.

[1] Twistlas$^{®}$ is a registered trademark of TRUMPF GmbH + Co. KG, Johann-Maus-Str. 2, 71254 Ditzingen, Germany. http://www.trumpf.com

4.5.4 Laser Welding of Plastics

Plastics come in two types, thermoplastics and thermosetting plastics. The thermoplastics can be melted at fairly low temperatures, whereas the thermosetting plastics, such as Bakelite®[2], do not melt but usually char when heated sufficiently. Welding thermosetting plastics is not possible by laser or any heating method; they are usually glued in some way. However thermoplastics such as polythene (listed in Table 4.9 with their properties and possible weld combinations) can be and often are welded by heating. The current methods of joining thermoplastics are hot iron welding, hot gas welding, extrusion welding, vibration or frictional welding (ultrasonic or linear vibration welding), electromagnetic heat source, resistive implant welding, dielectric welding, laser welding and, of course, adhesives and metal fasteners.

4.5.4.1 Conduction Welding of Plastics

Conduction welding of plastics, in the style of metal welding, only works on very thin plastic films, partly owing to the poor thermal conductivity of plastics (see Table 4.9). For example, 0.1 mm thick polyethylene film can be seam-welded at $100 \, \text{m} \, \text{min}^{-1}$ with a 900 W CO_2 laser [49] Likewise, polyethylene sheeting for making plastic bags can be cut and welded simultaneously with a CO_2 laser provided the beam is shaped to cause heating for welding on one side of the cut and to have a sharp intense power density for cutting on the other side. This can be achieved by arranging a coma focus by tilting the lens or using a diffractive optic element [50]. Attempts to weld plastic drums (4-mm-thick polyethylene) this way were only made to work by using the laser as a form of hotplate and pushing the sticky edges together after heating.

4.5.4.2 Transmission Welding of Plastics

A major advance took place in 1985 with the invention of transmission welding of plastics, in which the top layer of plastic was transparent to the radiation and the lower layer was absorbing. This meant the radiation had to be capable of transmission as is the case with the near-infrared lasers such as Nd:YAG and diode lasers operating in the range 750–900 nm but not with CO_2 radiation at 10,600 nm. Such welds usually had a transparent top layer and a black plastic lower layer that was loaded with carbon black as the absorber. In 1997 Mercedes found a black plastic that could transmit the radiation and its example of a keyless entry device for its cars was the first reported case of mass production using transmission laser welding [49].

In 1998 The Welding Institute patented its Clearweld®[3] process, which was a major advance in laser welding of plastics. In this process two pieces of plastic both transparent to the laser radiation can be lap-welded by applying a layer of absorbing material at the interface between them. The subtlety is that the absorbing layer is of a material that absorbs over a narrow absorption band near the laser wavelength, has little absorption

---

[2] Bakelite® is a registered trademark of Bakelite AG, Gennaer Strasse 2–4, 5862 Iserlohn, Germany.

[3] Clearweld® is a registered trademark of Gentex Corporation. http://www.clearweld.com

Table 4.9 Properties of thermoplastics that have been welded by laser [62, 73]

Plastic	Melting point (°C)	Density (kg m⁻³)	Thermal conductivity (J m⁻¹ K⁻¹)	Specific heat (kJ kg⁻¹ °C⁻¹)	Thermal diffusivity (m² s⁻¹)	Maximum recommended service temperature (°C)	Minimum operating temperature (°C)	Glass temperature (°C)	Plastics to which it can be welded by laser
MDPE and HDPE	120–130	940–960	0.32	2.3	0.14	80	−200 to −30		–
LDPE	105–115	910–920	0.32	2.3	0.15	120	−200 to −30		–
ABS	80–100	1,040–1,060	0.13–0.2	1.46	0.08	65	–		ABS/polycarbonate; acrylic; polycarbonate
Acrylic (PMMA)	130	1,170–1,200	0.2	1.46	0.11	60–93	–	105	ABS
Nylon 12, PA12	190–350	1,100	0.25	2.4	0.09	100–121	−30		PA12
Polycarbonate	267	1,200–1,220	0.19–0.22	–	–	150	–	145	ABS; ABS/polycarbonate; polycarbonate
PET	260	1,370	0.41	1.0	0.29	75	–	69	Possibly polycarbonate
Polypropylene	160	855	0.13	1.9	0.08	90	−10		Possibly HDPE
Polystyrene	240	1,050	0.13	1.2	0.103	95	–	95	Modified PPO
PVC	180	1,380	0.28	–	–	87	−10	81	Possibly acrylic

MDPE medium-density polyethylene, *HDPE* high-density polyethylene, *LDPE* low-density polyethylene, *PMMA* poly(methyl methacrylate), *PET* poly(ethylene terephthalate), *PPO* poly(phenylene oxide)

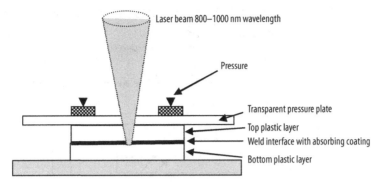

Figure 4.35 The transmission welding process for transparent plastics using the Clearweld® technique

in the visible waveband (400–700 nm, *i.e.*, you should not be able to see it), has good solubility in the host material, has good stability during application and should not degrade to a coloured product. There are not many chemicals that can do that. The ones currently used are cyanines (absorption peak 785 nm), squarylium (absorption peak 800 nm) and croconium (absorption peak 820 nm) absorbers. These are fairly complex organic molecules and are also costly.

The process is illustrated in Figure 4.35. The Clearweld® absorber is applied as a dilute solution by inkjet printing, needle deposition, spraying or as a film. The absorbing molecules are held in a solvent solution. On depositing, they may agglomerate or even crystallise, which will reduce their efficacy. Using a film layer, the dispersion of the absorber is better.

When a laser is shone onto a sandwich, as shown in Figure 4.35, the absorbing layer heats up sufficiently to melt both interfaces, creating a weld and HAZ of around 0.1 mm thickness. The extent of the heating is determined by the laser power, traverse speed, spot size and the absorptivity of the layer. This absorptivity is a function of the quantity of the absorber; at least 21% absorption is necessary for success. The process of absorption by these organic molecules is by excitation of an electron from the highest occupied molecular orbital of the chromophore into the first vacant antibonding orbital, promoting the absorber to the first excited singlet state. This energy increase is considerable (around 150–300 kJ mol^{-1}). The absorber molecule loses this energy most commonly by internal conversion to a vibrational state, followed by vibrational relaxation (around 10^{-13} s). The vibrational energy is transferred to the surrounding molecules. Other deactivation routes such as fluorescence are generally less important. One interesting variation is that a two-wavelength diode laser operating at 808 and 940 nm can be used to simultaneously weld a triple-layer assembly, creating welds where the appropriate absorber for that wavelength has been deposited (Figure 4.36).

A rule of thumb in the plastic welding industry is that any plastic which can be ultrasonically welded can also be welded by a laser, with an appropriate wavelength (800–1,064 nm) [51]. There are many solutions available for welding of coloured polymers. The degree of complexity increases with the combination, starting from transparent/black at the lower end of complexity, and increasing in the following order: trans-

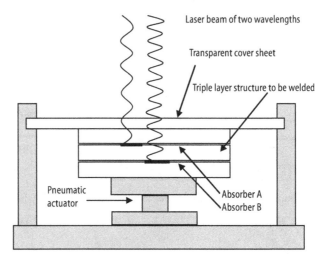

Figure 4.36 A three-layer stack being welded using two different absorbers and a laser beam having two wavelengths

Table 4.10 Measured absorptivity and HAZ for various deposition methods of the absorber in Clearweld® welding of plastics [49]

Method	Absorber deposited	Absorptivity (%)	Applied power (W)	Weld speed (m min^{-1})	Adjusted laser energy[a] (J mm^{-2})	HAZ depth (mm)
Ink jet	70	8	150	0.2	0.72	0.52
	70	8	150	0.6	0.24	0.21
Needle	275	37	100	3	0.15	0.13
	275	37	100	6	0.07	0.08
Spray	120	27	100	3	0.11	0.11
	120	27	100	6	0.05	0.08
Film	166	68	40	3	0.11	0.09
	166	68	40	6	0.05	0.06

[a] Adjusted to allow for the percentage absorbed by the absorber

parent/black, black/black, colour/black, colour 1/colour 2, colour 1/colour 1, transparent/transparent and white/white. Selection of the proper coating on the second layer is very important to promote absorption on the second layer.

Typical welding performance data are given in Table 4.10. The absorptivity of the different deposition methods is given in Table 4.11.

The weld requires firm clamping such as 400 N for 3 mm thick plastic [52]. This need for clamping is possibly the Achilles heal of the process since if one is clamping then it is as easy to use heated bars to make the weld, but possibly this is not so quick. However, the clamping for laser welding can be done by using a transparent plastic or glass plate to press on the stack (as shown in Figures 4.35, 4.36) or a transmissive ball through which the laser passes and is focused; the ball rolls along the required seam as shown in Figure 4.37 – this technique is known as Globo welding. With the glass

Table 4.11 Calculated welding performance using the Clearweld® technique with different dispensing methods for the absorber. The data are for lap welding 3 mm thick PMMA, though the thickness of the transparent plastic is not strictly relevant [49]

Method of dispensing absorber	Amount of absorber deposited $(\mathrm{nl\,mm^{-3}})$	Laser power (W)	Maximum speed below which failure would be in the parent material $(\mathrm{m\,min^{-1}})$	Calculated amount of absorber required $(\mathrm{nl\,mm^{-2}})$
Ink jet	5	150	0.43	6
	10	150	0.72	8
	19	150	0.9	13
Needle	85 (ink A)	100	4.40	88
	85 (ink B)	100	20.0[a]	83
	85 (ink C)	100	20.0[a]	76
Spray	60	100	6.7	39
	120	100	12.0	180
	240	100	20.0[a]	243

[a] Maximum traverse speed used

plate method and a powerful enough laser, the beam can be shaped by optics (*e.g.*, ring optics [53]) or masks to allow a complete circular weld or other shape to be done in one pulse. Alternatively the weld could be made in a quasi-simultaneous manner by scanning in the required pattern.

In transmission welding of plastics:

- the welds are strong, watertight and without any surface disruption, since the melt zone is only at the interface with the absorber;
- it is not necessary to have direct contact with the welding tool;
- the process is fast;
- there is greater design flexibility; and
- it is a clean process with no fume of dust.

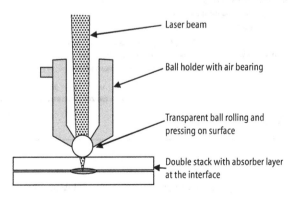

Figure 4.37 The principles of the Globo technique for welding plastics

4.6 Applications for Laser Welding in General

The laser has a certain thickness range over which it competes, as shown in Figure 4.38. Within this range, if the productivity of the laser can be used, then it will usually compete successfully, as in the costed example in Section 4.7.

The commercial advantages of the laser for welding lie in:

- the narrow HAZ, therefore low distortion and the possibility of welding near heat-sensitive material, such as electronics and plastics;
- the narrow fusion zone, which is cosmetically attractive;
- high speed; and
- ease of monitoring and automating.

From these advantages have sprung a multitude of applications with the future prospect of considerably more as the prices of lasers fall and the knowledge of production engineers rises. A sample of these applications can be listed as follows.

1. Hermetically sealing electronic capsules, which is possible owing to the low HAZ.
2. Welding of transmission systems for cars – taking advantage of the low distortion and the possibility of focusing near to potentially magnetic material (but note the cracking problems discussed in Section 4.4.7.1).
3. Welding the end plate on a piston assembly for a car, while a nylon washer is nearby.
4. Welding transformer laminates to reduce hum – the smaller weld zone reduces eddy losses.
5. Welding bimetallic saw blades.
6. Welding of stamped mufflers.

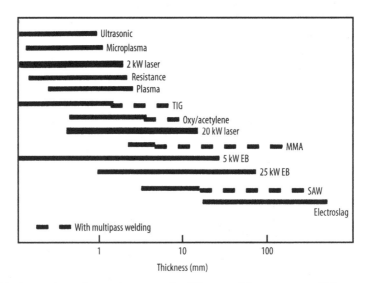

Figure 4.38 Comparison of operating ranges for different welding processes. *TIG* tungsten–inert gas, *MMA* manual metal arc, *SAW* submerged arc welding

7. Welding of cooker tops, made of two stamped sheets.
8. Welding of fire-extinguisher cylinders [54].
9. Flare welding of thick pipes.
10. Welding of complex shapes prior to pressing [55].
11. Repair of nuclear boiler tubes from the inside. There were nine internal welding units working in Japan in 1991 [55].
12. Laser soldering, which is fast becoming a major process in the electronics industry [56–58] particularly with the development of diffractive optic elements for multiple soldered points in a single shot.
13. Welding layered shaving blades.
14. Multiple welds in TV tubes [59] – this application uses 15 fibres placed at different locations in an assembly jig to make 15 spot welds sequentially from one Nd:YAG laser. The commercial advantage is the time saved in locating the spot welder.
15. Strip welding in continuous mills – the laser is replacing flash welding in some installations.
16. Sheet metal products such as washing machines and heat exchangers – one technique being used on some heat exchangers and aircraft parts is to weld a flat pack of two or more layers in an appropriate pattern and then blow the shape up with compressed gas or hydraulically.
17. The laser is being used more and more for three-dimensional welding of car and aircraft components – this is a new production area since most other welding processes cannot be controlled well enough for three-dimensional manipulation and monitoring.
18. Welding aluminium spacer bars for double-glazing units.
19. Numerous automobile applications, some of which are shown in Figure 4.39 [60, 61].
20. Welding of polymers and plastics – there are numerous and increasing numbers of applications for plastic welding. They include welding spectacles, diving suits,

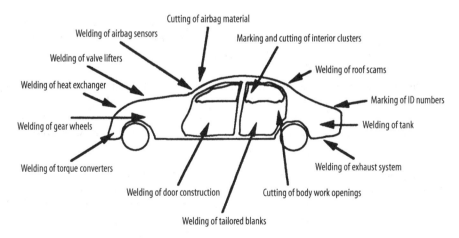

Figure 4.39 Some of the many ways the laser is being used in car production

waders, outdoor footwear, tents, parachutes and soon carpets, bookbinding and injection moulding [62–64]. For wetsuits the process is to discharge polymer through a nozzle into which a 200 W Nd:YAG laser beam is coupled. The beam gently heats the polymer and the joint area allowing the polymer to flow into the fabric. The process uses no toxic chemicals and is 100 times faster in making wetsuits than conventional methods since the seam is finished in one pass with no extra passes, stitching or sealing.

21. Welding fabrics. Fabrics made with some content of thermoplastic fibre can be welded instead of stitched, but there must be some element of compression. Alternative methods for joining fabrics apart from stitching are heat sealing, dielectric or high-frequency welding and ultrasonic welding.
    Applications include outdoor garments, air bags, inflatable buildings or airships, upholstery, foam-backed automobile fabrics, mattress fittings such as information labels, welding fabric to plastic-coated drawer fronts, sealing of medical equipment, manufacture of lithographic processing tanks and other watertight plastic containers [62] and the manufacture of inflation lugs to be fitted to plastic blow-up equipment. The list is growing by the day.

22. Smart garments. Consideration is being given to the manufacture of smart or advanced garments whose manufacture can be automated, thus saving labour costs. Some early work on the control and monitoring of such a process has been published [65]. Smart or advanced garments can include smart elements based on electrical or optical phenomena by including electrical conductors or optical fibres in the fabric; for example, a rescue service kit fitted with electronic sensing for temperature, blood, *etc.*, could be made capable of sending messages to some control centre in the event of an accident.

23. Underwater laser welding is being considered as one of the better methods for deep-sea divers – it is difficult to maintain an electric arc or flames at high pressures. The laser beam can be passed down a fibre for several kilometres if need be. Some of the recent work shows that the process works and that shroud gas is not strictly necessary since the high-power beam makes its own "keyhole" through the surrounding water [66].

24. Structural panels – since the laser can weld well and fast it is possible to consider welding panels consisting of a top and a bottom plate joined by some form of welded corrugation between. This welding can be speeded up by using the flare welding technique (Figure 4.24).

25. Shipbuilding – a vast change is taking place in the technology for shipbuilding which has come with the laser. The laser can be used to weld thick plate with less distortion and HAZ, which makes for greater precision and less time lost beating panels flat again. It can also be used to make lightweight floor and hatch cover panels as described in application 24 above [67].

Standing out from all these applications is that of tailored blanking for the pressed components of a car. This process is spawning not just a new process but whole new industries dedicated to manufacturing tailored blanks on a just-in-time basis. It is possibly illustrative to look at this one in more detail [68].

Sealing costs 3%
Welding costs 12%

Stamping costs 13%

Purchasing costs 24%

Material costs 28%

Figure 4.40 Cost breakdown for a car body in white

A typical car body consists of over 300 pieces, with different thicknesses, treatments and compositions. The cost of assembling them is divided as shown in Figure 4.40. The tailored blank process reduces the cost of material by increasing the yield from the supplied coils from approximately 40 to 65% (see Figure 4.42). It also gives:

- weight reduction;
- lower assembly costs;
- reduced number of parts;
- improved fatigue resistance;
- reduced number of overlapped joints;
- improved corrosion behaviour;
- lower tooling costs;
- improved crash resistance;
- lower press shop costs; and
- fewer sealing operations.

The overall saving is measured in dollars per car door alone. Such savings mean that this is the way in which complex sheet metal products are likely to be treated in the future.

The process itself is illustrated in Figure 4.41. To make this "side member panel" the materials are either galvanised (part E) or thinner (part C) and of various grades of steel.

A detailed study made by Toyota Motor Company, who pioneered this process in the 1980s [68], showed the need for low hardness and therefore low-carbon steels and that the weld bead must be at least 80% of the thickness of the butt joint for reasons of strength and deformability. This meant that the gap tolerance was only 0.15 mm for 0.8-mm-thick steel for autogenous welds. This limit could be extended by using a filler wire or powder – this extra material could help with obtaining the correct metallurgy of the weldment, to control the hardness. The material for the pieces of the blank is trimmed from rolls as shown in Figure 4.42. These parts are assembled on a special jig in which, by flexing the sheets, they can be finally pressed together to create a firm

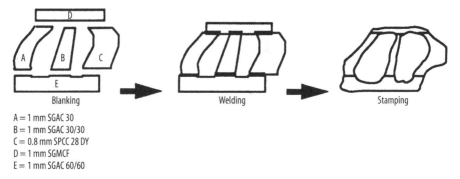

A = 1 mm SGAC 30
B = 1 mm SGAC 30/30
C = 0.8 mm SPCC 28 DY
D = 1 mm SGMCF
E = 1 mm SGAC 60/60

Figure 4.41 The sequence in the manufacture of tailored blanks [67]

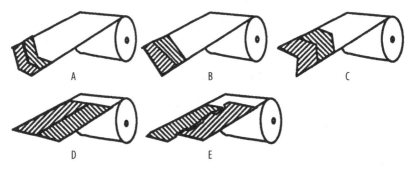

Figure 4.42 The cutting layout for tailored blanking [68]

compression at the joint line. The assembly is then laser-welded swiftly with a high-quality beam, to avoid too much HAZ. During welding the weld bead is monitored by observing the shape on the weld surface using a projected line beam from a He–Ne laser. The weldment is then ground flat and stacked. The blank stack is then transported to the car assembly plant, where it is pressed to the required shape.

4.7 Costed Example

The cost of laser welding is made up of capital and operating elements. The relative capital cost of a laser facility is listed in Table 4.12.

If, for the sake of this example, we cost the capital as an operating cost at 20% per year for 1,800 h per year, we can then compare the approximate cost of a 1 m weld in 3 mm mild steel made by a manual metal arc or laser as shown in Table 4.13.

The actual numbers are debatable, but the difference is striking – it relies upon keeping the laser working. The capital depreciation is fairly sensitive to the fraction of the year that the equipment is working; therefore, much of this work is done by job shops or in plant with a need for all-year operation. A similar conclusion is drawn by the cost analysis of FIAT and COMAU [69].

Table 4.12 Relative capital cost of a laser facility

Process	Capital cost in relative units based on cost for manual metal arc welding being 1
Oxy/fuel	0.2
Resistance butt weld	0.5–10
Manual metal arc	1
TIG	2
Metal–inert gas (300 A)	2
Submerged arc	10
Electroslag	20+
Microplasma	20+
Electron beam (7.5–20 kW)	10–450
Friction	4–100
Laser (2 kW)	100+

Table 4.13 Comparison of the welding costs for 1 m of weld by manual metal arc welding and laser

	Manual metal arc	Laser
System	300-A set	2-kW CO_2 laser plus workstation
Capital cost (£)	850	150,000
Consumables	1 m of 4-mm flux-coated mild steel rod	Gases at £5 per hour
Welding speed	1 mm s^{-1}	10 mm s^{-1}
Process time	1,000 s	100 s
Capital depreciation at 20% per year from 1,800 h per year (£)	0.026	0.46
Consumables (see above) (£)	0.5	0.11
Labour: £24 per hour manual metal arc; £20 per hour laser (£)	6.66	0.55
Power at 0.06 kWh at 4 kW for manual metal arc and 10 kW for laser (£)	0.066	0.016
Clean-up costs at 40% arc time (£)	2.20	
Total cost/m weld (£)	8.95	1.13

As a general rule:

A laser working for one shift per day is just paying for itself; if it is working for two shifts per day, it is distinctly profitable; if it is working for three shifts per day, you will probably find the owner in the Bahamas or some such place.

An interesting thought with which to finish this chapter!

Questions

1. Describe what happens within the keyhole during laser welding.

   a. Give a diagram of the melt pool and keyhole.
   b. Describe how the beam may be absorbed.

    c. Describe the nature of the material within the keyhole and how it may react with the incoming beam.

    d. What are the observed flow patterns within the weld pool? What are the forces causing this movement?

2. In what ways does the plasma affect the process?
3. How can the plasma be used to control the focal position of the beam?
4. Why is it possible to use CO_2 shielding gas when welding using a Nd:YAG laser and why is this not recommended for CO_2 laser welding?
5. List the advantages of a keyhole welding process in comparison with TIG, resistance and ultrasonic welding.
6. A theoretical model of the laser welding process gives the relationship $P = 0.483 \times \rho v w t C_p T_m$, where ρ is the density of the solid metal ($kg\,m^{-3}$), v is the welding speed ($m\,s^{-1}$), w is the weld width (m), t is the material thickness (m), C_p is the specific heat of the material being welded ($J\,kg^{-1}\,{}^{\circ}C^{-1}$) and T_m is the melting point of the material (${}^{\circ}C$).

    a. What was the basis of this model?

    b. What welding speed would one expect with a 5 kW laser welding 2 mm thick stainless steel with a beam focus that gives a 1.5 mm-wide weld?

The density of stainless steel is $8,030\,kg\,m^{-3}$, the specific heat is $500\,J\,kg^{-1}\,{}^{\circ}C^{-1}$ and the melting point is $1,450\,{}^{\circ}C$.

References

[1] Beyer E, Behler K, Herziger G (1990) Influence of laser beam polarisation in welding. In: Industrial laser annual handbook 1990. PennWell Books, Tulsa, pp 157–160
[2] Arata Y (1987) Challenge of laser advanced materials processing. In: Proceedings of the conference on laser advanced material processing (LAMP'87), Osaka, May 1987. High Temperature Society of Japan, pp 3–11
[3] Matsunawa A (2002) Science of laser welding-mechanisms of keyhole and pool dynamics. In: ICALEO 2002 proceedings, Phoenix, October 2002. LIA, Orlando, paper 101
[4] Greses J, Barlow CY, Steen WM, Hilton PA (2001) Spectroscopic studies of plume/plasma in different gas environments. In: ICALEO 2001 proceedings, Jacksonville, October 2001. LIA, Orlando, paper 808
[5] Greses J, Hilton PA, Barlow CY, Steen WM (2002) Plume attenuation under high power Nd:YAG laser welding. In: ICALEO 2002 proceedings, Phoenix, October 2002. LIA, Orlando, paper 808
[6] Mazumder J (1983) Laser welding. In: Bass M (ed) Laser material processing. North-Holland, Amsterdam, pp 113–200
[7] PennWell Books (1990) Industrial laser annual handbook 1990. PennWell Books, Tulsa, pp 7–15
[8] Kugler T, Naeem M (2002) Material processing with super modulation. In: ICALEO 2002 proceedings, Phoenix, October 2002. LIA, Orlando, paper 506
[9] Holtz R (2002) Optimized laser applications with lamp pumped pulsed Nd:YAG lasers. In: ICALEO 2002 proceedings, Phoenix, October 2002. LIA, Orlando, paper M409
[10] Katayama S, Wu Y, Matsunawa A (2001) Laser weldability of Zn coated steels. In: ICALEO 2001 proceedings, Jacksonville, October 2001. LIA, Orlando, paper P520
[11] Tsukamoto S, Kawaguchi I, Arakane G, Honda H (2001) Suppression of porosity using pulse modulation of laser power in 20 kW CO_2 laser welding. In: ICALEO 2001 proceedings, Jacksonville, October 2001. LIA, Orlando, paper 1702

[12] LASAG AG (1997) LASAG KLS brochure LASAG AG Switzerland (Headquarters): C.F.L. Lohner-strasse 24, 3602 Thun, Switzerland

[13] Popov S (2006) IPG Laser GmbH fibre lasers – driving material processing markets. In: Proceedings of AILU workshop on fibre lasers – future of laser material processing, Cranfield, 8 March 2006

[14] Lewis S, Naeem M (2006) 100W CW fibre laser vs pulsed Nd:YAG for micro joining. In: Proceedings of AILU workshop on fibre lasers – future of laser material processing, Cranfield, 8 March 2006

[15] Duley WW(1996) UV lasers affects and applications in material science. Cambridge University Press, Cambridge

[16] Schlessinger L, Wright J (1979) Inverse-bremsstrahlung absorption rate in an intense laser field. Phys Rev A Gen Phys 20:1934–1945

[17] Raizer YP (1965) Breakdown and heating of gases with a laser light pulse. Sov Phys JETP 21:1009

[18] Ducharme R, Kapadia PD, Dowden JM, Hilton P, Riches S, Jones IA (1997) An analysis of the laser material interaction in the welding of steel using a CW CO laser. In: ICALEO'96 proceedings, October–November 1996. LIA, Orlando, pp D10–D20

[19] Matsunawa A (1982) Role of surface tension in fusion welding, part I. J Weld Res Inst 11(2):145–154

[20] Matsunawa A (1983) Role of surface tension in fusion welding, part II. J Weld Res Inst 12(1):123–132

[21] Matsunawa A (1984) Role of surface tension in fusion welding, part III. J Weld Res Inst 13(1):147–156

[22] Albright CE, Chiang S (1988) High speed laser welding discontinuities. In: ICALEO'88 proceedings, Santa Clara, October–November 1988. Springer, Berlin/IFS, Kempston, pp 207–213

[23] Wilgoss RA, Megaw JHPC, Clarke JN (1979) Assessing the laser for power plant welding. Weld Met Fabr Mar 117

[24] Seaman FD (1977) Role of shielding gas in high power CO_2 CW laser welding. SME technical paper no MR77-982. Society of Manufacturing Engineering, Dearborn

[25] Sepold G, Rothe R, Teske K (1987) Laser beam pressure welding – a new technique. In: Proceedings of the conference on laser advanced material processing LAMP'87, Osaka, May 1987. High Temperature Society of Japan, pp 151–156

[26] Duhamel R (1996) Restrained joint laser welding. Industrial Laser Review Aug 3–4

[27] Norris IM (1989) High power laser welding of structural steels-current status. In: Proceedings of the conference on advances in joining and cutting processes 89, Harrogate, October 1989. The Welding Institute, Great Abington, paper 55

[28] Shannon G, Steen WM (1996) Laser welding with coaxial powder fill nozzle for sheet and thick section welding. In: ICALEO'96 proceedings, Orlando, October–November 1996. LIA, Orlando, pp 20–27

[29] Akhter R (1990) The laser welding of zinc coated steel. PhD thesis, University of London

[30] Katayama S, Wu Y, Matsunawa A (2001) Laser weldability of zinc-coated steels. In: ICALEO 2001 proceedings, Jacksonville, October–November 2001. LIA, Orlando, paper P520

[31] Akhter R, Steen WM (1991) The gap model for welding zinc coated steel sheet. In: Proceedings of the conference on laser systems applications in industry, Turin, 7–9 November 1990. IATA

[32] Tzeng Y-F (1996) Pulsed laser welding of zinc coated steel. PhD thesis, Liverpool University

[33] Nonhof CJ (1988) Materials processing with Nd:YAG lasers. Electrochemical Publications, Ayr, p 192

[34] Alexander J, Steen WM (1980) Effects of process variables on arc augmented laser welding. In: Proceedings of Optica '80 conference, Budapest, Hungary, November 1980

[35] Jørgensen M (1980) Increasing energy absorption in laser welding. Met Constr 12(2):88

[36] Akhter R, Watkins KG, Steen WM (1990) Modifications of the composition of laser welds in electro-galvanised steel and the effects on corrosion properties. J Mater Manuf Process 5(4):67–68

[37] Oakley PJ (1982) 2 and 5 kW fast axial flow carbon dioxide laser material processing. In: ICALEO'82 proceedings. LIA, Orlando, pp 121–128

[38] Glasstone S (1953) Textbook of physical chemistry. Macmillan, London, p 450

[39] Duley WW, Mueller RE (1990) Laser penetration welding in low gravity environment. In: Proceedings of the XXII ICHMT international conference on manufacturing and material processing, Dubrovnik, Yugoslavia, August 1990, pp 1309–1319

[40] Steen WM, Eboo M (1979) Arc augmented laser welding. Construction III(7):332–336

[41] Steen WM (1980) Arc augmented laser processing of materials. J Appl Phys 51(11):5636–5641

[42] Walz C, Seefeld T, Sepold G (2001) Process stability and design of seam geometry during hybrid welding. In: ICALEO 2001 proceedings, Jacksonville, October 2001. LIA, Orlando, paper 305

[43] Engström H, Nilsson K, Flinkfeldt J (2001) Laser hybrid welding of high strength steels. In: ICALEO 2001 proceedings, Jacksonville, October 2001. LIA, Orlando, paper 303

[44] Gu H, Mueller R (2001) Hybrid welding of galvanised steel sheet. In: ICALEO 2001 proceedings, Jacksonville, October 2001. LIA, Orlando, paper 304

[45] Trentmann G (1997) Laser welding doubles up to tackle aluminium. Europhotonics Jun–Jul 49–50

[46] O'Neill W, Steen WM (1988) Infra red absorption by metallic surfaces as a result of powerful u/v pulses. In: ICALEO'88 proceedings, Santa Clara, October–November 1988. Springer, Berlin/LIA, Orlando, pp 90–97

[47] Bonss S, Brenner B, Beyer E (2000) Hybrid welding with a CO_2 and diode laser. Industrial Laser User 21:26–28

[48] Narikiyo T, Miura H, Fujinaga S, Ohmori A, Inoue K (1999) Combination of two Nd:YAG laser beams and their welding characteristics. J Laser Appl 11(2):91–95

[49] Hilton PA, Jones IA, Kennish Y (2002) Transmission laser welding of plastics. In: Proceedings of the conference LAMP

[50] Cole CF, Noden SC, Tyrer JR, Hilton PA (1998) The application of diffractive optical elements in high power laser materials processing. In: ICALEO'98 proceedings, November 1998. LIA, Orlando, pp A84–A93

[51] Anscombe N (2004) Laser welding penetrates the plastics market. Photonics Spectra Sep 60–66

[52] Jones I, Rostami S (2003) Laser welding of plastics - process selection software. In: ICALEO 2003 proceedings, Jacksonville. LIA, Orlando, paper 601

[53] Xu G,Tsuboi A,Ogawa T, Ikeda T (2008) Super-short times laser welding of thermoplastic resins using a ring beam optics. J Laser Appl 20(2):116–121

[54] Industrial Laser Review (1996) Joining fire extinguishers. Industrial Laser Review Apr 3

[55] Matsunawa A (1991) Present and future trends of laser materials processing in Japan. In: Proc SPIE vol 1502, pp 60–71, Industrial and Scientific Uses of High-Power Lasres, Jean P. Billon and Eduardo Fabre (eds)

[56] Lau KH, Man HC (1995) Excimer laser soldering for fine pitch surface mounted assembly. In: ICALEO'95 proceedings, San Diego, November 1995. Springer, Berlin/LIA, Orlando, pp 15–24

[57] Adachi A, Hirota J, Hoshinouchi S (1995) Fluxless soldering with laser assembly of TCP. In: ICALEO'95 proceedings, San Diego, November 1995. Springer, Berlin/LIA, Orlando, pp 35–41

[58] Brandner M, Seibold G, Chang C, Dausinger F, Hugel H (2000) Soldering with solid state and diode lasers: energy coupling, temperature rise and process window. J Laser Appl 12(5):194–199

[59] Laser Focus World (1985) Fibres multiplex industrial YAG laser beams, lasers and applications. Laser Focus World 21(6):8

[60] Roessler DM (1990) New laser processing developments in the automotive industry. In: Industrial laser annual handbook. PennWell Books, Tulsa, pp 109–127

[61] Sinar R (1997) On the road to better automotive production. Laser Power Beam Processing Mar 6–9

[62] Warwick M, Gordon M (2006) Application studies using through transmission laser welding of polymers. In: Proceedings of the conference joining plastics 2006. NPL, London

[63] Opto Laser Europe (1997) Laser welding polymers enter mass production. Opto Laser Europe Jul 15–17

[64] DeMais R (1995) Laser enhanced bonding produces strong seams. Laser Focus World Aug 32–33

[65] Jones I, Rudlin J (2006) Process monitoring methods in laser welding of plastics. In: Proceedings of the conference joining plastics 2006. NPL, London

[66] McNaught W, Deans WF, Watson J (1997) High power laser welding in hyperbaric and water environments. J Laser Appl 9:129–136

[67] Pantsar H, Salminen A, Kujanpaa V (2001) Manufacturing procedure and cost analysis of laser welded all steel sandwich panels. In: ICALEO 2001 proceedings, Jacksonville, October 2001. LIA, Orlando, paper 1709

[68] Azuma K, Ikemoto K (1993) Laser welding technology for joining different sheet metals for one piece stamping.In: Laser applications for mechanical industry proceedings, NATO ASI, Erice, Sicily. Kluwer, Dordrecht, pp 219–233

[69] Marinoni G, Maccagno A, Rabino E (1989) Technical and economic comparison of laser technology with conventional technologies of welding. In: Steen WM (ed) Proceedings of the 6th international conferences on lasers in manufacturing (LIM6), Birmingham, UK, May 1989. IFS, Kempston, pp 105–120

[70] Matsunawa A, Kim J-D (2004) Basic understanding on beam-plasma interaction in laser welding. In: Proceedings of PICALO 2004, Melbourne, Australia. LIA, Orlando, paper 401

[71] Mizutani M, Kayayama S Keyhole behaviour and pressure distribution during laser irradiation on molten metal. In: ICALEO 2003 proceedings, Jacksonville. LIA, Orlando, paper 1004

[72] Burrows G, Croxford N, Hoult AP, Ireland CLM, Weedon TM (1988) Welding characteristics of a 2 kW YAG laser. Proc SPIE 1021:159–166

[73] Perry RH (1963) Perry's chemical engineers handbook, 4th edn, McGraw-Hill, New York

"There are still unknown depths to laser welding."

5 Theory, Mathematical Modelling and Simulation

There is safety in numbers

Jane Austen, Emma II,i

We haven't the money so we've got to think

Lord Rutherford (1871–1937) attributed in Prof. R.V. Jones's 1962
Brunel Lecture, 14 February 1962

5.1 Introduction

The processes to be modelled are described in their various forms in this book. There are several aspects that represent a challenge to any would-be modeller. They are shown in Figure 5.1. Some of the models that have been developed are listed in Table 5.1.

At first, the challenge of modelling looks insurmountable. Moreover, some phenomena, such as free surface deformation, are shrouded under a plasma and are difficult to measure for model validation. Observing the complication, one school says "why bother?" and another school says "why not?" and plunges in with glee regardless of direction. On the other hand, intelligent application of boundary conditions can make the mathematical challenge more amenable and the recent ready availability of fast com-

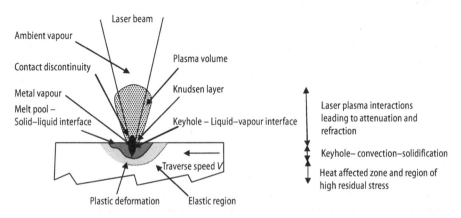

Figure 5.1 Some of the physical interactions during laser material processing

Table 5.1 Models discussed in this chapter showing the sections in which they feature

Heating models	Section	Heat + flow models	Section	Heat + flow + plasma models	Section
Analytical models					
Stationary models					
Lumped heat capacity	3.3.3.2.1	Basic surface tension	5.9.2	Basic equations for plasma	5.10
1D heat flow	5.3	Navier–Stokes flow	5.9.3.2		
Instantaneous point source	5.4.1				
Continuous point source	5.4.2				
Stationary Gaussian source	5.4.3				
Moving models					
Moving point source	5.5				
Hypersurface line source	5.6.1				
Moving Gaussian source	5.6.2				
Moving line source	5.7				
Moving line source arounda cylinder	5.7.2				
Moving point–line source	5.7.3				
3D semi-infinite plate	5.8.1				
3D transient model for finite slab	5.8.2				
Numerical models					
3D thermal models	5.9.1	Self-consistent 2D	5.9.3.1	Flow with vaporisation	5.9.6
		Self-consistent 3D	5.9.3.2	Drilling	Chapter 3
		Effect of Maragoni and Prandtl numbers	5.9.3.3	Mass additions	5.9.7
		Surface deformation	5.9.5	Surface alloying	5.9.7.1
				Surface cladding	5.9.7.2
				Ablation	5.10
				Particle size distribution during ablation	5.10

puter power has made modelling of many complicated processes possible. There are even software packages which make it too simple for even the uninitiated to have a go. Such packets are SIMPLE, SOLA®[1], ANSYS®[2], FLUENT®[3], FIDAP®[4], ABAQUS®[5], *etc.*

---

[1] SOLA® is a registered trademark of the Montreal Exchange. http://www.sola-x.ca

[2] ANSYS® is a registered trademark of SAS IP Inc., a wholly owned subsidiary of ANSYS Inc. http://www.ansys.com

[3] FLUENT® is a registered trademark of Fluent Inc. http://www.fluent.com

[4] FIDAP® is a registered trademark of Fluent Inc. http://www.fluent.com

[5] ABAQUS® is a registered trademark ABAQUS Inc., in the USA and other countries. http://www.abaqus.com

The target with modelling has three levels:

1. semiquantitative understanding of the process mechanisms for the design of experiments and display of results – dimensional analysis, order of magnitude calculations;
2. parametric understanding for control purposes – empirical and statistical charts, analytical models; and
3. detailed understanding to analyse the precise process mechanisms for the purpose of prediction, process improvement and the pursuit of knowledge – analytical and numerical models.

The first two we have seen being used in the text of this book, such as the lumped heat capacity model (Equation 3.2) in Chapter 3 and the semi-empirical relationship for depth of hardening during transformation hardening shown in Section 6.2.1, and no doubt we have an instinctive appreciation for these methods.

Most experimental and theoretical results are likely to be reported using dimensionless groups because of the economy of expression achieved this way.

The dimensional analysis of heat flow or any other physical problem implies a total knowledge of all the variables involved but no knowledge of how they are related. For any equation describing a physical relationship we know that the units must balance and that therefore there is a restriction on the number of ways the variables can be related. The way to take advantage of this is to arrange for the variables to be grouped as dimensionless groups, in which case the units will automatically balance, and we are left with a reduced number of variables – the groups. Once that has been done, then some experiments must be performed to show how the groups are related. Buckingham's Π theorem [1] tells us how many groups to expect: the number of independent groups, $n = i - r$, where i is the number of variables and r is the greatest number of these which will not form a dimensionless group (usually the same as the number of basic dimensions, except where there is some unusual symmetry). The usual dimensionless groups involved in heat transfer problems are:

- the Fourier number, $F = \alpha t / x^2$, a form of dimensionless time;
- the Péclet number, $Pe = vD/\alpha$, a form of dimensionless velocity (ratio of convection to conduction);
- the Reynolds number, $Re = \rho v D/\mu$, another form of dimensionless velocity (ratio of viscous to inertial forces);
- the Weber number, $V^2 \rho L/g_c \sigma$, the ratio of inertial forces to surface tension forces;
- dimensionless temperature, $T^* = (T - T_0)/(T_1 - T_0)$;
- dimensionless power, $P/k\pi DT$, there are a few variations; and
- dimensionless distance, x/D.

α is the thermal diffusivity ($m^2\,s^{-1}$), t is the time (s), x is the distance (m), v is the velocity ($m\,s^{-1}$), ρ is the density ($kg\,m^{-3}$), μ is the viscosity ($kg\,m^{-1}\,s^{-1}$), P is the incident absorbed power (W), k is the thermal conductivity ($W\,m^{-1}\,K^{-1}$), D is a characteristic distance, such as the beam diameter (m), T is the temperature (K) and g_c is the gravitational constant.

The third type of model involves complex mathematics or computing and is used as a means of inspiring awe or as a real tool of great importance. The main objective of mathematically rigorous models is to understand the physics of the process and include all possible phenomena to estimate their effect on the process and their importance. This enables one to include complicated phenomena, such as convection and discontinuities at the liquid–vapour or solid–liquid interfaces. It is extremely important to processes such as laser processing, which often include all the four phase changes. It is this type of model that is discussed in particular in this chapter. It should be remembered that the model is only a tool and as such has no significance other than in its use. Mathematical modelling can be akin to astrology, in that one can become mesmerised by the calculations and forget the underlying principles and objectives; the model has a sanctity of its own! (It can also become a "black hole" for research time.)

The questions we would like answered by a model and which are difficult or impossible to answer by experiment are as follows.

1. How did the microstructure in a laser weld or surface treatment arise? For this we need to know thermal gradients G, rates of solidification R, time above certain temperatures, stresses and stirring action or extent of mixing.
2. How did the weld bead form? There is a need to understand waves and movements in the weld pool and surface solidification processes.
3. What is the cause of cracking or distortion? We need to know the stress history and residual stresses.
4. What temperatures are likely in the environment of the laser event? We need to know the depth of penetration of isotherms within thermally sensitive materials.
5. What is the optimum operating region? We need to be able to forecast experimental results, describe the effects of parameters not normally separable from others (*e.g.*, the effect of the thermal conductivity is impossible to find by experiment since the density, *etc.*, will also change if the material is changed) and forecast development opportunities.

This information cannot be found from experiment, although the model results can be checked by experiment, implying that the reasoning leading to the result is probably correct.

5.2 What is a Model?

A mathematical model simply predicts future conditions on the basis of the present understanding of the phenomena. Figure 5.2 illustrates the simple concept of a model which can be as elementary as the conversion of the temperature from degrees Fahrenheit to degrees Celsius.

The model in this case is the equation $C = \frac{5}{9}(F - 32)$.

The accuracy and precision of a model depends on whether all the appropriate physics has been included. For example, pressure will effect the equation. It is important that all the physics are included for accuracy. Precision will depend on the level of approximation during the calculation. To develop a mathematical model, the following questions need answering to make the problem tractable.

Figure 5.2 Simple concept of a mathematical model

1. Why is the model being built? For example, there is a need to know the cooling rate to establish the reason for the observed microstructure, or some such answer.
2. What is being modelled? Is it a transport process for heat, mass or momentum flow or all three?
3. How well is the physical process understood? This is critical in determining the appropriate rate equations to use:

   a. choice of rate process;
   b. boundary conditions/initial conditions; or
   c. independent/dependent variables.

4. What relationships are there to describe the process quantitatively? Usually after identifying a control volume, there are three conservation equations – mass, heat or energy and momentum, sometimes there is also a fourth, vorticity. There are also flux equations such as Fourier's first law for heat flow, Fick's first law for mass by diffusion and Newton's law of viscosity. Armed with these relationships, the questions become:

   a. Is it tractable?
   b. What simplifications are possible, particularly concerning the boundary conditions?

5. What solution techniques can be used?

   a. transform to use an already-available solution (exact);
   b. approximate analytical solution in limiting cases (approximation); or
   c. numerical methods.

6. How can the solution be checked?

   a. look at limiting cases;
   b. examine boundary conditions; or
   c. check against measured values.

5.2.1 Derivation of Fourier's Second Law

Nearly all models in unsteady-state heat transfer have to solve Fourier's second law.[6] So this is our starting point.

---

[6] Fourier's first law is $q = -k\mathrm{d}T/\mathrm{d}x$, where q is the heat flow per unit area. Baron Jean Baptiste Joseph Fourier (1768–1830) was a French mathematician famous for his representation of

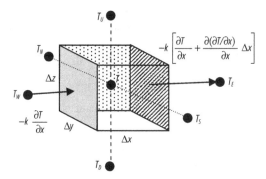

Figure 5.3 Heat flow through a differential element. Subscripts represent north (N), south (S), east (E), west (W), up (U) and down (D)

Consider the differential element shown in Figure 5.3. The heat balance on the element is

$$\text{Heat in} - \text{heat out} = \text{heat accumulated} + \text{heat generated}.$$

The difference between the heat in and the heat out depends upon the different rates of conduction and convection.

Conduction. Consider only the x direction (the other directions being analogous): heat flow by conduction in the x direction is given by:

heat accumulated $= (\text{Heat in} + \text{Heat out})$

$$= \left\{ -k\frac{\partial T}{\partial x} + k\left[\frac{\partial T}{\partial x} + \left(\frac{\partial(\partial T/\partial x)}{\partial x} \right) \delta x \right] \right\} \delta y \delta z = k\frac{\partial^2 T}{\partial^2 x}\delta x \delta y \delta z.$$

The total conduction in all three dimensions is

$$= k\left[\frac{\partial^2 T}{\partial^2 x} + \frac{\partial^2 T}{\partial^2 y} + \frac{\partial^2 T}{\partial^2 z} \right] \delta x \delta y \delta z,$$

which in vector notation is written as

$$k\nabla^2 T \delta x \delta y \delta z.$$

convection. This equals

$$\text{Heat in} - \text{heat out} = \rho C_p U_x \delta y \delta z T - \rho C_p U_x \delta y \delta z \left(T + \frac{\partial T}{\partial x}(x) \right)$$

$$= -\rho C_p U_x \frac{\partial T}{\partial x}\delta x \delta y \delta z,$$

---

trigonometric functions as a series and his masterpiece "Theorie Analytique de la Chaleur" from 1822, translated by A. Freeman in 1872, was considered one of the most important books of the nineteenth century. Through his association with Napoleon, he was Governor of Lower Egypt and later Prefect of Isere, where he wrote his work on heat. Though orphaned at 8 years old and of humble origins, he was made a baron before he died. His life in Egypt led him to believe that heat was healthy and so he spent much time swathed in blankets!

(*i.e.*, a positive heat loss for a negative thermal gradient), which in vector notation is written as

$$-\rho C_p U_x \nabla T \delta x \delta y \delta z.$$

Accumulation. This equals

$$\rho C_p \frac{\partial T}{\partial t} \delta x \delta y \delta z.$$

Generation. This equals

$$= H \delta x \delta y \delta z.$$

Thus, the *total balance* (divide throughout by $\delta x \delta y \delta z$) is

$$k \nabla^2 T - \rho C_p U \nabla T - \rho C_p \frac{\partial T}{\partial t} = -H,$$

or

$$-\frac{U}{\alpha} \nabla T + \nabla^2 T - \frac{1}{\alpha} \frac{\partial T}{\partial t} = -\frac{H}{k}. \tag{5.1}$$

This is the basic equation to be solved. (The definitions of the symbols are listed at the end of the chapter.)

5.3 Analytical Models with One-dimensional Heat Flow

If the heat flows in only one direction and there is no convection or heat generation, the basic equation becomes (from Fourier's second law), as just derived,

$$\frac{\partial^2 T}{\partial z^2} = \frac{1}{\alpha} \frac{\partial T}{\partial t}. \tag{5.2}$$

If it is assumed that there is a constant extended surface heat input and constant thermal properties, with no radiant heat loss or melting, then the boundary conditions are as follows.
    At $z = 0$,

$$\text{Surface power density,} \quad F_0 = \left\{ \frac{P_{tot}(1 - r_f)}{A} \right\} = -k \left[\frac{\partial t}{\partial z} \right]_{surf}.$$

At $z = \infty$,

$$\frac{\partial T}{\partial z} = 0.$$

At $t = 0$,

$$T = T_0 .$$

The solution is

$$T_{z,t} = \frac{2F_0}{k} \left\{ (\alpha t)^{1/2} \text{ierfc} \left[\frac{z}{2\sqrt{at}} \right] \right\} , \tag{5.3}$$

where the "integral of the complimentary error function", ierfc, means

$$\text{ierfc}(u) = \frac{e^{-u^2}}{\sqrt{\pi}} - u \left[1 - \text{erf}(u) \right] . \tag{5.4}$$

Since the error function, $\text{erf}(u)$, is not freely available, as it is with logarithms, and it is difficult to manipulate algebraically, it can be substituted by a polynomial with an accuracy of one part in 2.5×10^{-5}, which is usually sufficient for our needs. The polynomial is [2]

$$erf(u) = 1 - \left(a_1 b + a_2 b^2 + a_3 b^3 \right) e^{-u^2} ,$$

where $b = (1 + cu)^{-1}$, $a_1 = 0.3480242$, $a_2 = -0.0958798$, $a_3 = 0.7478556$ and $c = 0.47047$

Incidentally, the error function $\text{erf}(x)$ is defined as

$$\text{erf}(x) = \left(\frac{2}{\sqrt{\pi}} \right) \int_0^x e^{-\xi^2} d\xi .$$

This leads to the differential of the error function

$$\frac{d \left[\text{erf}(x) \right]}{dx} = \left(\frac{2}{\sqrt{\pi}} \right) e^{-x^2} ,$$

and the values at extremes are $erf(0) = 0$ and $erf(\infty) = 1$.

If the power is turned off, then the material will cool for $t > t_1$ according to the relationship

$$T_{z,t} = \frac{2F_0}{k} \sqrt{\alpha} \left[\sqrt{t} \, \text{ierfc} \left(\frac{z}{2\sqrt{\alpha t}} \right) - \sqrt{t_0 - t_1} \, \text{ierfc} \left(\frac{z}{2\sqrt{\alpha(t - t_1)}} \right) \right] , \tag{5.5}$$

where the variables are listed at the end of the chapter.

This model gives a feel for the extent of the conduction process and applies when the flow is in one direction, which will be reasonably correct when the heat source is large compared with the depth considered or when the flow is in a rod or some such geometry (see also Gregson [3, 4]).

Table 5.2 Heating and cooling times as a function of power for a melt depth of 0.025 mm in nickel

Power density (W cm^{-2})	Surface temperature (°C)	Melting time (s)	Cooling time (s)
550,000	2,730	6.71×10^{-5}	3.05×10^{-5}
200,000	1,960	2.64×10^{-4}	2.47×10^{-5}
50,000	1,590	2.77×10^{-3}	2.17×10^{-5}
5,000	1,469	0.236	2.17×10^{-5}
500	1,456	23.1	2.10×10^{-5}

However, this calculation has not allowed any variation or concept of beam size or mode structure, speed is only simulated by time on and no allowance is made for workpiece thickness. Some typical solutions from Brienan and Kear [6] are given in Table 5.2.

The solutions from this model are useful. Brienan and Kear developed graphs of the cooling rate, thermal gradient and solidification rates expected with this form of heating. They are shown for pure nickel in Figures 5.4–5.7.

Incidentally, Equation 5.3 shows that the bulk of the heat conduction is within $[z/2(\alpha t)^{1/2}] = 1$, *i.e.*, Fourier number of 1. This is so since $2\text{ierfc}(1) = 0.1005$ and $2\text{ierfc}(0) = 1.1284$, which is the value at $z = 0$. So for any time t, some 90% of the temperature change lies within this depth. This gives some interesting and simple estimates of the heat lost by conduction: for example, the heat lost in conduction per unit area is approximately $z_{F=1} \rho C_p \left(\frac{T_{surf}}{2} \right)$.

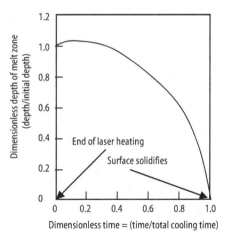

Figure 5.4 Melt depth history for pure nickel with 500,000 W cm^{-2} absorbed power, initial melt depth 1.2 mm and maximum temperature 2,038 °C [3, 4]

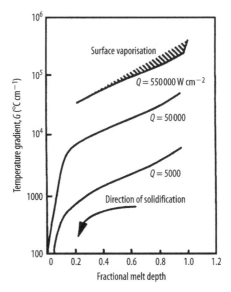

Figure 5.5 Transient behaviour of the temperature gradient at the melt front for pure nickel; initial depth 0.025 mm [3, 4]

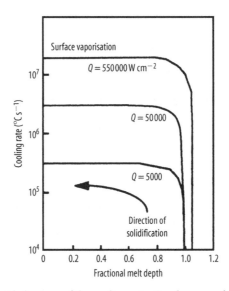

Figure 5.6 Transient behaviour of the cooling rate. Conditions as for Figure 5.5 [3, 4]

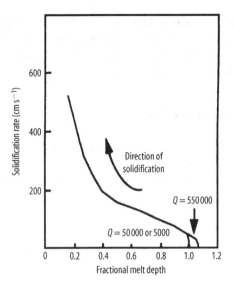

Figure 5.7 Transient behaviour of the solidification rate. Conditions as for Figure 5.5 [3, 4]

5.4 Analytical Models for a Stationary Point Source

5.4.1 The Instantaneous Point Source

The differential equation for the conduction of heat in a stationary medium, assuming no convection or radiation, is

$$\left[\frac{\partial^2 T}{\partial x^2} + \frac{\partial^2 T}{\partial y^2} + \frac{\partial^2 T}{\partial z^2}\right] = \frac{1}{\alpha}\frac{\partial T}{\partial t}.$$

This is satisfied for the case of an instantaneous point source of energy $Q\rho C$ by [7]

$$T = \frac{Q}{8(\pi\alpha t)^{2/3}}\exp\left\{-\left[(x-x')^2 + (y-y')^2 + (z-z')^2\right]/4\alpha t\right\}. \qquad (5.6)$$

As $t \to 0$, this expression tends to zero at all points except (x', y', z'), where it becomes infinite. Also, the total quantity of heat in the infinite region is

$$\iiint_{-\infty}^{+\infty} \rho C T \mathrm{d}x\mathrm{d}y\mathrm{d}z = \frac{Q\rho C}{8(\pi\alpha t)^{3/2}} \int_{-\infty}^{+\infty}\exp\left[\frac{-(x-x')^2}{4\alpha t}\right]\mathrm{d}x$$

$$\times \int_{-\infty}^{+\infty}\exp\left[\frac{-(y-y')^2}{4\alpha t}\right]\mathrm{d}y \int_{-\infty}^{+\infty}\exp\left[\frac{-(z-z')^2}{4\alpha t}\right]\mathrm{d}z$$

$$= Q\rho C,$$

which is as it should be.

Thus, the solution in Equation 5.6 may be interpreted as the temperature in an infinite solid due to a quantity of heat $Q\rho C$ instantaneously generated at $t = 0$ at the point (x', y', z'). Q is seen to be the temperature to which a unit volume of the material would be raised by the instantaneous point source.

5.4.2 The Continuous Point Source

Since heat is not a vector quantity, the effects from different heat sources can be added. Thus, if heat is liberated at the rate of $\varphi(t)\rho C$ per unit time from $t = 0$ to $t = t'$ at the point (x', y', z'), the temperature at (x, y, z) at time t is found by integrating Equation 5.6 over that time period:

$$T(x, y, z, t) = \frac{1}{(8\pi\alpha)^{3/2}} \int_0^t \varphi(t') e^{\frac{-r^2}{4\alpha(t-t')'}} \frac{dt}{(t - t')^{3/2}},$$

where $r^2 = (x - x')^2 + (y - y')^2 + (z - z')^2$.
   If $\varphi(t)$ is constant and equal to q, we have

$$T = \frac{q}{4(\pi\alpha)^{3/2}} \int_{1/\sqrt{t}}^{\infty} e^{\frac{-r^2\tau^2}{4\alpha}} d\tau,$$

where $\tau = (t - t')^{-1/2}$

$$= \frac{q}{4\pi\alpha r} \text{erfc}\left\{ \frac{r}{\sqrt{4\alpha t}} \right\}. \tag{5.7}$$

As $t \rightarrow \infty$ this reduces to $T = q/4\pi\alpha r$, a steady temperature distribution in which a constant supply of heat is continually introduced at (x', y', z') and spreads outwards into an infinite solid.

5.4.3 Sources Other than Point Sources

By integrating this point source solution over an area, one can calculate the heating from line sources, disc sources or Gaussian sources or any other definable distribution. Carslaw and Jaeger's book "Conduction of Heat in Solids" [7] is a collection of solutions for nearly every conceivable geometry likewise Dowden's book "The Mathematics of Thermal Modelling: An Introduction of the Theory of Laser Material Processing" [5].
   For example, the solution for the surface central point under a stationary Gaussian source can be shown to be

$$T_{\text{contGauss}}(0, 0, t) = \frac{2P(1 - r_f)D}{\pi D^2 k \sqrt{T}} \tan^{-1}\left[\frac{2(\sqrt{\alpha t})}{D} \right], \tag{5.8}$$

from which the equilibrium temperature of that spot would be

$$T(0,0,\infty) = \frac{P(1 - r_f)D}{D^2 k\sqrt{\pi}},$$

giving,

$$\frac{T\pi kD}{P(1 - r_f)} = \sqrt{\pi} = 1.77. \tag{5.9}$$

This is an interesting rule of thumb for calculating the maximum possible temperature if the only loss mechanism is conduction.

5.5 Analytical Models for a Moving Point Source

By integrating the point source solution over time and moving it by making $x = (x_0 + vt)$, Rosenthal [8] developed his well-known fundamental welding equations. He actually had three equations: the one-dimensional solution for a melting welding rod; the two-dimensional solution for a moving line source (simulating a keyhole on thin plate welding); and the three-dimensional moving point source (simulating thick plate welding, which can also be applied for laser surface treatments).

The solution for the moving point source assumes a semi-infinite workpiece, no radiant loss, no melting and constant thermal properties over the temperature range concerned. The solution is [5]

$$T - T_0 = \frac{Q}{2\pi k} \frac{e^{-v(x-R)/2\alpha}}{R}. \tag{5.10}$$

For the boundary conditions,

$$\frac{\partial T}{\partial x} \to 0 \text{ for } x \to \infty; \quad \frac{\partial T}{\partial y} \to 0 \text{ for } y \to \infty; \quad \frac{\partial T}{\partial z} \to 0 \text{ for } z \to \infty,$$

and $\frac{\partial T}{\partial R} 4\pi R^2 k \to Q$ for $R \to 0$, where

$$R = \sqrt{x^2 + y^2 + z^2}.$$

From this, the rate of cooling for the centreline surface spot ($y = 0, x > 0, z = 0$) can be derived as [9, 10]

$$\frac{\partial T}{\partial t} = 2\pi k \left[\frac{v}{Q}\right](T - T_0)^2. \tag{5.11}$$

It is therefore possible to estimate the expected quench rates for various surface treatments, as shown in Figure 5.8 using a 2 kW laser on steel, $k = 52 \text{ W m}^{-1} \text{ K}^{-1}$; melting

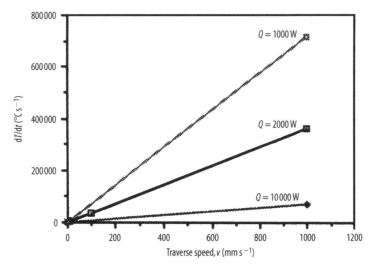

Figure 5.8 Estimated cooling rates from Equation 5.11 for steel; $k = 52\ \mathrm{W\ m^{-1}\ K^{-1}}$ melting point $1{,}500\,^\circ\mathrm{C}$

point $1{,}500\,^\circ\mathrm{C}$. These are fast cooling rates; faster rates can be achieved by pulsing; a 1-ps pulse would give a quench rate of around $10^{13}\ \mathrm{K\ s^{-1}}$.

5.6 Alternative Surface Heating Models

5.6.1 The Ashby–Shercliffe Model: The Moving Hypersurface Line Source

Many models have been published. They are mostly based on the fundamental solution for a point source or line source. A significant one for laser processing is that of Ashby and Shercliffe [11]. They derived a solution, as a development of the Ashby–Easterling hypersurface point source [12], in which the heat source is a moving finite line source situated above the surface and parallel to it. The advantage of this is that it allows a pseudo-beam diameter to be considered. The assumptions are as follows: the heat source is a line source of finite width in the y direction and is infinitesimally small in the x or welding direction; the workpiece is homogeneous and isotropic, having constant thermal properties; no account is taken of latent heat effects, radiation or surface convection heat losses; and the work piece is semi-infinite.

The solution for the temperature variation with time is

$$T - T_0 = \frac{Aq/v}{2\pi k\left[t\left(t + t_0\right)\right]^{1/2}}\exp\left[-\frac{1}{4\alpha}\left\{\frac{(z + z_0)^2}{t} + \frac{y^2}{t + t_0}\right\}\right]. \tag{5.12}$$

The cooling rate is

$$\frac{dT}{dt} = \frac{T - T_0}{t}\left\{\frac{(z + z_0)^2}{4\alpha t} - \frac{1}{2}\left(\frac{2t + t_0}{t + t_0}\right)\right\}. \tag{5.13}$$

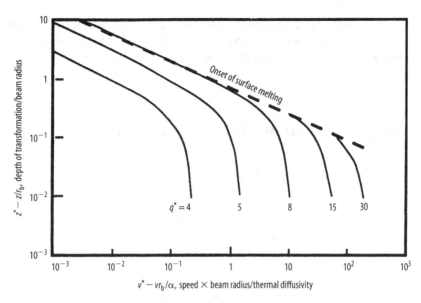

Figure 5.9 The Ashby–Shercliffe [11] "master plot" for predicting laser transformation depth of hardening. $q^* = AP/r_b k\,(T_{trans} - T_0)$, a dimensionless beam energy. T_{trans} is the transformation temperature (K)

All this can be plotted in their "master plot" as shown in Figure 5.9. This model is convenient to use and fits transformation hardening problems quite well.

5.6.2 The Davis *et al.* Model: The Moving Gaussian Source

First solved by Cline and Anthony [13], the moving Gaussian surface heat source problem was solved by Davis *et al.* [14] and they obtained a complex equation – as the reader might expect. However, they then derived an expression for the depth of hardness assuming constant thermal properties and no surface heat losses by convection or radiation; they also assumed that there were no latent heat effects.

Their solution for the depth of hardness, d, is

$$d = 0.76D \left[\frac{1}{\frac{\alpha'C}{q} + \pi^{1/4}\left(\frac{\alpha'C}{q}\right)^{1/2}} - \frac{1}{\frac{\alpha'C}{q_{min}} + \pi^{1/4}\left(\frac{\alpha'C}{q_{min}}\right)^{1/2}} \left(\frac{q_{min}}{q}\right)^3 \right], \qquad (5.14)$$

where

$$\alpha' = \frac{\rho u D C}{2k}; \quad q = \left[\frac{2P(1-r_f)}{D\pi^{3/2}k(T_c - T_0)}\right]; \quad q_{min} = (0.40528 + 0.21586\alpha C)^{1/2},$$

where T_c is the transformation temperature (°C), q_{min} is the minimum absorbed power for which any hardening can occur and $C = C_\infty - 0.4646\,(C_\infty - C_0)\,(\alpha C_0)^{1/2}$.

The constants vary with the material. For En8 (~ AISI 1040) they are $C_\infty = 1.599$ and $C_0 = 0.503$. Although this expression may look at first sight to be a bit formidable, it is simple to solve on a personal computer. The results have been shown to fit experimental findings very well [15].

5.7 Analytical Keyhole Models – Line Source Solution

5.7.1 Line Source on the Axis of the Keyhole

The Rosenthal solution for the moving line source [8] assumes that the energy is absorbed uniformly along a line in the depth direction. The assumptions are as for the other Rosenthal models. Figure 5.10 illustrates the heat flow pattern. The solution simulates single-pass welding in thin materials or a fully penetrating keyhole weld. It is [8]

$$T - T_0 = \frac{Q}{2\pi k g} e^{+\frac{vx}{2\alpha}} K_0 \left[\frac{vR}{2\alpha} \right]. \qquad (5.15)$$

The boundary conditions are

$$\frac{\partial T}{\partial x} = 0 \text{ as } x \to \infty; \quad \frac{\partial T}{\partial y} = 0 \text{ as } y \to \infty; \frac{\partial T}{\partial z} = 0$$

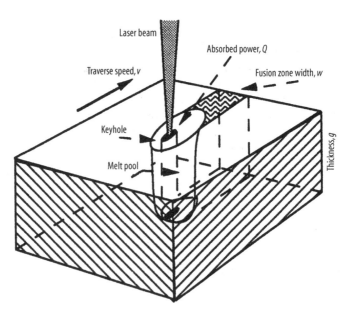

Figure 5.10 Approximate geometry of the heat flow pattern from a line source forming a keyhole weld

and

$$-\frac{\partial T}{\partial R}2\pi Rkg \rightarrow Q \text{ as } R \rightarrow 0,$$

where the symbols are explained at the end of the chapter.

Bessel functions are sometimes known as cylinder functions and occur in diffusion problems with cylindrical symmetry and on many other occasions. Bessel functions of the first kind, known as a "Bessel function", are

$$J_p(z) = \sum_{n=0}^{\infty}\left[(-1)^n\left(\frac{z}{2}\right)^{p+2n}\right]/[n!\Gamma(p+n+1)],$$

where $\Gamma(p+n+1) = (p+n)!$

Bessel functions of the second kind are known as Neumann functions,

$$Y_p(z) = \frac{\cos(p\pi)J_p(z) - J_p(z)}{\sin p\pi}.$$

Such functions have tables available but are difficult to solve.

From Equation 5.15, an asymptotic solution for the rate of cooling can be derived for the centreline location, $x > 0$, $y = 0$ and the Péclet number $vx/2a \gg 1$ (*i.e.*, at high speeds) as

$$\frac{\partial T}{\partial t} = 2\pi k\rho C\left[\frac{vg}{Q}\right]^2 (T - T_0)^3. \tag{5.16}$$

It can be seen that the relationships derived so far are between the dimensionless groups $vR/2\alpha$ and $Q/[2\pi(T - T_0)kg]$. Swifthook and Gick [16] observed that there were analytical solutions for the Bessel function at high or low speeds, and hence produced the plot shown in Figure 5.11. The limit for high speed was mentioned in Section 4.4.5. The limit for low speed is not so straightforward to understand. This graph is useful in predicting required powers and speeds for welding.

Example 5.1. A laser beam is required to weld 10-mm-thick 304 stainless steel at 10 mm s^{-1}. The usual weld width for the laser being considered is 1.5 mm. What laser power is required assuming a 90 % absorption in this keyhole welding process?

The thermal properties of 304 stainless steel are as follows: thermal diffusivity α = 0.49×10^{-5} m^2 s^{-1}, thermal conductivity k = 100 W m^{-1} K^{-1} and melting point T_m = 1,527 °C.

The solution is as follows. $Y = vw/\alpha = (0.01 \times 0.0015)/0.49 \times 10^{-5} = 3.06$. From Figure 5.11 or from the equation, $Y = 0.483X$,

$$X = 7.0 = Q/gkT = Q/0.01 \times 100 \times 1527.$$

Therefore Q = 10.6 kW and with 90 % transfer efficiency the required power is 10.6/ 0.9 = 11.8 kW.

This is a large power, but then it is capable of welding thick stainless steel at a high speed. If we only had a 2 kW laser then the calculation would show what speed we could expect. That is an exercise for the reader! One can see that such sums assume that the

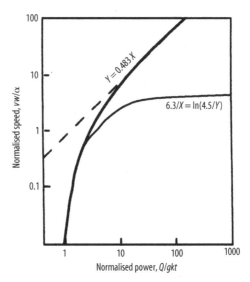

Figure 5.11 Normalised speed versus normalised power. A solution of the moving line source problem [16]

penetration is possible and that the beam can be focused to a fine spot – a line. It can also be seen that any calculation of the temperature near or within the experimental beam diameter will be badly wrong. These limitations can be overcome to some extent by using a cylindrical heat source as in Bunting and Cornfield's solution [17], or by using finite-element or finite-difference numerical methods.

5.7.2 Line Source Around the Surface of a Cylinder: One-dimensional Transient Model for Cylindrical Bodies

Sandven [18] presented a model which predicts the temperature distribution in the vicinity of a moving ring-shaped laser spot around the periphery of the outer surface of a cylinder or the inner surface of a hollow cylinder. Sandven [18] developed his model on the basis of the flat-plate solution provided by Carslaw and Jaeger [7] and assumed that the temperature distribution for cylindrical bodies can be approximated by a product solution of the form

$$T = \theta g I, \tag{5.17}$$

where θ depends on the cylindrical geometry of the work piece, g is the wall thickness and I is the analytical solution for an equivalent solution on a flat plate.

The expression for a cylindrical work piece is

$$T = \left(1 \pm 0.43\sqrt{\phi}\right) \frac{2Q_0\alpha}{\pi k g V} \int_{x-B}^{x+B} e^u g K_0 \left(z^2 + u^2\right)^{1/2} du, \tag{5.18}$$

where the plus sign is used for heat flow into a cylinder and the minus sign is used for heat flow out of a hollow cylinder; the other symbols are explained at the end of the chapter.

The integral part of Equation 5.18 must be evaluated numerically for $z > 0$. Graphical solutions for $z = 0$ for various values of B were provided by Sandven [18] to estimate an approximate hardened depth.

5.7.3 Analytical Moving Point–Line Source

Since temperatures can be added, it is relatively simple to add the point and line source solutions, and so simulate the Fresnel absorption by the line source and the plasma absorption by the point source located at any specified point along the line. Such a solution has been derived by Dowden [5], p. 154, Steen *et al.* [19].

Figure 5.12 shows the fit of an experimental fusion zone to the calculated zone allowing one to speculate on the proportion of power in the plasma or Fresnel absorption. This line of reasoning was taken further by Dowden *et al.* [20], who found, by arguing that the energy absorbed by the inverse bremsstrahlung process was approximately equal to that lost by the ablation process, that the depth of penetration in welding would therefore be determined mainly by the Fresnel absorption and the heat conduction losses. Taking an average value of the Fresnel absorption coefficient, R' (0.79), they were able to derive an expression for the radius of the keyhole, r:

$$-2(1 - R')[P_0/r_0][r/r_0]^{(1-2R')}(dr/dz) = 2\pi k[T_v - T_0]\text{cyl}(r/2L),$$

where the symbols are explained at the end of the chapter.

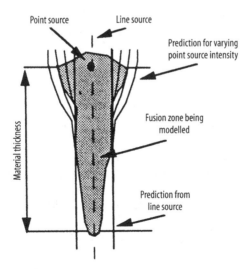

Figure 5.12 The solution achieved with the point–line source model [19]

This first-order differential equation has a boundary condition $r(0) = r_0$. By integrating from $z = 0$ to d, one can find the depth of penetration. One of the surprising findings from this work, amongst many gems, is that the radius at the full depth of penetration is $r_d = r_0/3$, a fact which coincides geometrically with the position of incidence of the first reflection off the keyhole wall from the top of the keyhole!

5.8 Three-dimensional Models

5.8.1 Three-dimensional Model for a Semi-infinite Plate

The thermal analysis for laser heating reported by Cline and Anthony [13] seems to be the most realistic analytical model reported so far. They used a Gaussian heat source and determined the three-dimensional temperature distribution by solving the following equation:

$$\frac{\partial T}{\partial t} - \alpha \nabla^2 T = \overline{Q}/c_p, \tag{5.19}$$

where $\overline{Q}$ is the power absorbed per unit volume and c_p is the specific heat per unit volume. They used a coordinate system fixed at the workpiece and superimposed the known Green function solution for the thermal distribution of a unit point source to represent the Gaussian distribution. The expression developed to determine the temperature distribution is as follows:

$$T(x, y, z) = P(c_p \alpha R)^{-1} f(x, y, zV), \tag{5.20}$$

where f is a distribution function,

$$f = \int_0^\infty \frac{\exp\left(-\overline{H}\right)}{(2\pi^3)^{1/2}\left(1+u^2\right)} d\mu, \tag{5.21}$$

and

$$\overline{H} = \frac{\left(X + \frac{\tau u^2}{2}\right) + Y^2}{2\left(1+u^2\right)} + \frac{Z^2}{2\mu^2}, \tag{5.22}$$

where $\mu^2 = 2\alpha t'/R$, $\tau = VR/\alpha$; $X = x/R$, $Y = y/R$ and $Z = z/R$, where P is the total power, R is the beam radius, t' is the earlier time when the laser was at (x', y') and v is the uniform velocity.

The cooling rate can also be calculated using the following expression:

$$\frac{\partial T}{\partial t} = V\left[x/r^2 + (V/2\alpha)(1 + x/r)\right] T, \tag{5.23}$$

where $r = \sqrt{x^2 + y^2 + z^2}$.

5.8.2 Three-dimensional Transient Model for Finite Slabs

In the context of laser chemical vapour deposition, Kar and Mazumder [21] solved the three-dimensional transient heat conduction equation with temperature-dependent thermophysical properties by considering both convection and radiation losses of energy at the boundaries. The geometrical configuration for this problem is given in Figure 5.13, where the laser beam moves in the x direction at a constant velocity, U. They solved the following equation

$$\rho C_p \frac{\partial T}{\partial t} = \nabla \left[k(t) \nabla T(x, y, z, t) \right], \tag{5.24}$$

and obtained the following expression for the temperature distribution:

$$T^* = \frac{1}{(\bar{q}+1)\,\overline{T}_a^{*\bar{q}}} \left\{ -\overline{T}_a^{*\overline{q+1}} + (\bar{q}+1)\,t'^* \right\} + \overline{T}_a^*, \tag{5.25}$$

where

$$T'^* \left(x^*, y^*, z^*, t^*\right) = \sum_{l=0}^{\infty} \frac{K_{lx}(x^*)}{N_{lx}} \left[\sum_{m=0}^{\infty} \frac{K_{my}(y^*)}{N_{my}} \left\{ \overline{T}^{iv} \left(\lambda_{lx}, \lambda_{my}, z^*, t^*\right) \right. \right.$$
$$\left. \left. + \sum_{n=0}^{\infty} \left(\psi_1(0)e^{-\lambda_{lml}^2 t^*} + \psi_2(t^*)\right) \frac{K_{nz}(z^*)}{N_{nz}} \right\} \right] - T_c \overline{T}^{iv} \left(\lambda_{lx}, \lambda_{my}, z^*, t^*\right)$$
$$= f_{lm}(t^*)\,(Z_0 z^* + 1)\,/\Delta_z,$$

where $\Delta_z = z_0\,(1 - z_H) - z_H$ (the symbols are explained at the end of the chapter).

Equation 5.25 is used to study the effect of various process parameters on chemically reactive zones and to determine the operating conditions for laser chemical vapour deposition. The chemically reactive zone is defined as the zone where the temperature is more than or equal to that at which the chemical reaction that generates the film-forming material takes place. A critical scanning speed can be identified and calculated at which the two boundary curves of the chemically reactive zone will collapse into one, giving rise to the narrowest possible film deposition region, as illustrated in Figure 5.14.

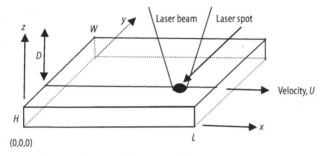

Figure 5.13 Geometric configuration of the substrate and the relative position of the laser spot

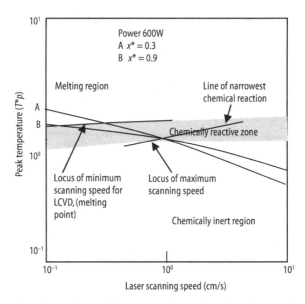

Figure 5.14 Examples of the theoretical process limits that have been calculated from the model [21]

5.9 Numerical Modelling

There are two main numerical techniques, finite difference and finite element. Both solve the basic conservation equations such as Fourier's second law for all internal points and then have special equations for the boundaries. A finite-difference model starts by dividing up the space to be considered into small boxes, whereas the finite-element model divides the space with linked lines.

Numerical methods remove many of the limitations that apply to the analytical methods. This is because the volume to be analysed is divided into small pieces (control volumes), each of which could have its own values of thermal conductivity, density, surface tension or heat input. For example, the heat source does not have to be concentrated in a point, line or plane. Temperature-dependent thermophysical properties and real boundary conditions may be included. In spite of the inherent advantages in numerical methods, only a few numerical models for heat flow in laser processing have been developed so far. This is partly because of the complexity in building such models and partly because the model does not easily show general trends. An analytical solution would show, for example, the temperature of the surface is proportional to $1/\sqrt{\text{time}}$. This would not be so apparent from a numerical solution, which is specific for a given time. To get a trend one would have to run the model several times, like an actual experiment. On the other hand a numerical solution can include more detail physics of the problem and thus reveal the effect of different phenomena on the process.

The way in which a problem can be introduced to the computer is by forming the defining differential equations into finite-difference form. For example, the heat balance

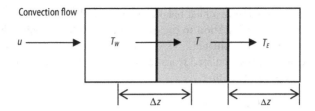

Figure 5.15 Notation for Equation 5.26 for the finite-difference equation

on the element shown in Figure 5.15, which is either heating up or cooling down, is given by Fourier's second law as

$$\alpha \frac{\partial^2 T}{\partial z^2} = \frac{\partial T}{\partial t} - u \frac{\partial T}{\partial z}.$$

This can be expressed in finite-difference form as

$$\alpha \left\{ \frac{[(T_w - T)/\Delta z] - [(T - T_E)/\Delta z]}{\Delta z} \right\} \approx \frac{T' - T}{\Delta t} - u \left(\frac{T_w - T_E}{2\Delta z} \right).$$

Therefore, the temperature, T', after one time interval, Δt, is

$$T' = \left[u \left(\frac{T_w + T_E}{2\Delta z} \right) + \alpha \left(\frac{T_w + T_E - 2T}{\Delta z^2} \right) \right] \Delta t + T. \tag{5.26}$$

The three-dimensional version of this equation is solved for all internal points to find the temperature after one time interval, Δt. This is done all over the whole matrix. An instability may arise in that the two gradient terms are centred on time $t = t$, whereas the variation of temperature with time term is centred on $t = t + \Delta t/2$. The Crank–Nicholson method [22] is sometimes used to overcome this problem, which could otherwise be a source of computing inefficiency.

If the problem is one of a quasi-steady-state nature, such as the weld pool being stationary relative to the laser beam in a moving substrate, then the time term may be dropped and the solution is found through a process known as "relaxation". In this form of calculation, used by Mazumder and Steen [23], the heat terms should add to zero. If they do not, then the steady-state temperature field is slowly relaxed, i.e., changed, by a calculated amount until they do.

If there is melting, then the latent heat is usually treated as an abnormal specific heat. Henry et al. [24] arranged for stages in melting to be considered by letting the specific heat C_p be a step function of the form

$$C_p(T) = \begin{array}{l} C_{p0} + \frac{\Delta H_m}{\Delta T_m} \text{ from } T_m \le T \le T_m + \Delta T'_m \\ \\ C_{p0} + \frac{\Delta H_v}{\Delta T_v} \text{ from } T_v \le T \le T_v + \Delta T'_v \end{array}, \tag{5.27}$$

where C_{p0} is the normal constant specific heat, ΔH_m and ΔH_v are the latent heats of fusion and boiling, respectively, and $\Delta T'_m$ and $\Delta T'_v$ are the temperature bands over which the transition occurs.

The boundary points have a heat balance similar to that of the internal points, but there has to be some modification to the gradients extending outside the zone, where the temperatures are not calculated.

The surface points are calculated from the virtual surface temperature gradient that would give the imposed heat flux:

$$k \left(\frac{\partial T}{\partial x} \right)_{x,y,1} = P_{x,y} \left(1 - r_f \right) - \left(h_c + h_r \right) \left(T_{surf} - T_a \right) , \tag{5.28}$$

where T_a is the ambient temperature. The value of P_{xy} depends upon which power distribution is chosen. For example, if the mode structure is Gaussian, TEM00, then P_{xy} is defined as

$$P_{x,y} = \frac{P_{tot}}{r_b^2 \pi} \exp \left(\frac{-2r^2}{r_b^2} \right)$$

where $r^2 = x^2 + y^2$.

Boiling of the surface can be allowed for by assuming that if the boiling point is reached at a certain matrix point, then that point will disappear, allowing the power to fall on the matrix point beneath it, with some plasma absorption loss accounted for by the Beer–Lambert absorption law, or Bouguer's exponential law [25]:

$$P_{x,y} = P_0 e^{-\beta \Delta z} , \tag{5.29}$$

where β is an absorption coefficient with units of inverse metres.

The side points are considered to be sufficiently far away that $\partial T / \partial x \approx 0$ and $\partial T / \partial y \approx 0$: the base points for thick substrates are assumed to have $\partial T / \partial z \approx 0$; but the base points for thinner substrates are calculated from the estimated z gradient at the full depth of g,

$$\left(\frac{\partial T}{\partial z} \right)_{x,y,g} = - \frac{\left(h_c + h_r \right) \left(T_{surf} - T_a \right)}{k} . \tag{5.30}$$

These finite-difference equations link the individual boxes and so the computer iterates around the complete matrix until the answer has converged sufficiently.

Hopefully, this quick summary has shown the reader not already familiar with these techniques how very versatile numerical models can be and that almost any physical phenomenon can be added and considered. This is clearly the strength of this technique as a tool for questioning process mechanisms. However, since the average model requires 0.5 MB of memory and several minutes on a fast computer, it is not a model useful for everyday prediction or control work. Some examples of the sort of solutions one can obtain from this form of analysis are shown in Figures 4.34 and 5.16–5.18. On the basis of this approach to the problem, a number of situations have been examined, which further illustrate the freedom in calculating made possible by numerical techniques.

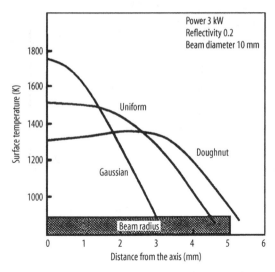

Figure 5.16 Finite-difference solution for the effect of mode structure on the surface temperature distribution [57]

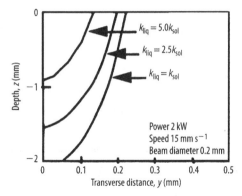

Figure 5.17 Finite-difference solution illustrating the effect on the melt pool size of the effective thermal conductivity of the melt pool. The effective conductivity is made up of the material conductivity and a convective element or eddy conductivity. This calculation shows how significant the pool stirring action could be on the shape and size of the melt pool [58]

5.9.1 Three-dimensional Thermal Model

The numerical solution to the three-dimensional heat transfer model for laser materials processing on a semi-infinite workpiece was developed by Mazumder and Steen [23] and was later modified by Chande and Mazumder [26]. It allows for temperature-dependent thermophysical properties, spatial distribution of the heat source (Gaussian, uniform or any other known distribution), radiative heat losses, latent heat of transformation and the removal of material by boiling.

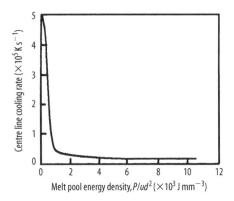

Figure 5.18 A finite-difference solution for the centreline cooling rate versus the melt pool energy density. The cooling rate will determine the structure of the solidification process (see Chapter 6 and [58])

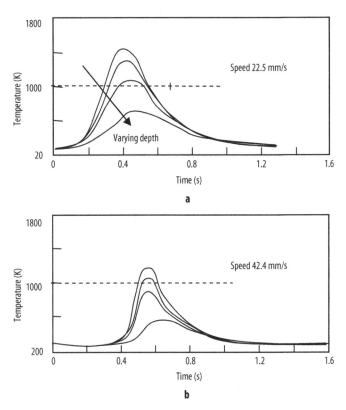

Figure 5.19 Theoretically predicted thermal cycle time during laser heating of En8 steel (EN8 is similar to AISI 1040 steel) (power 2 kW, beam radius 3 mm, reflectivity 0.4): **a** speed 22.5 mm s^{-1}, and **b** speed 42.4 mm s^{-1}. (After [27])

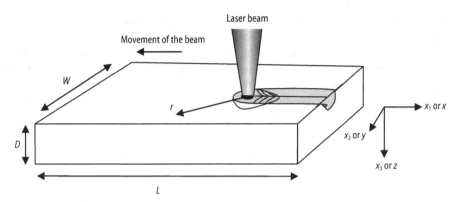

Figure 5.20 The configuration modelled for the two-dimensional self-consistent case

Courtney and Steen [27] used the nonboiling version of this model to predict the depths of the heat-treated zone and thermal cycle time, as shown in Figures 5.19 and 5.20. Knowledge of thermal cycle time is essential to calculate the carbon diffusion distance, cooling rates, time at temperature and other thermal aspects of microstructure formation.

5.9.2 Flow Within the Melt Pool – Convection

For processes such as melt quenching, alloying and welding, convection in the melt pool plays a major role in the process physics. It is also the single most important factor influencing the geometry of the pool, including pool shape, undercut and ripples, and can result in defects such as variable penetration, porosity and lack of fusion. Convection is also primarily responsible for mixing, and therefore affects the composition of the melt pool. The heat transfer and thus the cooling rate is greatly enhanced in the presence of convection [26]. The homogeneity of solute redistribution during laser surface alloying, as reported by Chande and Mazumder [28], can only be explained by the presence of convection currents within the pool.

Surface-tension-driven flow has been identified as responsible for the convection within the molten pool. Anthony and Cline [29] did an analysis on the surface-tension-driven flow within the molten pool. It was essentially a one-dimensional problem and the flow field did not couple to the energy equation. Therefore, no additional information was obtained as far as the heat transfer process was concerned. Oreper *et al.* [30] developed a two-dimensional convection model which included buoyancy, electromagnetic and surface tension forces. Numerical solutions based on a specified pool profile were obtained. It was found that the surface tension gradient is usually the dominant factor. However, the solid–liquid interface is not known *a priori* as they had to assume.

5.9.3 Pool Shape

The pool shape is, however, a part of the problem to be solved. In fact, the solid–liquid interface, *i.e.*, the pool shape, is a piece of information that is of great interest. A two-dimensional transient model for the laser surface melted pool was first developed by Chan *et al.* [31, 32]. It showed as expected that surface tension was responsible for the fluid flow and convection. The cooling rate at the edge of the pool was found to be higher than that at the bottom of the pool below the centreline, which agrees with the experimental findings that the microstructure is finer at the edge of the pool than at the bottom of the pool. It also predicted that the recirculating velocity is of 1 or 2 orders of magnitude higher than that of the scanning velocity.

5.9.3.1 Two-dimensional Self-consistent Transient Model for a Surface-tension-driven Pool

A two-dimensional transient self-consistent (*i.e.*, the solid–liquid interface position is a part of the solution) model for a laser-melted pool was first developed by Chan *et al.* [33]. The geometrical configuration of the process is given in Figure 5.20. The basic assumptions of this model were as follows:

1. A laser beam having a defined power distribution strikes the surface of an opaque material of infinite width, thickness and length.
2. Only part of the energy is absorbed.
3. Absorbed energy induces surface-tension-driven flow owing to the high temperature gradient.
4. The liquid metal is considered to be Newtonian so that the Navier–Stokes equation is applicable.
5. All properties of the liquid metal and solid metal are constant, independent of temperature (except the surface tension). This allows simplifications of the model; however, variable properties can be treated with slight modifications. The dependence of surface tension on temperature is assumed to be linear.
6. The latent heat of fusion is neglected since the energy liberated is small compared with total enthalpy change associated with temperature differences.
7. Thermal conductivity is assumed to be the same for both liquid and solid phases for the sake of simplicity of the model.
8. The surface of the melt pool is assumed to be flat to simplify the surface boundary condition, and hence surface rippling is neglected.

The governing equations are as for all these models based on conservation of mass, energy and momentum for each difference control volume. The fluid flow is governed by a balance between viscosity and surface-tension-driven surface flow. The boundary of the pool is defined by the velocities being zero and the temperature at the melting point. This type of problem is generally known as a Stefan problem in the literature [34] (Stefan problems involve moving surfaces and assume that the last point of the solid phase has the same temperature as the first point of the liquid phase). An iterative scheme is used to solve for the interface.

The relevant equations are given in [35].

The SOLA® computer program [36] was employed. The basic method of the algorithm is presented in Hirt's report [36].

This model provides details of the flow field and heat transfer on a plane perpendicular to the scanning direction. The flow behaviour predicted by this model will be valid

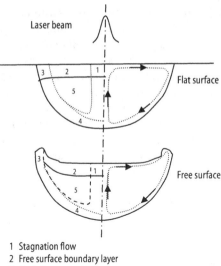

Laser beam

Flat surface

Free surface

1 Stagnation flow
2 Free surface boundary layer
3 Cooled corner region
4 Solid–liquid interface boundary layer region
5 Isothermal inviscid core

Figure 5.21 General features and various regions of the flow [33, 35, 37]

for regions 1, 2 and 5 in Figure 5.21. However, the front-to-back motion and its effect on heat transfer also play important roles in the physical process, especially for the fact that the recirculating velocity is so much higher than the scanning velocity. Such a motion cannot be obtained from two-dimensional models: a three dimensional model is required.

5.9.3.2 Three-dimensional Self-consistent Quasi-steady-state Model

Recently developed three-dimensional models [33, 35, 37] have provided better understanding of the flow behaviour. From these better accuracy of the magnitude of flow is obtained and the results should be valid for regions 1, 2, 4 and 5 in Figure 5.21.

Kou and Wang [37] developed a three-dimensional model for convection in a laser-melted pool using both surface-tension-driven and buoyancy-driven flow.

The semi-implicit method for pressure linked equation algorithm [38] was used to calculate the velocity field. A conjugate heat transfer method was used at the solid–liquid interface for the calculation. This model predicted a velocity profile and magnitude (on the order of metres per second) similar to that of the transient two-dimensional model developed by Chan *et al.* [32]. The conjugate heat transfer method uses discontinuous viscosity variation at the solid–liquid interface; therefore, the flow behaviour at the solid–liquid interface (Figure 5.21, region 4) will not be realistic unless an extremely fine grid is used. However, this model reconfirmed that surface tension induced by a temperature gradient significantly affected the flow behaviour and pool shape.

Chan *et al.* [33, 35] took a step-by-step approach to construct three-dimensional models for laser-melted pools. Initially, an axisymmetric stationary spot source was considered [33]. Subsequently, a perturbation solution based on low scanning velocity was sought for the basic axisymmetric case [35]. The advantage of seeking a perturbation solution, as it turns out, is that the three-dimensional flow is modelled by two sets of two-dimensional equations, which provide considerable computing advantages. However, the perturbation solution is valid only for low scanning velocities. Another model, which is a full three-dimensional numerical solution of the Navier–Stokes equations, was developed using a point-by-point partially vectorised iteration scheme. This model was then modified to accommodate a free surface [35]. Such models were used to study the effect of convection on pool geometry, cooling rate and solute redistribution.

The assumptions of the materials properties for the quasi-steady-state three-dimensional models are similar to those of a two-dimensional model. These models also assume a grey body radiative heat loss and a flat surface as the starting point.

5.9.3.3 Axisymmetric Model: Effect of Marangoni and Prandtl Numbers

A standard alternating direction iterative (ADI) [39] method is employed to solve the governing equation for the axisymmetric case. The effect of convection and the Prandtl number on the melt pool shape is studied using this model. The very existence of the flow field changes the mechanism of heat transfer from conduction to convection. It tends to move the higher-temperature fluid on the surface right underneath the beam sideways. This, in turn, melts down the solid at the edge of the molten pool, creating a wider pool. Because of the conservation of energy, the size of the molten pool must remain roughly unchanged, with or without convection. As a result, the molten pool becomes shallower. The effect of an increase in the Marangoni number on the pool shape was studied. The Marangoni number is a measure of the surface-tension-driven convection. An increase in the Marangoni number $\left[\gamma q r_0^2 / k\mu\alpha\right]$ implies an increase in the amount of heat being brought sideways. Consequently, the molten pool becomes wider and wider. The ranges of aspect ratios (width to depth ratio) for different Marangoni numbers are quite large, up to a 150 % increase as compared with the conduction case.

The effect of changing the Prandtl number was also studied [40, 41]. The Prandtl number is the ratio of the momentum diffusion to heat diffusion. To study its effect, the momentum diffusion must be kept constant and only the heat diffusion can be allowed to vary. This corresponds to keeping the Reynolds number (Re = Ma/Pr) constant and changing Pr and Ma. This, in some sense, implies that the flow field is being maintained while changing the thermal diffusion. Caution must be exercised because of the complexity of the process owing to the coupling of the energy and momentum equations. An increase in the Prandtl number with the Reynolds number being kept constant will increase the Marangoni number and hence the convective heat transfer. The aspect ratio will, therefore, increase with the Prandtl number.

5.9.4 Some Model Results

These models have been used to study surface temperature, velocity field, solute redistribution, temperature of the molten pool and cooling rate.

5.9.4.1 Surface Temperature

The temperature attains its maximum at the centre underneath the beam and decreases radially outwards. It is interesting to note that the temperature gradient, the driving force, has three distinct regimes. Its magnitude increases from zero at the centre in the radial direction roughly to the edge of the laser beam. It then decreases, and finally increases again as the flow approaches the edge of the molten pool. The fluid flow driven by the surface tension gradient becomes important in the second stage, where it tends to smooth out the temperature. Finally, the flow approaches the edge of the molten pool and turns downward.

5.9.4.2 Velocity Field

The variation of surface temperature due to surface heating induces a surface tension gradient pulling the molten materials radially outwards. The general pattern of the flow field is that the molten materials are going radially outwards on the surface. As the flow approaches the edge of the molten pool, it goes down and turns around. It then moves to the centre and comes up to complete the recirculation pattern. In addition to this recirculation, there is also the motion of the workpiece. Therefore, material enters the molten pool from the front portion of the pool and goes through the recirculation pattern, ultimately resolidifying on the trailing edge.

5.9.4.3 Solute Redistribution

To help the reader gain a qualitative understanding of the mechanism of solute redistribution, a particle trajectory is plotted in Figure 5.22 and experimental values are shown

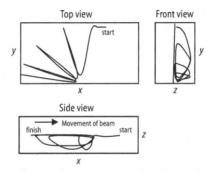

Figure 5.22 Trajectory of a particle as it enters, recirculates and ultimately freezes [40, 41]

in Figures 4.8–4.10. Figure 5.22 shows the particle trajectories in three views – front, side and top. The recirculating pattern of the particle can be clearly observed. This recirculating pattern implies that a particle will travel on a rather long path before it freezes into the resolidifying surface. Consequently, the molten materials can be well mixed. The composition within the melt pool is, therefore, uniform.

5.9.4.4 Temperature Field Within the Molten Pool

The isotherms in the vicinity of the solid–liquid interface are compressed near the surface of the pool and stretched apart near the bottom owing to the existence of the flow field. This distortion is due to recirculation of the molten material. The isotherms and the solid–liquid interface are asymmetrical owing to the motion of the workpiece.

5.9.4.5 Cooling Rate

The microstructure at the moment of solidifying is related to the cooling rate: the higher the cooling rate, the finer the microstructure. The cooling rate (*i.e.*, the temperature gradient) decreases from a maximum at the centreline to zero at the edge of the molten pool and also decreases from a maximum on the surface to a minimum at the bottom of the molten pool. Such a variation in cooling causes a variation of the resolidified microstructure. Further investigation of the solidification mechanism is required to fully predict the resulting microstructure.

5.9.5 Effect of Flow on Surface Deformation

Most of the work discussed so far assumes a flat surface. The effect of surface rippling induced by a surface tension gradient during laser surface melting and alloying was first studied by Anthony and Cline [29]. They essentially solved the Navier–Stokes equation for the one-dimensional steady-state condition with surface tension and buoyancy forces. They derived an algebraic relationship for the ripple height:

$$\Delta h = (3/2)\frac{T_{\mathrm{m}}}{\rho g h_0}\frac{\partial \sigma}{\partial T}, \tag{5.31}$$

where Δh is the height difference between the surface below the beam and the surface at the solidifying solid–liquid interface, T_{m} is the melting temperature of the metal and

$$h_0 = 10R \int_0^\infty \frac{\exp\left\{\left(-(r_0 D v)^2/r\mu^4\right)\left[2\left(1+\mu^2\right)\right]^{-1}\right\}}{(2\pi^3)^{1/2}\left(1+\mu^2\right)}\,\mathrm{d}\mu,$$

where r_0 is the radius of the laser beam, D is the thermal diffusivity of the liquid, v is the velocity of the laser beam and μ is the fluid viscosity. This work also estimated a critical velocity required to suppress surface rippling. However, this calculation will not be very realistic for melt pools with recirculating velocities.

Chan *et al.* [42, 43] modified the three-dimensional numerical model to study the forced surface deformation. The method shows the following trends. Thermocapillarity drives the surface fluid radially outwards at high velocities. These high velocities displace more mass from the central surface region than can be replaced by the recirculating flow, thus causing a depression. The displaced mass builds up at the solid–liquid interface, causing the surface to bulge upwards, where it is then forced downwards into the molten pool. For this calculation, the contact angle is unknown. The contact edge height, however, was defined equal to the initial surface height.

5.9.6 Model for Flow with Vaporisation

Processes such as deep penetration welding, cutting and drilling have vaporisation affecting the flow dynamics. One of the major problems of treating the vapour–liquid interface is the discontinuity at the interface and this makes application of continuum mechanics rather difficult. Theoretical understanding of such problems is in its infancy.

Recently, Chan and Mazumder [40] developed a one-dimensional steady-state model where a thin discontinuity of the order of a molecular mean free path thickness at the solid–liquid interface is incorporated using the "Knudsen layer" jump condition and the Mott–Smith type solution. In this model, the vaporisation process creates a recoil pressure which pushes the vapour away from the target and expels the liquid. The materials are, therefore, removed in both the vapour phase and the liquid phase. The removal rates of the materials are incorporated in the moving boundary immobilisation transformation. The vapour phase is assumed to be optically thin so that its absorption of the high-energy beam is negligible. However, this is a simplifying assumption at this initial stage. Rockstroh and Mazumder [41] have observed from their spectroscopic work that the beam does interact with the plasma. Owing to the simplification, closed-form analytical solutions are obtained for this analysis. The effect of heat source power on removal rates, vaporisation rates, liquid expulsion rate, surface temperature and Mach number were examined. Results were obtained for three different materials – aluminium, superalloy, and titanium and are shown in Section 3.3.3.2.3.

5.9.7 Mass Additions – Surface Alloying and Cladding

For processes such as laser surface alloying and cladding, mass transport is a very important phenomenon. A mathematical model for mass transport during laser surface alloying would help clarify several aspects of the problem. A set of processing conditions could be tested for uniformity of mixing and for the resultant average compositions in the liquid state. The mass flux necessary to obtain a desired average composition in the liquid state could be computed. Powder loss during alloying would then be estimable, knowing the actual mass flux during laser processing. The model could be used to simulate the effect of varying the method of supplying alloying elements. Having predicted an average liquid pool composition and measured the actual solid-state composition, one can calculate the effective solute partitioning coefficient C_s^* at the solid–liquid interface. This value could then be compared with the value determined from the equilibrium phase diagram to check if conditions of local equilibrium existed ahead of the solid–liquid interface. This would be a useful check as local equilibrium is assumed in predicting possible compositions during rapid solidification from equilibrium phase diagrams.

5.9.7.1 Two-dimensional Transient Model for Mass Transport in Laser Surface Alloying

Chande and Mazumder [44] used the momentum transfer model described earlier and coupled it with the diffusion equation to estimate the mass transfer during laser surface alloying,which was reasonable because convection dominates the process.

For mass transfer calculations, the following assumptions were made:

1. The effect of alloying on solute mass diffusivity was neglected because accurate high-temperature data were unavailable; mass transport by diffusion was negligible compared with that by convection ($u_0 d/D > 100$). This simplified the formulation.
2. Mass flux at the surface was constant and uniform across the width of the pool. This rate and distribution could be altered to allow for any other experimental condition, such as wire feed or nonuniform powder addition.
3. There was no transfer of alloying elements across the solid–liquid interface. This was a good assumption that was verified experimentally using the electron probe microanalysis technique.

5.9.7.2 One-dimensional Transient Model for Laser Cladding

Laser cladding is a process used for altering the surface properties of substrate materials. Owing to a high cooling rate in laser cladding, extended solid solutions are formed whenever a mixture of several elements is used as cladding powder. Usually, the extended solid solutions have a higher melting point and better oxidation and corrosion-resistance properties than the alloys with equilibrium compositions. To study the extension of solid solubility during laser cladding, a diffusion model was developed by Kar and Mazumder [45–48].

The cladding was assumed to melt almost instantaneously as soon as it is exposed to the laser beam and reaches a uniform temperature T_2. The method of the solution is given in [46].

5.10 Modelling Laser Ablation

The laser ablation process is used for producing thin films with nanocrystalline single-crystal structure as well as nanocrystalline particles. The simultaneous evaporation associated with laser ablation makes it a preferred process for the fabrication of multicomponent films of particles. However, the physics associated with the process is rather complicated, including a discontinuity at the liquid–vapour interface. The mathematical modelling of the laser ablation process to produce nanocrystalline particles to

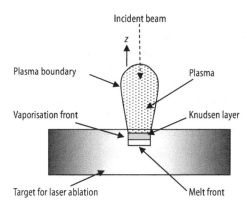

Figure 5.23 The process physics during laser ablation

provide quantitative understanding is discussed in [49–51]. One of the features of the model is the jump condition caused by the sonic emission of vapour and hence a discontinuity known as the Knudsen layer (see Figures 5.1, 5.23).

To understand the effects of various laser parameters on the diameters of the particles formed during laser ablation, the droplet-growth theory of Lifshitz and Slyozov [52] can be used to determine the particle size distribution in the plasma. According to this theory, the mean diameter of a cluster $D_c(t)$, is given by

$$D_c(t) = (32\alpha Dt/9)^{1/3}. \tag{5.32}$$

Here D is the mass diffusion coefficient and $\alpha = 2\sigma_t v' C_{0\infty}/(k_B T)$, where σ_t is the surface tension, v' is the molecular volume of the droplet material and $C_{0\infty}$ is the concentration of the saturated solution, which is defined in terms of the volume of the material dissolved in unit volume of the solution. Although Equation 5.32 was used in this study, it should be noted that the detailed condensation processes and the arrangement of chemical bonds are very important in the formation of the clusters. The model is described in [52]. From this study various observations were made:

1. More material melts and vaporises as the laser intensity or the ablation time increases. Only a very small amount of material vaporises during one laser pulse, but an extremely large vapour flux is generated.
2. The temperature and vapour flux at the vaporisation front increase rapidly during the first several nanoseconds of the pulse and then become constant near the end of the pulse.
3. The radial and axial velocities vary from a few to several hundred metres per second.
4. A weakly ionised plasma is formed, and the ion mass fraction increases as one moves towards the plasma boundary.
5. Nanoscale particles exist near the vaporisation front. The particle size increases as one moves towards the plasma boundary, or as the residence time of the particles in the plasma increases.

6. The energy losses due to conduction in the target, absorption and scattering in the plasma and vapour phases, conversion of thermal energy into the kinetic energies of the ablated particles, and the conduction, convection and radiation losses from the target surface are very important in the accurate determination of the ablation rates.
7. The trends for the theoretical and experimental results are similar, but the theoretical predictions give higher values than the experimental data, which could be due to the multidimensional energy losses and the lack of high temperatures.

5.11 Semiquantitative Models

There are a number semiquantitative models, such as that in Section 5.3 for calculating the HAZ, but the one by Klemens [53] should be noted. He was analysing the keyhole by assuming only radial heat flow, so a heat balance indicates that

$$P(r) = -k(\partial T/\partial r)2\pi r. \tag{5.33}$$

If the power is only absorbed by the keyhole plasma of radius r_c, then we know that for a "top hat" mode (uniform step function)

$$P(r) = W\pi r_c^2 = P, \tag{5.34}$$

where W is the power per unit area per unit depth (W m^{-2} m^{-1}).

Integrating Equation 5.33 and using Equation 5.34 as a boundary condition gives

$$(P/2\pi k)\left[\ln\left(r/r_c\right)\right] = T_c - T, \tag{5.35}$$

where T_c is the plasma temperature degrees Kelvin.

At the edge of the keyhole $r = r_0$ and $T = T_v$, the boiling point in degrees Kelvin. Therefore,

$$r_0/r_c = \exp\left[2\pi k\left(T_c - T_v\right)/P\right]. \tag{5.36}$$

Thus, we have some concept of the size and temperatures within the keyhole. There are other gems in Klemens's paper which show that there is likely to be a neck at the top of the keyhole.

Other semiquantitative arguments are presented in Steen and Courtney [54], showing that the transformation hardening depth is likely to be proportional to the group $P/(VD)^{1/2}$. The reasoning is as follows.

The depth of penetration of a given isotherm will be governed by the laws of heat conduction, which, as we have seen, involve the Fourier number. Thus, the depth of this isotherm will be given from a particular value of the group $\Delta x^2/\alpha t$. The value of the isotherm at that depth will be governed by the heat intensity, which is P/D. Now the time of heating from a moving disc source will be given by some value such as D/V.

Assuming the relationships are linear, we have

$$\Delta x = A \left(\frac{P}{D} \right) \left(\frac{\alpha D}{V} \right)^{1/2} ,$$

and therefore

$$\Delta x = f \left[\frac{P}{(DV)^{1/2}} \right] , \tag{5.37}$$

which fits the experimental findings quite well.

A further illustration is from Steen and Powell [55] on the movement of the solidification front beneath a laser clad. It shows the likelihood that the interface region will resolidify almost instantly after the formation of a fusion bond if the operating conditions are correct (see Section 6.5.1).

Olsen [56] calculated the melt thickness at the cut front during laser cutting by the following elegant reasoning. The molten film at the cut front is maintained by thermal conduction through the melt before being blown away by the assist gas passing through the kerf. Thus, there is a thickness described by conduction and another described by the drag/pressure forces.

The conduction thickness, t_{cond}, is found from Fourier's first law. The heat to melt, q (watts per unit depth), at the speed of cutting, u (m s^{-1}), is

$$q = uw\rho \left[(T_m - T_0) C_p + L_m \right] ,$$

where w is the width of the cut and the liquid–solid interface is assumed to be at the melting point, T_m.

From Fourier's first law this heat is supplied by conduction through the melt. This ignores convection on the principle that the temperature through the depth of the cut does not vary significantly since the surface of the melt is approximately at the boiling point throughout. Thus, assuming the film is thin and the temperature gradient is approximately linear,

$$q = uw\rho \left[(T_m - T_0) C_p + L_m \right] = -kw \frac{(T_v - T_m)}{t_{cond}} .$$

Hence

$$t_{cond} = \frac{k(T_v - T_m)}{\rho u \left[(T_m - T_0)C_p + L_m \right]} . \tag{5.38}$$

The momentum-controlled thickness, t_{mom}, can be calculated from the force moving the film vertically due to the pressure drop across the depth of the cut, ΔP, and the gaseous drag forces. A mass balance on the material to be expelled from the cut front of depth, h (m) and cutting speed, u (m s^{-1}) and the melt created of thickness t_{mom} (m) and flow rate down the cut front of v (m s^{-1}) is

$$\left[t_{mom} v \right] = hu ,$$

where h is the depth of the cut in metres.

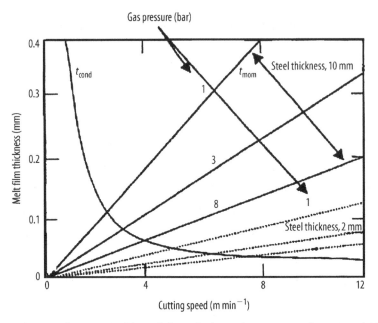

Figure 5.24 Theoretical estimates of the melt front thickness during cutting for various thicknesses of plate and assist gas pressures

Assuming the gaseous drag forces are negligible compared to the overall pressure drop through the cut depth, the velocity, v, is approximately given by the energy balance

$$v = \left(\frac{2\Delta P}{\rho}\right)^{1/2} ;$$

hence,

$$t_{\text{mom}} = hu\left(\frac{\rho}{2\Delta P}\right)^{1/2} . \tag{5.39}$$

These two thicknesses should always be the same and hence the film thickness for various processing conditions, such as variations in assist gas pressure, can be estimated from plots such as those in Figure 5.24.

5.12 Conclusions

In this chapter on mathematical modelling we have tried to present the concept of process modelling from a simple equation (in this case using a temperature-conversion calculation) through analysis based on amenable geometries to modelling complexities incorporating different process physics, including multiple dimensions, convection, free surface deformation, extended solid solution, vaporisation and discontinuity at the solid–liquid interface. The purpose of all these models is to give a deeper understanding of the process mechanisms. Those wishing to create a model have the choice either to select an existing model that they think is most suitable for the physics of their problem or to build a model from scratch using their own boundary conditions and solution techniques, some of which have been shown in this chapter, with more detailed descriptions available in the references given. At the minimum we hope this chapter will remove some of the mystery surrounding mathematical models and give the reader some concept of their accuracy and importance. At best we hope it will empower and enthuse the reader to have a go. Best of luck!

Nomenclature

a	Radius of cylinder	m
A	Cross-section area of beam	m^2
b	Half width of the heat source in the direction of motion	m
B	Dimensionless welding group	$(vb/2\alpha)$
C_p	Heat capacity at constant pressure	J kg^{-1} K^{-1}
cyl	Function based on Bessel functions	
d	Diameter of the laser beam	m
D	Characteristic distance	m
F_0, or F	Fourier number, $\alpha t/x^2$	
F_0	Absorbed power density	W m^{-2}
g	Plate thickness	m
g_c	Gravitational acceleration	m s^{-2}
h	Heat transfer coefficient	W m^{-2} K^{-1}
h_c	Convective heat transfer coefficient	W m^{-2} K^{-1}
h_r	Radiative heat transfer coefficient	W m^{-2} K^{-1}
H	Heat generated per unit volume	J m^{-3}
k	Thermal conductivity	W m^{-1} K^{-1}
$k^*(T^*)$	Nondimensional thermal conductivity	
K_0	Bessel function of the second kind and zero order	
L	Length of the substrate	m
L	Link intensity (k/ruC_p)	m
L_c	Characteristic length based on domain size	m
N_{lx}	Neumann function	
p	Pressure	N m^{-2}
P	Incident power	W
P_{tot}	Total delivered beam power	W
P_0	Power striking the surface	W
Pe	Péclet number (vD/α)	
Pr	Prandtl number $(C\mu/k$ or $\alpha\mu/\rho)$	

$\overline{q}$	Slope of the $\log k(T)$ versus T graph	
q	Net heat flux distribution	$\mathrm{W\,m^{-2}}$
Q	Total power from laser	K ($Q\rho C$, the point source energy)
Q	Heat input per unit volume $[p(1-r_f)]$	W
Q_0	Power density	$\mathrm{W\,m^{-2}}$
r	Radial distance from the centre of the laser beam	m
r_f	Reflectivity	
r_0	Radius of laser beam or keyhole	m
r, z	Cylindrical coordinates	m
R	Radial distance	m
Re	Reynolds number ($\rho u D/\mu$)	
t	Time variable	s
t_0	Time on	s
t_1	Time off	s
t^*	Normalized time variable	
T	Temperature	K
T_0	Initial temperature	K
u	Integration variable	
U	Translational speed of the substrate with respect to the laser beam	$\mathrm{m\,s^{-1}}$
v	Traverse speed	$\mathrm{m\,s^{-1}}$
V	Velocity of source in the x direction	$\mathrm{m\,s^{-1}}$
W	Width of the substrate,	m
x	Length	m
x^*	Normalised longitudinal position	
x_1, x_2, x_3	Rectangular coordinates	m
X	Dimensionless welding group (vx/α)	
y	Transversal position	m
y^*	Normalised transversal position	
z	Depth	m
z^*	Normalised depth	
z_0	Height of the hypersurface line source above the substrate	m
Z	$v/2\alpha$	
∇	The three-dimensional del operator	$\mathrm{m^{-2}}$

Greek Symbols

α	Thermal diffusivity	$m^2\,s^{-1}$
β	Beer–Lambert absorption coefficient	m^{-1}
γ	Surface tension	$N\,m^{-1}$
γ'	Temperature coefficient of surface tension	$N\,m^{-1}\,K^{-1}$
ε	Emissivity	
θ	Angular variable in cylindrical coordinates,	rad
λ	Latent heat of fusion	$J\,kg^{-1}$
μ	Dynamic viscosity	$kg\,m^{-1}\,s^{-1}$
ν	Kinematic viscosity	$m^2\,s^{-1}$
ρ	Density	$kg\,m^{-3}$
σ	Stefan–Boltzmann constant	$W\,m^{-2}\,K^{-4}$
$\varphi(t)$	Rate of heat liberation in a continuous point source	K
τ	Laser-substrate interaction time, $(t - t')^{-1/2}$	s
ω	Relaxation factor	

Subscripts

l	Liquid
m	Melting
s	Solid
∞	Ambient
trans	Transformation temperature

Superscripts

*	Dimensionless quantities

Questions

1. What effect would double the thermal diffusivity have on the expected depth of transformation hardening when surface heating with a 1-cm-diameter laser beam?
2. What is the maximum possible surface temperature at the centre of a stationary Gaussian beam of 5 W power and 3 mm diameter spot size incident on a slab of glass whose reflectivity is 0.5 and whose thermal conductivity is $1.3\ W\,m^{-1}\,K^{-1}$?
3. In one-dimensional heat flow how does the temperature vary with time?
4. Using the line source model, how fast could 3 mm stainless steel be welded with a 2 kW laser capable of giving a 1 mm wide keyhole weld? The melting point of stainless steel is $1{,}527\,^{\circ}C$, the thermal diffusivity is $0.49 \times 10^{-5}\ m^2\,s^{-1}$ and the thermal conductivity is $100\ W\,m^{-1}\,K^{-1}$.

References

[1] Perry RH, Green D (eds) (1973) Perry's chemical engineering handbook, 5th edn. McGraw-Hill, New York, pp 2–114

[2] Abramowitz M, Stegun LA (eds) (1964) Handbook of mathematical functions. National Bureau of Standards, applied mathematics series no. 55. US GPO, Washington

[3] Gregson V (1983) Laser heat treatment. In: Bass M (ed) Laser materials processing. North-Holland, Amsterdam, pp 201–23

[4] Carslaw HS, Jaeger JC (1959) Conduction of heat in solids, 2nd edn. Oxford University Press, Oxford, pp 11–12

[5] Dowden MJ (2001) Mathematics of Thermal Modelling: An Introduction to the Theory of Laser Material Processing. Chapman and Hall/CRC, Boca Raton, Fl

[6] Breinan EM, Kear BH (1983) Rapid solidification laser processing at high power density. In: Bass M (ed) Laser material processing. North-Holland, Amsterdam, pp 235–295

[7] Carslaw HS, Jaeger JC (1959) Conduction of heat in solids, 2nd edn. Oxford University Press, Oxford

[8] Rosenthal D (1946) The theory of moving sources of heat and its application to metal treatments. Trans ASME 48:849–866

[9] Adams CM (1958) Cooling rates and peak temperatures in fusion welding. Weld J 37:210S–215S

[10] Jhaveri P, Moffatt WG, Adams CM Jr (1962) The effect of plate thickness and radiation on heat flow in welding and cutting. Weld J 41(1):12s–16s

[11] Ashby MF, Shercliff HR (1986) Master plots for predicting the case depth in laser surface treatments. CUED/C-MAT/TR134. Engineering Department, Cambridge University

[12] Ashby MF, Easterling KE (1984) The transformation hardening of steel surfaces by laser beams. Acta Metall 32:195–1948

[13] Cline HE, Anthony TR (1977) Heat treating and melting material with a scanning laser or electron beam. J Appl Phys 48:3895

[14] Davis M, Kapadia P, Dowden J, Steen WM, Courtney CHG (1986) Heat hardening of metal surfaces with a scanning laser beam. J Phys D Appl Phys 19:1981–1997

[15] Bradly JR (1988) A simplified correlation between laser processing parameters and hardened depth in steels. J Phys D Appl Phys 21:834–637

[16] Swifthook DT, Gick AEF (1973) Penetration welding with lasers. Weld Res Suppl Nov 492s–498s

[17] Bunting KA, Cornfield G (1975) Towards a general theory of cutting: a relationship between power density and speed. Trans ASME J Heat Transf 97:116–122

[18] Sandven OA (1979) Heat flow in cylindrical bodies during laser transformation hardening. Proc SPIE 198:138–143

[19] Steen WM, Dowden J, Davis M, Kapadia P (1988) A point line source model of laser keyhole welding. J Phys D Appl Phys 21:1255–1260

[20] Dowden J, Ducharme R, Kapadia P, Clucas A, Steen WM (1996) A simplified mathematical model for the continuous laser welding of thick sheets. Report EU194. The Welding Institute, Great Abington

[21] Kar A, Mazumder J (1989) Three-dimensional transient thermal analysis for laser chemical vapor deposition on uniformly moving finite slabs. J Appl Phys 65:2923–2934

[22] Crank J, Nicholson P (1947) A practical method for numerical evaluation of solutions of partial differential equations of the heat conduction type. Proc Camb Philos Soc 43:50–67

[23] Mazumder J, Steen WM (1980) Heat transfer model for CW laser materials processing. J Appl Phys 51:941–947

[24] Henry P, Chance T, Lipscombe K, Mazumder J, Steen WM (1982) Modelling laser heating effects. In: ICALEO'82 proceedings, Boston, September 1982. LIA, Tulsa, paper 4B-2

[25] Jenkins FA, White HE (1951) Fundamentals of optics, 2nd edn. McGraw-Hill, London, p 197

[26] Chande T, Mazumder J (1981) Heat flow during CW laser materials processing. In: Mukherjee K, Mazumder J (eds) Lasers in metallurgy. The Metallurgical Society of the AIME, Warrendale, pp 165–194

[27] Courtney CGH, Steen WM (1979) Surface heat treatment of En8 steel using a 2 KW continuous-wave CO_2 laser. Met Technol 6:456–462

[28] Chande T, Mazumder J (1982) Mass transport in laser surface alloying: iron-nickel system. Appl Phys Lett 41:42–43

[29] Anthony TR, Cline HF (1977) Surface rippling induced by surface tension gradients during laser surface melting and alloying. J Appl Phys 48:3888–3894

[30] Oreper GM, Eagar TW, Szekely J (1983) Convection in arc weld pools. Weld J 62:307–312

[31] Chan CL, Mazumder J, Chen MM (1983) A model for surface tension driven fluid flow in laser surface alloying. In: Metzbower EA, Copley SM (eds) Second international conference on applications of lasers in materials processing, Los Angeles, 24–26 January. ASM, Metals Park, pp 150–158

[32] Chan CL, Mazumder J, Chen MM (1984) A two-dimensional transient model for convection in laser melted pools. Metall Trans 15A:2175–2184

[33] Chan CL, Mazumder J, Chen MH (1987) A three-dimensional axisymmetric model for convection in laser melted pool. Mater Sci Eng 3:306–311

[34] Viskanta R (1984) Phase change heat transfer. In: Lane GA (ed) Solar heat storage: latent heat materials. CRC, New York, 153–222

[35] Chan CL, Zehr R, Mazumder J, Chen MM (1986) Three-dimensional model for laser weld pool. In: Kuo B, Mehrabian R (eds) Proceedings of the 3rd Engineering Foundation conference on modeling and control of casting and welding processes. TMS-AIME, Warrendale, pp 229–246

[36] Hirt CW, Nichols BD, Romero NC (1975) A numerical algorithm for transient fluid flows, UC-34 and UC-79d, April 1975. University of California Berkley

[37] Kou S, Wang JH (1986) Three-dimensional convection in laser melted pools. Metall Trans 17A:2265–2270

[38] Patankar SV, Spalding DB (1972) A calculation procedure for heat, mass and momentum transfer in three-dimensional perabolic flows. Int J Heat Mass Transf 15:1787–1806

[39] Carnahan B, Luther HA, Wilkes JO (1969) Applied numerical methods. Wiley, New York

[40] Chan CL, Mazumder J (1987) One-dimensional steady-state model for damage by vaporization and liquid expulsion during laser materials interaction. J Appl Phys 62:4579–4586

[41] Rockstroh TJ, Mazumder J (1987) Spectroscopic studies of plasma during laser materials interaction. J Appl Phys 61:917–923

[42] Chan CL, Chen MM, Mazumder J (1988) Asymptotic solution for thermocapillary flow at high and low Prandtl numbers due to concentrated surface heating, ASME J Heat Transf 110:140–146

[43] Chan CL, Chen MM, Mazumder J (1985) Convection in the central region of a deep melt pool due to intense, non-uniform surface heating. In: ASME/AIChE national heat transfer conference, Denver 4–7 August. ASME, New York, 85-HT-21

[44] Chande T, Mazumder J (1985) Two-dimensional transient model for mass transport in laser surface alloying. J Appl Phys 57:2226–2232

[45] Kar A, Mazumder J (1987) One-dimensional diffusion model for extended solid solution in laser cladding. J Appl Phys 61:2645–2655

[46] Kar A, Mazumder J (1988) One-dimensional finite-medium diffusion model for extended solid solution in laser cladding of Hf on nickel. Acta Metall 36:701–712

[47] Kar A, Mazumder J (1989) Extended solid solution and nonequilibrium phase diagram for Ni-Al alloy formed during laser cladding. Metall Trans A 20:363–371

[48] Li LJ, Mazumder J (1984) A study of the mechanism of laser cladding processes. In: Mukherjee K, Mazumder J (eds) Laser processing of materials. The Metallurgical Society of AIME, Los Angeles, pp 35–50

[49] Kozlov, IM, Romanov GS, Stankevich YA, Teterev AV (1989) Heat Transf Sov Res 21:222

[50] Zel'dovich YB, Raizer YP(1966) Physics of shock waves and high-temperature hydrodynamic phenomena, vol 1 (translation edited by Hayes WD, Probstein RF). Academic, New York, pp 166, 265, 444, 195

[51] Lifshitz IM, Slyozov VV (1961) J Phys Chem Solids 19:35

[52] Kar A, Mazumder J (1994) Mathematical model for laser ablation to generate nanoscale and submicrometer-size particles. Phys Rev E 49(1):410–419,

[53] Klemens PG (1976) Heat balance and flow conditions for electron beam and laser welding. J Appl Phys 47(5):2165

[54] Steen WM, Courtney GHC (1979) Surface heat treatment of En8 steel using a 2 kW continuous wave CO_2 laser. Met Technol 6(12):456

[55] Steen WM, Powell J (1990) Theoretical modeling of the laser cladding process. In: Proceedings of the European scientific lens workshop on mathematical simulation, Lisbon 1989. Sprechsaal, Coburg, pp 143–160
[56] Olsen FO (1994) Fundamental mechanisms of cutting front formation in laser cutting. Proc SPIE 2207:402
[57] Sharp MC (1986) Mathematical modeling on continuous wave CO_2 laser processing of materials. PhD thesis, University of London
[58] Steen WM, Mazumder J (1984) Mathematical modeling of laser/material interactions. Report AFOSR-82-0076 GRA vol 84, no 17, 17 Aug 1984

"A Fast Fourier Transformation!"

6 Laser Surface Treatment

> Beauty is only skin deep, but it is only the skin you see
>
> *Fifteenth century proverb amended by A. Price, 44 Vintage xix (1978)*

6.1 Introduction

The laser has some unique properties for surface heating. The electromagnetic radiation of a laser beam is absorbed within the first few atomic layers for opaque materials, such as metals, and there are no associated hot gas jets or eddy currents and there is even no radiation spillage outside the optically defined beam area. In fact the applied energy can be placed precisely on the surface only where it is needed. Thus, it is a true surface heater and a unique tool for surface engineering. The range of possible processes with the laser is illustrated in Figure 6.1. Common advantages of laser surfacing compared with alternative processes are:

- chemical cleanliness;
- controlled thermal penetration and therefore distortion;
- controlled thermal profile and therefore shape and location of the HAZ;
- less after-machining, if any, is required;
- remote noncontact processing is usually possible; and
- relatively easy to automate.

Surface treatment is a subject of considerable interest at present because it seems to offer the chance to save strategic materials or to allow improved components with idealised surfaces and bulk properties. These ambitions are real and possible, but soon our maturity will make us realise that only particular parts of surfaces are vulnerable to corrosion or wear and that there is no need to cover great areas. Where great areas are required to be covered, for example, for appearance, paint will probably be most cost-effective, or where large-area coverage by metals is required, then electroplating is likely to be the winner; however, for discrete areas the laser has few competitors and can give a wide variety of treatments as discussed here.

Currently the uses of lasers in surface treatment include:

- surface heating for *transformation hardening* or annealing;
- *scabbling*, the removal of the surface of concrete or stone;
- surface *melting* for homogenisation, microstructure refinement, generation of rapid solidification structures and surface sealing;

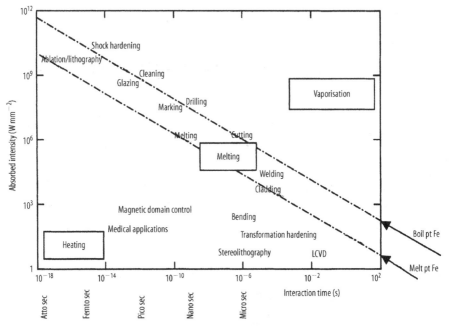

Figure 6.1 Range of laser processes mapped against power density and interaction time. The *diagonal lines* represent lines of constant temperature for the boiling and melting point of iron. (See Equation 5.3 for the dependence on $t^{1/2}$.)

- surface *alloying* for improvement of corrosion, wear or cosmetic properties;
- surface *cladding* for similar reasons as well as changing thermal properties such as melting point or thermal conductivity; this process has progressed into *direct laser casting* for low-volume manufacture of three-dimensional components (see Chapter 7);
- surface *texturing* for improved paint appearance;
- surface *roughening*: for enhanced glue adhesion;
- *plating* by laser chemical vapour deposition (LCVD) or laser physical vapour deposition; enhanced plating for localised plating by electrolysis or cementation or improved deposition rates;
- noncontact *bending* (see Chapter 9);
- magnetic *domain control*;
- *stereolithography* and other forms of layer manufacture;
- *paint stripping and cleaning* (see Chapter 10);
- *laser marking*;
- *micromachining*; and
- *shock hardening*.

It can be seen that these processes range from the low-power-density processes of transformation hardening, bending, scabbling and LCVD, which rely on surface heating without surface melting, to processes involving surface melting requiring higher power

Figure 6.2 A blown powder laser clad being formed

densities to overcome latent heat effects and larger conduction heat losses. These melting processes include simple surface melting to achieve greater homogenisation or very rapid self-quench processes as in laser glazing for the formation of metallic glasses, which is possible in certain alloys. Melting processes also include those where a material is added either with a view to mixing it into the melt pool, as in surface alloying and particle injection, or with a view to fusing on a thin surface melt, as in cladding (Figure 6.2). If very short pulses of great power intensity strike a surface, they are able to cause instant ablation, as in cleaning, and to send mechanical shock waves, originating from sudden thermal stress, through the material, resulting in surface hardening similar to shot peening but with a greater depth of treatment. The processes listed above are now discussed in turn.

6.2 Laser Heat Treatment

The initial goal of laser heat treatment [1] was selective surface hardening for wear reduction; it is now also used to change metallurgical and mechanical properties. There are many competing processes in the large subject of surface heat treatment (Figure 6.3). The laser usually competes successfully owing to reduced distortion and high productivity. Practical uses of laser heat treatment include:

* hardness increase;
* strength increase;
* reduced friction;
* wear reduction [2];
* increase in fatigue life;
* surface carbide creation;
* creation of unique geometrical wear patterns; and
* tempering is also possible.

Laser heat treatment is used on titanium, some aluminium alloys, steels with sufficient carbon content to allow hardening and cast irons with a pearlite structure. The process arrangement is illustrated in Figure 6.4. An absorbing coating is usually applied

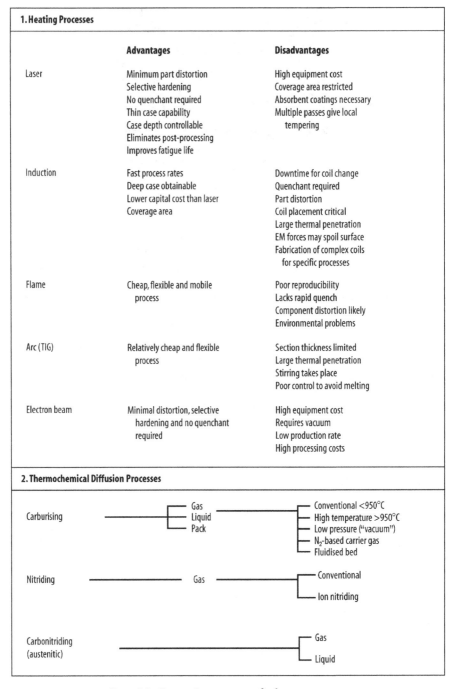

1. Heating Processes

	Advantages	**Disadvantages**
Laser	Minimum part distortion Selective hardening No quenchant required Thin case capability Case depth controllable Eliminates post-processing Improves fatigue life	High equipment cost Coverage area restricted Absorbent coatings necessary Multiple passes give local tempering
Induction	Fast process rates Deep case obtainable Lower capital cost than laser Coverage area	Downtime for coil change Quenchant required Part distortion Coil placement critical Large thermal penetration EM forces may spoil surface Fabrication of complex coils for specific processes
Flame	Cheap, flexible and mobile process	Poor reproducibility Lacks rapid quench Component distortion likely Environmental problems
Arc (TIG)	Relatively cheap and flexible process	Section thickness limited Large thermal penetration Stirring takes place Poor control to avoid melting

Electron beam	Minimal distortion, selective
	   hardening and no quenchant
	   required
		High equipment cost
		Requires vacuum
		Low production rate
		High processing costs

2. Thermochemical Diffusion Processes

Carburising — Gas / Liquid / Pack — Conventional <950°C / High temperature >950°C / Low pressure ("vacuum") / N_2-based carrier gas / Fluidised bed

Nitriding — Gas — Conventional / Ion nitriding

Carbonitriding (austenitic) — Gas / Liquid

Figure 6.3 Competing processes for heat treatment

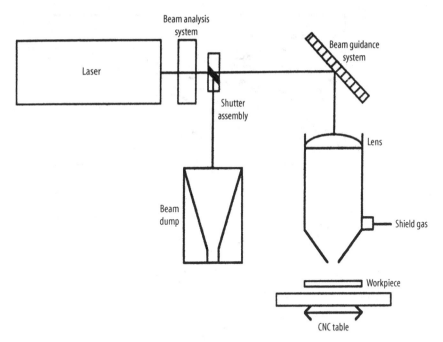

Figure 6.4 Experimental arrangement for laser heat treatment

Table 6.1 Typical values of the reflectivity of various surfaces to 10.6-μm radiation at normal angles of incidence

Surface type	Reflectivity (%)		
	Direct	Diffuse	Total
Sandpaper-roughened (1 μm)	90.0	2.7	92.7
Sandblasted (19 μm)	17.3	14.5	31.8
Sandblasted (50 μm)	1.8	20	21.8
Oxidised	1.4	9.1	10.5
Graphite	19.1	3.6	22.7
Molybdenum sulphide	5.5	4.5	10.0
Dispersion paint	0.9	0.9	1.8
Plaka® [1] paint	0.9	1.8	2.7

[1] Plaka® is a registered trademark of Pelikan. http://www.pelikan.com

to the metal surface to avoid unnecessary power loss by reflection. Some typical coatings with their average reflectivity values are listed in Table 6.1. The absorption coefficient can also be increased by allowing a polarised beam with the electric vector in the plane of incidence to be reflected at the Brewster angle (approximately 80° for metals – a glancing angle) [3]. This leads to a unique process for transformation hardening inside small holes – such as valve guides. The variation of reflectivity with angle of incidence was noted in Section 2.3.4. As the beam moves over an area of the metal surface, the temperature starts to rise and thermal energy is conducted into the metal component. Temperatures must rise to values that are more than the critical transformation

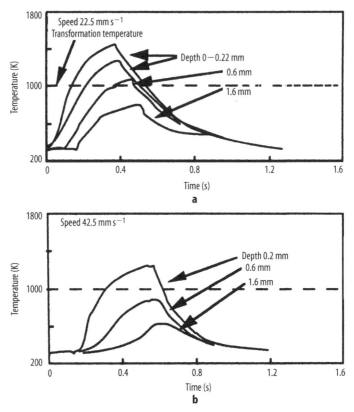

Figure 6.5 Theoretically predicted thermal cycles during laser heating of EN8 steel (EN8 is similar to AISI 1040 steel) (power 2 kW, beam radius 3 mm and reflectivity 0.4) [7]

temperature (Ac1, Figure 6.6) but less than the melt temperature. After the beam has passed, cooling occurs by quenching from the bulk of the material which has hardly been heated by this fast surface heating process. The thermal and structural histories are illustrated in Figures 6.5 and 6.6. The process is discussed in Section 6.2.3, but in essence for transformable alloys there is a phase change on heating which starts at the Ac1 temperature in Figure 6.6a and is complete at the Ac3 temperature. This new structure is unable to transform back again on rapid cooling owing to diffusion, which occurs while at the higher temperatures. The species diffusing is usually carbon. The result is a structure under some form of stress and hence unable to allow dislocations to flow. Such a structure has the property of being hard.

The laser beam is defocused or oscillated to cover an area such that the average power density is 10^3–10^4 W mm^{-2}. Using these power densities, a relative motion between the workpiece and the beam of 5 and 50 mm s^{-1} will result in surface hardening. If surface melting occurs and this is not desired, relative motion should be increased. A decrease in power density will produce the same effect. If no hardening or shallow hardening occurs, but deeper hardening was desired, relative motion should be decreased; an in-

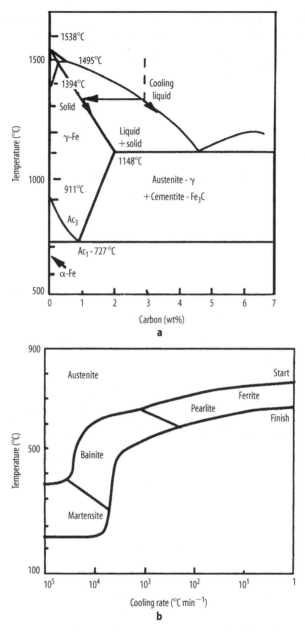

Figure 6.6 **a** The iron cementite (Fe–Fe₃C) system – a summary of stable and metastable equilibria to 7 wt% carbon [112], and **b** continuous cooling transformation diagram for a 0.38 wt% carbon steel [113] Analysis: C 0.38; Si 0.20; Mn 0.70; P 0.02; S 0.02

crease in power density will produce the same effect. The depth of hardening depends upon thermal diffusion and hence the heating time (D/V, where D is the spot size on the workpiece and V is the traverse speed), as well as the temperature, which is dependent on the specific energy (P/DV).

6.2.1 Heat Flow

The ideal power distribution is one which gives a uniform temperature over the area to be treated. This requires a dimpled power distribution since the heating effect is dependent on the edge cooling and surface heating, *i.e.*, P/D and not P/D^2, where P is the incident absorbed power. Methods of spreading the beam to simulate this are illustrated in Figure 6.7. They include:

1. defocused high-power multimode beams (top-hat mode);
2. one- or two-axis scanning beams (dithered zigzag mode);
3. kaleidoscopes;
4. segmented mirrors; and
5. special optics (axicon lenses, toric mirrors and diffractive optic elements).

Most of these are used in laser heat treatment. They all generate a reasonably uniform distribution of power over the central region of the beam path. The temperature distribution with depth during the temporal duration of the irradiation can be represented by equations derived from simple, but idealised, models of one-dimensional heat transfer. A simple test to determine if this representation can be used is to examine the cross-section of a heat-treated sample as in Figure 6.8. If the bottom of the hardened zone is flat and parallel to the surface under the central part of the cross-section, then the one-dimensional analysis will predict the temperatures in the heated material with reasonable accuracy (as discussed in Chapter 5). The edges of the cross-section are regions where the problem is two-dimensional and the one-dimensional heat flow model will not accurately predict the induced temperatures. Whether the edge or central model is dominant is determined by the processing speed and beam diameter expressed as the Peclet number ($Dv\rho C/k$) [4]. Transformation hardening with no surface melting is the simplest process to model mathematically [1]. There are no unknown convection or latent heat terms since there is no melt pool and surface heat losses follow the normal rules of convection and radiation [5,6]. An empirical relationship between $P/(DV)$ [1,2] and the depth of hardness was found by Steen and Courtney [7], as was noted in Section 5.9. The theoretical fit for this parameter, as calculated by Sharp and Steen [4] using a finite-difference model, is shown in Figure 6.9. The spread due to the Peclet number effect is shown to be slight. For En8 the Courtney fit was found to be

$$d = -0.10975 + 3.0\frac{P}{(DV)^{1/2}} \, . \tag{6.1}$$

Although the one-dimensional distribution is useful for approximate predictions, if more exact thermal distributions are required, then calculation must be made via numerical techniques such as finite-difference models [8]. This would be the case if allowance is to be made for variations in beam energy distribution, edge effects, finite

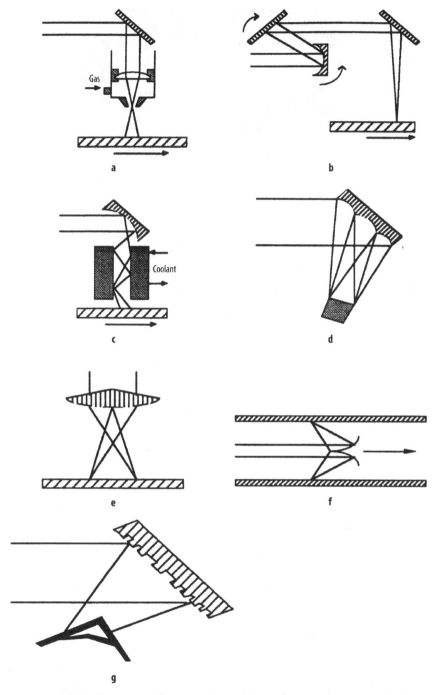

Figure 6.7 Methods of beam spreading: **a** unfocused beam, **b** rastored beam, **c** kaleidoscope, **d** beam integrator (segmented mirror), **e** axicon lens, **f** toric mirror, and **g** diffractive optic element

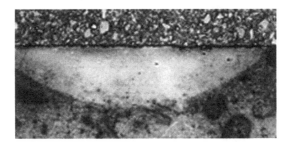

Figure 6.8 Microstructure of laser-transformation-hardened En24 ($\sim$ SAE 4340) steel. Power 1.6 kW, traverse speed 15 mm s^{-1}, beam diameter 6 mm. Composition (wt%): C 0.36–0.44; Si 0.1–0.35; Mn 0.45–0.7; P 0.035 max; S 0.04 max; Cr 1.0–1.4; Mo 0.2–0.35; Ni 1.3–1.7

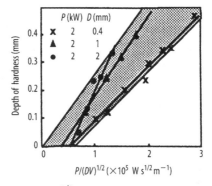

Figure 6.9 Theoretical plot of $P/(DV)^{1/2}$ for the depth of the 1,960 K isotherm for various P, V and D values. (From Sharp and Steen [4])

part size and particular geometries. It is also true if the hardened width is an essential part of the answer sought.

A greatly improved model was suggested by Ashby and Shercliffe [9], who solved the problem of a moving finite line source placed at a specified distance above the surface – to simulate a beam diameter. They produced the master plot illustrated in Figure 5.7. This plot has been shown to fit the results of transformation hardening quite well.

6.2.2 Mass Flow by Diffusion

In transformation hardening of steels, the parent structure consists of a nonhomogeneous distribution of carbon, *e.g.*, pearlite and ferrite, which upon heating above the phase transformation temperature, Ac1 temperature (723 °C for many steels), starts to diffuse to achieve homogeneity within the austenite phase. The rate of diffusion is described by equations similar to the equation for heat flow, but is usually much slower, *i.e.,*

$$\frac{\delta c}{\delta t} = D_{AB}\left[\frac{\delta^2 c}{\delta x^2} + \frac{\delta^2 c}{\delta y^2} + \frac{\delta^2 c}{\delta z^2}\right]. \qquad (6.2)$$

The diffusivity of carbon in austenite is approximately

$$D = 1 \times 10^{-5} e^{-9.0/T} \text{ m}^2 \text{ s}^{-1}$$

and in ferrite it is

$$D = 6 \times 10^{-5} e^{-5.3/T} \text{ m}^2 \text{ s}^{-1}.$$

When austenitisation has occurred, the carbon moves by diffusion down concentration gradients. The time for diffusion within the austenitic lattice varies with position within the laser-treated zone [4] (Figure 6.5). In laser transformation hardened zones there is always a region around the edges, if not throughout, where the carbon has not fully diffused and the resulting structure is a nonhomogeneous martensite: it is not yet known whether this nonhomogeneous martensite is preferable to homogeneous martensite. It would be expected that the higher carbon levels in certain regions would lead to higher hardness levels and therefore better overall wear resistance – as is observed in some laser-treated samples.

6.2.3 Mechanism of the Transformation Process

6.2.3.1 Steels

On rapid heating, pearlite colonies first transform to austenite. Then carbon diffuses outwards from these transformed zones into the surrounding ferrite, increasing the volume of high-carbon austenite. On rapid cooling, these regions of austenite which have more than a certain amount of carbon (*e.g.*, 0.05%) will quench to martensite if the cooling rate is sufficiently fast, although retained austenite may be found if the carbon content is above a certain value (more than 1.0%). The required rate of cooling is indicated by constant cooling curves, such as shown in Figure 6.6b. In laser transformation hardening the cooling rate is usually in excess of $1{,}000\,^{\circ}\text{C}\,\text{s}^{-1}$, which means that most steels will self-quench to martensite not bainite or pearlite.

The transformation of the pearlite is thought to proceed by diffusion from the cementite plates into the ferrite plates, possibly starting from one end of a pearlite colony (Figure 6.10). This time-dependent process does not take long but is sufficient to necessitate some heating above the austenitising temperature, Ac1, to allow it to proceed to any extent during laser treatment. The superheat, and therefore the extent of the diffusion process, is thus slightly affected by the prior size of the pearlite colonies. These colonies, on transformation, become austenite having 0.8% carbon. Carbon diffuses down the concentration gradient into the ferrite regions where there is virtually no carbon. The ferrite regions may also have transformed to the face centred cubic structure of austenite. The extent of the homogeneity of the resultant martensite will depend

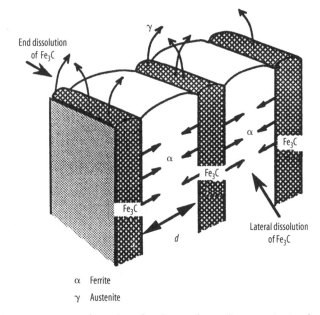

α Ferrite
γ Austenite

Figure 6.10 Routes for carbon dissolution during homogenisation [21]

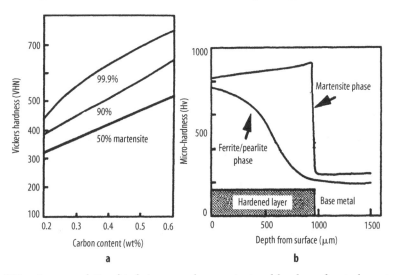

Figure 6.11 a Average relationship between carbon content and hardness for steels containing different amounts of untempered martensite [114], and **b** microhardness distribution in a non-homogeneous hardened specimen [115]

upon the size of the prior ferrite regions and the processing conditions – in particular, the interaction time (beam diameter/traverse speed). The hardness depends upon the carbon content (Figure 6.11).

The metallurgical changes which occur in laser-treated steels are similar to those for furnace-treated steels. However, the more rapid heating and quenching of the laser process results in variations in the type of martensite, particularly its fineness, the amount of retained austenite and carbide precipitation as well as the homogeneity of the hardened zone. The transformed zone is also more highly restrained, resulting in higher compressive stresses opposing the approximately 4% volume increase associated with martensitic phase changes.

6.2.3.2 Cast Irons

Ferritic grey cast iron consists of ferrite and graphite regions. As such it is difficult to harden by the laser because the diffusion time is too short. Typically the diffusion distance from the graphite is 0.1 mm for a 5 mm beam travelling at 20 mm s^{-1}. Thus, all that is formed is a hard crust around the graphite flakes or nodules. These can still give impressive wear properties although no change in the overall hardness value would be observed.

Pearlitic cast iron, formed by moderately fast cooling, consists of pearlite and graphite. In this case laser transformation hardening is successful in achieving very high hardness levels, as for 0.8% carbon or higher irons. With cast irons there is a fairly narrow window between transforming and melting. The irons are important for their ease of casting; hence, they have low melting points, whereas their Ac1 temperature is approximately constant as for all Fe/C alloys (Figure 6.6a).

Laser transformation hardening of spheroidal graphite cast iron may result in preferential melting around the graphite nodules owing to the lowering of the melting point as the carbon diffuses away from the graphite.

6.2.3.3 Silicon

Annealing of amorphous silicon has been used in the fabrication of active liquid crystal display panels and other electronic components [10].

6.2.4 Properties of Transformed Steels

6.2.4.1 Hardness

This depends upon the carbon content (Figure 6.11a). It has been found that the hardness value may be slightly higher than that found for induction hardening. This difference is probably due to the shallower zone in the laser process allowing a faster quench and therefore greater restraint, and hence higher residual compressive stress. A typical hardness profile is shown in Figure 6.12 for a carburised 20 CrMo steel.

Overlapping successive tracks induces a thermal experience in the neighbouring tracks so that there is some back-tempering. This is not necessarily undesirable since the softer stripe allows space for oil and wear debris. The extent of the hardness variation is illustrated in Figure 6.13. Patterned hardened surfaces have not received too much

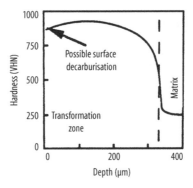

Figure 6.12 Variation of microhardness with depth for a 20 CrMo steel

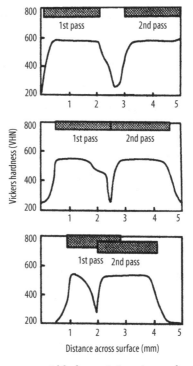

Figure 6.13 Surface hardness versus width for variations in overlap between successive passes. (From Bergmann [21])

attention as wear surfaces, mainly because prior to the introduction of the laser they were difficult to make. The laser can make patterned surfaces easily and therefore opens a whole new study in tribology.

The tempering of steels by a laser is used in the production of reinforcing wire for tyres. The wire is passed at speed through a shaped laser beam.

Table 6.2 Experimental results of rotational wear resistance for transformation hardened SK5 (AISI WI) steel

Property	Method of transformation hardening	
	Laser	Induction
Hardness	HRC 64–67	HRC 60–63
Case depth (mm)	0.7–0.9	2–3
Load (kg mm^{-2})	101	101
Scuffing	No occurrence	Slight
Wear loss	0.5	1.0

6.2.4.2 Fatigue

In steels and cast irons there is a residual compressive stress on transformation hardening due to the volume expansion on the formation of martensite (approximately 4% for 0.3 wt% carbon steel). This effect is particularly pronounced in the shallower hardened zones formed with the laser due to the greater restraint for such treatment. Fatigue cracks are generally initiated at the surface by tensile stresses; thus, the fatigue load must be sufficient to overcome this residual compressive stress before a crack can propagate. Improved fatigue life with laser heat treatment compared with induction hardening has been reported [1, 11].

6.2.4.3 Wear Resistance

Wear resistance has also been found to improve with laser treatment compared with oil or water quench. On SK5 steel (~ AISI 1080, however Carbon can vary between 0.65 to 1.4%) the pin-on-disc wear resistance of laser-treated surfaces was found to be twice that of an induction-hardened surface (Table 6.2).

6.2.4.4 Distortion

Owing to the reduced thermal load and penetration possible with laser treatment there is less distortion compared with that from flames or induction hardening. This is often the reason for using the laser. For example, the 12×12 mm^2 rail in a cash dispenser has an asymmetric parabolic groove 3 mm wide which is hardened by a 2.5 kW CO_2 laser with a 3 mm spot size in a single pass. Alternative processes would have the additional costs of straightening this part [12].

6.3 Laser Surface Melting

For surface melting the experimental arrangement is similar to that for transformation hardening, as shown in Figure 6.4, except that in this case a focused or near-focused beam is used. The surface to be melted is shrouded by an inert gas. The competing processes are listed in Figure 6.14. The main characteristics are:

Process	Characteristics
Laser	High capital cost, localised heating of sample, rapid solidification of melt zone to give fine recrystallised grain structure and good homogeneity. Controllable surface roughness Power density $10^2 - 10^4$ W cm^{-2}
Flame	Cheap capital cost, poor reproducibility, no fast quench available, environmental problems and sample distortion A flexible and mobile process
Plasma	Medium capital cost, very low heat input to sample
TIG	Limited section thickness, electromagnetic stirring, weld bead may be rough, large thermal penetration and medium heat input
Induction	Cheaper than laser, large thermal penetration, electromagnetic force may spoil surface, fast processing rate, deep case possible and fast area coverage

Figure 6.14 Competing surface melting processes

- moderate to rapid solidification rates producing fine near-homogeneous structures;
- little thermal penetration, nor surface hot gases, resulting in little distortion and the possibility of operating near thermally sensitive materials;
- surface finishes of around 25 μm are fairly easily obtained, signifying reduced work after processing; and
- process flexibility, due to software control and possibilities in automation.

The main areas of variation in processing centre around controlling the reflectivity, shaping the beam and shrouding the melt pool. Reflectivity is difficult to control owing to the melting process itself causing variations in the surface reflectivity. The initial reflectivity can be controlled in the same manner as for transformation hardening, by having an antireflection coating, but this is usually removed by the melting process; however, once the material becomes hot, the reflectivity is reduced, owing to the increased phonon concentration.

The reflectivity varies with the angle of incidence [13] and surface films play a significant role. The addition of a small amount of oxygen to the shroud gas has a notable effect on reflectivity [14] (see Figure 4.29). A surface plasma will initially help to couple the beam into the surface. If the plasma leaves the surface, then it will block or defocus the beam. Optical feedback systems, such as a reflective dome around the interaction region (see Figure 6.28) [15, 16], can increase the laser coupling by around 40%. Optical methods vary according to the method used to produce the required spot size or beam shape which may be required to control the flow in the melt pool, as well as for the method used to protect the optics from sputter and fume.

There are three metallurgical areas of considerable interest: cast irons, tool steels and certain deep eutectics which can form metallic glasses at high quench rates. All are essentially nonhomogeneous materials which can be homogenised by laser surface melting.

There are two reasons why laser surface melting is not widely used in industry:

1. If surface melting is required, then surface alloying is almost the same process and offers the possibility of vastly improved hardness, wear or corrosion properties.
2. The very high hardnesses achieved with cast irons and tool steels by laser surface melting are associated with some surface movement and hence may require some further surface finishing after treatment. This is not so easy to effect with the high hardnesses obtained.

The products of laser surface melting of some important engineering materials are as follows:

- *Cast iron* [17]. This commonly used engineering material usually consists of an inhomogeneous structure of ferrite and graphite in various forms (flakes, spheres, *etc.*) On surface melting with a laser, the hardening effects come from changes of graphite to cementite and austenite to martensite [17–20]. The precise value of the hardness depends on the extent of the carbon dissolution from the graphite giving a variation of hardness and structure with processing speed. The result is usually a very hard surface on one of the cheaper metals and this can be achieved by a simple, fast process. Figure 6.15 shows the melt interface for laser surface melted flake graphite cast iron. The structure varies from Fe_3C dendrites in the ledeburitic fusion zone through high dissolved carbon (around 1 wt%), giving retained austenite with some martensite to full martensite and partially dissolved graphite flakes. A form of trip through the iron carbon phase diagram! Figures 6.16 and 6.17 illustrate the variation in hardness and structure with traverse speed. The high hardness at slow speeds in Figure 6.16 for SG iron is due to nearly all the carbon dissolving and giving a ledeburitic white iron structure. The second peak at higher speeds, with only a small amount of carbon dissolution, is due to a martensitic structure. The intermediate

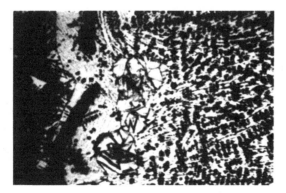

Figure 6.15 Micrograph of the melt interface for laser surface melted flake graphite cast iron (×150)

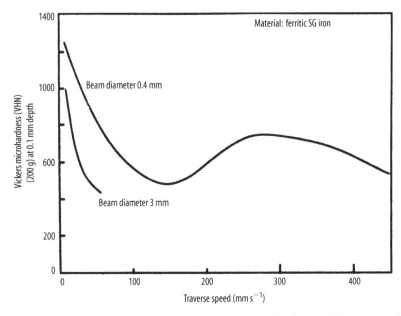

Figure 6.16 Variation of microhardness with scanning speed for ferritic SG cast iron. (From Hawkes *et al.* [18])

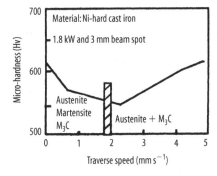

Figure 6.17 Variation in microhardness with scanning speed for nickel-hard cast iron [17]

low-hardness region is due to retained austenite. The improved wear properties are illustrated in Figure 6.18, whereas the fatigue properties are usually worse owing to residual tensile stress in nonmartensitic materials [21].

As an application example, surface melting of GGG60 cast iron ($\sim$ ASTM A536) camshafts using a 6 kW CO_2 laser is being commercialised in the automobile industry [22]. Conventionally, this surface melting has been practised using TIG remelting and clear chill casting methods. The laser process gives the same surface hardness as these processes but a finer ledeburitic microstructure, which gives a greater lifetime in test engines.

The surface of the laser-treated cams is smoother than that obtained with the competition since the melting is done with a line source of the same width as the cam in

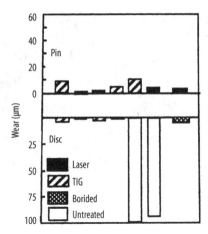

Figure 6.18 Comparison of the wear of surface-melted SG iron prepared by different processes. (From Bergmann [21])

a single pass. This means less after-machining by a factor of 2. Owing to this, the hardened layer does not need to be as deep as the TIG layer (1.5 mm for TIG, 0.5 mm for the laser) and hence the thermal load is reduced. This, in turn, gives a faster quench and sharper top profile, as well as less distortion and therefore straightening requirements. The straightening work load is reduced by a factor of 5–10. Process speeds using a 6 kW CO_2 laser allow the treatment of a four-cylinder, two-valve camshaft at 60 s per shaft. The whole process includes some preheat, which is required to avoid microcracks. Add to these advantages the ease of automation and control, which usually leads to a reduction in reject rate, and the process becomes very attractive.

Stainless steel. Fine structures are produced in both martensitic and austenitic stainless steels as expected from the high values of the cooling rate, GR. Without the phase expansion associated with the martensitic transformation, austenitic steels have a residual tensile stress, whereas single tracks of martensitic steel are usually under compression, which becomes tensile when they are annealed by overlapping. The residual tension adversely affects the stress corrosion properties and the pitting potential [23]. Lumsden *et al.* [24] found that laser melting and rapid solidification had differing effects on the pitting behaviour of a series of ferritic steels of composition Fe–13Cr–xMo, where x varies from 0 to 5%. Unless the molybdenum concentration was higher than 3.5%, laser melting had a deleterious effect or no effect on the pitting potential. The 5% alloy had a large increase in the pitting potential compared with the untreated alloy. Improved corrosion resistance of sensitised stainless steels has been noted by many workers owing to the finer structure reducing the tendency for intergranular and end grain corrosion [25, 26].

- *Titanium.* Titanium and its various alloys can take up a variety of crystal forms. In laser surface melting rapid quench structures are formed which have highly dislocated fine structures (Figure 6.19). The process must be carefully shrouded owing to the activity of titanium with oxygen [27, 28].
- *Tool and special steels.* These materials are usually hardened by a fairly long process of solution treatment to dissolve the carbides, followed by a controlled quench to

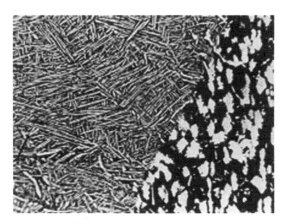

Figure 6.19 Micrograph showing the fine basket weave structure produced in laser surface melting IMI550 [27] (P = 1.6 kW, V = 200 mm s^{-1}, D = 0.5 mm) (×100)

give a fine dispersion of carbides. These carbides do not temper as easily as marten-site; hence, these steels have a high hot hardness and are suitable for tools. In laser surface melting this dissolution is accomplished very swiftly, producing a very hard, fine carbide dispersion with high hot hardness properties. The problem with the ap-plication of this process in production is that the laser melt track will have a surface waviness of around 10–25 μm and the track is very hard to machine.

- *Glazing grout and cement* [29–31]. The surface of certain cements can be melted without cracking and spallation to create a sealed surface which can be more easily cleaned as may be required in medical areas or radiation hazard zones. Lawrence *et al.* [30, 31] found that when using a diode laser at 820 nm wavelength, surface sealing occurred, but when using a CO_2 laser at 10.6 μm, spallation occurred (see Section 6.16). It is thought that the difference may be due to the different penetration depths of these two wavelengths in concrete – 470 ± 22 μm for CO_2 laser radiation and 177 ± 15 μm for diode laser radiation. These affect the stress fields at depth and hence the spalling. They also showed that the diode laser could densify Al_2O_3-based refractories, making them more corrosion-resistant at higher temperatures [32].

In all materials there is a tendency for cracking to occur if the hardness is high. Usually this can be avoided if some preheat is applied. As a rule of thumb, the required preheat is around 1°C per Vickers hardness number. This indicates a preheat of around 500°C for low-carbon steel, 650°C for 0.7 wt% carbon steel and 700°C for tool steels.

6.3.1 Solidification Mechanisms

6.3.2 Style of Solidification

Solidification [33] will proceed as either a stable planar front or as an unstable front leading to dendrites or cells. The process which ill occur depends on the occurrence of constitutional supercooling (Figure 6.20). Constitutional supercooling is caused by the

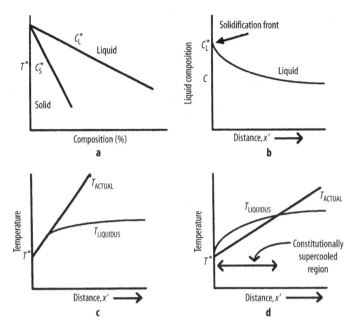

Figure 6.20 Constitutional supercooling in alloy solidification [33]: **a** phase diagram, **b** solute-enriched layer in front of a liquid–solid interface, **c** stable interface, and **d** unstable interface

thermal gradient being less steep than the melting point gradient, which is the result of partition effects taking place at the solidification front giving rise to composition variation in this region. The result is that the liquidus has a higher temperature than the actual temperature and hence there is supercooling ahead of the solidification front.

Consider a mass balance on the solidification front. The gradient of the solute in the liquid at the solidification interface is

$$\left[\frac{dC_L}{dx}\right]_{x=0} = -\frac{R}{D_L}C_L^*(1-k) . \tag{6.3}$$

Constitutional supercooling is absent when the actual temperature gradient in the liquid at the interface $G \geq (dT_L/dx)_{x=0}$: now the value of this gradient is

$$\left[\frac{dT_L}{dx}\right]_{x=0} = m_L\left[\frac{dC_L}{dx}\right]_{x=0} , \tag{6.4}$$

where C_L is the liquidus composition, x is the distance from the interface (m), T_L is the liquidus temperature (°C), R is the rate of solidification (m s^{-1}), D_L is the diffusivity (m^2 s^{-1}), C_L^* is the liquidus composition in equilibrium with solidus composition C_S^*, k is the partition coefficient, m_L is the slope of the liquidus (dT_L/dC_L) and G is the thermal gradient (°C m^{-1}).

Combining these two equations and letting $C_S^* = kC_L^*$ – the equilibrium condition – we obtain the following general constitutional supercooling criterion.

There is no constitutional supercooling if

$$\frac{G}{R} \geq -\frac{m_L C_S^* (1-k)}{k D_L}.$$ (6.5)

The ratio G/R should be large for a stable planar front solidification mechanism. Figure 6.21 illustrates this equation and further introduces the concept of "absolute stability" when the solidification rate, R, is so large that there is insufficient time for diffusion.

6.3.2.1 Scale of Solidification Structure

If the dendritic or cellular structure is sufficiently fine, then it is possible to approximate the liquid between the cells as being like a small stirred tank whose composition will be determined by the rate of diffusion out of the cell depleting the concentration of the cell, in fact Fick's second law:

$$D_L \frac{\delta^2 C_L}{\delta y^2} = \frac{\delta C_L}{\delta t}.$$ (6.6)

Now

$$\frac{dC_L}{dt} = \left(\frac{dC_L}{dt}\right)\left(\frac{dT}{dx}\right)\left(\frac{dx}{dt}\right) = -\left(\frac{GR}{m_L}\right).$$

Substituting for dC_L/dt and integrating with respect to y across the cell width λ, we obtain

$$\left[\frac{\delta C_L}{\delta y}\right]_{y=0} = -\frac{GR\lambda}{m_L D_L} \quad \text{and} \quad \Delta C_{L_{max}} = -\frac{GR\lambda^2}{2 m_L D_L}.$$ (6.7)

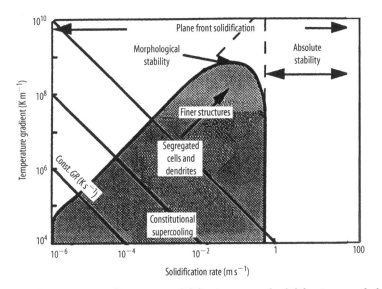

Figure 6.21 Temperature gradient versus solidification rate and solidification morphology

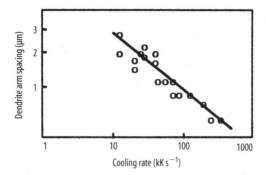

Figure 6.22 The logarithm of dendrite arm spacing versus the logarithm of the cooling rate. The *line* has a slope of approximately −0.5 as expected from Equation 6.7

We observe that the parameter GR is related inversely to the square of the cell spacing, λ. GR is the cooling rate in degrees Celsius per second. In laser surface melting extremely high cooling rates can be achieved (around $10^6 \, ^\circ C \, s^{-1}$) and therefore finer structures result, as illustrated in Figure 6.22.

6.3.2.2 Material Flow Within the Melt Pool

There are many forces acting on the melt pool as shown in Figure 6.23 [34]. One of the largest is that from the variation in surface tension, σ, due to the steep thermal gradients:

$$\text{Surface shear force} = \frac{\delta\sigma}{\delta x} = \frac{\delta\alpha}{\delta T}\frac{\delta T}{\delta x} . \qquad (6.8)$$

Consider the following example. For nickel the variation of surface tension with temperature $d\sigma/dT = 0.38 \times 10^{-3} \, J \, ^\circ C^{-1} \, m^{-2}$. For laser processing the thermal gradient

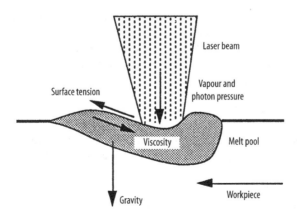

Figure 6.23 Forces on the melt pool

is of the order of $dT/dx \sim 2.5 \times 10^6\,°C\,m^{-1}$. Therefore, the shear force is $0.38 \times 2.5 \times 10^3\,N\,m^{-2} = 10^3\,Pa = 0.01$ atm.

This is not an insignificant force but one which if applied to a small surface film of area A would work out to be of the order of $10G$ on the surface layer, a force similar to that in an airgun! It is perhaps a good thing there is some viscosity to restrain the motion.

(From Newton's law of motion $F = ma$. The mass of a film 1 mm thick and of density $\rho = 10,000\,kg\,m^{-3}$ is $m = \rho Ad = A \times 10$ kg. The force on area A is $F = A \times 10^3$. Therefore, acceleration $a = 10^3/10 = 10^2\,m\,s^{-2}$. With gravitational acceleration $g = 10\,m\,s^{-2}$, this acceleration is of the order of $10G$.)

Chan *et al.* [8] modelled this flow by solving the Navier–Stokes equation together with the heat flow equation. Their calculations suggest that the melt pool rotates approximately five times before solidifying. The marker experiments of Takeda *et al.* [35] (see Section 6.5.2, Fig. 6.34) also indicate that a very rapid mixing takes place within the melt pool, which is of great complexity owing to microeddies that are difficult to model.

6.4 Laser Surface Alloying

Surface alloying with a laser is similar to laser surface melting except that another material is injected into the melt pool. Laser surface alloying is also similar to surface cladding in that if the cladding process is performed with excess power, then surface alloying will result. It is therefore one extreme of surface cladding. The main characteristics of the process are as follows:

- The alloyed region shows a fine microstructure with nearly homogeneous mixing throughout the melt region. Inhomogeneities are only seen in very fast melt tracks (around $0.5\,m\,s^{-1}$).
- Most materials can be alloyed into most substrates. The high quench rate ensures that segregation is minimal [36]. Some surface alloys can only be prepared via a rapid surface quench, *e.g.*, Fe–Cr–C–Mn [36].
- The thickness of the treated zone can be 1–$2,000$ μm. Very thin, very fast quenched alloy regions can be made using Q-switched Nd:YAG lasers.
- Some loss of the more volatile components can be expected [37].
- Other characteristics are as for laser surface melting.

6.4.1 Process Variations

The variations in processing are similar to those for surface melting except that an alloy ingredient has to be added. The alloy can be placed in the melt zone by:

- electroplating [38];
- vacuum evaporation;
- preplaced powder coating [39];
- thin foil application;

- ion implantation;
- diffusion, *e.g.*, boronising [40];
- powder blowing [41];
- wire feed; and
- reactive gas shroud [42], *e.g.*, C_2H_2 in Ar or just N_2.

Laser surface alloying is capable of producing a wide variety of surface alloys. The high solidification rate even allows some metastable alloys to be formed in the surface. All this can be done by a noncontact method which is relatively easy to automate. The competing processes are shown in Figure 6.24. The laser offers precision in the placement of the alloy, good adhesion and vastly improved processing speeds. Provided the speed is lower than a certain figure (*e.g.*, 70 mm s^{-1} for 2 kW power), then the mixing is good and uniform. Some alloys suffer from cracking and porosity, which may put restrictions on shrouding and preheat. The surface profile can be quite smooth, particularly for Ni, with a small ripple of around 10 μm.

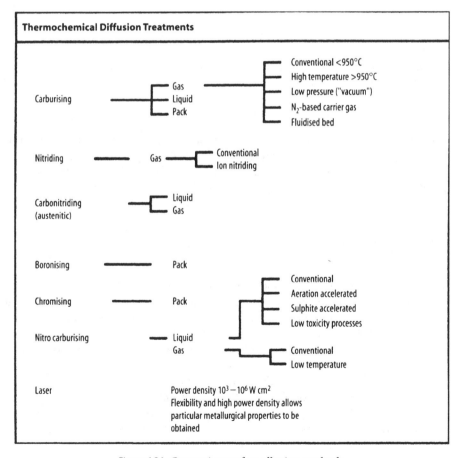

Figure 6.24 Competing surface alloying methods

6.4.2 Applications

Titanium: Titanium can be readily surface-alloyed by carbon or nitrogen. The latter can be supplied by having a nitrogen shroud gas [21,43]. One of the beauties of these processes is that the hard carbide or nitride solidifies first as a dendrite which would be hard to remove. The colour effects on titanium are starting to attract the attention of the art world.

Nitinol, NiTi: Nearly equiatomic NiTi alloy is an ideal biomedical candidate for things such as stents, because of its unique properties, such as superplasticity, the shape memory effect and radiopacity. To reduce the danger of nickel release, the alloy can be laser gas nitrided to produce a sealed corrosion-resistant surface [44].

Cast Iron: Surface alloying with chromium, with silicon and with carbon are all possible methods to make relatively cheap cast irons into superficially exotic irons. A study has been made by Steen *et al.* [17].

Steel: For alloying of steel numerous systems have been explored: chromium by melting chromium plate [38]; molybdenum [45]; boron [40]; nickel [24,44].

Stainless steel: The carbon alloying of stainless steel by melting preplaced powder has been studied by Marsden *et al.* [46].

Aluminium: Surface hardening of aluminium by alloying with silicon, carbon, nitrogen and nickel has been shown to be possible by Walker *et al.* [39] and others. Superalloys have been alloyed with chromium by Tien *et al.* [47].

Coinage: Laser surface alloying could become important in making machine-readable coinage or other metallic objects [48, 49].

Surface alloying has many advantages and great flexibility. Applied by laser, the process offers the possibility of surface compositional changes with very little distortion and surface upset. This has thus put engineers in the position where they could have the material they require for the surface and the material they require for the bulk. The problem of the choice is exhausting to contemplate!

6.5 Laser Cladding

The aim of most cladding operations is to overlay one metal with another to form a sound interfacial bond or weld without diluting the cladding metal with substrate material. In this situation dilution is generally considered to be contamination of the cladding which degrades its mechanical or corrosion-resistance properties. There are many cladding processes as shown in Figure 6.25. Thick-section cladding (more than 0.25 mm) is frequently carried out by welding methods: substantial melting of the substrate is produced and therefore dilution can be a major problem. Dilution is observed in TIG, oxyacetylene flame or plasma surface welding processes in which the melt pool is well stirred by electromagnetic, Marangoni and convective forces. This dilution necessitates laying down thicker clad layers to achieve the required clad property, but does have the advantage of a good interfacial bond. Negligible dilution is achieved in other cladding processes which rely on either forge bonding or diffusion bonding: forge

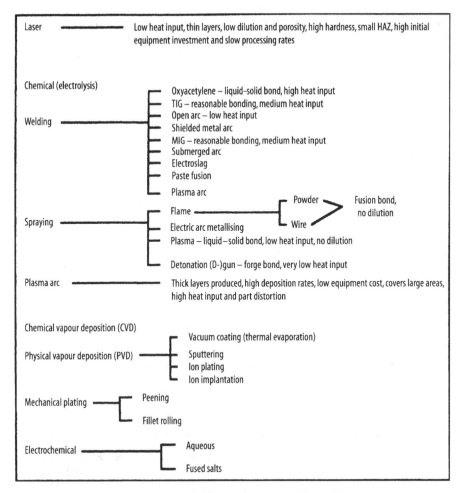

Figure 6.25 Competing cladding techniques. *MIG* metal–inert gas

bonds are made through the impact of high-speed particles with the substrate (*e.g.*, D-gun) or clad layer and diffusion bonding occurs between a solid and a liquid phase as in cladding by vacuum fusion. The fusion bond is usually the strongest and most resistant to thermal and mechanical shock, provided brittle intermetallics are not formed. A comparative study of the dilution, distortion, wear and other properties of clad layers made with a laser, plasma, vacuum furnace, TIG or oxyacetylene flame has been made by Monson and Steen [50, 51]. The particular advantage of the laser is its ability to heat and clad in specified areas alone.

Among the laser cladding routes are those which melt preplaced powder [52], or blown powder [41], those which decompose vapour by pyrolysis [56], or photolysis [57] as in LCVD, those which are based upon local vaporisation as in laser physical vapour deposition or sputtering and those which are based on enhanced electroplating or cementation [55]. These latter three processes are discussed separately in Sections 6.9–6.11.

The three most common methods of supplying the cladding material are:

- preplacement of cladding material as powder on the substrate;
- inert gas propulsion of material as powder into a laser-generated molten pool; and
- wire feed.

6.5.1 Laser Cladding with Preplaced Powder

Cladding with preplaced powder is the simplest method provided the powder can be made to stick until melted, even while the area is being shrouded in inert gas. Some form of binder is usually used, which is often an alcohol. The preplaced powder method involves scanning a defocused or rastered laser beam over a powder bed, which is consequently melted and welded to the underlying substrate. Minimal dilution effects were observed for a wide range of processing parameters.

Theoretical modelling of movement in the molten front [56] has shown that the melt progresses relatively swiftly through the thermally isolated powder bed until it reaches the interface with the substrate. At this point the thermal load increases owing to the good thermal contact with the high thermal conductivity substrate, causing resolidification. The results of the model are shown in Figure 6.26. This figure illustrates why a large operating region for achieving low dilution exists, but it also shows that only a small part of this region gives a fusion bond. In fact, it would be very difficult to achieve a low-dilution fusion bond by this method and avoid serious dilution.

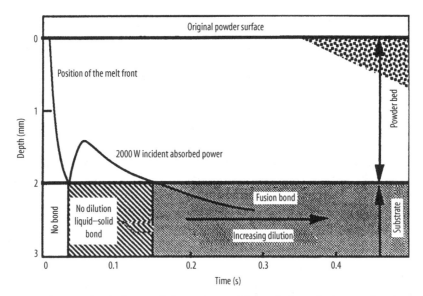

Figure 6.26 Theoretical calculation of the position of the melt front during preplaced powder cladding [51]

6.5.2 Blown Powder Laser Cladding

The main interest in blown powder laser cladding [41] is because it is one of the few cladding techniques which has a well-defined heated region, a fusion bond with low dilution and which is adaptable to automatic processing. The arrangement for cladding by blown powder is illustrated in Figure 6.27. In this figure an optical feedback system is shown. A reflective dome such as this has been shown by Weerasinghe and Steen [41] to recover around 40% of the delivered power. This is necessary when cladding surfaces of variable reflectivity such as machined and shot blasted surfaces as shown in Figure 6.28. The process took a great step forward with the invention of the coaxial powder feed illustrated in Figure 6.29 [57, 58]. With side-blown powder there is a directional effect on the clad bead shape and the alignment of the powder stream with the melt pool is critical. The coaxial system avoids both these problems and behaves almost like a "metal pencil", capable of writing in metal whatever is required. It is important in the design of these nozzles that the powder does not meet the beam until it is outside the nozzle, to avoid clogging.

It has been found by both Lin [59] and Hayhurst *et al.* [60] that the powder stream can be mildly focused with a coaxial nozzle. Lin found, using a circumferential powder feed, that the best focus occurs when the central nozzle is slightly proud of the outer nozzle, thus encouraging some Couette flow of the powder – as when a finger is placed in the stream from a tap. Hayhurst *et al.* used four independent powder streams at 90° to each other, which on colliding merged into a single central stream. Lin also mea-

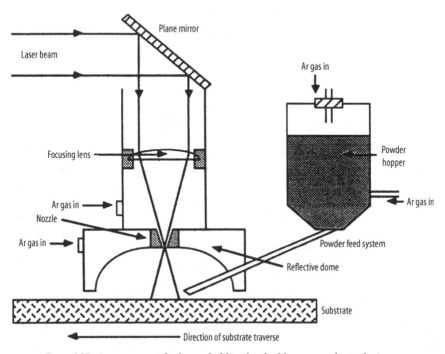

Figure 6.27 Arrangement for laser cladding by the blown-powder technique

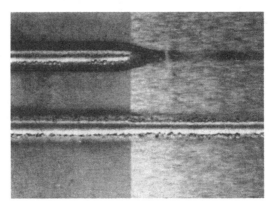

Figure 6.28 The *left-hand side* of the substrate was shot-blasted. The *right-hand side* had a ground finish. The *upper track* was made without the reflective dome; the *lower track* was made using the dome to recycle reflected energy [41]

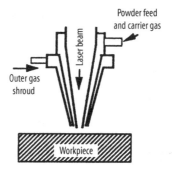

Figure 6.29 A coaxial powder feed nozzle

sured the powder temperature as it fell in the expanding beam. This showed a peak a few centimetres from the nozzle at the point where the growing powder preheat and the diminishing power density caused a maximum. This could be the optimal point for cladding since the powder may have melted but the substrate may not have. Gilkes *et al.* [61,62] in trying to clad aluminium onto iron for making bearings did not want the iron to melt for fear of making a brittle intermetallic. Hence, they were trying to operate in the zone where the powder had melted. However, on turning up the laser power, the maximum temperature in the powder stream rose to the point where the aluminium formed a plasma – with considerable noise – and they had invented a new process, that of "laser plasma deposition", which was capable of cladding aluminium onto iron. Gedda *et al.* [63] calculated the energy loss in the powder stream [62] and found that the powder absorbed approximately 2% of the beam and scattered or reflected another 9% for the conditions used. The remaining energy suffered 50% reflection off the workpiece, 1% re-radiation from the hot workpiece and 40% absorption by the process, of which 10% was used in melting the powder and 30% in heating the workpiece.

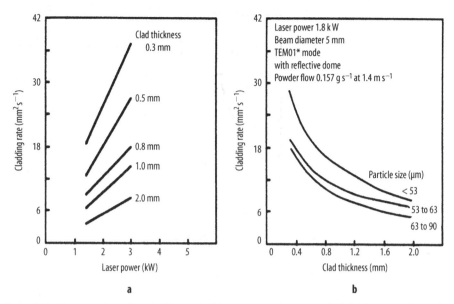

Figure 6.30 The variation of typical laser cladding rates with power, clad thickness and powder particle size using the blown-powder process with the reflective dome [41]

The powder feeder must have a very steady feed rate since every ripple in the flow is printed in the clad track. Feeding powder at rates of $0.1\,\text{g s}^{-1}$ or so is not a simple engineering feat. The pulse of screw feeders has to be ironed out with some capacitance in the delivery lines or fluidisation of the outlet from the screw. The powder must flow properly and not backup and avalanche. Thus, it must be dry and free-flowing. Pinkerton and Li [64] found that the powder shape played a small part in the process. They found that gas-atomised powder clad faster but water-atomised powder (which was therefore more irregular in shape) gave a better clad owing to fewer surface-active agents.

Blown powder cladding can have the low dilution associated with forge-bonded processes but the good surface strength and low porosity associated with the welding processes. The covering rate for laser powers greater than $5\,\text{kW}$ is attractive (Figure 6.30) and when consideration of powder costs and after-machining costs are taken into account, the process becomes economically comparable with other processes for covering large areas. However, the strength of this process is its ability to cover very small areas with precision, since the heat is highly localised and not associated with any hot jets. Thus, cladding takes place only where it is needed and can be done near thin walls or heat-sensitive parts. These are the niche characteristics for the applications of laser cladding.

Blown powder cladding is essentially conducted over a small melt pool area, which is travelling over the surface of the substrate. The thermal penetration can be controlled by the speed, power or spot size. It can be low and hence for thin clad layers distortion can be minimal (Figure 6.31). The clad layer will usually have a residual tensile stress that may reduce its ultimate tensile strength by approximately 50% [65].

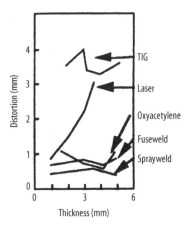

Figure 6.31 Hard-facing thickness versus the distortion of a standard-size sample for coatings produced by various techniques [116]

Area coverage is achieved by overlapping the clad tracks. There are three basic cross-sections of clad tracks [66] (Figure 6.32). For cladding without interrun porosity, the angle α (Figure 6.32c) must be acute, as in Figure 6.32a; this is defined by the aspect ratio of the track. For Colmonoy[1] Wallex[2] PC6 (Stellite[3] 6) powder, the aspect ratio (width/height) should be greater than 5 to avoid interrun porosity [67]. The parameter PVD/m^2 was estimated in Steen [67] to be correlated to the aspect ratio. This is plotted in Figure 6.33 as one limit on the operating region. Dilution (Figure 6.32b) represents another limit on the operating region. It is caused by excess energy above that needed to melt the powder. A term of the form P/mD based on an energy balance will have a maximum value before dilution sets in. For the Colmonoy® Wallex® PC6 this value was found to be 2,500 J g^{-1} mm^{-1}. This is also plotted in Figure 6.33. The final boundary is due to the need for a certain amount of energy to create a continuous track. This is defined by another energy balance term (P/DV) representing the minimum energy

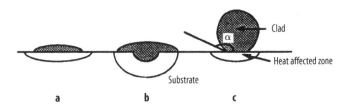

Figure 6.32 The three basic cross-section profiles for single-track clad beads

---

[1] Colmonoy® is a registered trademark of Wall Colmonoy, 550 Sand Sage Road, Los Lunas, New Mexico, 87031-4844, USA. http://www.wallcolmonoy.com

[2] Wallex® is a registered trademark of Wall Colmonoy, 550 Sand Sage Road, Los Lunas, New Mexico, 87031-4844, USA. http://www.wallcolmonoy.com

[3] Stellite® is a registered trademark of Deloro Stellite Holdings Corporation, St.Louis, USA.

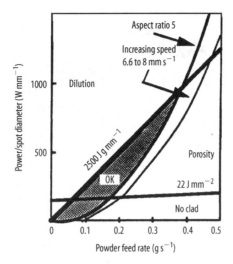

Figure 6.33 The operating window for blown-powder laser cladding

required for cladding before the track starts to be discontinuous; in the case of the Colmonoy® alloy it was $22\,\mathrm{J\,mm^{-2}}$. The terms in these parameters are as follows: P is the power absorbed after reflection and scattering (W), D is the spot size on the substrate surface (mm); V is the traverse speed $(\mathrm{mm\,s^{-1}})$ and m is the mass flow of powder trapped in the clad $(\mathrm{g\,s^{-1}})$.

There is a relatively large operating region for making fusion-bonded, low-dilution clad tracks as illustrated in Figure 6.33. It is a function of speed. It may be partly due to the phenomenon described in Steen [67] and first observed by Takeda *et al.* [68]. Takeda *et al.* placed some tin in the path of a clad track and then examined the track under a scanning electron microscope to see to where the tin had flowed, thus mapping the melt pool. For one track they stopped at the moment the tin was struck leaving a melt pool shape as illustrated in Figure 6.34. It appears that the molten leading edge of the melt pool is quickly covered by the new molten clad, which grows sufficiently fast to allow the interface region to resolidify (by conduction to the substrate) before any considerable dilution has occurred. This may account for this usefully large operating region. Another feature of this marker system was to observe that the tin had moved very fast from the front of the pool over the top to the back at a speed measured in metres per second. This movement is expected from the Marangoni effect. It also explains why it is possible to clad successfully on dirty or rusty surfaces – not recommended, but it is possible!

The quality of the bond in blown powder cladding is most likely to be a fusion bond because the powder will usually arrive hot but solid, and hence will not stick unless the surface has melted a little, thus ensuring a fusion bond. The powder would melt quickly under the incoming radiation once it was static.

For higher clad coverage rates, larger spots and higher powers would be required; also, the laser process can be enhanced by joining it to a plasma spray in a hybrid process [69].

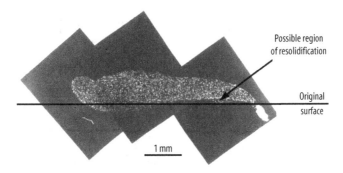

Figure 6.34 Scanning electron microscope micrograph of a clad longitudinal section, showing the location of tin which was picked up from the clad leading edge [68]

One variation on the process, invented by Fellowes *et al.* [70], is the bithermal laser cladding process. By this technique the powder can be fed into the laser beam focal spot, where it will melt instantly. It is then allowed to fall molten down the beam onto the substrate, which can be situated at a lower power density, achieved both by distance and tilting if need be. Fellowes *et al.* wished to coat steel and INCONEL® with silica, but in later work aluminium was coated with chromium in this way to achieve 78% Cr/Al alloy [71]. Direct cladding with chromium (melting point 1,875 °C) onto aluminium (melting point 660 °C) would generate a huge pool into which the chromium would sink if it could penetrate the Al_2O_3 skin. In the bithermal process the skin can be removed by a pulse from an auxiliary excimer laser firing from the side onto the tilted plate.

6.5.3 Applications

This almost unique precision cladding technique has a flexibility that is only just beginning to be understood. Not only can it clad with good fusion-bonded cladding in localised areas, but it allows build-up of the clad to make precision castings (see Chapter 7). The high rate of mixing in the clad melt pool causes homogenisation of the melt, for speeds below a certain value, and hence alloys can be formed *in situ* from cheaper ingredients or alloy systems can be rapidly analysed using this process [72]. The cooling rates are fairly swift in the small melt pool. This generates finer microstructures than competing processes. Monson and Steen [50, 51] calculated the cooling rate from secondary dendrite arm spacing (see Section 6.3.1) for various processes when cladding Colmonoy® 6. The results are shown in Table 6.3.

Table 6.3 Observed cooling rates for Colmonoy® 6 alloy [51]

Process	Cooling rate ($°C\,s^{-1}$)
Oxyacetylene	9.0
TIG	45.7
Plasma-transferred arc	111.6
Laser	3,045.2

Preheating affects the cooling rate. For example, the laser cooling rate fell to 236 °C s^{-1} when there was a preheat of 750 °C on a similar alloy, Colmonoy® 5. This helps when cladding crack-sensitive alloys. These high cooling rates may allow different levels of solid solubility and thus may affect the expectation from the equilibrium phase diagram.

The main applications of laser cladding are for corrosion or wear resistance (both adhesive and abrasive wear), but more recently the applications include a form of reverse machining, whereby the laser can put material on rather than have it removed, as with a lathe. Thus, a major application is in the repair business for refurbishment or salvage of high-value components or components that have been overmachined [58]. In one example the laser is used to rebuild the splines on drive shafts [73] by completely removing the worn splines and rebuilding with alloy steel by laser cladding. Similarly, bearings with oil seals used in the mining industry are resurfaced by laser cladding, the laser being particularly applicable since the heat input is low and hence there is little thermal distortion of these precision parts. Turbine blades can be rebuilt at their tips or leading edges [74].

For wear resistance the shroud interlock between turbine blades has been hard-faced with Triballoy® [4] to reduce the wear that may occur while the cold engine is heating up to fill the expansion gap between the blades [75]. There are numerous hard-facing applications and alloy systems that have been laser-clad: the cladding of aluminium with iron [76]; mild steel with NiCoCrB alloy for enhancing cavitation erosion resistance [77]; HASTELLOY® onto stainless steel [78]; and many others.

For biological properties hydroxyapatite has been laser clad onto a titanium prosthesis using a Nd:YAG laser to create a coating which allows growth of tissue between the prosthesis and the bone [79], thus avoiding the necessity for cements and the inevitable decay of the support bone. ZrO_2 has been added into the clad mix and found to enhance the strength of such coatings [70].

For alloy scanning, Sexton *et al.* [72] did an analysis of the Co:Al:Fe system by way of illustration of a technique for rapid alloy scanning using a triple powder feed system and varying the feed rate from the hoppers along the length of a single track. The high mixing within the clad pool gave alloy samples of varying composition along the track. They also invented a series of microtests to examine the basic properties of these alloys and hence could plot approximate ternary diagrams.

6.6 Particle Injection

This process is similar to laser cladding by the blown powder route except that the particles blown or projected into the laser melt pool do not entirely melt [81, 82]. This creates a structure similar to a Macadam road (Macadam is a type of road surface made of even conglomerate with a binder. It was pioneered by John Loudon McAdam around 1820) (Figure 6.35). The main advantages for this process are improved hardness and wear resistance with reduced friction coefficients in some systems. Process variations

---

[4] Triballoy® is a registered trademark of Deloro Stellite Holdings Corporation, St. Louis, USA.

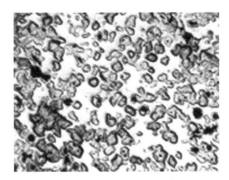

Figure 6.35 Macrograph showing a laser clad made with mixed powder of Stellite® and TiC. The TiC has not melted and forms a type of macadam road

centre around the particle delivery system, delivery pressure (vacuum or atmospheric) and the gas shrouding systems.

To achieve a good surface layer the hard embedded particles must be wetted by the metal matrix and they must have strong bonding to it. It is also desirable that the particles do not suffer too much dissolution while lying in the melt pool. These requirements mean the particle and the surface must be clean and the level of superheat must be kept as low as possible, compatible with the wetting condition.

This process is still at the laboratory stage but shows considerable promise in hardening aluminium and its alloys by the injection of TiC, SiC, WC or Al_2O_3 particles. It has also been applied to stainless steel. In the injection of WC into a Ni/Co alloy, Gassmann [83] used uncoated particles and a particle volume percentage greater than 40 to reduce Marangoni stirring and hence the rate of dissolution.

6.7 Laser-assisted Cold Spray Process

This process is a combination of the cold spray process, in which high-velocity particles impact on a surface and forge-weld to the surface, and laser cladding, where the particles are melted on arrival. By feeding of the high velocity particles down a laser beam they are slightly heated and therefore softened on the way to impact. On arrival they are further heated by the laser irradiation and so consolidated. The result is a lower-temperature cladding process [84]. In a further development of the process O'Neill blows powder at around 400 m/s onto a spot laser heated to aproximately 600 C by the laser, achieving a good bond, dense clad and higher deposition rate than with standard cold spray.

6.8 Surface Texturing

The rolls in temper mills are textured to dull the surface of sheet steels, which improves the grip in a press and the flow of paint on the final surface. There are several techniques

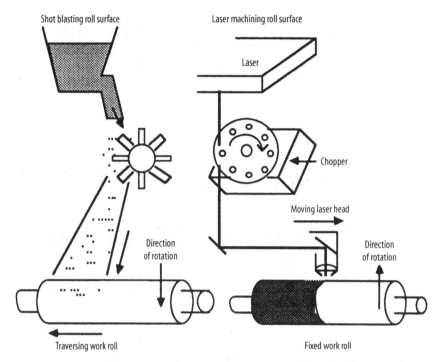

Figure 6.36 General arrangement for texturing tempering rolls by shot blasting or laser

for doing this [85, 86]. The conventional technique is to shot-blast the surface. This, however, gives a random roughness which exhibits a waviness in the finished paint surface. This can be avoided by a regular patterned roughness. The pattern can be put down by electric discharge machining, electron beam or laser (either CO_2 or Nd:YAG) methods. The results and the method with the CO_2 process are seen in Figures 6.36 and 6.37. The roughness must be higher than approximately 1 μm for press formability and the waviness should be as low as possible (Figure 6.38). It can be seen that the laser process is quite successful. It is necessary that the patterned roughness is very regular and uniform. The electric discharge machining approach lacks placement precision and the Nd:YAG laser does not give sufficiently consistent high-quality spots owing to the variations within the YAG rod. The electron beam method requires a vacuum chamber and special cleaning of the roll prior to work. On the whole, the CO_2 laser technique comes out on top when a comparison is made [86]. The CO_2 laser process requires some preheat of the roll surface prior to creating the crater to achieve an optimum crater shape. The shape is amazingly important to avoid the indent showing through the paint, yet to achieve a good grip. This preheat is achieved by a fast-rotating chopper system (Figure 6.39) in which the beam is reflected off the chopper blade onto another mirror which delivers the out-of-focus beam to the area to be treated. As the chopper rotates, the beam will pass between the blades and create the crater at the spot which has just been heated. Crater formation can be done at five to 30,000 craters per second with a 2.5 kW CO_2 laser. The focused beam has a power intensity of 12–65 MW cm^{-2} with pulse energies of the order of 25–125 mJ per pulse for 10–50 μs.

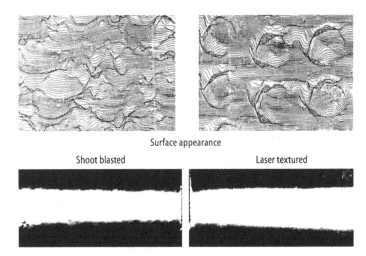

Surface appearance

Shoot blasted   Laser textured

Figure 6.37 Appearance of a strip light reflected in a painted surface [85]

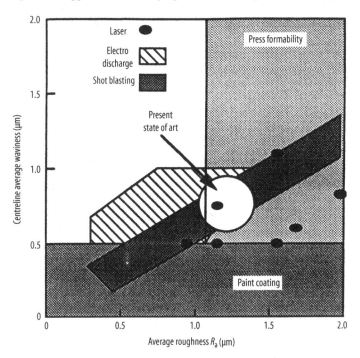

Figure 6.38 Surface roughness versus surface waviness for a painted surface of press steel. (From Shibata [85])

Another texturing process has been developed by Y.H. Han (private communication) at KIMM, South Korea. He varnished a roll with photosetting monomer and then scanned a pattern onto the roll using a low-powered ultraviolet He–Cd laser. After the unexposed monomer had been washed off the roll, the roll was etched to make an em-

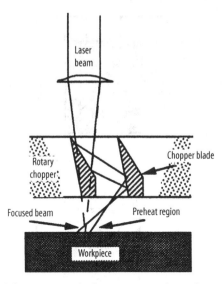

Figure 6.39 Beam delivery system for laser texturing for preheat and cratering [62]

bossing roll for paper or steel. A variation is to do the same process on a stainless steel mesh roll; the result is a screen for printing textiles. This could be a major breakthrough in the technique for textile printing rather than working from standard silk screens with problems in turnaround and storage. See also Chapter 8 on micromachining for the lithography of solar panels and LCD screens.

6.9 Enhanced Electroplating

The irradiation by a laser beam of a substrate used as a cathode during electrolysis causes a drastic modification of the electrodeposition process in the irradiated region. Interesting aspects of the process are:

- the possibility of rapid maskless patterning;
- the possibility of enhanced plating rate on selected areas; and
- the possibility of modifying the structure of electrodeposited coatings.

The first attempt to use the laser in combination with electroplating is thought – possibly mythically – to have been done by someone interested in using a laser Doppler anemometer to measure the flow at the cathode during plating. Instead of measuring flow rate, he found a new process! Possibly this story belongs to the IBM Thomas J. Watson Research Center, since Von Gutfeld et al. [87] give it the credit for the invention. From these experiments it was shown that the plating rate increased in the irradiated zone by up to 1,000 times!

Three processes have emerged: laser-enhanced plating; laser-induced electroless deposition or immersion plating; and laser-assisted etching.

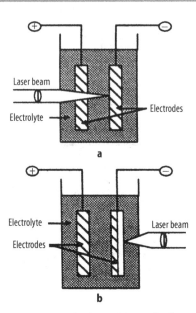

Figure 6.40 a A directly irradiated cathode, and **b** a cathode irradiated from the back

Laser-enhanced electroplating of nickel, copper and gold has been reported by Von Gutfeld *et al.* [87]. Power densities of $10–10^4$ W mm^{-2} are used. The laser used depends upon the transmissivity of the electrolyte and the absorption properties of the cathode. Thus, argon ion lasers (λ = 514.5 nm) are used for copper and nickel solutions, whereas krypton ion lasers (λ = 647.1 nm) or Nd:YAG lasers (λ = 1,060 nm) are used for yellow gold electrolytes. The power can be directed onto the front of the cathode or through a transparent cathode (Figure 6.40). A light pen has been developed by IBM in which a jet of electrolyte with a laser beam waveguided down it impinges on the area to be treated. The result is a dome-shaped deposit around 0.5 mm wide. The potential applications are maskless printing or repair of circuit boards.

Laser-induced electroless deposition has been tried for a number of systems, *e.g.*, copper from copper sulphate/hydrochloric acid [88]. This process poses the possibility of printing circuits on nonconductors such as ceramics.

6.10 Laser Chemical Vapour Deposition

Blowing thermally sensitive vapour onto a laser-generated hot spot can cause a deposit to be formed by pyrolysis. The rate of deposition is controlled by chemical reaction rates (Arrhenius equation) up to certain deposition rates dependent on the surface temperature; above these temperatures the process is controlled by mass transport. Under mass transfer control the quality of the deposit falls markedly from a smooth sheet to a rough surface and ultimately to powder. The rates are illustrated in Figure 6.41 [89]. The rate of deposition is usually slow, being a few micrometres per minute for most processes

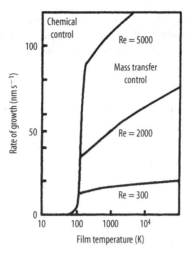

Figure 6.41 Temperature dependence of the growth rate of CoO deposit during LCVD of cobalt acetylacetonate from a jet of concentration 0.3 ppm. (From Steen [89])

owing to the need to avoid mass transfer control. Alternatively the vapour could be directly broken by the photons in a process of photolysis. This is particularly relevant to processing with the excimer laser. Leön et al. [90, 91] have deposited SiO_2 in this manner.

6.11 Laser Physical Vapour Deposition

The laser beam can be directed onto a target situated in a vacuum chamber. The target evaporates and the vapour condenses on the substrate among other areas. This process has the advantage of extreme cleanliness in the heating technique. It is possible using an excimer laser pulse to ablate the surface of the target, resulting in a deposit on the substrate which is of the same composition as the target material with no difference due to vapour pressures. Such a process has been described for the deposition of superconducting alloys [92], calcium hydroxyapatite for medical implants [93] and TiN on hydrogen-terminated silicon (100) [94].

6.12 Noncontact Bending

Noncontact bending [95, 96] is included in laser surface treatments since the process involves the generation of surface stress and plastic deformation. By passing a laser beam over a surface, one can create sufficient stress to cause plastic flow that results in a controlled bend on cooling. The process is one of precision bending with no springback and only software to control the direction and speed of the laser beam – a form of virtual tooling. This should appear to most readers like a magic wand that can make sheet metal curl up. The industrial significance is just beginning to be appreciated, but the process deserves a separate chapter and is therefore presented in Chapter 9.

6.13 Magnetic Domain Control

In magnetic domain control [97] the laser induces thermal stress fields, which cause the generation of subdomains leading to a significant decrease in eddy current losses in transformers (Figure 6.42). Electrical machines and transformers contain metal sheets having particular magnetic properties. These are planned to guide the magnetic field with only minimal losses. High-power transformers exclusively contain grain-oriented electrical sheets of an efficiency of up to 99%.

A CW Nd:YAG laser has been used by Neiheisel [98] to refine the magnetic domain size in transformer steel, whilst leaving the insulating coating on the steel intact. In the laser domain refinement process [97], a high-power focused beam is scanned rapidly $(100\,\mathrm{m\,s^{-1}})$ across the surface of 3% silicon–iron, as used in electrical transformers. The material after treatment shows no visible surface change; however, a reduction in the magnetic core loss has occurred.

A thermal shock is believed to be imparted to the microstructure which causes slip-plane dislocations to form, thereby producing new magnetic domain wall boundaries.

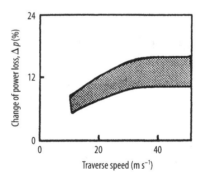

Figure 6.42 Reduction in core loss versus laser traverse speed. (After Gillner *et al.* [97])

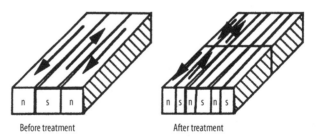

Before treatment   After treatment

Figure 6.43 The principles of domain refinement

By adjustment of the spacing of the scanned laser lines, the energy lost owing to moving the domain walls back and forth under the action of the applied AC field in the transformer is minimised. The laser lines restrict the length of the domains, which also acts to control the width of the domains. Thus, by adjustment of the spacing of the laser lines, the domain sizes can be controlled, *i.e.*, refined. The process is illustrated in Figure 6.43.

6.14 Laser Cleaning and Paint Stripping

A short (approximately nanosecond), sharp (approximately megawatt) burst of radiation can remove superficial material and hence clean [99–101]. The mechanism is debatable but could be based on one or all of the following causes: evaporation, photochemical bond breaking, mechanical pressure from the photons, the expansion shock wave in the substrate or recoil pressure from a plasma. Whatever the mechanism, the advantages are significant, including the facts that:

- it is a quiet process;
- it does not require solvents or abrasive particles;
- fume is easily collected by standard vacuum cleaning methods;
- it sterilises as it cleans;
- metals cleaned by laser may not need flux for subsequent soldering; and
- the process is easily monitored through the spectrum of the plasma generated.

With these advantages and the huge potential industry for cleaning and paint stripping, this process merits a separate chapter and is therefore presented in Chapter 10.

6.15 Surface Roughening

The adhesion of glue to a surface previously blasted with a pulse from an excimer laser is considerably stronger than without this treatment. Olfert *et al.* [102] used a XeCl excimer laser with a power density of $13\,\mathrm{J\,cm^{-2}}$ and a pulse length of approximately 30 ns on galvanised and galvannealed steel sheets. Maybe this is the beginning of a new technique for the car industry!

6.16 Scabbling

Large heavy rotary claws can scrape off the surface of a road or the concrete surface in a waterworks or nuclear power plant prior to resurfacing or decommissioning. These machines can do unnecessary damage to concrete tanks and create considerable dust, which, in the nuclear example, would be a problem. It has been found that the laser used in an unfocused condition (spot size greater than 10 mm) will generate a stress condition in the concrete surface such that it breaks and flies off. The process is quite spectacular, but it overcomes the problems listed above [103, 104]. High-speed filming of the process has shown that the break occurs at the leading edge of the laser-generated hot spot. The cracking has some semblance to that from a Hertzian stress zone. This process has the potential for more environmentally friendly civil engineering activities since it is relatively quiet. It is also one of the few options for removing polluted surface material without causing excessive dust or water waste. At present it is a little slow. The process has been elegantly modelled by Dowden [104].

6.17 Micromachining

Ultrashort pulses of a few nanoseconds and powers up to gigawatts has meant that the laser can ablate material with very little heating, if any. Femtosecond pulses with gigawatt power are thought to create such intense electric fields at the focus that they can ionise the solid material in the extremely short time of the pulse length. The material then flies apart through Coulomb forces. In the case of very short wavelengths, and therefore energetic photons, the energy of a photon can break chemical bonds without necessarily heating; this is a further method for "cold" ablation. Some doubt is cast on these theories by the experiments of Hess [105], who measured the velocity of particles of solid benzene being ejected from the surface of transparent NaCl cooled by liquid nitrogen. He also measured the ablation yield as a function of fluence. He concluded that the heat balance suggested that the process was photothermal in origin and the process appears "cold" because there is no time for conduction to occur over measurable distances. Whatever the answer, a precise method of ablating material without a significant HAZ is very valuable to the electronics industry. There was a whole section in the International Congress on Applications of Lasers and Electro-Optics conferences in both 2001 and 2002 dedicated to micromachining.

Ablation is used to drill fine specialist holes. These holes can now be drilled by an excimer laser in ceramics and plastics in a variety of shapes, even rifled holes of a few micrometres diameter. The laser is an ablation tool and coupled to a computer is capable of precision engineering on a scale hitherto unobtainable. For example, Q-switched Nd:YAG lasers are used to trim quartz oscillators for watches [106]. The oscillators are tiny tuning forks ranging in size from $2 \times 5\,mm^2$ to $5 \times 15\,mm^2$ and which are about $100\,\mu m$ thick. They are mounted in a watch in which a weak magnetic field drives the oscillator to vibrate at its resonance frequency. The signal derived from this frequency is used to drive all the watch circuits, including the signal driving the tuning fork; thus, the tuning fork has to be accurate. The quartz for the fork is cut into slices and machined to the shape required, but each fork will differ slightly. The mass of each fork is then increased by a metal overlay, the resonance frequency is measured and a computer program determines how much material must be removed. The laser then ablates the correct mass. Ablation and micromachining are now the basis of a growing industry and are therefore discussed in a separate chapter, Chapter 8.

6.18 Laser Marking

Short pulses from a Q-switched Nd:YAG laser, excimer or a transversely excited atmospheric pressure (TEA) laser can make a mark by removing a layer, modifying the surface morphology, causing a local reaction or activating a colour centre [107]. The mark thus made is then shaped into a pattern by direct writing, projecting through a mask or scan-writing a dot matrix.

Examples include the removal of a paint layer on anodised aluminium, surface melting, porosity or oxide formation and the change in colour of a pigment such as TiO_2 on heating.

Mask projection must have a specially homogenised and uniform power distribution in the laser beam. The speed is impressive but the flexibility is low, requiring masks to be changed. One hundred similar images per second have been achieved in marking Teflon[®5] cabling. Integrated circuits are printed this way using very carefully stabilised excimer lasers and precision stepper systems. One new development being considered is the use of LCDs as a software-controlled mask that can be rapidly changed.

Direct writing is performed by steering a laser beam with galvanometer-driven mirrors. The laser is usually a Q-switched Nd:YAG laser of approximately 100 W average power with short pulses of 100–200 ns, giving peak powers of 30–100 kW, and a high pulse rate of 0.1–100 kHz. Good pulse shapes are only possible at present in the range 1–20 kHz. One of the main development areas for direct writing concerns the software driving the galvanometer steering mirrors. The system also usually has large-aperture flat-field focusing optics. The print rate is of the order of 70 characters per second with traverse speeds of 3 m s^{-1}. The line thickness is dependent on the beam diameter, but is of the order of 50–100 µm; it can be as little as 15 µm with special optics, allowing character heights of 50 µm.

Marks to avoid counterfeiting can be made internally by nonlinear effects using femtosecond laser pulses. The primary packaging may be organic (polymer) or inorganic (glass, fused silica or quartz). Nd:YAG lasers with nanosecond pulses can create microcracks 5–100 µm in diameter: these are unacceptable for packaging, *e.g.*, bottles which may shatter, although they are very beautiful for the gift trade in large blocks of glass. Using a femtosecond laser, the process is one of ablation without cracking, extremely fine definition and almost invisible marks that can only be read by special optics. It is, however, totally tamper-proof [108].

There are many competing technologies, as shown in Table 6.4, but the simplicity of using the laser, the permanence of the mark and the speed of application have opened up a considerable appetite within industry for marking for security, quality control, identification or legal reasons.

Some of the main advantages and disadvantages of the laser are shown in Table 6.5.

The following list shows some of the current areas in which laser marking is being used: keypads for mobile phones or PCs; tags for goods and cattle; bar codes for anything from electronic items to retail goods; calibration marks on tools and gauges; serial numbers on cars and machine tools; identification numbers on cable and production

---

[5] Teflon® is a registered trademark of E.I. du Pont de Nemours and Company. http://www.dupont.com

Table 6.4 A comparison of marking technologies [107]

Process	Permanence	Throughput	Flexibility	Costs	Quality	Material disposal	Maintenance
Laser	+	+	+	−	+	+	0
Printing	0	−	−	+	+	+	−
Ink jet	0	+	+	+	+	+	−
Chemical etching	+	−	−	+	+	−	+
Label/name plate	−	+	+	+	+	+	0
Pantograph	+	−	+	−	0	+	+
Silk screen	−	+	−	−	+	+	−
Metal stamp	+	+	−	−	−	−	0
Moulded/ embossed	+	+	−	−	+	−	0
Engraving	+	0	−	0	0	−	−

+ is a point of merit, − is a disadvantage, 0 is average value

Table 6.5 Typical advantages and disadvantages of laser marking

Advantages	Disadvantages
Flexible in design and application	High investment costs
Fully automatable	Need for skilled operators
Noncontact technology, no clamping	Thermal load on workpiece
Good readability	
Good text permanence	

lines; artistic designs for buttons and plaques; hallmarking gold and silver items. This list is probably endless!

6.19 Shock Hardening

Laser shock hardening [109, 110] or "laser shot peening" has until recently been viewed as a curiosity. It is now, however, an industrial process and for good reasons. The purpose of shot peening is to create a compressive stress in the surface and thus increase the fatigue strength of the article. The compressive stresses are induced by cold-working the surface with impacting shot. The magnitude of the compressive stress induced this way can be as high as 60% ultimate tensile strength. The depth of treatment depends on the shot energy and the material, but is generally of the order of 1 mm maximum and 0.25 mm average. With this depth under compression, the balancing tensile stress is borne by the material below the surface layer. For thin materials this could be a problem, inducing fatigue beneath the surface. Thus, the compressive layer thickness should be no greater than 10% of the workpiece thickness. The process of shot peening is also used to refine grain structures, close pores in certain cast or sintered products and produce improvements in fretting, galling and hydrogen embrittlement stress corrosion cracking and intergranular corrosion. It will also eliminate residual tensile stresses that result from various metal-forming techniques such as electric discharge machining, turning, milling, broaching, grinding, chemical milling and laser cutting.

The same or better compressive stress can be created by a powerful radiant blast, but it must be a short pulse (10–100 ns) and of good beam quality. This was one of the breakthroughs made by Lawrence Livermore National Laboratory in understanding this process. A powerful master oscillator power amplifier system based on Nd:glass was built to give 100 J pulses at 6 Hz and 10–100 ns pulses. It is essential that the pulse is short or the result will be a drilled hole [111]. The average power of the laser was only 600 W. To achieve a uniform phase front on impact over an area of 1 cm^2, the laser incorporated stimulated Brillouin scattering reflectors. This machine can laser shot peen 100 mm^2 per pulse or 1 m^2 per hour. The part to be hardened is coated with an absorbing layer and the plasma generated is confined by a surface layer of water. The general arrangement is illustrated in Figure 6.44. The impact generates pressures of the order of 1,000–10,000 atm. Since this happens over a considerable area (1 cm^2), the depth of treatment can be as much as 3 mm, compared with 1 mm for shot peening, and there is little surface damage.

The advantages of laser shot peening are:

- there is less surface roughening than conventional shot peening;
- there are no embedded particles;
- it can harden right into corners where shot could not reach;
- there is no material to recycle, collect, grade and clean, as there is with shot peening;

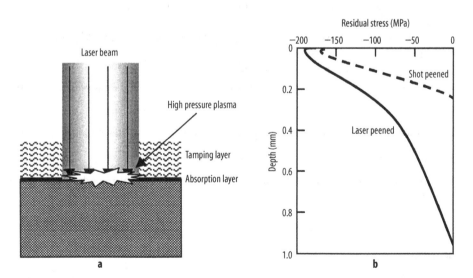

Figure 6.44 a The laser peening process. Short pulses of 10–100 ns and 100 J cm^2 create plasma pressures of 1,000–10,000 atm, and **b** comparison of depth of hardening between shot peening and laser peening. (Data taken from Hammersley *et al.* [109] for INCONEL® 718)

- there is little wear on the equipment; and
- no contaminated pellets are created when treating radioactive material.

6.20 Conclusions

Optical energy is an ideal form of energy for surface treatment. The uses of laser surface treatment include surface heating, bending, scabbling, melting, alloying, cladding, texturing, roughening, marking, cleaning, shock hardening and stereolithography and layered manufacturing processes, which will be discussed in the next chapter. The advantages offered by the laser are the highly localised, clean nature of the process, low distortion and high quality of finish. It is thus not surprising to find that laser processing of surfaces is a subject currently enjoying much research and industrial interest. With the development of highly automated workstations and lasers which are cheaper, more powerful, reliable and compact, surface treatment by lasers is set to be the fashion of the next decade or so.

Questions

1. What is the mechanism for laser surface hardening? What are the main controlling parameters?
2. How would you reduce the surface reflectivity?
3. How could a laser beam be used to uniformly irradiate an area?
4. Why do dendrites form?
5. What would you have to do to get a fine metallurgical structure? What parameters affect the size of the dendrites?
6. How could the material for laser surface alloying be applied?
7. Sketch the expected flow structure within a laser-generated melt pool. What are the main forces acting on the pool?
8. What processes compete with laser cladding?
9. Is there a difference in clad quality between preplaced powder and blown powder cladding?
10. What is the point of surface texturing?
11. Describe the process of laser shock hardening. List some of the advantages it has over shot peening.

References

[1] Gregson VG (1984) Laser heat treatment. In: Bass M (ed) Laser materials processing. North-Holland, Amsterdam, chap 4
[2] Bell T (1982) Surface heat treatment of steel to combat wear. Metallurgia 49(3):103–111

[3] Dausinger F, Beck M, Rudlaff T, Wahl T (1988) On coupling mechanisms in laser processes. In: Hugel H (ed) Proceedings of the 5th international conference on lasers in manufacturing (LIM5), Stuttgart, September 1988. IFS, Kempston, pp 177–186

[4] Sharp M, Steen WM (1985) Investigating process parameters for laser transformation hardening. In: Proceedings of the 1st international conference on surface engineering, 1985, Cambridge, UK, paper 3

[5] Mazumder J, Steen WM (1980) Heat transfer model for CW laser material processing. J Appl Phys 51(3):941–946

[6] Ashby MF, Easterling KE (1984) The transformation hardening of steel surfaces by laser beam. Acta Metall 32:1935–1948

[7] Steen WM, Courtney C (1979) Laser surface treatment of En8 steel using a 2 W CO_2 laser. Met Technol 6(12):456

[8] Chan C, Mazumder J, Chen MM (1984) A two dimension transient model for convection in a laser melted pool. Metall Trans A 15:2175–2184

[9] Ashby MF, Shercliff HR (1986) Master plots for predicting the case depth in laser surface treatments. Document CUED/C-Mat/TR 134. English Department, Cambridge University

[10] Laser Focus World (1996) Excimer laser beam several inches across is used to anneal amorphous silicon in fabrication of AMLCD panels. Laser Focus World May 101

[11] Mazumder J (1983) Laser heat treatment: the state of the art. J Met 35(5):18–26

[12] Morley J (1995) CO_2 lasers go to the bank. Photonics Spectra Jul 20–21

[13] Olsen FO (1982) Laser material processing at the Technical University of Denmark. In: Proceedings of Materialbearbeitung mit CO-Hochleist, Stuttgart, April 1982, paper 2

[14] Jørgensen M (1980) Increasing energy absorption in laser welding. Met Constr 12(2):88

[15] La Rocca AV (1982) Laser applications in manufacturing. Sci Am Mar 80–87

[16] Weerasinghe VM, Steen WM (1983) Laser cladding by powder injection. In: Kimmitt M (ed) Proceedings of the conference lasers in manufacturing 1, Bedford, UK. IFS, Kempston, pp 125–132

[17] Steen WM, Chen ZD, West DRF (1987) Laser surface melting of cast irons and alloy cast irons. In: Belforte D, Levitt M (eds) Industrial laser annual handbook 1987. LIA, Orlando, pp 80–96

[18] Hawkes IC, Steen WM, West DRF (1983) Laser surface hardening of S.G. cast iron. Metallurgia 50(2):68–73

[19] Trafford DNH, Bell T, Megaw JHPC, Bransden AD (1983) Laser treatment of grey iron. Met Technol 10(2):69–77

[20] Hawkes IC, Walker AM, Steen WM, West DRF (1984) Applications of laser surface melting and alloying to alloys based on the Fe–C system. In: Mukherjee K, Mazumder J (eds) Proceedings of lasers in metallurgy 2, Los Angeles, USA, February 1984. ASM, Metals Park, pp 169–182

[21] Bergmann HW (1986) Laser surface melting of iron-base alloys. In: Draper CW, Mazzoldi P (eds) Proceedings of NATO Advanced Study Institute on laser surface treatment of metals, San Miniato, Italy, 2–13 September 1985. Nijhoff, Dordrecht, pp 351–368

[22] Mordike S (1993) Laser surface remelting of camshafts. Lasers Eng H.2.:S43/60

[23] Lamb M, Steen WM, West DRF (1984) Structure and residual stresses in two laser surface melted stainless steels. In: Proceedings of the conference Stainless Steels'84, Gothenburg, Sweden, September 1984

[24] Lumsden JB, Gnanamuthu DS, Moores RJ (1984) Intergranular corrosion of steels and alloys. In: McCafferty E, Clayton CR, Oudar J (eds) Fundamental aspects of corrosion protection by surface modification. Electrochemical Society, Pennington, p 122

[25] Anthony TR, Cline HE (1977) Heat treating and melting material with a scanning laser or electron beam. J Appl Phys 48:3895–3900

[26] Jeng Y, Quale BE, Modern PJ, Steen WM, Bastow BD (1993) Laser surface treatment to improve intergranular corrosion resistance of 18/13 Nb 304L in nitric acid. Corros Sci 35:1289

[27] Folkes JA (1985) Laser surface melting and alloying of titanium alloys. PhD thesis, University of London

[28] Folkes JA, Henry P, Lipscombe K, Steen WM, West DRF (1984) Laser surface melting and alloying of titanium alloys. In: Proceedings of the 5th international conference on titanium, Munich, September, 1984

[29] Lawrence J, Li L (2000) Surface glazing of concrete using a 2.5 kW diode laser and the effects of large beam geometry. Opt Laser Technol 31:583–591

[30] Lawrence J, Li L, Spencer JT (1998) A two-stage ceramic tile grout sealing process using a high power diode laser – part I: grout development and materials characteristics. Opt Laser Technol 30:205–214

[31] Lawrence J, Li L, Spencer JT (1998) A two-stage ceramic tile grout sealing process using a high power diode laser – part II: mechanical and chemical properties. Opt Laser Technol 30:215–223

[32] Li L (2000) The advances and characteristics of high power diode laser material processing. Opt Lasers Eng 34:231–253

[33] Flemings MC (1974) Solidification processing. McGraw-Hill, New York

[34] Hawkes IC, Lamb M, Steen WM, West DRF (1983) Surface topography and fluid flow in laser surface melting. In: Proceedings of CISFFEL, vol 1. Commissariat á l'Energie Atomique, Gif-sur-Yvette, pp 125–132

[35] Takeda T, Steen WM, West DRF (1984) Laser cladding with mixed powder feed. In: ICALEO'84 proceedings, Boston, November 1984. LIA, Orlando, pp 151–158

[36] Eiholzer E, Cusano C, Mazumder J (1985) Wear properties of laser alloyed Fe–Cr–Mn–C alloys. In: ICALEO'85 proceedings, November 1985. LIA, Orlando, p 8

[37] Blake A, Mazumder J (1982) Control of composition during laser welding of Al–Mg alloy using plasma suppression technique. In: ICALEO'82 proceedings. LIA, Orlando, pp 33–50

[38] Christodoulou G, Steen WM (1984) Laser surface treatment of chromium electroplate on medium carbon steel. In: Metzbower EA (ed) Proceedings of the 4 h international conference on lasers in material processing, Los Angeles, January 1983. ASM, Metals Park, pp 116–126

[39] Walker AM, West DRF, Steen WM (1983) Laser surface alloying ferrous materials with carbon. In: Waidelich W (ed) Proceedings of Laser'83 Optoelectronik conference, Munich, June 1983, pp 322–326

[40] Lamb M, Man C, Steen WM, West DRF (1983) The properties of laser surface melted stainless steel and boronised mild steel. In: Proceedings of CISFFEL, vol. 1. Commissariat á l'Energie Atomique, Gif-sur-Yvette, pp 227–234

[41] Weerasinghe VM, Steen WM (1984) Laser cladding with pneumatic powder delivery. In: Metzbower EA (ed) Proceedings of the 4th International conference on lasers in material processing, Los Angeles, January 1983. ASM, Metals Park, pp 166–175

[42] Walker AM, Folkes J, Steen WM, West DRF (1985) The laser surface alloying of titanium substrates with carbon and nitrogen. Surf Eng 1(1):23–29

[43] Folkes J, West DRF, Steen WM (1986) Laser surface melting and alloying of titanium. In: Draper CW, Mazzoldi P (eds) Proceedings of NATO Advanced Study Institute on laser surface treatment of metals, San Miniato, Italy, 2–13 September 1985. Nijhoff, Dordrecht, pp 451–460

[44] Man HC, Cui ZD, Yue TM (2002) Surface characteristics and corrosion behaviour of laser surface nitrided NiTi shape memory alloy for biomedical applications. J LaserAppl 14(4):242–247

[45] Tucker TR, Clauer AH, Ream SL, Walkers CT (1982) Rapidly solidified microstructures in surface layers of laser alloyed molybdenum on Fe–C substrates. In: Proceedings of the conference on rapidly solidified amorphous and crystalline alloys, Boston, MA, November 1981. Elsevier, New York, pp 541–545

[46] Marsden C, West DRF, Steen WM (1986) Laser surface alloying of stainless steel with carbon. In: Draper CW, Mazzoldi P (eds) Proceedings of NATO Advanced Study Institute on laser surface treatment of metals, San Miniato, Italy, 2–13 September 1985. Nijhoff, Dordrecht, pp 461–474

[47] Tien JK, Sanchez JM, Jarrett RT (1983) Outlook for conservation of chromium in superalloys. In: Proceedings of technical aspects of critical materials used by the steel industry, vol. II-B. National Bureau of Standards, Washington, p 30

[48] Liu Z, Watkins KG, Steen WM (1999) An analysis of broken-layer characteristics in laser alloying for coinage applications. J Laser Appl 11(3):136–142

[49] Liu Z, Pirch N, Gasser A, Watkins KG, Hatherley PG (2001) Effect of beam width on melt characteristics in large area laser surface alloying. J Laser Appl 13(6):231–238

[50] Monson PJE, Steen WM (1990) A comparison of laser hardfacing with conventional processes. Surf Eng 6(3):185–194

[51] Monson PJE (1988) Laser hardfacing. PhD thesis, University of London

[52] Powell J, Henry PS, Steen WM (1988) Laser cladding with preplaced powder: analysis of thermal cycling and dilution effects. Surf Eng 4(2):141–149

[53] Steen WM (1978) Surface coating with a laser. In: Proceedings of the conference on advances in coating techniques. TWI, Great Abington, pp 175–187

[54] Jardieu de Maleissye J (1986) Laser induced decomposition of molecules related to photochemical decomposition. In: Draper CW, Mazzoldi P (eds) Proceedings of the NATO Advanced Study Institute on laser surface treatment of metals, San Miniato, Italy, 2–13 September 1985. Nijhoff, Dordrecht, pp 555–566

[55] Roos JR, Celis JP, VanVooren W (1986) Combined use of laser radiation and electroplating. In: Draper CW, Mazzoldi P (eds) Proceedings of the NATO Advanced Study Institute on laser surface treatment of metals, San Miniato, Italy, 2–13 September 1985. Nijhoff, Dordrecht, pp 577–590

[56] Powell J, Henry PS, Steen WM (1988) Laser cladding with preplaced powder: analysis of thermal cycling and dilution effects. Surf Eng 4(2):141–149

[57] Lin J, Steen WM (1996) Design characteristics and development of a nozzle for coaxial laser cladding. In: ICALEO'96 proceedings, Detroit, October 1996. LIA, Orlando, pp 27–36

[58] Azer MA (1995) Laser powder welding: key to component production, refurbishment and salvage. Photonics Spectra Oct 122–127

[59] Lin J (2000) Laser attenuation of the focused powder stream in coaxial laser cladding. J Laser Appl 12(1):28–33

[60] Hayhurst P, Tuominen J, Mantyla T, Vuoristo P (2002) Coaxial laser cladding nozzle for use with a high power diode laser. In: ICALEO 2002 proceedings, Phoenix, October 2002. LIA, Orlando, paper 1201

[61] Gilkes J (1999) Intermetallic free laser deposition of Al alloy onto mild steel for the bearing industry. PhD thesis, Liverpool University, 1999

[62] Gilkes J, Brown WP, Watkins KG, Shannon GJ, Steen WM (1997) Aluminium flame deposition. In: ICALEO'97 proceedings, San Diego, November 1997. LIA, Orlando, pp F27–36

[63] Gedda H, Powell J, Wahlstrom G, Li W-B, Engstrom H, Magnussen C (2002) Energy redistribution during CO_2 laser cladding. J Laser Appl 14(2):78–82

[64] Pinkerton A, Li L (2002) A comparative study of multi-layer laser deposition using water and gas atomised 316L stainless steel powder. In: ICALEO 2002 proceedings, Phoenix, October 2002. LIA, Orlando, paper 1208

[65] Kahlen F-J, Kar A (2001) Residual stress in laser-deposited metal parts. J Laser Appl 13(2):60–69

[66] Steen WM, Weerasinghe VM, Monson PJE (1986) Some aspects of the formation of laser clad tracks. Proc SPIE 650:226–234

[67] Steen WM (1988) Laser surface cladding. In: Proceedings of the Indo-US workshop on principles of solidification and materials processing, SOLPROS, Hyderabad, India, January 1986. Office of Naval Research, American Institute for Biological Sciences, Washington, pp 163–178

[68] Takeda T, Steen WM, West DRF (1985) In situ laser alloy formation by laser cladding. In: Kimmit MF (ed) Proceedings of LIM 2 conference, Birmingham, UK. IFS, Kempston, pp 85–96

[69] Nowotny S (2002) Laser based hybrid techniques for surface coating. In: ICALEO 2002 proceedings, Phoenix, October 2002. LIA, Orlando, paper 603

[70] Fellowes FCJ, Steen WM, Coley KS (1989) Ceramic coatings for high temperature corrosion resistance by laser processing. In: Proceedings of the 2nd international conference on surface engineering with high energy density beams, Lisbon, Portugal, September 1989. IFHT, Aachen, pp 435–445

[71] Liu Z, Watkins KG, Steen WM, Vilar R, Ferreira MG (1997) Dual wavelength laser beam alloying of aluminium alloy for enhanced corrosion resistance. J Laser Appl 9(4):197–204

[72] Sexton CL, Steen WM, Watkins KG (1993) Triple hopper powder feed system for variable composition laser cladding. In: ICALEO'93 proceedings, Orlando, October–November 1993. LIA, Orlando, pp 824–834

[73] Anderson T (2002) Practical aspects of laser cladding with high power lasers. In: ICALEO 2002 proceedings, October 2002, Phoenix. LIA, Orlando, paper 1505

[74] Katuria YP (2000) Some aspects of laser surface cladding in the turbine industry. Surf Coat Technol 132:262–269

[75] McIntyre RM (1983) Laser hard surfacing of turbine blade shroud interlocks. In: Metzbower EA (ed) Lasers material processing. ASM, Metals Park, pp 230–240

[76] Carroll JW, Liu Y, Mazumder J (2001) Laser surface alloying of aluminium with iron. In: ICALEO'01 proceedings, Jacksonville, October 2001. LIA, Orlando, paper P537

[77] Kwok CT, Cheng FT, Man HC (2001) Laser hardfacing of 1050 mild steel using NiCoCrB for enhancing cavitation erosion-corrosion resistance. In: ICALEO'01 proceedings, Jacksonville, October 2001. LIA, Orlando, paper P500

[78] Henry M, Fearon E, Watkins KG, Dearden G (2001) Laser cladding of Hastalloy to critical surfaces of stainless steel components. In: ICALEO'01 proceedings, Jacksonville, October 2001. LIA, Orlando, paper 1003

[79] Lusquiños F, Pou J, Arias JL, Boutinguiza M, Leön B, Pérez-Amor M (2001) Calcium phosphate coatings obtained by laser cladding. In: ICALEO'01 proceedings, Jacksonville, October 2001. LIA, Orlando, paper P523

[80] Kim J-D, Liu J (2001) Effect of zirconia on laser cladding of hydroxyapatite bioceramic on Ti-alloy. In: ICALEO'01 proceedings, Jacksonville, October 2001. LIA, Orlando, paper P558

[81] Ayers JD (1981) Particulate composite surfaces by laser processing. In: Mukherjee K, Mazumder J (eds) Lasers in metallurgy. Metallurgical Society of AIME, Warrendale, pp 115–126

[82] Cooper KP, Ayer JD (1988) Surface modification by laser melt/particle injection process. In: Proceedings of laser surface modification conference, New Orleans, LA, 1988, pp 15–131

[83] Gassmann RC (1995) Laser cladding of WC/WC2-CoCrC, WC/WC2-NiBSi composites for enhanced wear resistance. Mater Sci Technol 11(5):520–526

[84] Cockburn A, Bray M, O'Neill W (2008) The laser assisted cold spray process. Laser User (53):30–31

[85] Shibata K (1992) Recent trends in laser material processing in the Japanese automotive industry. In: Proceedings of the European conference on laser applications technology ECLAT'92, Oct 12–15 1992, Göttingen, Germany, pp 1–10

[86] Hector LG, Sheu S (1993) Focussed energy beam work roll surface texturing science and technology. J Mater Process Manuf Sci 2:63–117

[87] Von Gutfeld RJ, Tynan EE, Melcher RL, Blum SE (1979) Laser enhanced electroplating and maskless pattern generation. Appl Phys Lett 35:651–653

[88] Al-Sufi AK, Eichler HJ, Salk J (1983) Laser induced copper plating. J Appl Phys 54:3629–3631

[89] Steen WM (1976) Discrete deposition of cobalt oxide on glass using a laser. PhD thesis, University of London

[90] Leön B, Klumpp A, Pérez-Amor M, Sigmund H (1991) Excimer laser deposition of silica films – a comparison between two methods. J Appl Surf Sci46:210

[91] Szorenyi T, Gonzales P, Fernandez D, Pou J, Leön B, Péres-Amor M (1991) Gas mixture dependency of the LCVD of silica films using an ArF laser. J Appl Surf Sci46:206–209

[92] Ding MQ, Rees JA, Steen WM (1989) Plasma assisted laser evaporation of superconducting YBaCuO thin films. In: IPAT 89–ion and plasma assisted techniques–7th international conference; Geneva, Switzerland, May 1989, pp 421–425

[93] Carts YA (1994) Laser based coatings fight catheter infections. Laser Focus World Dec 26–27

[94] Lu YF, Wang HD, Ren ZM, Chong TC, Low TS, Wu XW, Cheng BA, Zhou WZ (1999) Pulsed laser deposition of TiN thin film on silicon (100) at different temperatures. J Laser Appl 11(4):169–173

[95] Namba Y (1987) Laser forming of metals and alloys. In: Proceedings of conference LAMP'87, Osaka, May 1987, pp 601–606

[96] Vollertsen F (1994) Mechanisms and models for laser forming. In: Geiger M, Vollertsen P (eds) Laser assisted net shape engineering (LANE 1994). Meisenbach, Bamberg, pp 345–360

[97] Gillner A, Wissenbach K, Beyer E, Vitr G (1998) Reducing core loss of high grain oriented electrical steel by laser scribing. In: Hugel H (ed) Proceedings of the 5th international conference on lasers in manufacturing (LIM5), September 1988, Stuttgart. IFS, Kempston, pp 137–144

[98] Neiheisel GL (1984) Laser magnetic domain refinement. In: ICALEO'84 proceedings, Boston, November 198. LIA, Orlando, pp 102–111

[99] Lee JM (2002) Lasers and cleaning process. Hanrimwon, Seoul

[100] Asmus JF (1978) Light cleaning: laser technology for surface preparation in the arts. Technol Conserv 3(3):14–189

[101] Watkins KG (2000) Mechanisms of laser cleaning. Proc SPIE 3888:165–174

[102] Olfert M, Mueller RK, Duley WW, North T, Hood J, Sakai D (1996) Enhancement of adhesion in coated steels through excimer laser surfacing. J Laser Appl 8(2):79–87

[103] Johnston E, Shannon GJ, Spencer J, Steen WM (1997) Laser surface treatment of concrete. In: ICALEO'97 proceedings, Orlando, November 1997. LIA, Orlando, pp A210–218

[104] Dowden JM (2001) The mathematics of thermal modelling – an introduction to the theory of laser material processing. Chapman and Hall/CRC, London

[105] Hess P (1993) New insights into laser induced polymer ablation. Lambda Highlights 42:1–3

[106] Gitin M (1996) Diode pumped lasers in nick of time. Photonics Spectra May 122

[107] Kaplan AFH (1998) Laser marking. In: Schuöcker D (ed) Handbook of Eurolaser Academy, vol 2. Chapman and Hall, London, chap 7

[108] Mottay E (2008) Femtosecond pulses combat counterfeiting. OLE Apr 17–19

[109] Hammersley G, Hackel LA, Harris F (2000) Surface pre-stressing to improve fatigue strength of components by laser shot peening. Opt Lasers Eng 34:327–337

[110] Clauer H, Holbrook JH, Fairand BP (1981) Effects of Laser induced shock waves on metals. In: Meyers MA, Murr LE (eds) Shock waves and high strain rate phenomena in metals. Plenum, New York, pp 675–702

[111] Sarady I (1991) Application of pulsed laser for phase transformation and welding. PhD thesis, Lulea University

[112] Kubaschewski O (1982) Iron-binary phase diagrams. Springer, Berlin

[113] Atkins M (1980) Atlas of continuous cooling transformation diagrams for engineering steels. ASM, Metals Park

[114] Boyer HE, Gall TL (eds) (1978) Properties and selection: irons and steels. In: Metals handbook, 9th edn, vol 1. ASM, Metals Park

[115] Zhu L (1991) Surface modification of materials using high powered lasers and arc image intensifier. PhD thesis, Liverpool University

[116] Chande T, Mazumder J (1985) Two dimensional, transient model for mass transport in laser surface alloying. J Appl Phys 57(6):2226–2232

"You said it was a fast process so do this lot by 12 o'clock"

7 Rapid Prototyping and Low-volume Manufacture

If a picture is worth a thousand words, what's a model worth?

7.1 Introduction

In recent years two things have happened. The first is the astounding growth of computer memories, which now allow the design of "three-dimensional" models which can be rotated and visualised, surfaces to be accurately placed in a three-dimensional coordinate system, surface tangents and normals to be calculated and the model to be sliced. The second is the invention of the laser and methods for accurately guiding the finely focused beam. The result of combining the two is the growth of a new industrial sector in rapid construction of models, prototypes or manufactured parts for small batch runs of up to a few hundred pieces. The construction can be done within a small number of working hours, fully automatically using a powerful computer, a very flexible fabrication system and some after-treatment techniques.

Today it is just possible to start the day with a sketch of the new design of, for example, a sauce bottle, pass the sketch to a computer programmer who "draws" it into a CAD software package, by lunch. In the afternoon the CAD design is automatically sliced and the sliced data are fed to a fabricator, who builds the first model by the evening. As part of the design development cycle, it is far easier with a model, rather than with a drawing, to assess whether the design is attractive, easy to handle, stands firmly on a table, holds the correct volume and looks good alongside other objects with which it might normally be expected to be seen. If there is some fault in the design, it is usually fairly straightforward to adjust the design.

This design style saves months in the design process. Some have argued that there is an 80% saving in time and cost with these rapid prototyping techniques [1]. This saving in time is critical in the fashion-sensitive markets of today. There are some 25 different rapid prototyping methods available today, of which 70% are based on the application of a laser [2]. This is a major breakthrough in design technology. It has all happened since 1984. The chronology of four of the main processes is shown in Figure 7.1 [3]. Naturally the sales of equipment have been growing exponentially since 1990, even though current prices are very high (around £140,000 to 500,000 for a stereolithography machine).

This chapter aims to describe the range of laser-based processes which have been developed with some comment on their build rate, accuracy and methods for quantity manufacture.

Date	SLA	LOM	SLS	FDM	LDC
1993			Sinter station 200	FDM 1000 Patent granted	Invented
1992			SLS 125 β test	3D modeller β test	
1991		LOM-1015	SLS 125	3D modeller	
1990	SLA-500				
1989	SLA-250			Invented	
1988	SLA-1 β testing				
1987	SLA-1 Patent granted	Patent granted	Invented		
1986	Patent granted	Invented			
1985					
1984	Patent application				

Figure 7.1 History of rapid prototyping. *SLA* stereolithography, *LOM* laminated-object manufacture, *SLS* selective laser sintering, *FDM* fused-deposition modelling, *LDC* laser direct casting. (After Murphy [3])

7.2 Range of Processes

7.2.1 Styles of Manufacture

Manufacturing processes [3, 4] can be divided into those which are formative, subtractive or additive. Formative processes include stamping, forging, drawing, rolling or extruding. Subtractive processes include turning, milling, grinding, EDM, ECM, water-jet erosion and laser machining. Additive processes are a recent addition to the manufacturing scene and include the processes described in this chapter. A Venn diagram showing the relationship between these processes is given in Figure 7.2 [3].

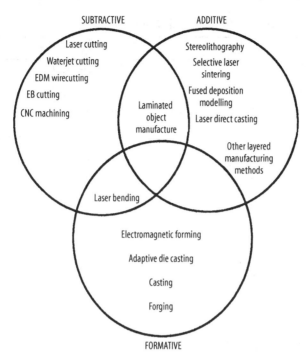

Figure 7.2 Automated fabrication techniques. *EDM* electric discharge machining, *EB* electron beam

7.2.2 Classification of Rapid Prototyping Techniques by Material

The starting material can be solid or liquid. The liquid can be made *in situ* by melting or can be formed into a solid *in situ* by polymerisation as shown in Table 7.1 [2].

The main rapid prototyping techniques are based on stereolithography, selective laser sintering, laminated-object manufacture and fused-deposition techniques, with laser direct casting or Direct metal Deposition (DMD) poised ready to enter the fray.

7.3 Computer Aided Design File Manipulation

The build process starts by designing the part in a computer. The basic design stages within the computer are illustrated in Figure 7.3. The objective is to deliver to the fabricating machine readable data on each slice of the part to be built. To build three-dimensional objects from the data held in a computer design, the internal and external surfaces of the object must be precisely defined. Since most model-building strategies available today are based on layer overlays, the computer model must be capable of being sliced within the computer program to yield data for the beam-steering mirrors or x/y stages. Thus, CAD systems which only reproduce blueprint drawings are inadequate, whereas those with wireframe, surface or solid modelling capabilities are

Table 7.1 Classification of rapid prototyping processes

Feed	Basic process	Process name
Solid	Gluing sheets	Laminated-object manufacture (Helisys, Hydronetics)
	Melting and solidification	Fused-deposition manufacturing (Stratasys)
Powder	1 component	Selective laser sintering (DTM, Hydronetics, Westinghouse), laser direct casting, DMD
	1 component and binder	3D printing, ballistic particle manufacture
Liquid	Liquid polymerisation	Single-wavelength lamps, stereolithography (Cubital, solid ground curing) Laser beams, stereolithography (3D Systems™, Quadrax, Grapp, DuPont™, Laser Fare, Sony/D-Mec, Mitsui E&S, Mitsubishi/CMET, EOS), holography (Quadtec)

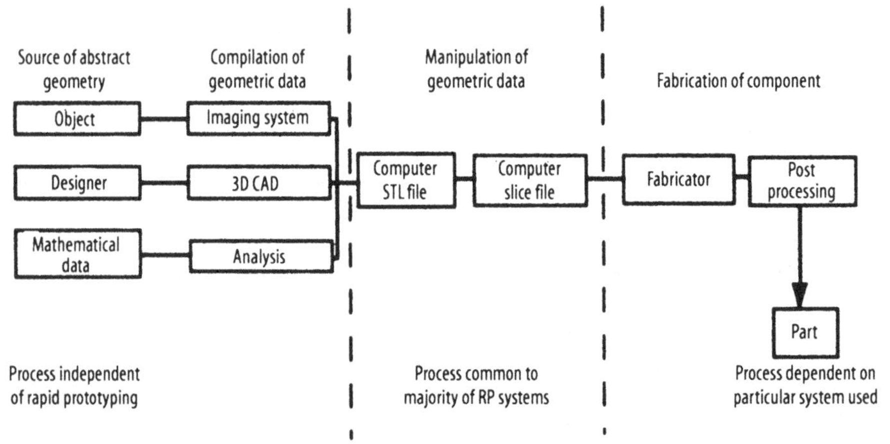

Figure 7.3 The basic stages in rapid prototyping processes. (*CAD* – Computer Aided Design)

adequate. The surface must be defined by a set of points with three-dimensional coordinates. This can be done by scanning an actual object and feeding the scanned data into the computer as a "cloud" of points over which another program will draw connecting tangents and so form a surface; constructing the object within the computer usually from sets of basic objects available in the CAD package; or through a mathematical formula of the surface. However it is done, a CAD is created which has a defined surface. This surface is sometimes defined by drawing over it a set of triangles with flat faces. The coordinates of the vertices of these triangles are stored in a stereolithographic (STL) file (STL stands for Standard Template Library) as a set of numbers. These triangles must mate up against each other to avoid holes. They must also have all vertices mating with other vertices – the vertex to vertex rule illustrated in Figure 7.4. Very long thin triangles should be avoided. There are programs available which will automatically

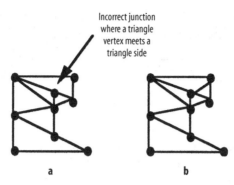

Incorrect junction
where a triangle
vertex meets a
triangle side

a   b

Figure 7.4 The vertex to vertex rule: **a** incorrect, **b** correct [4]

correct for these errors. The STL file is sliced to give the shape of the object at different levels parallel to a base plane which has to be defined.

Some CAD packages, notably including the tomographic scan systems (CAT) in medicine, will allow direct slicing without using the STL file. Although the STL file is currently accepted as an industry standard, it has problems with large and complex models. Recently new six axis software, DMDCAD, has been developed by the POM group working with Tata Consulting Services (TCS) to read directly from the solid CAD model to create tool paths for the deposition process, avoiding the need to make an STL file (www.pomgroup.com). The new software saves considerable design time.

The STL file or directly sliced data of a section of the model are then translated by a further software package into a set of instructions to drive a CNC machine (G-Code) or galvanometer mirrors or some such guidance system on the fabricator.

7.4 Layered Manufacturing Issues

7.4.1 General

Currently the only commercial rapid-build technique is to construct a three-dimensional model layer by layer, although concepts of directly printing real holographic images are being considered. For example, two-photon polymerisation using femtosecond pulses that are not absorbed by the monomer but are by two photons in a nonlinear reaction allows resolutions finer than the diffraction limit and the casting of holographic images as objects in three dimensions instead of building by layer. In building an object by layers, there are a number of problems: the steps at the edge of each layer; the layer thickness to be chosen; the accuracy of each layer as a function of shrinkage and process variables; and the orientation of the layering. These issues are discussed in the next sections.

7.4.2 Stair Stepping

Figure 7.5 illustrates the problem. Stair stepping is unavoidable with most techniques since each deposited layer is likely to have an edge of a different slope from the intended

Figure 7.5 Stair stepping can be arranged to favour the most important surfaces

surface shape. Some attempts at angled cutting by laser for laminates in laminated-object manufacture are being tested but angles are limited to 45° [5]. Infilling the steps with filler or smoothing the steps by sanding is usually done if it is necessary. Smoothing the steps by laser has been tried but the etch rate is uniform all over and hence the step is not removed but is merely machined further into the object!

7.4.3 Layer Thickness Selection

The stair stepping is reduced with finer steps but the computing penalty is large. The best strategy is to have a variable step height depending on the slope of the object and the capability of the fabrication process.

7.4.4 Accuracy

Different processes have different problems but shrinkage, distortion and curling are common. These problems are usually associated with cooling or phase changes in the material used. A further source of inaccuracy is the stereolithography program, which puts facets onto curves.

7.4.5 Part Orientation

The build time is a function of the number of layers built; hence, the larger the steps the better. The orientation can be arranged so that larger steps are not particularly detrimental.

By varying the orientation, one can put the "steps" on any face. Hence, the important faces should be built in the slicing plane. The principle is illustrated in Figure 7.5.

7.4.6 Support Structures

Overhanging parts need to be supported in many rapid prototyping processes, particularly if one is casting in a vat of liquid (stereolithography process). Some processes are not dependent on support structures; selective laser sintering, for example, has the support necessary from the powder bed.

Another use for support structures is as a constraint to part distortion, particularly for large thin fins.

7.5 Individual Processes

7.5.1 Stereolithography

Stereolithography [6] or three-dimensional printing is a process by which intricate models are directly constructed from a CAD package by polymerising a plastic monomer. Ultraviolet light of wavelength 325 nm from a low-power He–Cd laser shines onto a pool of liquid monomer, causing it to selectively polymerise and set. The process relies on a scanned laser beam to selectively harden successive thin layers of photopolymer, building each layer on top of the previous layer until a three-dimensional part has been constructed. The photopolymer used for stereolithography has a very thick consistency and quickly hardens upon exposure to specific wavelengths of light.

7.5.1.1 Model Design

The first step in creating a model by stereolithography is to perform the design function on a three-dimensional CAD system. The CAD image is constructed as either a solid or a surface model complete with final wall thicknesses and interior details. Next, the solid model image is orientated on the screen into the position which best facilitates construction by stereolithography. A supporting structure made of thin cross-webs may be required for elements of the model which overhang, because the stereolithography process constructs the model layer by layer beginning from the bottom surface. The use of support structures may be minimised by optimising the orientation of the part. If the model is to be used for investment casting later, then the solid parts may be designed with a honeycomb structure (QuickCast™[1] [7], see Section 7.6.2). After the model has been made, the support structure will be broken away or cut off.

7.5.1.2 Model Slicing and Data Preparation

On completion of the design, the electronic model, in the form of an STL file, is transferred to the stereolithography computer, where it is electronically sliced into thin layers. The thickness of the individual layers is selected by the operator. The layers are generally thinner in areas requiring the greatest accuracy. The computer then breaks each layer into a gridwork of vector data that will later be used to guide the direction and velocity of the laser beam.

7.5.1.3 Model Creation by Stereolithography

The stereolithography process is illustrated in Figure 7.6. It begins by lowering a platen to a precise depth into a vat of photopolymer. The distance the platen is lowered corresponds to the thickness of the model layer to be hardened. A wiper ensures the plate is uniformly covered and that no waves or bubbles occur on the monomer surface. Next, the laser beam is activated and guided in the x- and y-axes by two computer-controlled,

---

[1] QuickCast™ is a trademark of 3D Systems Inc. http://www.3dsystems.com

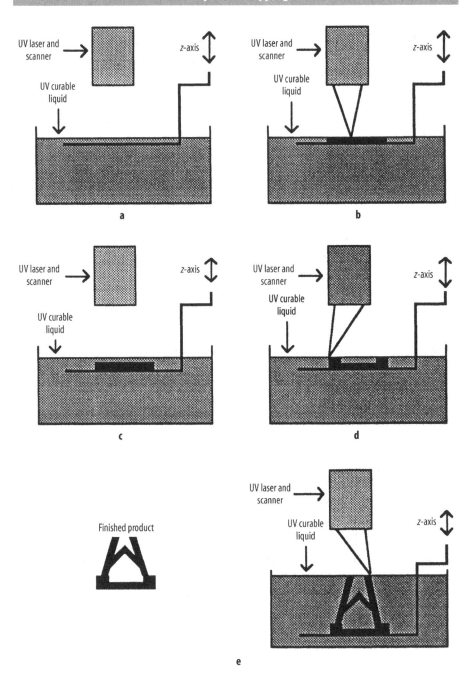

Figure 7.6 The sequence for making a part by stereolithography [6]: **a** the platen is lowered to a preset depth for the slice being cast, **b** the initial layer is polymerised by ultraviolet (*UV*) light to form the bottom layer from the CAD image to be cast, **c** and **d** the cycle is repeated for each layer, until **e** the finished part is produced

galvanometer-driven scanning mirrors. The motion of the laser beam selectively hardens the photopolymer in the areas corresponding to the first slice of the model. The laser is pulsed and hence the track consists of a series of overlapping cones of hardness. The spot size is of the order of 0.2 mm, with a depth of typically 0.5 mm, depending on the laser fluence. The first layer solidified becomes the bottom layer of the model. The platen is again lowered a precise distance into the vat and the laser draws the second layer on top of the first one. This layer adheres to the first because the depth of penetration of each pulse is greater than the thickness of the layer and hence overcures the prior layer. This process is repeated as many times as is necessary to recreate the entire object layer by layer. When completed, the platen is raised from the vat and the model is ready for removal of the support structure and postcuring. A typical processing time would be around 1 h, with a further 2 h for postcuring, to make a three-dimensional model in around 3 h with variations for complexity.

The photopolymers used are composed of a photoinitiator and liquid monomers – usually with epoxy, vinyl ether or acrylate functional groups. On exposure to ultraviolet radiation, the photoinitiators are excited and a small percentage of these molecules transform to a reactive species which can stimulate photopolymerisation through the formation of free radicals. The resins differ in their curing time and final physical properties. This means some selection is required prior to casting.

7.5.1.4 Postcuring

Maximum polymer hardness is not attained directly from the stereolithography process and therefore some postcuring is usually required. This is accomplished by exposing the model to high-intensity ultraviolet light in a special curing unit. Exposure times vary from 30 min to 2 h depending on part geometry.

Postcuring is also used as a method of quickly hardening thick walls and large volumes in models. Large volumes are more quickly created by entrapping liquid polymer within the model structure to be hardened later during the postcuring cycle.

7.5.1.5 Process Variations

3D Systems™[2] owns the original patent for stereolithography, which is very well covered. However, Quadrax, Grapp, DuPont™[3], Laser Fare, Sony/D-Mec, Mitsui E & S, Mitsubishi/CMET and EOS all have variations on the central theme:

- EOS uses a flat-field lens system to ensure the beam is always normal to the vat.
- Mitsubishi moves the beam on a flying optics x/y stage instead of using galvanometer mirrors.
- Mitsui E & S in its Colamm-300 system casts beneath the platen to avoid surface levelling problems and hence its process is a form of inverted stereolithography. This also removes the need for large vats of polymer and adds the possibility of changing polymers during build. Mitsui claims improved accuracy of ±0.1 mm with layer thicknesses down to 10 microns.

---

[2] 3D Systems™ is a trademark of 3D Systems Inc. http://www.3dsystems.com

[3] DuPont™ is a trademark of E.I. du Pont de Nemours and Company. http://www2.dupont.com

- DuPont™ has upgraded the software and control aspects.
- Sony uses ultraviolet-curable resins from Japan Synthetic Rubber based on acrylate urethane resins, which, owing to them being low viscosity materials, simplifies the levelling problems. Sony also has a more complex scanning system.

More strategic variations are being considered, such as printing whole layers in one shot using more powerful lasers or ultraviolet lamps and masks. The mask could be an liquid crystal display LCD and hence give a highly flexible image. Another variation is to print the real image from a hologram – a complete three-dimensional object in a single shot [8] – but there is still much to do to perfect that route.

7.5.1.6 Advantages and Disadvantages of Stereolithography

The advantages of stereolithography are:

- fairly good surface finish and accuracy;
- fairly good speed of casting;
- fully automatic;
- both internal and external surfaces can be modelled; and
- multiple parts can be cast at the same time depending on the size of the platen.

The disadvantages are:

- shrinkage and distortion may lead to separation and splitting;
- the surface finish depends on the slice thickness and the material;
- the finishing operations can be tedious and messy;
- large thin surfaces are difficult;
- the surface finish and the tolerance are not complementary and sometimes a choice has to be made;
- the polymers used have a high thermal coefficient of expansion and hence do not make good moulds;
- support structures are needed for overhanging parts, and this has to be built into the CAD design;
- the optimum tolerance at present is only ±0.2 mm; and
- the equipment is expensive.

7.5.2 Selective Laser Sintering

7.5.2.1 Process Description

The system [9] generates three-dimensional parts by selectively fusing thermoplastic, ceramic or metallic powders with the localised heat from an infrared laser, usually a CO_2 laser. The process, illustrated in Figure 7.7, is similar to stereolithography in that a thin layer is placed on a platen and a pattern is solidified on it; after which another layer is laid on top and the process is repeated until the object is built slice by slice. In this case the thin layer is a layer of fine powder which is spread over the surface

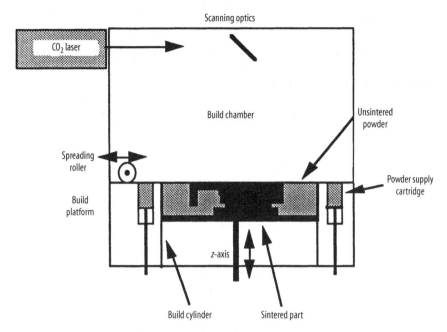

Figure 7.7 Selective laser sintering

to a precise depth by a plough. The laser beam from a 50 W CO_2 laser is then scanned or rastered by galvanometer-driven mirrors in the required x/y pattern taken from the STL file. To minimise the required laser energy and to speed up the process, the chamber is heated and may be filled with an inert gas, such as nitrogen, to avoid oxidation or burning of the powders. The completed article is withdrawn from the unsintered powder bed and bead-blasted to remove unsintered adhering particles. The unsintered particles can be recycled after sieving.

The raw material can be anything which can be sintered or will bond without melting. This includes polycarbonate, investment casting wax, nylon, glass-reinforced nylon (tiny nylon-coated glass spheres) and plastic-coated metallic powders (*e.g.*, carbon steel coated with a polymer).

7.5.2.2 Accuracy

The process is surprisingly accurate, ±0.05–0.25 mm. The build strategy does not normally require support structures for overhanging parts, except when using certain wax powders [10], but the article will be porous. For monomode powders (*i.e.*, powders of one size) the face centred cubic structure would suggest 26% porosity, but by mixing powder sizes, one can can reduce this. Porosity means the article may be friable. With some metallic bronze articles the porosity can be partially filled by heating in a vacuum and allowing solder to run into the article.

7.5.2.3 Variations

The original process was invented in 1986 at the University of Texas. Since then DTM has started marketing the process. Hydronetics and Westinghouse have two further processes which are variations of the central theme.

Three-dimensional printing is not a laser process, but uses the same equipment to lay a uniform layer of powder which is then scanned with a nozzle which lays down tracks of binder to form the layer of the object.

7.5.2.4 Advantages and Disadvantages of Selective Laser Sintering

The advantages of selective laser sintering are:

- any powder which can be sintered or will bond without melting can be used;
- the materials are nontoxic;
- it does not usually require support structures;
- the accuracy is in the range ±0.05–0.25 mm;
- it has fairly good speed, for example, a piece 305 mm × 380 mm can be cast at 12–25 mm h^{-1};
- the finished parts are in some cases functional and testable prototypes as well as conceptual models;
- wax models can be built in a few hours and used to make functional prototypes by investment casting; and
- it is a fully automatic process.

The disadvantages are:

- large flat sheets need tie-down supports to stop them curling;
- part accuracy depends on size and complexity;
- the surface finish is rough compared with that produced by stereolithography, the article is porous and hence of poor mechanical strength;
- the equipment is expensive;
- cavities are difficult to clean since the powder tends to cling and bead blasting is difficult in confined regions;
- shrinkage can create serious residual stresses which can cause distortion or cracking;
- temperature control is critical; and
- processing is best done in an inert atmosphere to avoid oxidation, fire or even explosions.

7.5.3 Laminated-object Manufacture

7.5.3.1 Process Description

This process [11] is marketed by Helisys. The building process is again layer by layer; however, in this case the layer is a sheet of paper, polyester film or other laminate, which

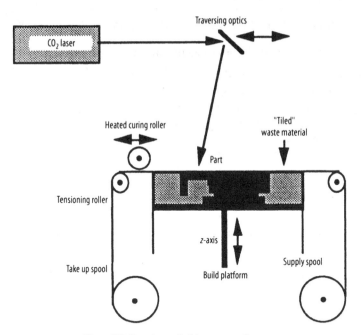

Figure 7.8 Laminated object manufacture

is spread over the previous layer and stuck with adhesives. A laser then cuts the required shape in the layer and also cuts the unwanted material into small squares, called "tiles", for later removal. The unwanted material acts as support for the article during the building process. Enclosed cavities have to have the tiles removed during building. The process is illustrated in Figure 7.8.

7.5.3.2 Accuracy

The xy accuracy is very high at ± 0.1 mm per 100-mm since it depends solely on the accuracy of the laser cutting system. There is no shrinkage. However, the z accuracy depends on the adhesive thickness and hence for a deep article the accuracy can be poor. There is shrinkage in that direction as the adhesive sets. The adhesive needs to be uniformly spread.

7.5.3.3 Variations

The laminates may be metal sheets or layers of sintered metal powders. The thickness steps may be large if the process also includes some edge treatment. In one process the sheets are metal and are laser-cut at an angle [5] or are welded [12] or clad.

Another process is "solid ground curing" (Cubital), where a layer of photocuring resin is exposed via a mask into the required shape. The untreated resin is then removed and replaced with wax. The set resin and wax sheet is then machined flat and the next layer is added on top, and so on. The object of the wax is to act as a support structure.

7.5.3.4 Advantages and Disadvantages of Laminated-object Manufacture

The advantages of laminated-object manufacture are:

- cheaper materials can be used than for the other processes;
- no support structure is required;
- composite structures of different materials can be made;
- tooling can be made this way (*e.g.*, punches, dies, *etc.*);
- low equipment cost;
- fully automatic process; and
- high accuracy ±0.1 mm.

The disadvantages are:

- part removal can be a delicate matter;
- closed surfaces and re-entrant faces cannot be made as a single piece;
- part strength is limited;
- paper can absorb moisture and distort or delaminate;
- the build rate is slow; and
- the structural strength of finlike parts is poor.

7.5.4 Laser Direct Casting or Direct Metal Deposition (DMD)

7.5.4.1 Process Description

This process [13–15] is based on laser cladding using a coaxial powder feed system, as shown in Figure 7.9. The part is built by layers as before; however, the build direction can be altered by tilting the build table. By this means there is no need for support structures. Any metal powder that is weldable can be used. The finished article can be made in the required material and hence this process is more a form of low-volume manufacture rather than one primarily for prototyping. The metallurgical structure has epitaxial growth from layer to layer – a form of directional solidification. The grain size is smaller than might be achieved by conventional casting, being more of a chill structure. The material can be fed as a powder or as a wire. The coaxial powder feed system [16] makes the powder process straightforward. The system behaves like a "metal pencil" laying down a track of approximately 1 mm depth and width determined by the spot size. The powder catchment efficiency is one of the design aspects of the process and may be as high as 89%, but is more usually 30–40%. It is unwise to recycle the unused powder because it contains oxides and irregular particles. The process needs to be shrouded for most metals. This can be done by gas jets in the powder supply system for less reactive metals. However, since the object is hot during the casting process, reactive metals, such as titanium, must be cast in an inert atmosphere. The work at Applied Research Laboratory, Pennsylvania State University [17], on the casting of titanium, uses a chamber as shown in Figure 7.10. The chamber is bottom-filled with argon, which is a dense gas.

The uniformity of the build height and width when constructing a wall is determined by traverse speed, power and powder feed rate. Thus, at the turn points where

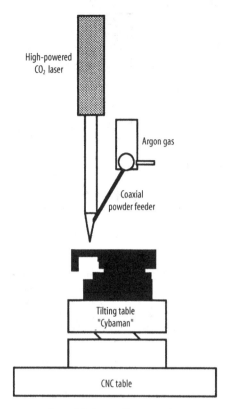

High-powered
CO_2 laser

Argon gas

Coaxial
powder feeder

Tilting table
"Cybaman"

CNC table

Figure 7.9 Laser direct casting

the CNC table may slow, or at overlapping points, there will be a rise in wall height or melt back unless precautions are taken. Mazumder *et al.* [14,15] used a system of height sensors which gave a feedback signal to turn the laser off at the peak points. This system achieved very good build accuracy of approximately ±0.04% as well as a smoother surface. Hands *et al.* [18] used a two-wavelength pyrometer to measure the temperature of the clad melt pool in a system similar to their focal control equipment (see Chapter 12). The argument for measuring the temperature to control the height is that the temperature of the pool varies as it rises up or falls down in the expanding laser beam. If the focus is set above the clad pool, then a fall in height would cause the power density to be lower and the pool temperature to fall. Thus, the pool would shrink and catch less powder and hence the level would fall further. This represents process instability. If, on the other hand, the focus is below the clad surface, the height control is not unstable and may be self-correcting. There is a further advantage to this arrangement in that any powder that rises up the beam by convection would not enter the focal region and would thus not form plasma that might interrupt the process. However, control of the temperature is another way in which the height during build can be automatically maintained. A smarter and simpler method is that of Fearon [19], who controlled the height by the flow of powder. This was achieved by arranging that if the height rose too

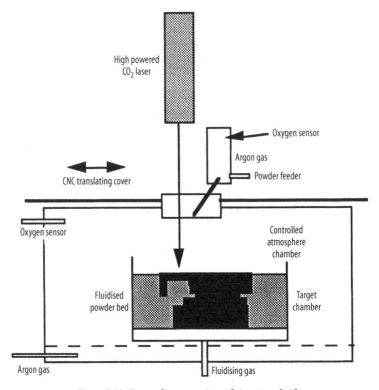

Figure 7.10 Laser direct casting of titanium [17]

far the powder flow would be absent in that region. With this level of control available, the fall-back position, of machining between layers, should be unnecessary.

The wall roughness is approximately 13–50 μm, whereas the valley-to-peak distances range between 75 and 275 μm [14]. This is similar to sand casting. It is affected by the powder purity; recycled powder will give a rougher wall.

A variation in the process has been proposed by Gedda *et al.* [20]. They blew the powder at rates such that it did not weld to the substrate, but instead used the substrate as a mould to form a casting – the "laser casting" process. Alternatively, copper chills can be put along the side of a wall that needs rebuilding and the new material is blown into the resulting mould to form a casting that is welded to the wall but not the copper chills – the "laser clad-cast" process. This may be useful for the repair of worn edges.

7.5.4.2 Accuracy

The accuracy is poor at present since there is considerable shrinkage on solidifying and cooling. The reproducibility is higher and hence the inaccuracies can, in part, be removed by altering the build path.

7.5.4.3 Variations

There is a family of fused-deposition methods, most of which do not use a laser. The Stratasys fused-deposition method passes a thermoplastic string into a heated extrusion head, where it is melted and applied as a track to build, as for laser direct casting. The plastic resolidifies reasonably quickly, in 0.1 s. The extrusion head is moved on an x/y stage. The extrusion rate is controlled by a precision pump. The process is quick, laying tracks at 23 m min^{-1}. The materials can be thermoplastic (nylon and acrylonitrile–butadiene–styrene-like plastics) or wax (machinable or investment casting waxes). It is a desktop process which can be safely operated in an office environment.

The ModelMaker™[4] building process is similar but uses two extrusion heads, one for thermoplastic and the other for wax for the support structures. The materials are ejected from inkjet nozzles as hot liquids which solidify on impact with the cooler build surface. Tracks are laid at approximately 3.5 m s^{-1}. Each plane is back-machined between layers.

7.6 Rapid Manufacturing Technologies

The build time for even these fast processes [4] is still several hours per part and in most of the processes described the article has been built in the wrong material for it to be functional. Thus, some route to convert the plastic model into the required functional material, which is usually metal, either as a one-off or as short batch runs is required. This can be achieved either by silicone rubber moulding for low-temperature materials or by investment or sand casting for high-temperature materials.

7.6.1 Silicone Rubber Moulding

The part is immersed in silicone rubber, which sets around it. The object is carefully cut free and the result is a cavity in silicone rubber which can be filled with a variety of plastics and coloured plastics to make functional or multiple copies of the original model. The original model may need a barrier coating to avoid contaminating the silicone and inhibiting the curing process. The making of a silicone mould usually takes several days.

7.6.2 Investment Casting

In investment casting the article is coated in ceramic by repeated dipping and drying until a ceramic shell develops. The shell can be mounted in a sand bed for support. The shell is then autoclaved to burn out or melt the pattern. Metal can then be poured into the cavity as in normal casting and the shell is broken away to reveal the cast object.

---

[4] ModelMaker™ is a trademark of 2Bot Corporation. http://www.2bot.com

Both the model and the mould are lost for each part cast. Thus, it is handy if the models are produced from a silicone mould from the original prototype.

Stereolithographic models are sometimes made with an internal structure like a honeycomb to cause them to soften and burn away in the investment process. If they were solid, they would expand and crack the casting. The form of the honeycomb pattern is critical because the structure must be able to drain successfully after the stereolithography process. The technique is known as QuickCast™ [7].

7.6.3 Sand Casting

In this process the model is buried in sand or sand and binder. The sand is set either cold (by blowing CO_2 into the casting) or hot (by sintering the resin-coated sand). The casting is broken open along joint lines and the model is removed. By joining the casting parts together again, one creates a casting cavity into which metal can be poured. For example, the casting of aluminium has been done this way [21]. This process is not easy if the internal structure is important.

7.6.4 Laser Direct Casting

As already noted, laser direct casting, direct metal deposition (DMD) or weld build techniques of whatever name create the functional part directly. They can also do it using a variety of materials or with internal structures that are unattainable by all previous methods. For example, the University of Michigan has a start-up company Precision Optical Manufacturing (POM) which uses this process, in which, among other things, it manufactures an H13 tool steel die with copper heat sinks and conformal cooling channels, which reduces the cooling time for an injection moulding by 10–40% [14,22]. It has also been estimated by the National Center for Manufacturing Studies, USA that the die can be produced in only 60% of the traditional time for making dies. There are also proposals for the manufacture of lightweight beams by this technique, the lightness being achieved both by the choice of material and by the internal structure of the beam; a comparison with bone structures comes to mind.

Mazumder's group at Michigan has also pioneered the possibility of designing the material itself using this technique. For example, by the choice of materials and by building complex wall structures – mixtures of voids and solids – one can construct materials with a negative Poisson ratio [23]. Design control over the Young's modulus, shear modulus, Poisson ratio and even the thermal expansion coefficient is possible this way by choice of suitable metal matrix composites. Such materials can be designed and built by laser direct casting processes – and the use of a huge computer databank. This is a heady vision of the large impact that the laser may have on the design of manufactured goods in the future. This process has removed a number of design constraints that we considered axiomatic in the past.

The manufacture of complex parts directly from a computer drawing should greatly reduce costs and time in getting a prototype or product ready for use. It is applicable to many industries. Mazumder *et al.* [14] suggested that it would be of use to the US

Navy as it would allow an aircraft carrier to proceed to sea with only barrels of powdered metal and CAD drawings of all spare parts together with the equipment to make the parts as required, without needing a large inventory of spare parts. Futuristic that may be, but stranger things have happened in the past. The automotive and electronics industries see the potential for the rapid fabrication of tooling and dies. These are potential applications in addition to those cited next for the other rapid prototyping processes.

7.6.5 Rapid Prototyping Tooling

Possibly, rapid prototyping is most useful in creating tools, as in the example just given. These may be die-casting and injection-moulding tools, press tools, EDM tools and "soft tools". Soft tools are tools which can make up to 20 or more parts to test the design of the tool itself. All these are achievable targets for this new manufacturing method.

7.7 Applications

Rapid prototyping is one of the most exciting developments in design techniques in the last 100 years. Models are far superior to drawings in clarifying conceptual problems. The ability to iterate a design within a given time is essential for optimising art or engineering artefacts. Important applications which have been proposed or demonstrated include:

- three-dimensional topographic maps from satellite data;
- architectural models complete with complex internal structures;
- templates and cams for mechanical controls;
- three-dimensional plots of scientific and engineering data;
- manufacturing-related models;
- geometric fits of parts for engines or factory layouts;
- medical models drawn directly from the tomographic data, such as personalised prostheses and dentures, reconstruction of bones and skulls after traffic accidents and many other examples [24];
- a new art form; and
- conservation of historical artefacts [25].

7.8 Conclusions

Rapid prototyping or low-volume manufacture is a new technological tool for the manufacturing industries. It has brought with it a range of striking advantages such as:

- *Reduced lead time*: from months and years to days and weeks for a given design cycle; this saving in time has been reckoned to be 80% by Tromans and Winpenny [1].

- *Reduced costs*: both through direct manufacturing savings due to reduced times and reduced wastage due to design faults and mistakes.
- *Improved quality of the product*: the design stage can be so much more careful and may have several iterations before production; these iterations would have been too costly or time-consuming to contemplate previously.

Since this manufacturing style matches so well with the fashion-sensitive markets of today, and since rapid prototyping or low-volume manufacturing is dominated by laser techniques, this could well be one of the major application areas of lasers in the future.

Questions

1. Write brief notes on four different rapid prototyping techniques.
2. What is meant by an STL file?
3. How is a three-dimensional image stored in a computer?
4. In layer build models what is the importance of part orientation?
5. Compare the advantages of stereolithography with those of selective laser sintering and laser direct casting.
6. How could a rapid prototype model be used for mass production?

References

[1] Tromans G, Winpenny D (1995) Rapid manufacturing. In: Proceedings of the 4th international conference on rapid prototyping and manufacturing, Belgirate, Italy, June 1995, pp 27–41

[2] Pera L, Marinsek G (1992) The role of the laser in rapid prototyping. In: Proceedings of the NATO Advanced Study Institute (ASI) laser applications for mechanical industry, Erice, Sicily, April 1992, pp 293–303

[3] Murphy M (1995) Rapid prototyping by laser surface cladding. PhD thesis, Liverpool University

[4] O'Neill W (1997) Rapid Prototyping'97. Notes for the Eurolaser Academy, Liverpool University

[5] Erasenthiran P, O'Neill W, Steen WM (1997) An investigation of normal and slant cutting using CW and pulsed laser for laminated object manufacturing techniques. Proc SPIE 3097:48–57

[6] Jacobs PF (1992) Rapid prototyping and manufacturing: fundamentals of stereolithography. Society of Manufacturing Engineers, Dearborn

[7] Jacobs PF(1995) Stereolithography accuracy, Quickcast™ and rapid tooling. In: ICALEO'95 proceedings, San Diego, November 1995. LIA, Orlando, p 194

[8] Obata K, Passinger S, Ostendorf A, Chichkov B (2007) Multi-focus system for two photon polymerisation using phase modulated holographic technique. In: ICALEO 2007 proceedings. LIA, Orlando, paper N107

[9] Juster NP (1994) Rapid prototyping using selective sintering process. Assem Autom 14(2):14–17

[10] Juster NP, Childs THC (1994) A comparison of rapid prototyping processes. In: Proceedings of the 3rd Euro conference on rapid prototyping and manufacturing, Nottingham, July 1994, pp 35–52

[11] Feygin M (1990) Laminated object manufacture. In: Proceedings of the national conference on rapid prototyping, OH, USA, 1990, pp 63–65

[12] Erasenthiran P, Ball R, O'Neill W, Steen WM, Jungreuthmayer C (1997) Laser step shaping for laminated object manufacturing parts. In: Fabbro R et al (ed) ICALEO'97 proceedings, San Diego, November 1997. LIA, Orlando, pp 17–20

[13] McClean MA, Shannon GJ, Steen WM (1997) Shaping by laser cladding. In: Geiger M, Vollertsen F (eds) Proceedings of conference LANE'97, Erlangen, Germany, September 1997, pp 115–127

[14] Mazumder J, Dutta D, Kikuchi N, Ghosh A (2000) Closed loop direct metal deposition: art to part. J Opt Lasers Eng 34:397–414

[15] Koch J, Mazumder J (1993) Rapid prototyping by laser cladding. In: ICALEO'93 proceedings, Orlando, November 1993. LIA, Orlando, pp 556–565

[16] Lin J, Steen WM (1996) Design characteristics and development of a nozzle for coaxial laser cladding. In: ICALEO'96 proceedings, Detroit, October 1996. LIA, Orlando, pp 27–36

[17] Arcella FG, Laurel MD, Witney EJ, Krantz MTS (1995) Laser forming near shapes in titanium. In: ICALEO'95 proceedings, San Diego, November 1995. LIA, Orlando, pp 178–183

[18] Hand DP, Fox MDT, Haran FM, Peters C, Morgan SA, McClean MA, Steen WM, Jones JDC (2000) Optical focus control system for laser welding and direct casting. J Opt Lasers Eng 34:415–427

[19] Fearon E (2003) Laser free form fabrication applied to the manufacture of metallic components. PhD thesis, Liverpool University

[20] Gedda H, Kaplan A, Powell J (2002) Laser casting and laser clad-casting: new processes for rapid proto-typing and production. In: ICALEO '02 proceedings, Phoenix, October 2002. LIA, Orlando, paper 1209

[21] Fussel PS (1995) Rapid production of aluminium artifacts via casting. In: Proceedings of the 4th Euro conference on rapid prototyping and manufacturing, Belgirate, Italy, June 1995, pp 69–81

[22] Koch J, Heyzner D, Mazumder J (1996) A metallurgical analysis of laser clad H13. In: ICALEO'96 proceedings, Detroit, October 1996. LIA, Orlando, paper A143–150

[23] Lakes R (1987) Foam structures with negative Poisson's ratio. Science 23:1038–1040

[24] Industrial Laser Review (1993) Selective laser sintering improves prosthesis design. Industrial Laser Review Feb 5

[25] Industrial Laser Review (1994) Technology resolves artefact dispute. Industrial Laser Review Apr 5

"Posselthwaite, I think we should keep this to ourselves."

8 Laser Ablative Processes – Macro- and Micromachining

It must be done like lightning

Ben Jonson (1573?–1637), Every Man in His Humour (1598) IV v

8.1 Introduction

The interest in ablation as a process has recently become important with the introduction of very powerful, ultrashort laser pulses of immense power but little energy. Such pulses can deliver energy at a rate that the material can only absorb by evaporating or flying apart. Thus, material is removed leaving very little heat-affected material. These lasers have become a form of unbluntable machine tool, which can operate on very fine structures causing very little chemical or mechanical damage.

The applications for such a tool are slowly being recognised. It has never been possible before to work with such cleanliness and precision except via chemical etching, which can hardly be described as clean. Chemical etching has its own set of problems and is limited in the materials on which it can be used. Thus, laser ablation has a growing niche market in surgical applications, lithography, micro-optical and electronic device manufacture, marking, cutting, scribing and cleaning.

This chapter discusses the physical phenomena that take place as a laser ablates a surface and finishes with some examples of applications. Three regimes that differ in their rates of delivery of energy are identified: slow, medium and fast. These are summarised in Table 8.1.

Table 8.1 Main regimes for ablation

	Rates of energy delivery		
	Slow	Medium	Fast
Main removal mechanisms	Thermal heating Photo-thermal effects Thermal stress	Photochemical Photophysical	Coulomb explosion Mechanical shock
Main interaction	Evaporation Spallation	Evaporation	Evaporation mechanical/ thermal/Coulomb

8.2 Basic Mechanisms During Short Radiant Interactions

8.2.1 Thermal Models

When photons arrive at a surface, their energy is so small compared with the energies of the atomic nuclei that they are only able to have a significant impact on small charged particles such as electrons. There is an analogy with waves arriving on a sandy beach: the waves can move the individual sand particles but not the beach or the cliffs. If, however, the wave gets to tsunami proportions, different things will happen and the beach can be reshaped. If the electron in the irradiated structure is free, it will move in the incoming electric field (which is the photon). This results in a secondary wave due to the electron changing direction or for a brief moment a high-speed electron. The same result will occur if the electron is held in the phonon structure, except the electron will lose its energy not by radiating but by nudging the internuclei bonding. For copper it takes approximately 10^{-17} s for the electron to lose its energy to the lattice. This time is known as the "electron cooling time". The lattice takes a further short period to heat up. The meaning of "heat up" is for the lattice to start vibrating, which we detect as temperature. This time is of the order of picoseconds and the time is called "the lattice heating time". There is a third time period of importance and that is the duration of the laser pulse. The overall picture is of an electric wave arriving at a surface and rattling all the electrically charged parts, first electrons move and then the structure starts to wobble. Since all this happens very quickly for most laser processes, it is of no practical interest to know what the fine mechanics of light absorption is. However, as pulse lengths shorten and powers rise, the wave gets higher and individual parts are perceived to act at different rates and not virtually all together.

There are three thermal model regimes:

1. Case 1: Long pulses compared with the relaxation times (more than 1 ms).
2. Case 2: Shorter pulses with thermal equilibrium still maintained but heat loss is now principally by evaporation (less than 1 ns).
3. Case 3: Ultrashort pulses that are short compared with the relaxation times of the structure. This leads to nonthermal equilibrium, with electrons and the lattice having different temperatures, direct evaporation and Coulomb effects (less than 1 ps).

8.2.1.1 Photothermal Ablation

Case 1 is the processing regime that has been discussed in the rest of this book. Heat flows by thermal conduction and material evaporates by boiling after first melting. However, this thermal heating can cause material removal by routes other than straight boiling, one such route is photothermal ablation.

If the structure is heated sufficiently that the vibrations can themselves break the weaker bonds, then the boiling point of the broken structure may be lower than that of the original structure and evaporation occurs without the melting point being reached; this is of particular relevance to polymers. Polymers consist of long macromolecules joined along their length by strong covalent bonds. These chains are bonded to each other by weaker bonds. To remove a whole chain would require the simultaneous breaking of many weak bonds. As the material is abruptly heated, so these bonds, both strong and weak, may break, thus reducing the molecular size and the boiling point of the material.

8.2.2 Nonthermal Models

8.2.2.1 Photochemical Ablation

In some cases the energy of the photon itself can be sufficient to directly break the chemical bonds without bothering with heating.

The energy of a photon is readily calculated from its value given by $h\nu$ or hc/λ (h is Planck's constant, ν is the radiant frequency, λ is the wavelength and c is the velocity of light). The comparative energy of the photons from different lasers and the chemical bond energy for different linkages are listed in Table 8.2

Thus, the photon energy from a KrF excimer laser, at 4.9 eV, is sufficient to directly break a C–C bond, whose energy is only 3.62 eV. On its own one photon may break a bond but the bond would almost instantly reform after the photon has passed. If, however, a whole avalanche of photons were to strike at the same time, then many bonds would be broken simultaneously, possibly resulting in permanent chemical change (as in sunbathing).

The flux of photons from a 1 GW pulse from a KrF laser whose photon energy is 4.9 eV (*i.e.*, $4.9 \times 1.602 \times 10^{-19}$ J) would have a flux of $10^9/7.8 \times 10^{-19}$, *i.e.*, 1.28×10^{27} photons per second.

The atomic spacing in a polymer is approximately $1°Å$ (10^{-10} m). This gives a bond density of 10^{20} m^{-2}, or 10^{14} mm^{-2}. If the beam is focused to a 0.2 mm-diameter spot (0.03 mm^2), then there would be $10^{14} \times 0.03 = 3 \times 10^{12}$ bonds within the irradiated zone. This gives $1.28 \times 10^{27}/3 \times 10^{12} = 4.2 \times 10^{14}$ photons per bond per second. Almost certainly all the bonds would have been affected within the pulse length of less than nanoseconds. For femtosecond pulses and shorter wavelengths this photon saturation becomes even more extreme.

Table 8.2 Photon energies and chemical bond energies

Photon energies Source	Wavelength (μm)	Frequency (Hz)	Energy (eV)
Cyclotron	0.1 (X-ray)	2.9×10^{15}	12.3
Excimer	0.249 (UV)	1.2×10^{15}	4.9
Argon	0.488 (blue)	6.1×10^{14}	2.53
Nd:YAG	1.06 (IR)	2.8×10^{14}	1.16
CO_2	10.64 (IR)	2.8×10^{13}	0.12

Chemical bond energies					
Type of bond C–C	C=C	C≡C	C–H	C–N	C–S
Energy (eV) 3.62	6.40	8.44	4.30	3.04	4.96

1 eV = 1.602×10^{-19} J

8.2.2.2 Mechanical Stresses

Explosive decomposition has been observed with photosensitive polymers and brittle material such as concrete (see Section 6.16). The stresses are due to thermal expansion or photochemical reactions. For example, the depolymerisation of PMMA can be accompanied with a 20% increase in volume; also smaller light fragments of gaseous materials such as CO, CN and CH_2 will cause inner pressure build-up [1].

By far the greatest effect comes from thermal expansion. For example, during heating the thermal strain is given by

$$\delta l = l\beta\Delta T,$$

where β is the coefficient of thermal expansion and ΔT is the temperature rise.

Or for sudden events leading to a nonlinear response,

$$\delta l = l\left[\beta\Delta T + \propto \Delta T^2\right].$$

(Allowance can be made for the Poisson effect within β if one is converting from volumetric expansion coefficients.)

Taking the simpler equation for an order of magnitude calculation of the acceleration of a particle away from a surface that is suddenly irradiated or for a particle which is suddenly heated, one can calculate the expansion of a particle of radius r during the heating time of the laser pulse, assuming a uniform particle temperature:

$$\delta r = r\beta\Delta T.$$

From this the rate of change of the radius, v, can be calculated:

$$\frac{\partial r}{\partial t} = r\beta\frac{\partial\Delta T}{\partial t} = v.$$

Putting approximate values into this equation ($\beta = 10^{-6}\,°C^{-1}$, $\Delta T \approx 2,000°C$ and pulse length $\delta t = 1\,ns$), we get $v = 2 \times 10^6 r$. For a 10-μm-radius particle this amounts to $2 \times 10^6 \times 10^{-5} = 20\,m\,s^{-1}$.

To get to $20\,m\,s^{-1}$ in $10^{-9}\,s$ requires an acceleration of $2 \times 10^{10}\,m\,s^{-2}$. This is a formidable acceleration for a small particle of density approximately $3,000\,kg\,m^{-3}$ and therefore of mass $(4/3)\pi r^3 \rho = 1.2 \times 10^{-11}\,kg$. Thus, the force (from Newton's law, $F = ma$) is $1.2 \times 10^{-11} \times 2 \times 10^{10} = 0.24\,N$. This force – a quarter the weight of an apple – on a tiny speck of dust is sufficient in many cases to cause it to fly away from the surface, in spite of weak van der Waals forces holding it down. This acceleration away from the surface due to photoablation was photographed by Charles Otis and Bodil Barren at the IBM Thomas J. Watson Research Center in Yorktown Heights [2, 3] (Figure 8.1).

This effect is enhanced for shorter pulses particularly if they are shorter than the time required to lose energy as sound. The knock can come from the expansion of either the particle or the substrate, or both. As with other waves the reflected wave is not necessarily to be ignored, particularly if some shape factor focuses the reflection on the surface.

These instantaneous stresses can result in defects forming within the structure. The spatial distribution of defects as a function of pulse length is shown in Figure 8.2.

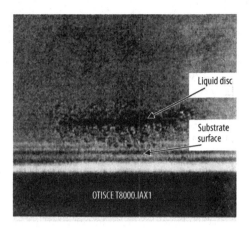

Figure 8.1 A CCD image taken after a time delay of 8.0 μms between the excitation pulse from a KrF excimer laser of 180 mJ cm^{-2} and the probe pulse. The excitation pulse irradiated the silicon sample, which was coated with 0.5-μm film of alcohol–water, vertically from above and the probe pulse was incident at a glancing angle, which gives a mirror image of the event. The size of the disc of water rising from the surface is the same as the beam diameter [2, 3]

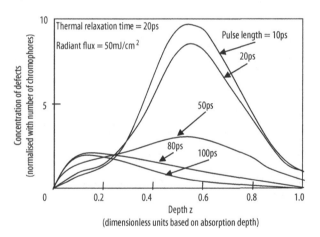

Figure 8.2 Spatial distribution of defects generated by acoustic waves due to transient thermal stresses; showing that defects mainly form with pulse lengths shorter than the thermal relaxation time. (From [1])

8.3 Case 2: Nanosecond Pulse Impact

Gross [4] modelled the impact of a 30 ns 400 kW peak power pulse on a metal surface from first principles. The model required around 100 parallel computers to process it! The results give some interesting numbers for parameters not easy or impossible to measure, and therefore an interesting glimpse into the probable activity during these very rapid events.

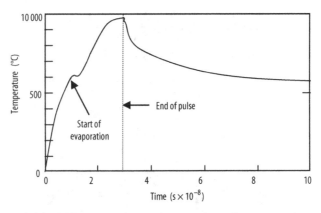

Figure 8.3 Detail of the calculated temperature history of a surface spot in the centre of a 30 ns pulse of 400 kW peak power showing a quadratic temperature rise with a glitch for the latent heat effect at the start of evaporation and a logarithmic cooling [4]

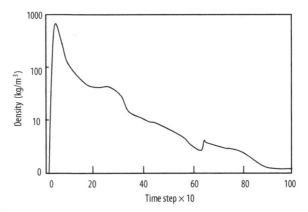

Figure 8.4 The calculated history of the plasma density generated by a 30-ns pulse of 400 kW. The irregular fall is thought to be caused by latent heat effects [4]

The peak temperature history (Figure 8.3) shows the normal quadratic rise in temperature during heating and logarithmic cooling. There are steps during heating caused by the latent heat effects. Plasma density is not a parameter that is easily measured but it can be calculated. During the pulse impact the density rises to almost that of a solid of $800\,\mathrm{kg\,m^{-3}}$ and then decays rapidly (Figure 8.4). The pressure at centre of impact also rises to very large values at the end of the pulse of several thousand atmospheres (Figure 8.5). This is the basis for laser shot peening.

After 61 ns, 31 ns after the end of the pulse, the mass flow within the impact region is still progressing and is essentially around the edges of the growing pit, with a nonuniform pressure across the forming hole (Figures 8.6 and 8.7). The gas pressure and density vary down the depth of the hole (Figure 8.8). The flow around the edges

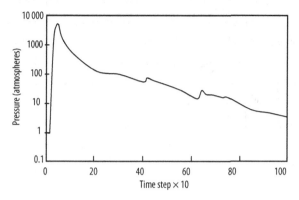

Figure 8.5 Calculated pressure history at the centre of a 30-ns, 400-kW pulse impact [4]

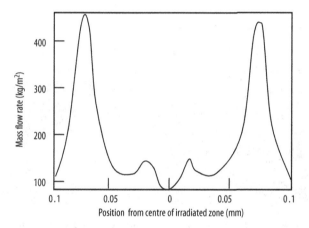

Figure 8.6 Mass flow across and within the impact region 61 ns after the start of a 30 ns pulse. Note the strong flow around the edges [4]

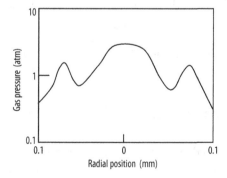

Figure 8.7 Calculated variation of pressure across the irradiated zone 61 ns after the start of a 30-ns, 400-kW pulse. The pressure is dying down considerably but still encourages a radial flow [4]

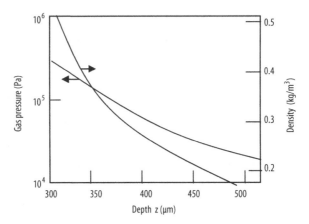

Figure 8.8 The variation of gas pressure and density down the depth of the hole after 61 ns, which is 31 ns after the end of the 30 ns pulse [4]

Figure 8.9 The flow from the edges of a forming hole captured after incomplete ejection from a short pulse of 511 nm radiation at fluences well below that for optimum drilling performance. The copper vapour laser beam was delivered by fibre and focused. (Photo courtesy of David Coutts, now at Physics Department, Macquarie University, Sydney, NSW, Australia)

of the hole and the implied flow from these calculations are seen in the appearance of a forming nanosecond impact crater (Figure 8.9).

8.4 Case 3: Ultrashort Pulses

Thermal concepts do not describe the processes taking place under femtosecond high-power pulses. Here the image is one of a colossal electromagnetic wave sufficient to rattle the entire electronic structure, strip electrons from atoms and leave them momentarily charged, during which time they repel each other by Coulomb forces [5]. This effectively removes material without significantly heating the surroundings. Other effects are impact stresses as already seen, plasma absorption and subsequent thermal and hydrodynamic movements of the target material.

The photon density is such that multiphoton events can occur. For example a 1 mJ pulse of 1 fs is a terawatt (10^{12} W) of power. This is a stream of, for example, 2.5 eV photons (blue light) of $10^{12}/2.5 \times 10^{-20} = 5 \times 10^{31}$ photons per second. These photons will be travelling at 3×10^8 m s^{-1} (speed of light), which means that there would be 1.6×10^{13} photons per angstrom. Apart from the problem of deciding how large a photon is, or where it is, this is a close-packed stream and likely to cause multiphoton events. If the wavelength of the blue light is 0.488 µm, then there would be 1.6×10^{17} photons within the length of one wavelength! This is a betting certainty that there will be some photons that are exactly in phase causing multiphoton events of energy 5, 7.5, 10 and even 12.5 eV (X-radiation is around 12.3 eV; X-rays are sometimes emitted from these events).

The mechanism of ionising by radiation depends on whether the photons are absorbed at the site, blocked prior to the site or transmitted through the site. If they are absorbed, then ionisation may occur directly by photoionisation – this is tricky since it implies precise multiphoton impact on an atom – or impact ionisation from a high-speed electron. This latter event is more likely since in this high density of photons it is probable that an electron may absorb two or more photons before impacting with the electronic structure of the atom (the electron cooling time is around 10^{-17} s for copper).

During the femtosecond-scale pulse the material may be in a form of solid-state plasma. What happens in the blindingly short time of a few femtoseconds? Firstly, with the binding electrons removed, the structure will start to fly apart – Coulomb explosion as described by Park [5]. He calculated the velocity of expansion as a function of the ionisation rate for a semiconductor and found that it is around 1,000 m s^{-1}.

In 10^{-13} s material travelling at 1,000 m s^{-1} would move approximately 10^{-10} m, almost the size of an atom. Is that sufficient to prevent recondensation? Apparently.

This physics is at the heart of the research into atomic fusion power generation in which a truly massive short pulse (petawatts 10^{15}) totally strips the nuclei of their electrons and, if the density can be maintained for long enough, allows the bare nuclei to be pushed together, resulting in a mass loss revealed as large quantities of energy.

8.5 Applications

How can we use this unbluntable machine tool, capable of very fine detailed work with little to no contamination of the workpiece by heat, chemicals or stress?

Ultrashort pulses behave in three major ways depending on their energy: low energy can cause melting; medium energy may also cause microcracking; high energy causes ablation.

8.5.1 Low-energy Pulses (Less than 150 nJ)

Low levels of energy can cause changes in the refractive index in glass and plastics possibly owing to the momentary melting at the focus and then resolidification with a higher density. These changes in refractive index can be used in optofluidic cells, including tunable photonic crystal waveguides, optical lenses and Bragg gratings [6, 7].

Microspheres of silica or polystyrene can be used as microlenses to create sub-wavelength-diameter holes. They also have the ability to assemble into a self-aligned pattern if they are made to flow over a flat plate. This has been used to make arrays of holes for solar cell platforms and microfiltration screens. The holes are made by a pulse of ultraviolet radiation shining vertically onto the sheet of spheres. If the radiation is incident at an angle, then slots and other patterns can be made [8]. This latter arrangement also avoids the spheres being ejected from the plate by the process. In another variation, polystyrene or silica microspheres of a few hundred nanometres are suspended in a fluid between a glass cover slip and a coated substrate (polyimide). A 532 nm laser traps the particles in a Bessel-beam optical trap (a ring-shaped beam) while a 355 nm pulsed laser beam illuminates the microspheres and is focused onto the translating substrate to produce features of the order of 100 nm – 3 times smaller than the wavelength of the patterning beam. Bessel beams, which have the property of reduced diffraction and long depth of focus, can be approximated by focusing a Gaussian beam through an axicon lens. Such beams are useful in optical tweezers systems. The Bessel beam separates the particle to be used and controls its position [9]. Moving the substrate and overlapping the spots can produce continuous line features with a positional accuracy of 30 nm for line widths of 263 nm.

Low-energy beams can be used for heating amorphous silicon to cause it to become microcrystalline and so alter its electrical properties.

8.5.2 Medium-energy Pulses (150–500 nJ)

Ultra short pulses when focused will usually give nonlinear events that have sharp power thresholds but so little energy that collateral thermal damage is negligible and the heat readily dissipates into the surrounding material. Ultrashort pulse applications include writing three-dimensional waveguides in silica, fabricating microfluidic devices and performing surgery inside living cells without puncturing the cell membrane. Medium energy levels of 150–500 nJ focused into glass may give a birefringent change owing to a chemical depletion in the irradiated layer leading to nanocracking, whose direction is fixed by the polarisation of the beam. These nanocracks can be erased and rewritten by light polarised in a different way; this is potentially useful for holographic storage [7].

8.5.3 High-energy Pulses (More than 500 nJ)

8.5.3.1 *Lithography*

The trends in most advanced technological applications are towards using multilayers as thin as possible (*e.g.*, submicron) and offering increased functionality involving complex multilayered structures, which are usually combinations of thin metals, metal oxides, ceramics or organic layers on glass, metal or polymer substrates [10]. Most of the layers are thermally and optically thin. Selective laser patterning of these layers is not a simple task, particularly to do so with industrial robustness. Nanosecond machining leads to thermal effects as discussed in this chapter and therefore could not be used for this application; however, machining with pulses of picosecond or less duration and powers just above the ablation threshold can achieve machining with no debris, fine focus and good depth control. Such pulses have peak intensities of more than 10^{14} W cm^{-2}, which will ionise and machine any material since the mechanism is based on nonlinear absorption. This also means that ablation only occurs where there is nonlinear photon absorption, thus allowing features below the diffraction limit of the radiation used. The pulse is so short there are no problems with interference from plasma and the debris is blown clear in what has been called "cold ablation". Examples of areas of application are:

1. Lab-on-a-chip engraving: These small devices, measured in square millimetres, have machined capillary flow channels and areas for electrodes, mixing and spectral analysis so that laboratory testing and analysis can be done quickly on minute quantities of material. The size of these devices may allow dozens of parallel tests simultaneously in twinned cells. They are used in cell analysis, DNA analysis, immunoassay based on antigen–antibody reactions, genomics and proteomics. Microfluidic devices sort cells and perform parallel chemical synthesis or screen for biological pathogens. Optofluidic devices include tunable photonic crystal waveguides, optical lenses and devices with a tunable refractive index [6].

2. Three-dimensional microfluidic structures can be made by using high-peak-power ultrashort pulses which vaporise the target material, such as glass, driving the material outwards from the focal point and leaving a void surrounded by an overdense layer. A series of pulses can carve a channel [7]. Such structures have been considered for making internal Fresnel zones for lenses and microfluidic channels. Exciting research with a product you cannot see with a naked eye!

3. Microarrays used in the pharmaceutical industry for high-throughput drug discovery.

4. Lab-on-a-chip fitted with electrodes for electrophoresis, made by drilling holes with an excimer laser and then filling them by a vapour deposition process with metallic conductors [10].

5. Etching miniaturised electrical circuits – often in single-shot machining on a moving conveyor using 308-nm excimer radiation. The metallised plastic strip may be 50 nm thick – these circuits do not carry high currents. The ultraviolet radiation passes through the plastic to strike the interface with the metal, causing the plastic at the interface to vaporise and completely remove the metal, leaving a thinned plastic and a good circuit. Feature sizes down to 10 μm can be made this way and areas up to 400 mm^2 have been processed with a single pulse energy of 1 J. This can generate 18,000 circuits per minute on a reel-to-reel arrangement at a pulse rate of 300 Hz with continuous feed since the pulse length is so short [10].

6. Submicron surface structuring: One simple method to structure a surface is to etch the interference pattern that is readily made between two beams of coherent radiation. This can give a regular array of lines spaced very tightly, such as 0.5 μm apart. Such structures can be used for plasmonic devices, Bragg reflectors and align-

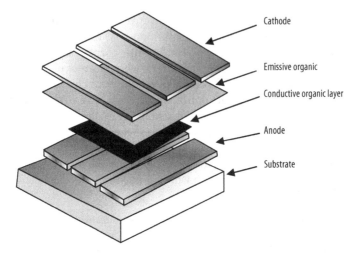

Cathode

Emissive organic

Conductive organic layer

Anode

Substrate

Figure 8.10 The structure of an organic light emitting diode stack (OLED) [14]

ment grooves for liquid crystals. Plasmonic devices are becoming the subject of intense research since they may hold the key to creating ultrafine structures some 10–20 times finer than the current Blu-ray Disc™[1] systems [11]. Plasmons are the quantised energy from plasma, just as photons are for electromagnetic radiation or phonons are for sound. When electrons oscillate within a metal surface or structure, they generate a very short wavelength radiation capable of being finely focused but not able to be transmitted much further than 100 nm or so. To focus this radiation requires minute Fresnel lenses some 10 μm in diameter. These lens structures can be etched by laser techniques.

7. Machining of OLEDs [12]. OLEDs are used in television screens, computer displays, small, portable system screens such as cell phones and PDAs, advertising, information and indication signs. OLEDs can also be used in light sources though they are not as bright as LED sources. OLEDs typically emit less light per area than inorganic solid-state-based LEDs, which are usually designed for use as point light sources. An OLED consists of many layers of different materials of 100–200 nm thickness (Figure 8.10). The patterning requires the removal of one or more layers without depositing debris. This is difficult but has been achieved with picosecond and femtosecond pulses. The competing technology is printing, which is limited in scale.

8.5.3.2 Stealth Dicing

Wafer dicing plays a critical role in the fabrication of semiconductor devices, devices which are becoming ever smaller and more complex. The standard methods of dicing

---

[1] Blu-ray Disc™ is a trademark of the Blu-ray Disc Association, 10 Universal City Plaza, T-100 Universal City, California, 91608, USA. http://www.blu-raydisc.com

are to use a diamond saw for silicon wafers thicker than 100 μm or laser ablation if they are thinner. In "stealth dicing" [13] internal perforations are made by focusing a pulsed laser beam at a wavelength that is transmitted through the wafer but which is absorbed by nonlinear processes at the focus – as in the internally etched blocks of glass in tourist shops. The internal perforation leaves the surface top and bottom pristine. The next step is to expand the tape on which the thin wafer sheet is stuck and cause the perforation to crack. There is no debris, surface cracking or thermal damage, unlike with competing processes. The technique can be used for speciality wafers where they may have other surface features, such as dye-attached films for adhesive stacking. Such additions make the traditional sawing or ablation processes more difficult and vulnerable to debris. In stealth dicing the internal perforation is made first, the dye is applied and the tape is pulled in that sequence. The film is then in the exact shape of the chip. Microelectromechanical (MEM) system devices can also be separated this way.

8.5.3.3 Fabricating Flat-panel Displays

Flat panel displays are another example of sophisticated multilayered structures as illustrated in Figure 8.11. Lasers are being used in many ways in their construction [14]. One example is the fabrication of colour filters, where the problem is to etch the contact pattern into the surface indium tin oxide layer without damaging the black mask material deeper in the sheet. Traditionally this etching has been done with chemicals, but this is messy and lacks precision. Laser dry etching of colour filters can be done by laser ablation with ultrashort pulses. It has also been done using a Q-switched diode-pumped solid-state laser incident at an angle of around 75° to enhance the absorption in the indium tin oxide layer and diminish the absorption in the lower black mask layer.

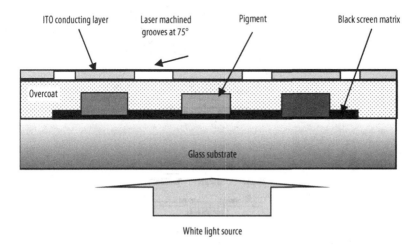

Figure 8.11 The construction of a colour filter for a flat panel. The indium tin oxide (*ITO*) conducting coating has to be precisely patterned by laser ablation but without damaging the screening black matrix [14]

8.5.3.4 Laser Transfer Technique

In building up a microelectronic circuit, traditionally one makes the circuit board and then mechanical pick and place robots place the devices. These robots cannot handle very small components (less than 1 mm square) or thin components (less than 50 µm thick). An alternative is to use the laser in a process named "lase and place" in which the component is mounted on a thin opaque plastic film on a laser-transparent support. When a laser is fired at the component through the transparent support, the plastic film ablates and effectively blows the component onto a receptor plate, which has been very precisely located to receive the circuit component (Figure 8.12 [15]). Some 100 devices per second can be mounted this way with a lateral precision of 25 µm for short travel distances of 100 µm.

A similar technique has been used to transfer OLED material onto a flat-panel display board. In this latter case the rigid glass donor plate is coated in molybdenum which is coated with the organic material for OLEDs. This is clamped against a substrate plate which has the essence of the circuit mounted on it – electrodes, *etc*. By scanning a laser in a precise pattern over the top donor plate, the beam heats the molybdenum layer and hence deposits the OLED in a pattern on the substrate plate. This can be repeated with other donor plates to achieve the complexity required of a high-quality flat-panel dis-

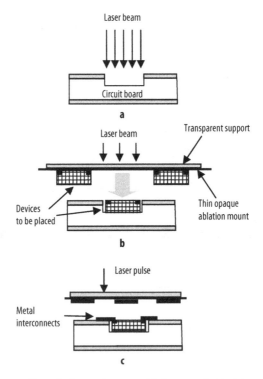

Figure 8.12 The laser transfer technique: **a** laser machining a pocket to take a device, **b** laser transfer of the device into the prepared pocket, and **c** laser transfer of interconnect circuitry [15]

play. The same technique can be used to repair broken conductors on panels. Tantalum is the metal usually transferred in this latter case.

8.5.3.5 Microcladding

Microcladding can be an alternative to electroplating, with a saving of materials. It has been done using wire feed or surface melting a printed or deposited paste. Using powder feed is an attractive alternative to making deposits of between 20–100 µm size, but there are serious problems with getting steady powder flow with 1–15 µm powder particles – which are little bigger than smoke. Applications for microcladding include making miniature tools such as hole stamps; depositing radio markers in gold, platinum or tantalum on stents, depositing electrical contacts and repair and modification of moulds [16].

8.5.3.6 Laser Print Forming

Laser print forming [17] is a process in which a tool made by scanning a focused ion beam or laser and possibly reproduced by polydimethylsiloxane is pressed at a low temperature of around 1–200 °C into a paste bed of nanoparticles which melt at a lower temperature than bulk material. The compacted structure after removal of the press tool is then sintered by laser. It has similarities to embossing.

8.5.3.7 Microstructures

Extreme ultraviolet lithography operating at 13.5 nm has been used to make the circuitry for single chips measuring a staggering 22 mm × 33 mm. Previously only part of such an area could be treated at a time. AMB in collaboration with IBM has developed a special photoresist for extreme ultraviolet lithography consisting of a polymer and photoacid generator that dissolves after exposure to extreme ultraviolet radiation. There are further additions of conductive and absorber films made from chromium nitride and tantalum nitride. New photoresists made of molecular glass are under consideration. The point being made is that simply having shorter wavelengths and finer focus is not the end of the road towards nanoelectronics [18].

8.6 Summary

Ultrashort laser pulses have opened the way for a new style of material processing, that of ablation, and a new scale of working, that of nanotechnology. Both are significant events in the progress of technology that is likely to enhance our quality of life.

Questions

1. a. Write brief notes on the four ways in which photons can remove material from a solid.

    b. How many photons would there be in the length of one wavelength for a femtosecond pulse containing 1 mJ of blue light (0.488-μm wavelength; 2.5 eV per photon)?

    c. Is it likely that nonlinear optical events will occur when using such a pulse? (1 eV = 10^{-20} J; velocity of light is 3×10^{8} m s^{-1}).

    d. What are the current applications of ultrashort laser pulses of the order of femtoseconds?

2. a. It has been suggested that there are three different thermal regimes that are generated by laser pulses of differing lengths. Briefly describe these differences.

    b. What is the difference between photothermal and photochemical ablation?

    c. Describe the main physical effects that occur when a focused 10 fs pulse of 1 mJ of blue light (0.488 μm wavelength) strikes a surface.

    d. What is the approximate expansion velocity of a 10 μm-diameter particle that is struck by a 1 ns laser pulse, given that the coefficient of thermal expansion is $10^{-6}\,^\circ\text{C}^{-1}$ and the temperature rise to the boiling point is 2,000 K?

References

[1] Anisimov SI, Bityurin NM, Luk'yanchuk BS (2003) Models for laser ablation. In: Peled A (ed) Photo-excited processes, diagnostics and applications (PEPDA). Kluwer, Dordrecht, pp 121–159

[2] Kelly R, Miotello A, Barren B, Otis CE (1992) Novel geometrical effects observed in debris when polymers are laser sputtered. Appl Phys Lett 60:2980–2982

[3] Zapka W (2002) The road to "steam laser cleaning. In: Luk'yanchuk B (ed) Laser cleaning. World Scientific, Singapore, pp 34–40

[4] Gross M (2004) Transient numerical simulation of laser material processing with focus on laser cutting. PhD thesis, Heriot-Watt University

[5] Park JH, Chan CL (2003) Ultra short pulse laser material removal by Coulomb explosion. In: ICALEO 2003 proceedings, Jacksonville, LIA, Orlando, paper M503

[6] Erickson D (2008) Optofuidics emerges from the laboratory. Photonics Spectra Feb 74–79

[7] PennWell Corporation (2007) Microspheres simplify nanopatterning processes. http://www.laserfocusworld/articles/325407

[8] Hetch J (2008) Ultrashort pulses write sharp tiny features and perform micro surgery. Laser Focus World Oct 71–74

[9] Overton G (2008) Direct writing nano-patterning uses optically trapped micro spheres. Laser Focus World Aug 25–27

[10] Delmdahl R (2008) Laser micro-structuring with excimer lasers. Laser User (53):13

[11] University of California Berkeley (2008) Denser computer chips possible with plasmonic lenses that fly. http://www.physorg.com/news143902428.html

[12] Kanarkis D (2008) Ultra fast laser patterning of OLED for solid state lighting. Laser User (53):26–27

[13] Lares M (2008) Laser dicing technique cuts wafers from the inside. Photonics Spectra Jan 77–79

[14] Washio K, Hitz B (2008) Fabricating flat panel displays with lasers. Photonics Spectra Jan 62–68

[15] Mathews SA, Charipar NA, Metkus K, Pique A (2007) Manufacturing microelectronics using 'lase-and-place'. Photonics Spectra Oct 70–75

[16] Jambor T, Wissenbach K (2008) Micro cladding with a fibre laser. Laser User (51):38–40

[17] Hu Q, Hu P, O'Neill W (2008) Laser assisted micro structure fabrication by using nano particles. Laser User (52):33–35

[18] Jeffries E (2008) Extreme UV lithography to power electronics. Materials World May 9

9

Laser Bending or Forming

The shape of things to come

after the book (1933) by H.G. Wells (1866–1946)

9.1 Introduction

When a material is heated, the atoms it contains vibrate; the hotter it is, the stronger the vibrations and hence the atoms push each other apart and the material expands. If it is restrained from expanding, the atoms rearrange themselves in what is described as elastic or plastic deformation. With elastic deformation the atoms can slide back into their original positions on cooling because they have not moved into a different energy well. With plastic deformation the atoms remain in their new positions on cooling because they have found new stable positions.

The laser can selectively heat any specified area, causing it to expand and if the expansion is restrained this can cause localised plastic deformation. On cooling this area that had suffered from plastic deformation will be too short and will thus come under tension from its surroundings. This tension in turn will cause the material to bend according to the normal laws of mechanics.

The process of thermal bending is not new [1]. It has been used in shaping plates for ship hulls and by blacksmiths for centuries. The process was manual and very labour-intensive. It required highly skilled labour using hoses and oxyacetylene torches, who would sit on stools or upturned buckets for hours at a time slowly coaxing the heavy steel sheets into the right shape. The shapes would be checked against wooden templates stored in lofts and manoeuvred by crane. After the shapes had been checked, chalk marks would be made to guide the next round of thermal straining. It has been estimated that 18,000 man-hours is required to make the three-dimensional plates for a US Navy destroyer [2].

The first references to the use of lasers as the heat source for bending was in the early 1980s with a report from Kitamura [3] and the first public publications of Namba [4, 5] and Scully [6]. Prior to that the process was described as "thermal distortion". Such is the world, there are usually two ways of viewing the same thing; from one view it is distortion and from another a precision manufacturing process with a brilliant future.

9.2 The Process Mechanisms

The laser can bend material by both thermal and nonthermal routes. Four thermal routes have been noted. They are the temperature gradient mechanism, the point source mechanism, the buckling mechanism and the upsetting mechanism [7].These are based on two different material responses: shortening action due to plastic flow on heating but not on cooling; and buckling, in which the material does not necessarily plastically deform by compression but by bending so that it springs up or down in a bulge. In the temperature gradient mechanism the concept that is useful to keep in mind is that the areas heated above a certain temperature are likely to be shortened by this process, and the bend is always towards the laser. Such a concept gives a first-order approximation as to the shape that will result. In the buckling mechanism greater bend angles can be achieved both towards and away from the laser.

The nonthermal bending routes are based on scribing a surface [8] or laser-induced shock waves which usually cause a microbend away from the laser [9, 10]. These mechanisms will now be discussed in turn.

9.2.1 The Thermal Gradient Mechanism

If the material is heated by a laser such that there is a steep thermal gradient through the thickness (see Figure 9.1, mechanism a) then the surface will be under compression, since it wishes to expand. However, the sheet will be restrained from bending by the stiffness of the rest of the material, which is still cold. Plastic flow will occur in the surface region only, provided the temperature is high enough to cause sufficient thermal strain. This plastic flow will not be recovered on cooling because:

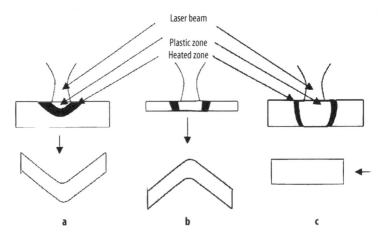

Figure 9.1 The three main thermal bending mechanisms: **a** temperature gradient, **b** buckling and **c** upsetting

1. on cooling the rest of the material will heat up a little by conduction, causing that region to expand, thus reducing the growing tension in the cooling region;
2. the mechanical properties of the cooling region also become stiffer as they cool; and
3. finally the area over which the stresses operate during cooling is the whole sheet instead of the much smaller zone of heating, which was of approximately spot size.

Thus, the plastic deformation formed on heating is mainly not recovered and hence the piece will bend towards the laser on cooling.

To create this thermal gradient means that a laser beam must traverse the workpiece moving at such a speed that the thermal depth, z, is small compared with the workpiece thickness, s_0. The thermal depth is approximately given by some constant value of the Fourier number ($z^2/\alpha t$, where α is the thermal diffusivity and t is the interaction time; $t = D/v$, the spot size D divided by the traverse speed v).

Hence, the temperature gradient mechanism will most probably be predominant if

$$\frac{\alpha D}{v s_0^2} \ll 1. \tag{9.1}$$

Initially, as the beam starts to traverse a sheet, there will be a bend away from the laser owing to the expansion of the small area under the beam. It is very slight and is hard to measure since the whole of the sheet is reluctant to bend owing to such a small spot being heated; nevertheless this counterbending diminishes the compressive stress in the surface region and hence reduces the ultimate bending angle. As the beam continues across the sheet, the initial region cools and contracts, causing a bend to form at that end of the traverse. It can be visualised that the whole stress pattern in the sheet is moving during the process, not by much but by something. This makes the bending process very slightly asymmetric. This asymmetry is further aggravated by the different levels of mechanical restraint at the edges and the middle of the sheet. Thus, in making a symmetric part it is necessary to consider the sequencing of traverses both in direction and location [11–13].

The amount of bending per pass is not very great, being approximately 1–3° (see Figure 9.2). It is constant for the first ten to 20 passes for a given material, laser power and beam size (see Section 9.3). After this number of passes the bending angle starts to fall per pass owing to work hardening and thickening of the material at the bent edge. The material thickens owing to the plastic deformation. This leads to one of the advantages that laser bending has over mechanical bending, in that it thickens the material on the bend rather than thinning it. This is particularly relevant when bending pressure piping.

9.2.2 The Point Source Mechanism

If the beam is stationary, then the heated zone is a spot rather than a line. If the pulse is brief, a thermal gradient will be created and the mechanism is similar to the temperature gradient mechanism (see Figure 9.3). The mechanical bend is different since the

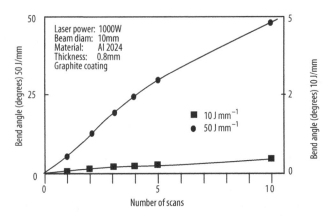

Figure 9.2 Bend angle versus number of passes at two line energies [11]

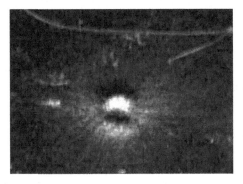

Figure 9.3 The surface "pucker" in aluminium made by firing 50 pulses at 1-s intervals from a 500-W laser focused to a spot size of 3-mm diameter [12]

effect is to produce a shortened spot, resulting in a pucker in the surface and a bend towards the laser along the line of least resistance – usually the smallest width. If the pulse is longer, then the heating is through the thickness and the buckling mechanism starts to predominate with a bulge upwards or downwards depending on some predilection of the material. This process is used for the adjustment of microcomponents [14]. The bending angle is usually very slight, of only one tenth to 1/100th of a degree [7].

9.2.3 The Buckling Mechanism

If there is little thermal gradient through the depth of the sheet, then the temperature gradient mechanism will not work since the top and bottom will be similarly affected. Instead, the expansion resulting from the through heating will cause a bulge to appear (see Figure 9.1, mechanism b). This bulge can move upwards or downwards depending on whether there is an initial bend, a residual stress or an applied stress. The centre of the bulge is hotter than the edges; hence, the centre can bend plastically but the edges

Table 9.1 Methods of ensuring the correct direction of bending when using the buckling mechanism

Cause of bias	Probability for bending direction relative to laser (%)	
	Towards	Away
Elastic prebending away (due to external forces)	0	100
Elastic prebending towards	100	0
Plastic prebending away	0	100
Plastic prebending towards	100	0
Relaxation of surface compressive stress (rolled sheet)	100	0
Counter bending due to initial temperature gradient	0	100

will tend to bend elastically. On cooling the plastic bend remains. The rate of bending is in the range 1–15° per pass. This is considerably greater than for the temperature gradient mechanism, partly owing to the fact that considerably more energy has been put into the sheet by the slowly moving beam of large diameter.

For this mechanism to be useful the direction of bending must be predictable. This can be achieved by ensuring there is a bias in one direction or the other. Table 9.1 lists some of the concepts.

The external force for elastic bending could be gas jets, gravity or an actual mechanical pressure. The plastic bending is a little more tricky since it may be present, but not apparent, owing to the way the sheet had been rolled or some such event and hence plastic prebending is not so reliable as the other techniques. It could also be achieved by an initial bend towards the laser by the temperature gradient mechanism. Rolled sheet has a compressive residual stress at the two surfaces and tension at the centre. When such a sheet is heated on one face, the compressive stress is annealed to some extent and hence the sheet will bend towards the heat source owing to the relaxation of the surface compression.

9.2.4 The Upsetting Mechanism

If the material geometry does not allow buckling owing to its thickness or section modulus, for example, a tube or extrusion, then buckling is restrained and is unlikely to happen (Figure 9.1, mechanism c). Thus, if the laser treatment produces a thermal field with no significant gradient, plastic deformation through the thickness will occur as the material is heated beyond a certain temperature, causing a thickening of the material along the heated zone. On cooling the plastic flow will not be restored since there is no force and the material is thus shorter than it was at the beginning. It is also thicker at the treated zone.

9.2.5 Laser-induced Shock Bending

Ultrashort laser pulses (less than 100 ps) can induce extreme pressures in the surface-generated plasma of more than 10 MPa. The resulting supersonic micro shock waves cause expansion of the material in the impact zone by a forging action [9,10]. The result is a microdeformation away from the laser of a few microradians. The bend is achieved with effectively no heating of the substrate; it is instantaneous and does not depend on cooling and therefore can be easily monitored; it can also give precision of nanometre accuracy for bends of up to 1°. It has been demonstrated to cause bending in difficult materials such as silicon. Typical operating parameters are 800 nm, 150 fs and 610 μJ, although it will work with 10 ps at 1,064 nm and 178 μJ.

9.3 Theoretical Models

The problem of modelling the bending process is centred on the difficulty of calculating the plastic flow during heating. Calculating the temperature field is well understood (see Chapter 5); calculating the stresses due to this temperature field is not so simple since the expansion of one part affects the stress in another and hence the expansion in that part, and so on. However, the stress field can be calculated, particularly by finite-element analysis. Having done so, one needs to compute the plastic strains using temperature-dependent stresses for the elastic limits and the plastic flow stress. Then the calculation needs to include the stresses during cooling and the recovery of some plastic flow. The resultant geometry has then to be figured to reveal the bending angle. Armed with a large computer and plenty of time, some have done this calculation, which is leading to a deeper understanding of this process.

However, for the general reader or busy engineer who simply wishes to know how the variables fit together to grasp the essence of the process, simpler models have been developed, principally by Vollertsen [7, 12–14]. Some of his models are summarised here. The general arrangement for bending is shown in Figure 9.4. The models assume that only two-dimensional stresses, transverse to the bend line, are relevant.

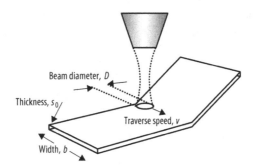

Figure 9.4 General arrangement for making a linear bend

9.3.1 Models for the Thermal Gradient Mechanism

9.3.1.1 The Two-layer Model or Trivial Model

If during surface heating by a laser it is assumed that the temperature field is a step function and all the thermal expansion in the heated layer is converted into plastic flow, and if it is also assumed that there is no straining of the compressed layer on cooling, then the resulting bend is determined solely by the geometry of the short piece at the surface and the longer piece beneath. These vast assumptions are illustrated in Figure 9.5.

From a heat balance on the surface element irradiated by the laser we get the temperature rise of the surface layer (the meaning of the symbols is given at the end of the chapter):

$$AP = \frac{\rho l v s_0 C_p \Delta T}{2};$$

therefore,

$$\Delta T = \frac{2AP}{\rho l v s_0 C_p}.$$

This causes thermal expansion of that layer by an amount Δl, where

$$\Delta l = \alpha_{th} \Delta T l = \frac{2AP \alpha_{th}}{\rho v s_0 C_p}. \tag{9.2}$$

From geometry and assuming the new lengths pivot about the centreline of the half-section, we have

$$\tan\left(\frac{\alpha_B}{2}\right) = \frac{\Delta l/2}{s_0/2} = \frac{\Delta l}{s_0}.$$

For small angles

$$\alpha_B = \frac{2\Delta l}{s_0} = \frac{4AP \alpha_{th}}{\rho v s_0^2 C_p}. \tag{9.3}$$

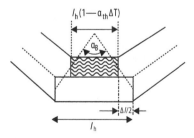

Figure 9.5 Two-layer or trivial model

This relationship has correctly identified many of the principal parameters involved in the process. The inverse dependency on the square of the section thickness is also in line with experiment, as is the linear dependence on the material properties of $(\alpha_{th}/\rho C_p)$ (see Figure 9.13) and to some extent the line energy (AP/v) (see Figure 9.12). The predicted value is, however, too large. This is to be expected since not all the expansion goes into plastic flow, nor is the temperature profile anything like a step function – except possibly for thin sections with high traverse speed.

9.3.1.2 The Two-beam Model

The two-layer or trivial model, just discussed, did not allow for the forces and moments within the bending sheet. It was also assumed that the heated layer was half the thickness of the section, thus not allowing for thermal conductivity. The two-beam model accounts for these but still relies on a layered structure, with the heated layer being uniformly heated and the lower layer not being heated. The effects of counterbending during the process are not allowed for, even though such a movement would reduce the compressive forces on the expanding heated layer and hence reduce the plastic flow and the resultant bending angle.

Assuming that the stress/strain field is two-dimensional, one can represent the bent sheet as shown in Figure 9.6. From this figure it can be seen that

$$\tan\left(\frac{\alpha_B}{2}\right) = \frac{l\left(\varepsilon_1 - \varepsilon_2\right)}{0.5s_0} \approx \frac{\alpha_B}{2}. \tag{9.4}$$

Now $\Delta l/l_h$ is the strain in the heated layer, ε_1, which during heating is made up of strain due to the bending force, bending moment and thermal expansion:

$$\varepsilon_1 = \frac{F}{A_1 E_1} - \frac{M_1}{E_1 I_1}z_1 + \alpha_{th}\Delta T.$$

If it is assumed that the thermal expansion part is entirely converted to a plastic deformation, then on cooling the value of the strain is

$$\varepsilon_1 = \frac{F}{A_1 E_1} - \frac{M_1}{E_1 I_1}z_1 - \alpha_{th}\Delta T. \tag{9.5}$$

Similarly the unheated layer has a strain given by

$$\varepsilon_2 = -\frac{F}{A_2 E_2} - \frac{M_2}{E_2 I_2}z_2. \tag{9.6}$$

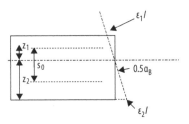

Figure 9.6 Strain in a bent beam

From standard mechanics for large curvatures it can be assumed that

$$\frac{M_1}{E_1 I_1} = \frac{M_2}{E_2 I_2} = \frac{M}{EI} = \frac{\alpha_B}{2l} . \tag{9.7}$$

Inserting Equations 9.5, 9.6 and 9.7 into Equation 9.4 gives

$$\alpha_B = \frac{4l}{s_0} \left[-\frac{F}{A_2 E_2} - \frac{\alpha_B z_2}{2l} - \frac{F}{A_1 E_1} + \frac{\alpha_B z_1}{2l} - \alpha_{th} \Delta T \right] . \tag{9.8}$$

The force is also related to the bending moment by

$$F = \frac{M}{2s_0} = \frac{\alpha_B EI}{s_0 l} . \tag{9.9}$$

Substituting $z_1 - z_2 = s_0/2$ and Equation 9.9 into Equation 9.8 yields

$$\alpha_B = \frac{4EI\alpha_B (E_1 A_1 + E_2 A_2)}{s_0^2 (E_1 A_1 E_2 A_2)} + \alpha_B + \frac{4l\alpha_{th}\Delta T}{s_0} . \tag{9.10}$$

The beam cross-section is defined as

$$A_1 = bs_1$$
$$A_2 = b[s_0 - s_1] , \tag{9.11}$$

and the moment of the rectangular beam, I, is

$$I = \frac{bs_0^3}{12} . \tag{9.12}$$

Substituting Equations 9.11 and 9.12 into Equation 9.10 gives

$$\alpha_B = \frac{12\alpha_{th}\Delta t l s_1 (s_0 - s_1)}{s_0^3} . \tag{9.13}$$

The temperature difference between the two layers, ΔT, can be estimated from the energy balance. The energy input $Q = 0.5\Delta t PA$, where the dwell time $\Delta t = D/v$; D is the beam diameter and v is the traverse speed.

The temperature rise from this heat input into the upper layer, which is of thickness s_1 and mass $lDs_1\rho$ is

$$\Delta T = \frac{Q}{C_p l Ds_1 \rho} = \frac{PA}{2C_p l s_1 \rho v} ;$$

hence

$$\Delta T l s_1 (s_0 - s_1) = \frac{PA(s_0 - s_1)}{2C_p v \rho} . \tag{9.14}$$

Substituting Equation 9.14 into Equation 9.13 gives

$$\alpha_B = \frac{6\alpha_{th} PA(s_0 - s_1)}{\rho C_p v s_0^3} . \tag{9.15}$$

For the case where $s_1 = s_0/2$, which is implied in Equation 9.12,

$$\alpha_B = \frac{3\alpha_{th}PA}{\rho C_p v s_0^2},$$
(9.16)

which is the same as that found in the simpler model except for the constant, which is now 3 instead of 4.

For the case where $s_1 = B(\alpha D/v)^{1/2}$, which by careful choice of the value of the constant B is approximately the penetration depth of the plastic deformation isotherm, we have

$$\alpha_B = \frac{6\alpha_{th}PA\left[s_0 - B\left(\alpha D/v\right)^{1/2}\right]}{\rho C_p v s_0^3}.$$
(9.17)

This has the interesting property of becoming zero when the isotherm is fully penetrating, a logical ending to the thermal gradient mechanism, but the onset of the buckling mechanism. It also has a maximum value with varying traverse speed, beyond which for higher velocities and using the arbitrary value of 0.8 for B the bending angles are similar to those found in practice. Equation 9.17 introduces the thermal diffusivity into the problem and thus has accounted approximately for the temperature field.

9.3.1.3 The Residual Stress Model

An alternative method for accounting for the temperature field is given by Vollertsen in his residual stress model [7]. He derived the expressions for the two cases when the recrystallisation temperature at the surface is not reached and when it is reached. The assumptions are that the region outside or cooler than a certain isotherm, T_G, does not plastically deform, whereas that within the isotherm has all its thermal expansion absorbed in plastic flow. When the temperature at the surface exceeds the recrystallisation temperature, the thermal expansion is calculated as though the temperature was at the recrystallisation temperature, i.e., any further heating above this temperature does not affect the extent of thermal expansion. Such an assumption allows somewhat for the reflow on cooling of very hot zones. The results are summarised below.

Case 1. Partial penetration and surface temperature below the recrystallisation temperature
(the symbols are explained at the end of the chapter)

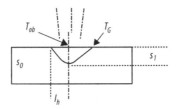

The bending angle is

$$\alpha_B = \frac{\varepsilon_{ob} I_h s_1}{s_0^3} \left(3\pi s_0 - 8 s_1\right), \tag{9.18}$$

where the surface strain, ε_{ob}, is

$$\varepsilon_{ob} = T_{ob} \alpha_{th} - \frac{k_f(T_{ob})}{E(T_{ob})}, \tag{9.19}$$

the depth of the temperature field, s_1, is

$$s_1 = \ln\left(\frac{4 T_G' r_n^{-2/3}}{3N}\right) \frac{(2 r_n \alpha t)^{1/2}}{2}, \tag{9.20}$$

and the width of the temperature field, l_h, is

$$l_h = r_n (4\alpha t)^{1/2} \left[-\frac{1}{2 r_n} \ln\left(\frac{4 T_G' r_n^{-2/3}}{3N}\right)\right]^{1/2}, \tag{9.21}$$

with

$$T_{ob} = \frac{3}{4} N r_n^{2/3} \quad \overline{T_{ob}} = \frac{T_{ob} + T_G}{2} \quad N = \frac{8 A P (\alpha t)^{1/2}}{\pi k D^2} \quad r_n = \frac{D}{4(\alpha t)^{1/2}}$$

$$T_G = T_G'(1 + \delta) \quad \delta = f(r_n, N, s_0, \alpha)$$

Case 2. The temperature field is fully penetrating and the surface temperature exceeds the recrystallisation temperature

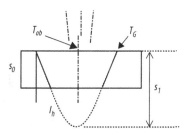

The bending angle, α_B, is

$$\alpha_B = \frac{\varepsilon_{ob} I_h}{s_1 s_0^3} \left[6 s_0^2 \left(s_1^2 - s_0^2\right)^{1/2} + 6 s_0 s_1^2 \arcsin\frac{s_0}{s_1} + 8 \left(s_1^2 - s_0^2\right)^{3/2} - 8 s_1^3\right], \tag{9.22}$$

and the other expressions are the same.

Such a model has made many of the assumptions of the simpler versions and hence it will only be approximate. It is thus debatable whether it is more trouble than it is worth, since the functional relationships between the major variables are given by the simpler models. If a detailed mathematical analysis is required, then possibly the way ahead is through finite-element analysis.

9.3.2 The Buckling Mechanism Model

If there is negligible temperature gradient through the thickness of the sheet, then the thermal gradient mechanism will not apply – there being no gradient. Instead the material may expand and bulge upwards or downwards to accommodate the extra length (Figure 9.7). The hottest part of this bulge may distort plastically, whereas the cooler part may distort only elastically, and hence on cooling the central part is bent, whereas the outer part recovers. This situation will occur if the sheet is thin compared with the heated spot size or the thermal conductivity is very high compared with the thickness, in fact if $(\alpha D/v)^{1/2} \gg s_0$.

Consider the buckling to be two-dimensional so that at stage 3 in Figure 9.7 a representation could be similar to that shown in Figure 9.8. The region at the top of the buckle, of length l_1, is assumed to bend plastically, whereas the region of length l_2 bends elastically. Both regions are represented as arcs of circles for geometric simplification. Since the elastic region recovers and is not permanent, the bending angle is determined by the bend of the plastic region, region 1.

From the geometry shown in Figure 9.8 the bending angle, α_B, is given by

$$\frac{\alpha_B}{2} = \frac{l_1}{r_1} = \frac{l_2}{r_2}.$$ (9.23)

Hence, we need to calculate r_2 and l_2 to give a value for α_B. The value of the radius, r_2, in the elastic region can be found from a balance of the bending moments in the two

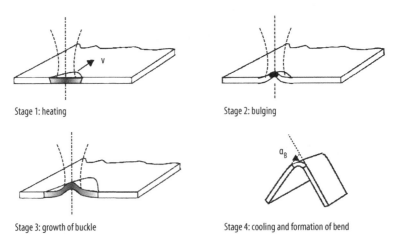

Stage 1: heating   Stage 2: bulging

Stage 3: growth of buckle   Stage 4: cooling and formation of bend

Figure 9.7 The stages in the development of a buckle. (After Vollertsen [7])

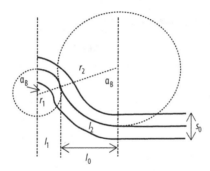

Figure 9.8 Two-dimensional representation of a buckle at stage 3

regions. The bending moment, M_{el}, is

$$M_{el} = \frac{Ebs_0^2}{12r_2},$$ (9.24)

where E is the elastic modulus, b is the width of the sheet s_0 is the sheet thickness.

The counterbalancing force from the plastic region, M_{pl}, comes from the resistance to flow:

$$M_{pl} = \frac{k_f(T_1)bs_0^2}{4},$$ (9.25)

where $k_f(T_1)$ is the flow stress in heated region.

Equating the two bending moments and solving for r_2, we get

$$r_2 = \frac{Es_0}{3k_f(T_1)}$$ (9.26)

l_2 = original length + the thermal extension in region 2 = $l_{02} + \Delta l$. (9.27)

Now

$$l_{02} = r_2 \sin\left(\frac{\alpha_B}{2}\right),$$ (9.28)

and

$$\Delta l = \alpha_{th}\Delta T_{av}l_h,$$ (9.29)

where ΔT_{av} is the average temperature of the total heated zone of transverse length l_h, longitudinal length Δx and thickness s_0. It can be estimated through a heat balance as before:

$$\Delta T_{av} = \frac{AP}{2C_p\rho s_0 v l_h}.$$ (9.30)

Substituting into Equation 9.29 gives

$$\Delta l = \frac{\alpha_{th} A P}{2 C_p \rho s_0 v} . \tag{9.31}$$

Thus, from Equations 9.27, 9.28 and 9.31 we get

$$l_2 = r_2 \sin\left(\frac{\alpha_B}{2}\right) + \frac{f' \alpha_{th} A P}{2 C_p \rho s_0 v} , \tag{9.32}$$

where f' is the fraction of the thermal expansion that causes an elongation in region 2. It has a value between 0 and 1, but since region 2 is far larger than region 1 it should be high, whereas because the temperature is less it should be low. Choosing an arbitrary value of 0.5 and substituting the value of l_2 from Equation 9.32 and the value of r_2 from Equation 9.26 into the equation for α_B (Equation 9.23), we get

$$\frac{\alpha_B}{2} = \sin\left(\frac{\alpha_B}{2}\right) + \frac{3 \alpha_{th} A P k_f (T_1)}{4 E s_0^2 C_p \rho v} . \tag{9.33}$$

Expanding the sine function by a Taylor series, we get

$$\sin(x) = \sum_{n=1}^{\infty} (-1)^{n+1} \frac{x^{2n-1}}{(2n-1)!} . \tag{9.34}$$

Thus,

$$\sin\left(\frac{\alpha_B}{2}\right) \approx \frac{\alpha_B}{2} - \frac{\alpha_B^2}{48} , \tag{9.35}$$

and hence the approximate equation for the angle of bending due to buckling, assuming a large heated area compared with the thickness of the sheet, is

$$\alpha_B = \left[\frac{36 \alpha_{th} k_f (T_1) A P}{C_p \rho E v s_0^2} \right]^{1/3} . \tag{9.36}$$

The validity of this equation should be limited to thin sheets with high values of the ratio of thermal conductivity to thickness and also to speeds that are fast enough to avoid overheating, which might reduce $k_f(T_1)$ to a low value and thus reduce the bending.

9.3.3 The Upsetting Mechanism Model

For this mechanism to occur there should be total restraint of the part, so that buckling cannot start, and there should be no thermal gradient. These conditions occur when bending extrusions or rigid sections or when processing flat sheets with certain traverse patterns, for example, radial paths on a plate, to make a bowl. The model is similar to the two-layer trivial model in that the compressed and plastically deformed layer is one side of the extrusion that is heated and the other layer is the other side of the extrusion.

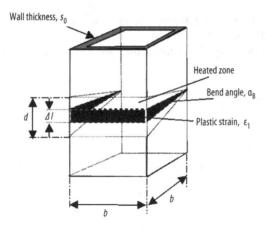

Figure 9.9 Bending of an extrusion by the upsetting mechanism

For a square-section tube of dimensions $b \times b$ (Figure 9.9), the bending angle would be

$$\tan\left(\frac{\alpha_B}{2}\right) = \frac{\varepsilon_1}{2b}.$$

For small angles this gives

$$\alpha_B = \frac{\varepsilon_1}{b}, \tag{9.37}$$

and

$$\varepsilon_1 = \frac{\alpha_{th} A P}{\rho v s_0 C_p}. \tag{9.38}$$

Substituting for ε_1 in Equation 9.37, we get

$$\alpha_B = \frac{\alpha_{th} A P}{\rho v b s_0 C_p}. \tag{9.39}$$

This represents a much smaller angle than found with the thermal gradient model owing to the factor of 1 instead of 4 and the larger value of b compared with the value of s_0.

9.4 Operating Characteristics

The experimental data confirm the general trend of the relationship between the bending angle and the operating parameters described by the equations derived in Section 9.3. The calculated values are, however, usually too high since the substantial assumptions did not allow for many secondary details such as thermal profiles, expansion of the "unheated" layer and full account of the elastic forces in the plastic regions. Nevertheless, they form a significant foundation for our understanding of the process.

9.4.1 Effect of Power

Bending can only start when the surface layer is sufficiently soft to plastically deform (approximately 600 °C for Ti–6Al–2Sn–4Zr–2Mo [15]). Thus, there is a threshold power required for a given traverse speed before the process will work. The bending angle then responds almost linearly with increase of power, as predicted. However, if the power exceeds a certain value, the bending angle starts to decrease owing to overheating of the upper layer, causing recrystallisation, and the heating of the lower layer that was assumed to be cool throughout. The expansion of the cool layer will reduce the overall bending angle by reducing the stress on the compressed layer in the temperature gradient mechanism route but represents the start of the buckling route. Melting and recrystallisation may affect the expansion coefficient. Cooling with air jets or water jets can improve the bending efficiency [16].

Figure 9.10 from the results of Vollertsen [7] shows these effects. In this case an aluminium alloy with a high thermal conductivity and low speed was bent by the buckling mechanism whereas the steel was bent via the temperature gradient mechanism.

9.4.2 Effect of Speed – "Line Energy"

The bending angle is expected to be inversely proportional to the speed for the thermal gradient mechanism and inversely proportional to the cube root of the speed during buckling. Vollertsen [7] and others have found a relationship nearer to $v^{-0.63}$. This is

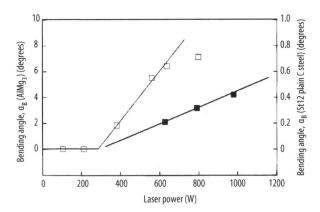

Figure 9.10 Influence of the laser power on the bending angle for AlMg$_3$ (*open squares*, 1 mm thick, speed 25 mm s^{-1}) bent by the buckling mechanism and plain carbon steel, St12 (*closed squares*, 2 mm thick, speed 83 mm s^{-1}) bent by the temperature gradient mechanism [7]

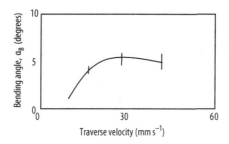

Figure 9.11 Bending angle versus velocity for a constant line energy of 33 J mm^{-1} and a single scan [17]. Laser power 250–1,300 W, beam diameter 10 mm

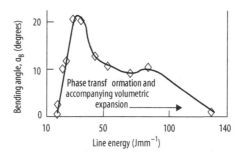

Figure 9.12 Bending angle versus line energy for α–β Ti alloy showing the effect of phase changes [21]. Laser power 1,300 W, beam diameter 10 mm, material Ti–6Al–4V, thickness 1 mm, graphite-coated, five scans

more in line with the model given in Equation 9.18, where an allowance is made for the varying thickness of the plastic region with speed. This depth is proportional to $v^{-0.5}$.

Since the speed is usually associated with the absorbed power, AP, the coupled energy per unit length, or "line energy" AP/v (J m^{-1}) has been looked at as a possible parameter for expressing the bending results. The data from Magee *et al.* [17] shown in Figure 9.11 suggest that above a certain velocity the bending angle is approximately constant for a fixed line energy. At the slower speeds the bending mechanism becomes less clear; it could well be a mix of buckling and the temperature gradient mechanism. This relationship changes as soon as other factors such as volumetric expansion due to phase changes occur (see Figure 9.12).

9.4.3 Effect of Material

The material properties are grouped as $\alpha_{\mathrm{th}}/(\rho C_p)$ in all the relationships derived in Section 9.3. A plot of this parameter against the bending angle justifies this prediction (Figure 9.13).

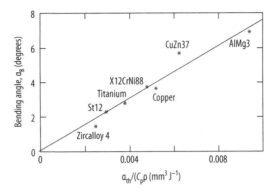

Figure 9.13 The linear relationship between the bending angle and the parameter [coefficient of thermal expansion/(specific heat × density)] [7]

9.4.4 Effect of Thickness – Thickening at the Bend

The predicted dependence of the bending angle on the thickness is given in Equation 9.2 for the temperature gradient mechanism as $\alpha_B = f(s_0^{-2})$ and from Equation 9.36 for buckling as $\alpha_B = f(s_0^{-2/3})$. These relationships have been found to hold fairly well.

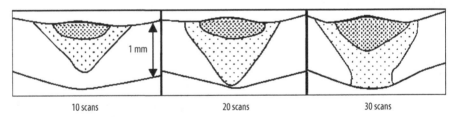

Figure 9.14 Thickening of 1-mm-thick α–β Ti. The contours show the heat-affected zone and the inner hardened regions. (Laser power 250 W, beam diameter 5 mm, traverse speed 15 mm s^{-1}, graphite-coated, with cooling between scans. (Taken from the micrograph in Magee *et al.* [17])

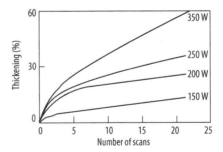

Figure 9.15 The percentage thickening of the bend line as a function of the number of passes [15]

The bending process is one of compression and plastic flow. This leads to a thickening of the bend line, which is one of the main differences between laser bending and mechanical bending (Figure 9.14). It is also one of the advantages of laser bending, particularly for pressure pipes and strength-sensitive articles. However, it does have the effect, together with work hardening, of reducing the bending angle per pass after ten or more passes (see Section 9.4.5). The percentage thickening per pass is almost linear (see Figure 9.15).

9.4.5 Effect of Plate Dimensions – Edge Effects

The resistance to counterbending varies with the distance from a free edge. Thus, the nearer a free edge is, the greater is the likelihood that a bend away from the laser will occur during heating. Such counterbending would reduce the stress on the plastic region and hence reduce the bending angle. This is a nuisance since it means the bending angle will vary across a sheet for a given power and speed. Some results from Magee *et al.* [18] are shown in Figures 9.16 and 9.17. The effect can be overcome by varying the speed or power during a traverse. For low speeds with high-conductivity aluminium alloy the effect is particularly marked (Figure 9.16), possibly associated with the onset of buckling. This reduced bending angle per pass near an edge has an effect on the expected bending angle as a function of the total sheet width, as shown in Figure 9.18 [7].

9.4.6 Effect of the Number of Passes

The bending angle would be expected to be the same for each pass and therefore the total bend would be proportional to the number of passes (see Figure 9.2). For the first

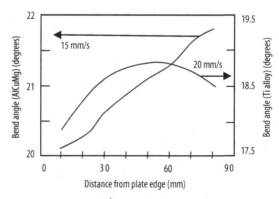

Figure 9.16 Bending angle versus distance from the plate edge for low traverse speeds [18]. Laser power 800 W, beam diameter 10 mm, dimensions $80 \times 80 \times 0.8$ mm^3, graphite-coated, ten scans

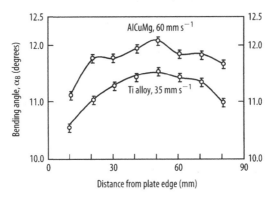

Figure 9.17 Bending angle versus distance from the plate edge for high traverse speeds [18]. Laser power 800 W, beam diameter 10 mm, dimensions $80 \times 80 \times 0.8$ mm^3, graphite-coated, ten scans

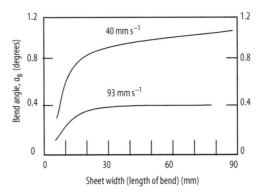

Figure 9.18 Bending angle versus sheet width for plain carbon steel with a beam diameter of 4.8 mm, a constant laser power and a single scan [7]

few passes this is so. After that the bending angle diminishes with the growing number of passes owing to:

1. the thickening of the section being bent;
2. strain hardening [19]; and
3. variation of the spot size owing to a changing angle of incidence as the bend grows [20].

For many metals, such as aluminium, the increased section modulus due to thickening is offset by the reduced hardness on heating, and hence the effect of multiple scans is diminished and an almost linear relationship is maintained. The results for alloys that harden, such as Ti–6Al–4V, are different. Some results from Magee *et al.* [21] are shown in Figure 9.19. For the lower line energy the total bending angle saturates since the energy delivered is insufficient to overcome the increased resistance required for bending. The higher line energy appears to be able to overcome this.

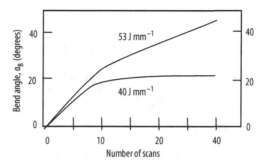

Figure 9.19 Variation of bend angle with number of passes for Ti–6Al–4V at different line energies, showing the effect of phase changes [21]. Beam diameter 10 mm, thickness 1 mm, graphite-coated

9.5 Applications

What has just been described is a remarkable bending process with the following characteristics:

1. noncontact bending;
2. bend seam easily directed by software changes;
3. bend direction can be towards or away from the laser beam or there can be no bend but only a shortening in a direction transverse to the beam path;
4. the bend itself is thicker and stronger than the parent sheet;
5. there is no spring-back;
6. the angle of the bend can be made with precision and repeatability; it can be sharp or curved depending on the spot size;
7. in-process sensing can be added to control the final angle;
8. the angle may vary very slightly along a single pass unless some variation in speed or power is included in the programme; and
9. the process is slow when making large bending angles; thus, a hybrid process based on roughing out the shape mechanically and finishing with the laser has been suggested as a future technique for large structures such as an aircraft wing manufacture [11, 22].

The current applications are few since the process is relatively new. However, there is a great deal of research taking place from which numerous potential applications arise. They include the following.

1. *Bending of pipes and extrusions* [5, 12, 23]. By control of the energy, the tube can be kept round and the outside wall radius thickness not thinned significantly. Compound bends in pressure pipes and flowing systems can be made this way. It may be used for making preforms for hydroforming. For large structures and when using highly defocused beams, there is some competition between the laser and high-powered arc lamps [24]. Apart from bending pipes, the pipes can also be shaped by reducing the diameter or creating wrinkles in the wall (Figure 9.20) [12].

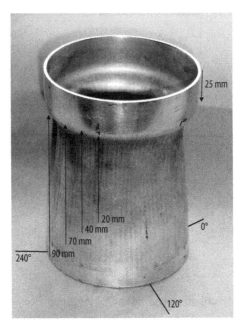

Figure 9.20 An upset pipe by Silve [12]. The 89 mm-diameter tube is of aluminium with a wall thickness of 1.8 mm. It was treated with a 600 W CO_2 laser beam. The beam diameter was 5 mm and the traverse speed was 30 mm s^{-1}. The scan pattern is shown as *arrows*. The tracks are all (except the short scan from the *top*) in the same direction for each pass starting at the *bottom* at different levels and moving to the *upper ridge*. The starting points are 120° apart. Three irradiations were made at each length (90, 70, 40, 20 mm). There was a 90-s cooling time after each circuit of the tube before shifting a 4° offset for the next circuit. The whole sequence was repeated three times. This meant nine scans at each zone

2. *A forming tool for astronauts* has been suggested by Namba [4]. It is obviously inconvenient to have heavy bending gear in space. The laser could be used to bend, cut, weld, clad, mark and clean, so it would be a useful tool to have in space – up there the easy availability of electricity and low pressures suggests a special space laser would not be too difficult to build.

3. *Adjustment of sealed electric contacts* [23,25]. The example here is the adjustment of relay switches mounted inside glass tubes. The reed switch is manufactured closed and then struck by a laser repeatedly until it just opens. The adjustment of electron guns can also be done this way.

4. *Straightening of rods* has been tried. In this process the laser beam is arranged as a line source and aligned along the edge of a thin rod that has to be straightened. The rod is then spun along its axis and the beam oscillates up and down the edge of the rod, just missing it if it is straight. Where it is bent the rod will be heated on the inside of the bend and eventually will straighten the rod by the buckling route.

5. *Curvature forming* and heat treatment of wiper blades [24,26]. A strip of spring steel coming from a production line at a speed of 1.3 m min^{-1} was curved using a 4 kW diode laser. The strip varied in thickness along its length but the curvature was

controlled by varying the laser power with the thickness measured by a laser. Then two opposing infrared lamps fully austinise the material to produce untempered martensite at Rockwell hardness (Rc) 58/60. To get the speed up to production rates of $10\,\mathrm{m\,min^{-1}}$ would require some 30–60 kW of power. This may be found by using arc lamps.

6. *Low-volume manufacture of prototype "pressed" sheet components.* The laser process does not require any hard tooling and hence to be able to make prototype bent sheet structures only requires the software to be designed. This application includes the concept for aircraft parts and prototype car parts. It would never be useful in the full-scale manufacture of cars or mass production owing to the slow speed of the process. However, for parts where only a few hundred per year are required, such as aircraft wings, laser bending is a serious option. Work on three-dimensional laser forming for correction of distortion and making designed shapes in aluminium structures has been carried out at the University of Liverpool, primarily aimed at applications in the aerospace industry [27].
A geometry-based control strategy for three-dimensional laser forming has been developed which involves iterative forming by multiple irradiation passes. The control strategy is based on using the error between the desired shape and the measured shape at each step to calculate an irradiation strategy based on following lines of constant angle or gradient of curvature [28].

7. *Artistic artefacts.* The laser offers a new bending tool capable of making strange shapes or making some shapes easily. For example, with only two tracks Silve made a sushi dish, shown in Figure 9.21. The dish was cut by laser to the required overall size, marked by laser and then bent by laser. The ornamental legs were added later. Silve [12, 23] developed an embryonic shape vocabulary in her thesis and subsequent publications.

8. *Micromanipulation of electronic components* [14]. This is one of the actual production applications by Philips in the Netherlands. The mount for a disc drive in a CD player is on a specially shaped sheet. Striking this sheet with a pulsed laser beam in one of three locations and repeating this as many times as is required allows adjustment to be made by bending up or down in the x, y or z axes. The laser route

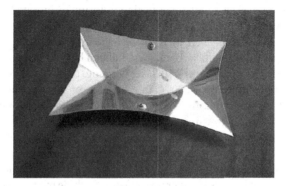

Figure 9.21 A laser cut, marked and bent sushi dish by Silve in aluminium and bronze

saves some 8 s per setting. This accurate final adjustment means that the previous manufacturing steps can be less rigid, thereby saving considerable time.

9. *Straightening of car body shells without spring-back* [29]. When fitting the car side to the roof there is often a gap, which has to be closed if welding is to take place. To close the gap means expensive clamping or mechanical bending, both of which risk damaging the sheet metal. The gap can be closed by laser bending without any spring-back.

10. *Straightening of welded sheets*. The angular distortion after welding can be reduced by some 50% using laser bending. However, it is not so easy to remove longitudinal curvatures [30].

11. *Bending difficult materials*. Since laser bending is essentially a compressive process even some brittle materials can be bent this way, such as cast iron [31]. Zirconia, glass and alumina cannot be formed this way owing to lack of plasticity [5].

12. *Bending thick-section plate for the shipbuilding industry*. The days of men sitting on buckets with a flame torch are numbered. Thick plates can be bent with powerful-enough lasers. The process could be fully automated and monitored. It would be very surprising if the shipbuilding industry does not take to this process with joy [6].

13. *Micro- and nanoadjustments*. The application for nonthermal bending via the laser-induced shock wave route includes micro- and nanoadjustments of microelectromechanical systems such as accelerometers that are built into automobile seat belts, miniature medical diagnostic labs-on-chips and 1-in. hard drives. It is also one of the very few techniques that can bend silicon without cracking it or heating it.

14. *Recent advances*.

    a. In 1997, Tam *et al.* [8] at IBM Almaden developed a laser curvature adjustment technique for controlling the curvature of magnetic head sliders in disk drives using a laser microbending technique. The aim was precise adjustment of the positive camber curvature of a slider to improve its tribological properties and allow reduced flying heights (below 25 nm) above the disk surface for increased disk storage density. The technique was not a thermal bending process, but involved scribing microscopic patterns on the reverse side of the slider, which, by inducing surface stress changes in the ceramic, produced a corresponding curvature change at the front-side air-bearing-surface. A Nd:vanadate laser was used and integrated with an optical monitoring technique for closed-loop control of slider curvature to within a few nanometres accuracy.

    b. Laser bending has been used for reshaping of wet-etched silicon microscale structures (50-μm-thick, 1-mm-wide silicon beams) using a CW Nd:YAG laser scanned across the beam. With laser bending, there is no need to heat the whole structure to more than 700 °C in a furnace, or for associated special forming tools. During the plastic deformation stage, dislocations are generated in the near-dislocation-free monocrystalline silicon that then affect thermomechanical and physical properties of the material and thus allow bending or reshaping to take place. The dislocations only occur on reaching the yield point of the material, unlike in the regions of elastic deformation. Applications of this laser microbending process under investigation include micromirrors for optical circuits and microscale grippers or "staples" for semiconductor chips [32, 33].

c. Research on laser forming of composites (fibre metal laminates) has been carried out at the University of Liverpool. Laser forming of multilayer aluminium/glass fibre laminate and aramid-reinforced aluminium laminates was achieved by applying the temperature gradient mechanism to a single aluminium outer layer, using an irradiation strategy designed to avoid thermal damage to the interlayers, which would otherwise result in structural degradation. The degree of localised forming along each irradiation line is limited to avoid delamination of the layers and preserve structural integrity [34, 35].

9.6 Conclusions

Laser bending is one of the latest processes to join the galaxy of growing applications for lasers. A noncontact, no spring-back, bending process must have a good future. It also offers the possibility of the "virtual" press tool and, since the laser can do many other processes besides bending, it could be the "super virtual manufacturing tool" – who knows?

Glossary

A	Absorptivity or area of cross-section	no units or m^2
b	Width of sheet	m
B	Constant	
C_p	Specific heat of workpiece	$J\,kg\,°C^{-1}$
D, d_1	Beam diameter	m
E	Elastic modulus	$N\,m^{-2}$
F	Bending force	N
f'	Fraction of the thermal expansion that causes an elongation in region 2	
I	Moment of inertia, section modulus	m^4
k	Thermal conductivity	$W\,m^{-1}\,K^{-1}$
$k_f(T_1)$	Flow stress in the heated region	$N\,m^{-2}$
$k_f(T_{ob})$	Flow stress in the surface region	$N\,m^{-2}$
l	Heated length along the bend seam	m
l_h	Transverse length	m
M	Bending moment	N m
P	Incident laser power	W
Q	Energy input	W
r	Radius of curvature	m
s_0	Sheet thickness	m
Δt	Heating time	s
ΔT	Temperature rise of the heated zone	°C
$\Delta T'$	Temperature difference between the two layers	°C
ΔT_{av}	Average temperature rise of the total heated zone	°C

v	Traverse speed	$\mathrm{m\,s^{-1}}$
Δx	Longitudinal length	m
z	Depth	m
α	Thermal diffusivity	$\mathrm{m^2\,s^{-1}}$
α_B	Bending angle	rad
α_{th}	Thermal expansion coefficient	$\mathrm{^\circ C^{-1}}$
ε_1	Strain of the heated layer	
ε_2	Strain of the unheated layer	
ε_{ob}	Surface strain	
ρ	Density of the workpiece being bent	$\mathrm{kg\,m^{-3}}$

Subscripts

1	Heated zone
2	Unheated zone
ob	Surface condition

Questions

1. List the ways a laser can be used to bend.
2. Derive an equation relating the main parameters involved in laser bending using the "trivial" argument.
3. How can a bend be arranged to be directed away from the laser during laser bending?
4. What is meant by shock bending?
5. What are the main advantages of laser bending over mechanical bending?

References

[1] Holt RE (1960) Flame straightening basics. Weld Eng Jun 49–53
[2] Peck DE, Jones G (2002) Line induced thermal strain forming. In: ICALEO'02 proceedings, Phoenix, October 2002. LIA, Orlando, paper 706
[3] Kitamura K (1983) Joint project on "Materials processing by high powered laser". Technical report JWESTP-8302. Japan Welding Engineering Society, pp 359–371
[4] Namba Y (1986) Laser forming in space. In: Wang CP (ed) Proceedings of the international conference on Lasers'85, Osaka, Japan. STS, McLean, pp 403–407
[5] Namba Y (1987) Laser forming of metals and alloys. In: Proceedings of LAMP'87, Osaka, Japan, 1987, pp 601–606
[6] Scully K (1987) Laser line heating. J Ship Prod 3:237–246
[7] Vollertsen F (1998) Forming, sintering and rapid prototyping. In: Shoucker D (ed) Handbook of the Eurolaser Academy. Chapman and Hall, London, chap 6
[8] Tam AC, Poon CC, Crawforth L (2001) Laser bending of ceramics and application to manufacturing magnetic head sliders in disk drives. Anal Sci 17:419–421
[9] Bechtold P, Roth S, Schmidt M (2008) Precise positional adjustments with laser induced shock waves. Photonics Spectra Jun 58–63

[10] Edwards KR, Carey C, Edwardson SP, Dearden G, Williams CJ, Watkins KG (2007) Laser peen forming for 2D shaping and adjustment of metallic components. In: 5th international conference on laser assisted net-shape engineering (LANE 2007), Erlangen, Germany, 2007, pp 569–580

[11] Magee J (1988) Laser forming of aerospace alloys. PhD thesis, Liverpool University

[12] Silve S (2000) Laser forming and creative metal work. PhD thesis, Buckinghamshire Chilterns University and Brunel University

[13] Edwardson SP, Watkins KG, Dearden G, French P, Magee J (2002) Strain gauge analysis of laser forming. In: ICALEO'02 proceedings, Phoenix, October 2002, LIA, Orlando, paper 703

[14] Hoving W (1997) Laser applications in micro-technology. In: Geiger M, Vollertsen F (eds) Proceedings of LANE'97, vol 2. Meisenbach Bamberg, pp 69–80

[15] Marya M, Edwards GR (2000) Factors affecting the laser bending of Ti–6Al–2Sn–4Zr–2Mo. J Laser Appl 12(4):149–159

[16] Li L, Chen Y, Lin S (2002) Characterisation of laser bending under different cooling conditions. In: ICALEO'02 proceedings, Phoenix, October 2002, LIA, Orlando, paper 704

[17] Magee J, Watkins KG, Steen WM, Calder NJ, Sidhu J, Kirby J (1998) Laser bending of high strength alloys. J Laser Appl 10(4):149–155

[18] Magee J, Watkins KG, Steen WM, Calder N, Sidhu J, Kirby J (1997) Edge effects in laser forming. In: Geiger M, Vollertsen F (eds) Proceedings of LANE'97, vol 2. Meisenbach Bamberg, pp 399–408

[19] Sprenger A, Vollertsen F, Steen WM, Watkins KG (1995) Influence of strain hardening on laser bending. Manuf Syst 24:215–221

[20] Edwardson SP, Abed E, Bartkowiak K, Dearden G, Watkins KG (2006) Geometrical influences on multipass laser forming. J Phys D Appl Phys 39(2):382–389

[21] Magee J, Sidhu J, Cooke RL (2000) A prototype laser forming system. J Opt Laser Eng 34:339–353

[22] Miyazaki T, Saito M, Yoshioka S, Tokunaga T, Misu T, Oba R (2001) Forming of thin plate with diode laser. In: ICALEO 2001 proceedings, Jacksonville. LIA, Orlando, paper G1603

[23] Silve S, Podschies B, Steen WM (1999) Laser forming – a new vocabulary for objects. In: ICALEO'99 proceedings, section F. LIA, Orlando, pp 87–96

[24] Peck DE, Jones G (2002) Line induced thermal strain forming. In: ICALEO'02 proceedings, Phoenix, October 2002. LIA, Orlando, paper 706

[25] Verhoeven ECH, de Bie WFP, Hoving W (2001) Laser adjustment of reed switches for micron accuracy in mass production. Ind Laser User 25:28–30

[26] Miyazaki T, Misu T, Yoshioka S (2002) Forming characteristics of thin metal plate with diode laser beam. In: ICALEO'02 proceedings, Phoenix, October 2002. LIA, Orlando, paper 702

[27] Dearden G, Edwardson SP, Abed E, Watkins KG (2006) Laser forming for the correction of distortion and design shape in aluminium structures. In: AILU Industrial Technology Programme, Photon'06 international conference on optics and photonics, Manchester, September 2006, Institute of Physics, London

[28] Abed E, Edwardson SP, Dearden G, Watkins KG, McBride R, Hand DP, Jones JDC, Moore AJ (2005) Closed loop 3-dimensional laser forming of developable surfaces. In: Vollertsen F (ed) International workshop on thermal forming (IWOTE'05). BIAS, Bremen, pp 239–255

[29] Geiger M, Vollertsen F, Deinzer G (1993) Flexible straightening of car body shells by laser forming. SAE Spec Publ 944:37–44

[30] Deinzer G, Vollertsen F (1994) Laserstrahlschweißen und -richten. Laser Optoelektron 26(3):S48–53

[31] Arnet H, Vollertsen F (1995) Extending laser bending for the generation of convex shapes. Proc Inst Mech Eng B J Eng Manuf 209:433–442

[32] Gärtner E, Frühauf J, Löschner U, Exner H (2001) Laser bending of etched silicon microstructures. Microsyst Technol 7:23–26

[33] Dearden G, Edwardson SP (2003) Some recent developments in two and three-dimensional laser forming for 'macro' and 'micro' applications. J Opt A Pure Appl Opt 5(4):S8–S15

[34] Edwardson SP, Dearden G, Watkins KG, Cantwell WJ (2004) Forming a new material. Ind Laser Solutions 19(3):16–20

[35] Edwardson SP, French PW, Dearden G, Watkins KG, Cantwell WJ (2005) Laser forming of fibre metal laminates. Lasers Eng 15:233–255

"But officer, I only shone my torch at it."

10 Laser Cleaning

I invented the laser eraser, which worked – and I even got a patent on it

Arthur L. Schawlow, Nobel laureate

What, after all, is a halo? It's only one more thing to keep clean

Christopher Fry, The Lady's Not for Burning

by K.G. Watkins, Liverpool University

10.1 Introduction

Laser cleaning is growing in importance, particularly in applications such as the removal of small debris particles from semiconductors and in art conservation. With the introduction of the Montreal protocol, which proposes long-term reduction on environmental and public health grounds in the use of organic solvents such as CFCs that are often used in industrial cleaning, it is to be expected that more generally available industrial embodiments of laser cleaning will emerge in the next few years.

Although the use of light in exerting a mechanical effect on surfaces was known long before the development of the laser and the precise origin of many laser applications is often unclear, it is possible to clearly attribute laser cleaning to three well-defined originators.

Arthur Schawlow (subsequently awarded the Nobel prize in physics in 1981 for his work on the development of laser spectroscopy) had collaborated at Bell Laboratories with Charles Townes in 1957–1958 on the development of the optical maser (laser) and by the mid 1960s had become a leading spokesperson for the future of lasers. Since possible military uses were being kept secret and were deemed to be unrealisable by many, the view was forming that the laser was an "Solution looking for a problem" and Schawlow was seeking to overcome this prejudice. As he told Suzanne Reiss in an interview series for University of California [1]:

> I did a lot of stuff to show that lasers were really not death rays. That's one reason I invented the laser eraser, which worked – and I even got a patent on it… People were talking about these death rays that you couldn't build and here was something you could build. If it had ever gone into mass production, it could have been practical to have one built into a typewriter. If you made a mistake, you bring it back to where it was typed, press the zap key, and off it would go… I didn't intend to patent it or try to make anything of it, but I just wanted an example of something you could do. I thought people might take up the

idea, but they didn't. First of all, IBM brought in the sticky tape for erasing and then word processors really took over.

Schawlow's patented prototype device with which the user would be able to invisibly correct typed and other printed documents with the aid of a small ruby laser is the first documented use of laser cleaning and a clear foreshadowing of the current-day use of lasers in the conservation of historic books and documents.

The second event in this history – which led to the use of lasers in art conservation – is equally well documented. In 1972 the Italian Petroleum Institute invited John F. Asmus from University of California, San Diego, to visit Venice to study laser holography for the recording of the city's decaying treasures, following serious flooding. During this work, Asmus (together with Italian collaborators including conservator Giancarlo Calcagno) was asked to observe the effect of the interaction of a focused ruby laser (which had until then been intended for holographic recording) with an encrusted stone statue. It was found that the darker encrustations were selectively removed from the surface, resulting in no apparent damage to the underlying, white stone. Asmus returned to the USA and began researching laser cleaning of art works, laying the basis for a powerful series of techniques, particularly in the use of pulsed ruby and pulsed Nd:YAG lasers [2–7]. Interestingly, the connection between the use of lasers in cleaning and in holographic recording of art works remains, with the two technologies developing side by side in the conservation world.

Finally, in the 1980s, back at Bell Laboratories (and working without knowledge of Asmus's work), a group working with Susan Allen of Florida State University developed techniques for the removal of small submicrometre debris particles from silicon wafers and other microelectronic devices [8,9]. This included a patented technique for the application of a thin layer of a water–alcohol mixture that significantly enhanced the laser cleaning efficiency as a result of the rapid evaporation of the liquid ("steam cleaning").

With some notable exceptions, the many developments of laser cleaning that are now currently available are traceable to the insight of these originators. The availability of new laser types and new operating ranges (*e.g.*, new wavelengths and shorter pulse lasers) means that this is an ongoing process with much that is still to be discovered.

Despite the rising volume of activity, particularly in the area of conservation, there has been insufficient consideration of the mechanisms involved in laser cleaning and how their understanding could lead to improved control and efficiency. Consideration of the mechanisms involved shows that laser cleaning is not in fact a single process but a range of processes that need to be carefully selected and optimised for each proposed application area [10, 11].

10.2 Mechanisms of Laser Cleaning

There are a range of mechanisms whereby radiation can remove material from a surface. They basically fall into three major groups; however, each process discussed below may use more than one mechanism:

1. evaporation processes – *selective vaporisation, ablation*;
2. impact processes – *spallation, evaporative pressure, photon pressure, dry and steam cleaning, laser shock cleaning*; and
3. vibration processes (based on acoustic or thermoelastic forces) – *transient thermal heating, angular laser cleaning*.

The mechanisms of the processes listed above are now discussed in turn.

10.2.1 Selective Vaporisation

From his work in Venice on the laser cleaning of stone and marble using a ruby laser [2–7], Asmus concluded that there are two principal cleaning mechanisms. In normal pulse mode (pulse duration approximately 1 μs to 1 ms), at relatively low laser intensity (10^3–10^5 W cm^{-2}) cleaning occurred as a result of the selective vaporisation of the surface contaminants compared with the underlying material, which remained almost wholly unaffected. This in turn occurred when the absorption coefficient of the darker encrustation was sufficiently large to lead to a temperature rise favouring vaporisation, whereas the absorption coefficient of the underlying material was sufficiently small to limit temperature rises to moderate values that did not allow the occurrence of cracking (as a result of differential thermal expansion), melting or vaporisation – conditions that are frequently obtained with dark encrustations on marble or stone. It was shown that a one-dimensional heating model could account for the difference in the surface temperature increase between two layers with different absorptivity to the incident laser energy.

The simplified one-dimensional model, shown diagrammatically in Figure 10.1, is sufficient to demonstrate the principle of selective removal of pollutant layers as a result of laser irradiation [7].

If a constant flux, F_0, is absorbed at the surface ($z = 0$) and there is no phase change in the material, the solution of the equation for heat flow in one dimension is (see Section 5.3)

$$T_{(0,t)} = \frac{2F_0}{\kappa}\left(\frac{\alpha t}{\pi}\right)^{1/2}, \tag{10.1}$$

where $T_{(0,t)}$ is the temperature at the surface after time t, κ is the thermal conductivity and α is the thermal diffusivity.

The following additional assumptions are implied:

1. The laser beam is uniform with no transverse variation in intensity.
2. The encrustation is uniform and planar.
3. The beam diameter is much larger than the encrustation thickness;.
4. The beam diameter is much larger than the thermal diffusion distance.
5. The beam energy is absorbed at the surface.

The thermal diffusivity is given by

$$\alpha = \kappa/\rho C_p, \tag{10.2}$$

where ρ is the density and C_p is the specific heat.

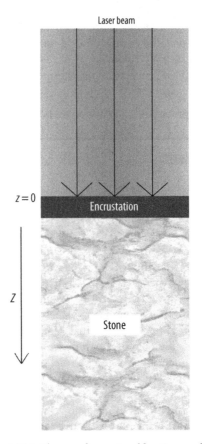

Laser beam

$z = 0$

Encrustation

z

Stone

Figure 10.1 The one-dimensional heating model

The distance, z, that a thermal wave will travel into the material is given approximately by

$$z = (\alpha t)^{1/2}. \tag{10.3}$$

For typical minerals $\alpha = 10^{-2}$ cm^2 s^{-1} and if t (the pulse length) is 10^{-3} s, Equation 10.3 shows that assumption 4 is valid and also that the thermal effect is strongly localised in the surface, since for this case

$$z = (10^{-2} \times 10^{-3})^{1/2} = 3 \times 10^{-3} \text{ (cm)}.$$

Selective vaporisation of surface layers depends on differences in the absorptivity of the substrate and the layer requiring removal, since

$$F_0 = (1 - R)I_0 = \beta I_0, \tag{10.4}$$

where R is the reflectance, β is the absorptivity and I_0 is the incident flux.

Then, from Equation 10.1,

$$T_{(0,t)} = \frac{2\beta I_0}{\kappa}\left[\frac{\alpha t}{\pi}\right]^{1/2}, \qquad (10.5)$$

and, if all other variables are held constant, different values of β for the substrate and the pollutant layer can result in significant differences in the surface temperature reached by the two layers.

For example, consider the removal of blackened layers from a white stone statue using a non-Q-switched (free-running) Nd:YAG laser. If values of the pulse length (t) (e.g., 10^{-3} s), κ for stone (e.g., $2\times10^{-2}\,\mathrm{J\,K^{-1}\,cm^{-1}\,s^{-1}}$) and α for stone (e.g., $10^{-2}\,\mathrm{cm^2\,s^{-1}}$) are assumed, Equation 10.5 has the simplified form

$$T(0) = C\beta I_0, \qquad (10.6)$$

where C is a constant. Plotting the combinations of β and I_0 required to produce a given temperature rise at the surface results in contours of constant temperature, as shown schematically in Figure 10.2.

One important limiting condition that is required for successful cleaning in this case is that the surface encrustation should be removed by vaporisation, whereas the melting temperature of the underlying substrate should not be exceeded. There may be a lower threshold of damage brought about by the specific nature of the laser–material interaction (e.g., stress-induced surface cracking), but clearly substrate melting or calcination is in all cases undesirable since morphological changes in the substrate will be implied in this case. Assuming that material constants such as κ and α are the same for both the encrustation and the substrate, Figure 10.2a shows that incident flux $I_0(1)$ is sufficient to raise the encrustation (with absorptivity 0.8) to its vaporisation temperature (T_1), whereas the same flux when incident on the cleaned substrate (with absorptivity 0.2) is insufficient to raise the temperature of the substrate to its melting temperature (T_2), which would require a flux $I_0(2)$. Laser cleaning under these conditions would be self-limiting in the sense that if the applied flux $I_0(1)$ were applied to the surface in a series of pulses, initial pulses would result in the vaporisation of the encrustation and, on removal of that layer, the temperature rise of the substrate would result in a temperature below the temperature required to melt the substrate [as shown schematically in Figure 10.3, $I_0(2) - I_0(1)$ would represent a safety margin for the successful cleaning of the object].

For a pulsed laser the result would be an initial series of pulses (Figure 10.3) in the presence of the absorbing coating layer in which the temperature of the layer was raised above the vaporisation temperature, followed, once the layer has been removed, by pulses resulting in a much smaller temperature rise as the laser interacts with the much less absorbing substrate – the so-called self-limiting effect. Indeed, this effect is often reported by practising conservators using lasers, the self-limiting effect being experienced as a reduction in the audible sound of the cleaning process.

Some combinations of encrustation and substrate are clearly susceptible to substrate damage by melting even if the incident flux $I_0(3)$ is chosen as the minimum required to raise the encrustation to its vaporisation temperature (T_3), as shown schematically

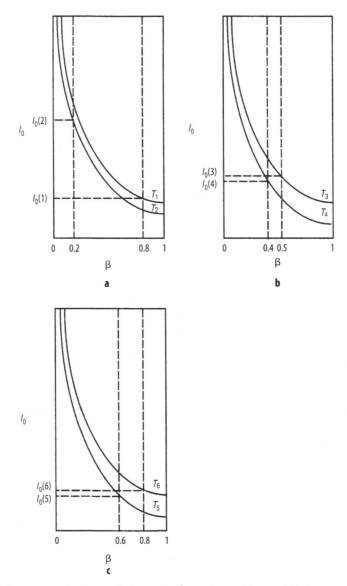

Figure 10.2 Temperature isotherms for intensity versus absorptivity combinations: **a** self-limiting, **b** condition of substrate melting, and **c** self-limiting condition for low vaporisation temperatures (see the text for details) [6]

in Figure 10.2b. Here the substrate melting temperature (T_4) would be reached by incident flux $I_0(4)$, which is exceeded by $I_0(3)$, since the absorptivity of the substrate (0.4) is large relative to that of the encrustation (0.5). Figure 10.2c shows the case where, although the absorptivities of the encrustation (0.6) and the substrate (0.8) represent an unfavourable combination, the encrustation can be removed by selective vaporisa-

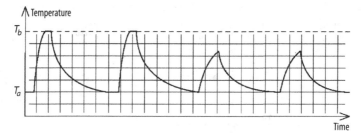

Figure 10.3 Pulse intensity versus time showing the basis for the self-limiting effect [6]

tion at incident flux $I_0(5)$ provided its vaporisation temperature (T_5) is sufficiently low compared with the melting temperature (T_6) of the substrate.

This schematic treatment, based on the stated simplifying assumptions, is sufficient to demonstrate the principle of selective vaporisation as a mechanism of laser cleaning and to outline the possibility of a self-limiting effect that can mitigate against substrate damage. Development of a more detailed understanding of this mechanism (which will depend on the determination of the role of the numerous elements and compounds in real encrusted layers) will allow more precise models to be developed.

10.2.2 Spallation

In practice, the selective vaporisation mechanism has been little applied in laser cleaning of art works, largely because of the relatively high temperature that is reached by the substrate despite the selective absorption effect and the relative slowness of the process. Shorter pulses, such as those delivered by Q-switched Nd:YAG lasers, induce very much less substrate heating and offer faster rates of contaminant-layer removal.

In Q-switched mode (pulse duration approximately 5–20 ns), a spallation (termed by Asmus "an ablation") mechanism was suggested to be responsible for the cleaning effect [6, 7]. At this high flux level (10^7–10^{10} W cm^{-2}) even relatively reflective surfaces absorb sufficient energy to reach the vaporisation temperature. High temperatures (typically 10^4–10^5 K) are produced in the vaporised material from the surface or the ambient gas, and at these temperatures this vapour becomes partially ionised and absorbs the laser energy strongly. The initial surface vaporisation stops as the target is shielded from the laser by the partially ionised ("plasma") vapour. As the pulse continues, the vapour is further heated and high pressures (1–100 kbar) can be produced, resulting in a shock wave which produces microscopic compression of the surface of the target material. When the laser pulse ends, the plasma expands away from the surface, the material surface relaxes and a thin surface layer (1–100 μm) is removed, resulting in spallation. Although cleaning is more rapid in this case, there is a greater propensity for damage to the material underlying the surface encrustation that requires removal. A schematic diagram showing interaction of giant pulse laser radiation with a solid surface is shown in Figure 10.4.

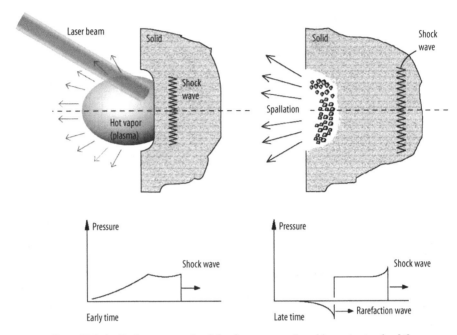

Figure 10.4 Spallation as a result of shock waves produced by a giant pulse [6]

For pulsed laser radiation, two relatively distinct regimes have been observed [12–18]: a laser surface combustion regime, which tends to be favoured at low intensity, and a laser surface detonation regime, which is favoured at higher intensity.

In addition to these two mechanisms proposed by Asmus, there are other laser–material interaction phenomena that can result in surface cleaning.

10.2.3 Transient Surface Heating

There is evidence that shock waves may be formed simply as a result of the very rapid heating and cooling of the irradiated surface as a result of interaction with a short laser pulse. The magnitude of the effect is sufficient that initially generated thermoelastic stresses can generate acoustic waves.

A number of workers have observed the effect of short-pulse lasers in generating thermoelastic stresses in the surface of solids (see also Section 12.2.3) [19–24]. Ready [25] clearly distinguishes such a transient surface heating mechanism from others where material is removed from the surface by evaporation. When laser radiation is deposited rapidly and absorbed by the surface, instead of the normal heating and contraction that would take place as a result of conventional thermal expansion at slower rates, the heated layer will exert a pressure on adjacent material and a compressive shock wave will travel through the material. For example, when using a Q-switched ruby laser, acoustic transducers measured elastic stress waves induced in glass targets with a travelling compressive stress wave moving into the material from the irradiated surface which was reflected as a tensile wave at a free surface in the material. If the magnitude of these induced stresses becomes sufficiently large so that the shear stress of the material is exceeded, removal of material may take place by physical fracture. However, of more interest from the laser cleaning point of view is the proposition now made here that the consequence of the rapid expansion and contraction of the treated surface could be the removal of a superficial deposit while at the same time the overall damage caused to the underlying substrate is negligible. This would mean that laser cleaning (*e.g.*, of polluted layers from the surface of stone) could take place with the presence of an audible acoustic wave but in the absence of plasma formation, or in a regime where plasma formation was not the main factor in determining the mechanism of cleaning. This could result in a mode of cleaning that is inherently less damaging to the substrate than the spallation mechanism. The evidence of this is given in the magnitude of the surface pressure that can be achieved as a result of the rapid thermal expansion and contraction effect.

White [24] modelling transient surface heating effects produced by repetitive laser pulses considered that temperature gradients both normal to and parallel to the surface are produced, and these result in thermal expansion which leads to strains in the body and stress waves which propagate away from the heated surface. The amplitude of the waves produced for a given absorbed power density depends on the elastic constraints applied to the heated surface. If this surface is unconstrained, the stress amplitude may be relatively small, but if the surface is constrained, for example, by contact with another body, the stresses induced and the wave amplitude may be very large. With use of a one-dimensional surface heating model, it was deduced that for Type 304 stainless steel with an absorbed power density of 10^4 W cm^{-2} and a frequency of 10^5 Hz, the peak temperature rise is 15.5°C and that a surface stress of 3.8×10^7 Pa (or 380 bar) would be produced for a constrained surface. This is 1.2×10^8 times the photon pressure (see later). For an unconstrained surface, the stress would be only 17 times the photon pressure. The influence of constraining the surface has been utilised in laser-based engineering processes such as laser shock hardening (see Section 6.19) and laser percussion cladding [26–29]. Here, it is found that very large mechanical stresses can be applied, sufficient to mechanically deform the surface of metallic components but only if the surface is constrained by a film of liquid such as water or a plate of glass. In both cases, the laser used is transmissive of the surface film. It has been observed that laser cleaning in the presence of a water film is intensified compared with cleaning in the absence of such a film [9].

The following question is raised: Is there a constraining effect during laser cleaning of the organic-type films encountered on stone? If these layers were sufficiently transmissive to the laser in use at its given wavelength, the mechanism of the generation of stress waves by transient surface heating may be a means of applying sufficient force to the surface to remove the coating layer. Acoustic waves would be produced as a result of the propagation of the elastic wave in the material. The mechanism would have a self-limiting effect since, once the coating had been removed, not only would the absorption coefficient of the substrate be less favourable, but the constraint would have been removed and the stresses produced in the surface would be very much lower, as modelled by White. These features, paralleling those posited by Asmus as a result of selective vaporisation, would be active in a regime where surface vaporisation was not an essential feature of the enabling mechanism.

Fernelius [30] attributes the discovery of a photoacoustical effect to Alexander Graham Bell in 1880 [31,32], who observed that a periodically interrupted light beam impinging on a solid generates a sound wave in the gas above the solid (pointing to a further, largely unrecognised, connection between Bell Laboratories and the history of laser cleaning). Under these conditions, the surface undergoes optical absorption and is heated by nonradiative transitions. The periodic heating generates thermal waves and stress waves in the material, which can be directly detected by attaching a transducer to the sample. Alternatively, these effects can be detected by having the sample in an enclosed cell and measuring the sound wave produced in the gas with a microphone. The sound wave is considered to be generated solely by the heat flow from the sample to the gas. Limiting regimes have been posited where the amplitude of the photoacoustic signal, S, fits an equation of the form

$$S = A\varphi^{-n},$$

where A is a constant, φ is the chopping frequency (pulse frequency) and n is an integer. For conditions where absorption is predominantly in the surface, $n = 1$. This could be created by a high absorption coefficient which does not allow light to penetrate into the sample or by a high repetition rate where the heat generated cannot be dissipated into the sample before the next pulse arrives. For the inverse conditions where there is penetration into the sample, $n = 3/2$.

Pulsed Q-switched ruby laser absorption in liquids was considered by Gournay [33]. The input fluence was 50–100 MW cm^{-2} with a pulse length of 10–50 ns with a surface temperature rise of the order of only 10 °C. Rates of change of temperature of 2×10^9 °C s^{-1} could be achieved. A 100 MW, 30 ns pulse on water produced pressures of 40 atm.

10.2.4 Evaporation Pressure

A further mechanism which can produce high-pressure pulses at an irradiated surface is the evaporation pressure. Aden *et al.* [34] and Knight [35] discuss the generation of a laser-induced shock wave as a result of the expansion of vapour away from a metal surface against the ambient gas (Figure 10.5). The concept is that there will be a region of compressed air between the advancing metal vapour and the ambient (uncompressed) air. The shock front would then be generated at the compressed air–ambient air interface. This mechanism would not require the absorption of laser energy in a plasma and would arise simply as a result of the high momentum of the evaporating material. High pressures are predicted in a model of this mechanism. Chan and Mazumder [36] state explicitly that for short, intense laser pulses, shock waves may be produced by the recoil pressure of rapid vaporisation of the material or by the interaction of the laser and the plasma, making it clear that these are two distinct processes. Ready [25] distinguishes recoil pressure pulses produced by evaporation and transient thermal heating effects as different mechanisms.

Phipps *et al.* [37] present a detailed account of experimental work on the pressure produced at metallic and nonmetallic targets under vacuum conditions where the in-

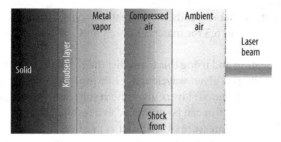

Figure 10.5 The rapid evaporation mechanism

tervention of an ambient gas (and hence the introduction of plasma blocking effects) is removed. This work could then be taken as being to some extent representative of conditions present in laser cleaning in a regime where plasma blocking effects are not present. A mechanical coupling coefficient, C_m, is defined as

$$C_m = P_a/I = M/W_L ,$$

where P_a is the ablation pressure, W_L is the laser energy, I is the incident laser intensity and M is the momentum imparted to the target.

An empirically determined relationship between C_m, the laser intensity, I, the laser wavelength, λ, and the laser pulse length, τ, is presented which is found to agree with experimentally determined values:

$$C_m = b\left[I\lambda(\tau)^{0.5}\right]^n ,$$

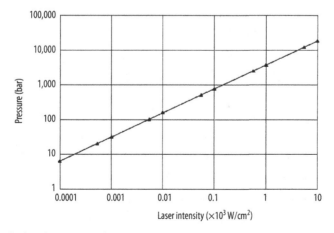

Figure 10.6 Calculated variation of evaporation pressure with laser intensity at a wavelength of 1.06 μm and for a pulse length of 10 ns

where b is a constant that depends on the material type, which takes on values of 5.6 for aluminium alloys and 6.5 for materials based on C–H bonds, and n is an empirical constant equal to −0.3.

Taking this equation and using this to predict the recoil pressure produced as a function of laser intensity at a given wavelength (1.06 μm) and pulse length (10 ns) for the case of organic (C–H bonded) layers gives the results shown in Figure 10.6. It can be seen that very large pressures can be produced as a result of this mechanism. For example, at 1×10^8 W cm^{-2} a pressure of 1,000 bar is predicted. At 1×10^7 W cm^{-2}, a pressure of some 200 bar is predicted. If linearity was retained at lower fluences, a cleaning regime that exploited this mechanism at a laser intensity of 1×10^5 W cm^{-2} would generate a pressure of 10 bar, which could still be sufficient for the removal of superficial organic deposits from stone or metal.

For a given laser intensity (1×10^9 W cm^2) and wavelength (1.06 μm) the pressure can be seen to increase with a decrease in pulse length (Figure 10.7). Similarly, at the

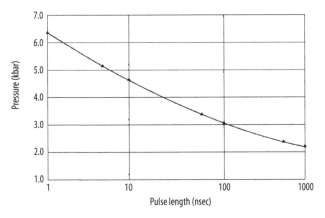

Figure 10.7 Calculated variation of evaporation pressure with pulse length at a wavelength of 1.06 μm and a laser intensity of 1 GW cm^{-2}

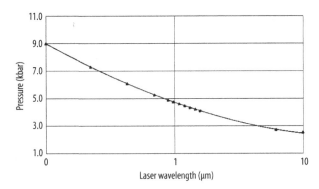

Figure 10.8 Calculated variation of evaporation pressure with laser wavelength for a pulse length of 10 ns and a laser intensity of 1 GW cm^{-2}

same intensity and for a given pulse length (10 ns), the pressure is found to increase with a decrease in laser wavelength (Figure 10.8).

Work at CNR, Florence, Italy, has developed special free-running Nd:YAG lasers in which the pulse length can be varied. It has been found that an optimum pulse length range (at about 20 µs – short free-running mode) exists for the selection of a version of the evaporation pressure mechanism which is particularly beneficial in art conservation of stone and metallic artefacts. This arises from concern that in many cases the mechanical effect induced by short Q-switched pulses may result in mechanical damage, particularly in fragile or friable objects, and hence the driving force has been to modify laser operation in order that intermediate-pulse-length regimes can be chosen in which both mechanical and heating effects are minimised while cleaning still takes place in an efficient manner [38, 39].

10.2.5 Photon Pressure

There will be a small pressure exerted on a surface simply as a result of the momentum of the arriving photons. Although the momentum of a photon is small, highly focused lasers are capable of providing a very high flux of photons. The resulting pressure was estimated in Section 2.1 by noting that

$$p = h/\lambda,$$

(where p is the photon pressure, h is Planck's constant and λ is the laser wavelength) and thus calculating the photon pressure exerted by a 1 kW CO_2 laser focused to a spot size of 0.1 mm diameter; the photon pressure found was 760 N m^{-2}. Using the same approach to calculate the variation of photon pressure with laser intensity for the case of a Q-switched Nd:YAG laser ($\lambda = 1.06$ µm, spot diameter 1 mm and pulse length 10 ns) gives the results shown in Figure 10.9. The pressure delivered to the surface is small but perhaps capable of moving small surface objects. This mechanism has been considered in the case of the removal of submicrometre particulates from

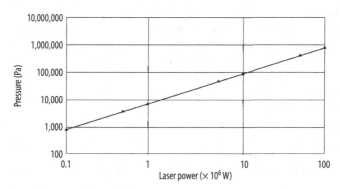

Figure 10.9 Calculated variation of photon pressure with laser peak power at a wavelength of 1.06 µm, for a pulse length of 10 ns and a spot size of 1 mm diameter

microelectronic components (with the additional assistance of a gas stream). However, it is unlikely that sufficient force would be applied for most laser cleaning purposes.

10.2.6 Ablation (Bond Breaking)

"Cold" ablation is an attractive mechanism in many applications of laser cleaning if it is the case that there is no (or little) attendant heating effect. Essentially, the energy, E, of a photon is dependent on its wavelength according to $E = h\nu$ (where h is Planck's constant and ν is the frequency). Hence, the energy per photon of an excimer laser (4.9 eV) is some 40 times greater than that of a CO_2 laser (0.12 eV). Thus, highly energetic ultraviolet wavelength lasers, such as excimer lasers, are capable of providing enough energy to directly break C–H bonds in organic materials. Since organic materials (including polymers) depend for their integrity on the presence of long-chain molecules based on single, double or triple C–H bonds, chain scission by a photon with sufficient energy can lead to the production of short chains that can be subsequently removed by any associated mechanical force induced by the short-pulse excimer laser. The chain-scission effect is a known deleterious factor in commercial plastics where strong sunlight has been found to cause mechanical degradation of plastics not containing ultraviolet stabilisers. In art conservation, the ablation effect produced by highly energetic photons has been applied to painting and rare manuscript conservation [40, 41]. It should be noted that the scission of double and triple C–H bonds requires photons of increasing energy and hence ultraviolet lasers of decreasing wavelength.

10.2.7 Dry and Steam Laser Cleaning

Work in the electronics industry on the effect of pulsed laser radiation on submicrometre surface particulates [8, 9] and published in the early 1990s has shown that the major part of the incident energy can be absorbed on a particular surface (the particulates or the substrate) depending on the wavelength of the laser used, offering potential for selective cleaning by control of the laser wavelength. Absorptivity of laser energy was found to be increased when the target was covered with a thin film of liquid, usually water, offering potential for water-enhanced laser cleaning. This effect was also observed in conservation [5] when the application of a thin layer of water by brush or steam blowing prior to laser treatment led to an enhanced rate of cleaning. A simple analysis shows that the adhesion forces between a substrate and a small particle (arising from van der Waals forces, the electrostatic double-layer force and capillary attraction in the presence of atmospheric moisture) are very large compared with gravitational forces. For the removal of the particle from the surface an acceleration that is inversely proportional to the square of the particle diameter is required. As a result the removal of very small (submicrometre) particles is difficult by conventional techniques.

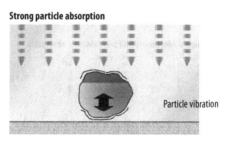

Figure 10.10 Mechanisms involved in "dry" laser cleaning for the removal of small particles from silicon wafer surfaces [9]

10.2.7.1 Dry Processing

For dry laser cleaning, two conditions are distinguished: strong substrate absorption and strong particle absorption (Figure 10.10):

1. *Strong substrate absorption.* Rapid pulsed heating of the dry substrate can lead to the ejection of micrometre and submicrometre particles as a result of the sudden expansion of the substrate surface. For example, 20-ns excimer laser pulses at a fluence of $350\,\mathrm{mJ\,cm^{-2}}$ removed $0.3\,\mu\mathrm{m}$ alumina particles from silicon surfaces. Although the expansion amplitude is small, the time is also very short, resulting in a strong acceleration.
2. *Strong particle absorption.* Twenty-nanosecond pulsed Nd:YAG laser radiation at a fluence of $650\,\mathrm{mJ\,cm^{-2}}$ was effective in the removal of micrometre-sized tungsten particles from a lithium niobate substrate as a result of selective absorption of the laser energy by the particles.

Particle removal can be enhanced by the presence of a liquid film [8, 9] subjected to different types of laser heating (Figure 10.11).

10.2.7.2 Wet Processing

Strong substrate absorption. For a short-wavelength laser such as an excimer laser and using short pulses (about 16 ns) the thermal diffusion distance in a material such as silicon is about $1\,\mu\mathrm{m}$ and in water is about $0.1\,\mu\mathrm{m}$. Hence, irradiation of silicon covered

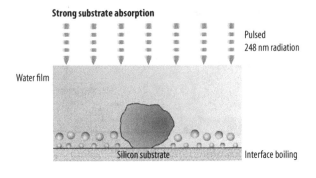

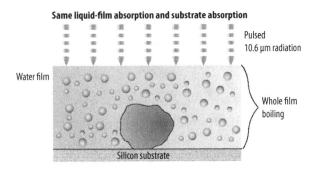

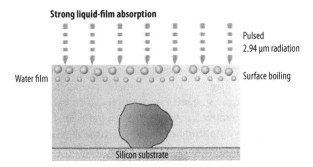

Figure 10.11 Mechanisms involved in "steam" laser cleaning for the removal of small particles from silicon wafer surfaces [9]

with a thin film of water results in very efficient heating of the liquid–substrate interface; superheating and explosive evaporation of the water leads to efficient particle removal.

Strong liquid film absorption. Removal of 0.2-µm gold particles from silicon in the presence of a water film by use of a Q-switched Er:YAG laser (2.94-µm wavelength) with 10-ns pulse length resulted in strong absorption in the water with a penetration depth of 0.8 µm and weak absorption in the silicon. Cleaning was less effective than in the case of strong substrate absorption since the peak temperature was achieved at the top surface of the liquid rather than at the substrate–liquid interface. In the area of art conservation, this approach has been taken further by Wolshbart at Duke University by attempting to exploit the coincidence of the major absorption peak of water and the Er:YAG laser wavelength for the removal of water-containing surface layers.

Joint liquid–substrate absorption. Use of a pulsed 10.6 μm CO_2 laser to remove alumina particles on silicon in the presence of a water layer resulted in particle removal but at a lower energy efficiency than in the case of strong substrate absorption. This was because the absorption depth of this laser in water is about 20 μm and hence, if the water layer is a few micrometres thick, only a fraction of the laser energy is absorbed by the water and this energy is distributed in the bulk of the water. The remaining energy penetrates too deep into the silicon to be useful for interface heating. Hence, more laser energy is required to create explosive evaporation of the liquid.

Tam *et al.* [42] conclude that the selection of laser process conditions to favour strong substrate absorption produces the most efficient steam cleaning technique. The ejection velocity of the exploding water droplets is estimated at 10^4 cm s^{-1}, causing the formation of a jet of water droplets and ejected particles that is visible up to 1 cm from the irradiated surface and which causes a shock pulse in the air which is audible as a characteristic snapping sound. The peak temperature is estimated to be about 370 °C and is accompanied by a high transient pressure (up to a few hundred atmospheres). However, since these conditions exist for only a very short time in pulsed laser operation, the extent of substrate damage is strictly limited.

Although no detailed account of laser steam cleaning applied to art restoration has appeared to date, it is known that this technique is highly effective in the removal of certain encrustations [5,12], for example, the removal of dark sulphurous encrustations from limestone sculpture using a Q-switched Nd:YAG laser where the water is applied as a thin surface layer by periodic brushing.

10.2.8 Angular Laser Cleaning

This technique [43,44] irradiates the stone surface at a glancing angle in contrast with typical laser cleaning, in which the laser beam is directed at a perpendicular angle of incidence to the target material (Figure 10.12). It has many advantages over conventional laser cleaning, including a dramatic improvement in cleaning efficiency. It was found that the cleaned area irradiated at a glancing angle of 20° was up to 10 times larger than the typical laser-cleaned area using a perpendicular angle. Hence, the speed at which contaminated areas are cleaned is greatly increased. Moreover, cleaning threshold fluences are reduced at glancing angles, therefore reducing risks of surface damage considerably in comparison with perpendicular laser cleaning. The principle of operation is shown in Figure 10.13. Access of the laser to the substrate is not shielded at a sufficiently oblique glancing angle and hence thermoelastic stresses near or beneath adherent dirt particles are higher compared with direct vertical radiation, thus in essence shaking the dirt off.

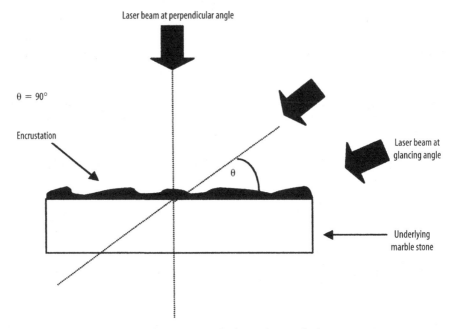

Figure 10.12 Angular laser cleaning [44]

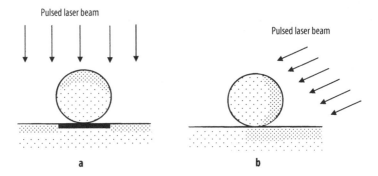

Figure 10.13 Principles of angular laser cleaning [44]

10.2.9 Laser Shock Cleaning

A plasma shock wave, produced by a breakdown of air due to an intense laser pulse, can be used for cleaning [44, 45]. The beam is directed parallel to the surface to avoid direct laser interaction with the target material and is focused a few millimetres above the area to be cleaned. The power density of the beam at the focal point is around 10^{12} W cm^{-2}. The gaseous constituents begin to break down and ionise; as a result a shock wave is produced which has an audible snapping sound. In air the typical peak pressure of the shock front for a spherically expanding plasma is estimated to be of the order of hundreds of megapascals. However, the precise value depends on the laser power den-

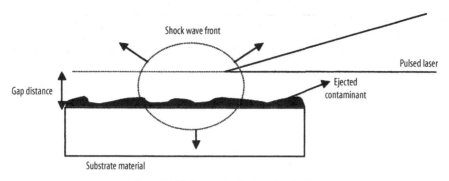

Figure 10.14 Laser shock cleaning [45]

sity and the distance from the shock wave (controlled by the gap distance established between the laser beam and the substrate surface). This new approach, illustrated in Figure 10.14, has unique characteristics. It is totally independent of the physical properties of the contaminants and the substrate, and hence does not encounter problems associated with other techniques, where the absorptivity of the substrate and that of the particles requiring removal to a given laser wavelength determine the efficiency of removal. Also, the risk of damage to the underlying substrate is minimised, as the incident beam does not come directly into contact with the workpiece. Again, as with angular laser cleaning, an increased cleaning efficiency is observed in terms of the size of the area cleaned and also speed at which the area is cleaned. It was found that the gap distance between the surface to be cleaned and the laser focus is critical in terms of successful cleaning, as this distance alters the pressure of the shock wave striking the surface.

10.3 An Overview of the Laser Cleaning Process

Figure 10.15 is an attempt to develop an overview of the various mechanisms just described and their place with respect to each other [11].

It should be recognised that this is an initial attempt to systematise this information and that the diagram will undoubtedly require updating in the future as a more exact understanding of the various mechanisms and their relations are developed. It is offered as much as a vehicle to promote discussion amongst laser cleaning practitioners at this stage as a definitive outcome. The guideline for vaporisation is intentionally schematic. There will be great variation in the actual values of the absorbed laser intensity and the interaction time required for vaporisation depending on the physical properties of the materials in question. Despite this, this method of presentation emphasises the diversity of the processes that could be involved in laser cleaning, the importance of the careful selection of intensity and interaction time and the possibility of the exploitation in cleaning by relatively low intensity processes that may not involve vaporisation (such as transient surface heating). In terms of acoustic emissions, the diagram points

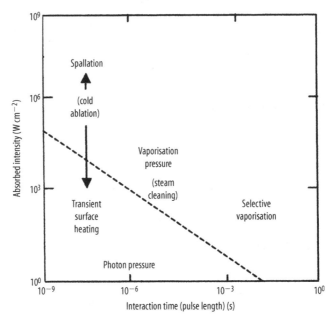

Figure 10.15 Absorbed intensity versus interaction time diagram showing regimes of candidate laser cleaning methods [11]

to three potential sources of acoustic waves and as a result the origin and significance of such signals and their relevance in cleaning should be carefully considered. The boundaries on the domains suggested by the diagram are not defined and hence the regimes suggested for the different mechanisms are indicative only. It may be the case that there is an overlap between domains such that more than one of these mechanisms may be simultaneously active under certain conditions (*e.g.*, transient heating combined with selective vaporisation or evaporation pressure combined with plasma detonation effects) and this would complicate the situation encountered in practical cleaning.

10.4 Practical Applications

It is possible to use the laser for cleaning, derusting, deoiling, depainting, deoxidising, degreasing and even the removal of ultrafine dirt [46]. For large-scale work it is used for paint stripping [47]. For example, using a 2 kW transversely excited atmospheric pressure (TEA) CO_2 laser, one can clean 8–10 m^2 per hour at less than $6 per square metre with a reduction in dangerous waste from chemical cleaners of the order of a factor of 10. These lasers usually work at approximately 9 J per pulse and 350 Hz and cost $40 per hour.

Despite good prospects for application in particle removal in the microelectronics sector and in more general cleaning tasks (such as graffiti removal), the main area of current usage of laser cleaning is in the area of the conservation of cultural heritage.

The organisation Lasers in the Conservation of Artworks (LACONA), which is currently supported by over 200 practising conservators and over 100 other scientists, has been in existence for 10 years and holds a successful biannual conference at which the latest developments in the area are shared. Lasers are currently being used for a wide range of conservation activity, which includes cleaning of marble and limestone sculpture [48, 49], restoration of paintings [40], cleaning and restoration of manuscripts [50] and conservation of a wide range of museum artefacts. In the restoration of paintings and coloured artefacts care has to be taken of the bleaching effects of strong radiation. The laser cleaning activity in picture restoration is often confined to the removal of the hard outer layer of varnish prior to finishing the art restoration with chemicals and soft brushes. Important developments have been made in the online monitoring of the cleaning process by laser induced spectroscopy techniques [41, 51] and in the application of neural network control systems [52, 53].

Laser cleaning may well replace chemical wet methods, which consume much water. A typical semiconductor plant uses hundreds of thousands of gallons of water per day. The cost of disposing of 100,000 gallons of fluoride-contaminated water runs at some $450,000, so there is obviously some commercial sense in finding a better way and the laser seems to offer just that.

The increasing availability of picosecond and femtosecond laser sources has emphasised the potential of these "ultrafast" devices in laser cleaning application [42, 54–57]. Because of the short interaction time compared with nanosecond laser sources, there is correspondingly less thermal effect on the target substrate, resulting in greater control of unwanted morphological changes there. Because of the initiation of multiphoton processes, nominally transparent targets may be processed [57] (see Chapter 8). At the same time, the cleaning rate is much reduced compared with nanosecond laser cleaning, resulting in a trade-off between the cleaning rate and limitation of unwanted effects. The possibility of wholly new mechanisms being operative in the ultrafast regime requires further investigation.

For further reading refer to the recent books edited by Lee [58] and Luk'yanchuk [59].

Questions

1. List the main methods whereby the laser can be used for cleaning.
2. What are the limiting factors on the incident laser intensity in selective evaporation?
3. How can laser cleaning be achieved without the radiation striking the surface?
4. What are the preferred conditions for laser cleaning by steam cleaning?
5. When is photon pressure significant?

References

[1] Reiss S, Schawlow AL (1996) Optics and laser spectroscopy. University of California interview series. University of California
[2] Asmus JF, Munk WH, Murphy CG (1973) Studies on the interaction of laser radiation with art artefacts. Proc Soc Photo-Opt Instrum Eng 41:72–76

[3] Asmus JF (1974) Laser consolidation tests. Final report. International Fund for Monuments, New York

[4] Asmus JF, Seracini M, Zetler MJ (1976) Surface morphology of laser-cleaned stone. Lithoclastia 1:23

[5] Asmus JF (1978) Light cleaning – laser technology for surface preparation in the arts. Technol Conserv 13:14

[6] Asmus JF (1986) More light for art conservation. IEEE Circuits Devices Mar 6–15

[7] Asmus JF (1987) Light for art conservation. Interdiscip Sci Rev 12:171

[8] Zapka W, Zeimlich W, Tam AC (1991) Efficient pulsed laser removal of 0.2 mm sized particles from a solid surface. Appl Phys Lett 58:2217

[9] Tam AC, Leung WP, Zapka W, Zeimlich W (1992) Laser cleaning techniques for the removal of surface particulates. J Appl Phys 71:3515

[10] Watkins KG (1997) A review of materials interaction during laser cleaning in art restoration. In: Proceedings of lasers in the conservation of artworks (LACONA 1), Crete, Greece, 4–6 October 1995. Mayer, Vienna, pp 7–15

[11] Watkins KG (2000) Mechanisms of laser cleaning. Proc SPIE 3888:165–174

[12] Von Allmen M, Blaser P, Affolter K, Sturmer E (1978) Absorption phenomena in metal drilling with Nd-lasers. J Quantum Electron QE-14:85

[13] Bonch-Breuvich AM, Imas YA, Romanov GS, Libenson MN, Mal'tser LN (1968) Effect of laser pulse on the reflecting power of a metal. Sov Phys Tech Phys 13:640

[14] Bergel'son VI, Golub AP, Loseva TV, Newchinov IV, Orlova TI, Popov SP, Svettsov VV (1974) Appearance of a layer absorbing laser radiation near the surface of a metal target. Sov J Quantum Electron 4:704

[15] Metz SA, Hettche LR, Stegman RL, Shreimpf JT (1975) Effect of beam intensity on target response to high intensity pulsed CO_2 laser radiation. J Appl Phys 46:1634

[16] Hettche LR, Tucker TR, Shreimpf JT, Stegman RL, Metz SA (1976) Mechanical response and thermal coupling of metallic targets to high intensity 1.06 μm laser radiation. J Appl Phys 47:1415

[17] Patel RS, Brewster MQ (1990) Effect of oxidation and plume formation on low power Nd-YAG laser metal interaction. Trans ASME J Heat Transf 112:170

[18] Bass M, Nasser MA, Swimm RT (1987) Impulse coupling to aluminium resulting from Nd:glass laser irradiation induced metal removal. J Appl Phys 61:1137

[19] White RM (1963) Elastic wave generation by electron bombardment or electromagnetic wave absorption. J Appl Phys 34:2123

[20] Percival CM (1967) Laser-generated stress waves in a dispersive elastic rod. J Appl Phys 38:5315

[21] Lee RE, White RM (1968) Excitation of surface elastic waves by transient surface heating. Appl Phys Lett 12:12

[22] Bushnell JC, McCloskey DJ (1968) Thermoelastic stress production in solids. J Appl Phys 39:5541

[23] Brienza MJ, DeMaria AJ (1967) Laser-induced microwave sound by surface heating. Appl Phys Lett 11:44

[24] White RM (1963) Generation of elastic waves by transient surface heating. J Appl Phys 34:3559

[25] Ready JF (1971) Effects of high-power laser radiation. Academic, New York, 116 ff

[26] Fabbro R, Fournier J, Ballard P, Devaux D, Virmont J (1990) Physical study of laser-produced plasma in confined geometry. J Appl Phys 68:775–784

[27] Peyre P, Fabbro R, Berthe R, Dubouchet C (1995) Laser shock processing of materials – physical processes involved and examples of applications. In: ICALEO 1995 proceedings. LIA, Orlando, pp 241–250

[28] Bubrujeaud B, Jeandin M (1994) Cladding by laser shock processing. J Mater Sci Lett 13:773–775

[29] Clauer AH, Fairand BP, Wilcox BA (1977) Laser shock hardening of weld zones in aluminium alloys. Metall Tran A 8:1871–1876

[30] Fernelius NC (1980) Photacoustical signal variations with chopping frequency for ZnSe laser windows. J Appl Phys 51:1756–1767

[31] Bell AG (1880) On the production and reproduction of sound by light: the photophone. J Sci Ser 3 20:305–324

[32] Bell AG (1881) Upon the production of sound by radiant energy. Philos Mag 11:510–528

[33] Gournay LS (1966) Conversion of electromagnetic to acoustic energy by surface heating. J Acoust Soc Am 40:1322–1326

[34] Aden M, Beyer E, Herziger G (1990) Laser-induced vaporisation of metal as a Riemann problem. J Appl Phys D 23:655–661

[35] Knight C (1979) Theoretical modeling of rapid vaporisation with back pressure. AIAA J 17:519–523

[36] Chan CL, Mazumder J (1987) One-dimensional steady-state model for damage by vaporisation and liquid expulsion due to laser-material interaction. J Appl Phys 62:4579–4596

[37] Phipps CR, Turner TP, Harrison RF, York GW, Osborne WZ, Anderson GK, Corlis XF, Hayes LC, Steele HS, Spiochi KC (1988) Impulse coupling to targets in vacuum by KrF, HF and CO_2 single-pulse lasers. J Appl Phys 64:1083–1096

[38] Salembeni R, Pini R, Siano S (2002) A variable pulse width Nd-YAG laser for conservation. J Cult Herit 4(Suppl 1):72–76

[39] Margheri F, Modo S, Massoti L, Mazzinghi P, Pini R, Siano S, Salembeni R (2000) Smart Clean: a new laser system with improved emission characteristics and transmission through long optical fibres. J Cult Herit 1:S119–S123

[40] Georgiou S, Zafiropolos V, Anglos D, Balas C, Tornari V, Fotakis C (1998) Excimer laser restoration of painted artworks: procedures, mechanisms, effects. Appl Surf Sci 127–129:738–745

[41] Tornari V, Zafiropolos V, Bonarou A, Vainos NA, Fotakis C (2000) Modern technology in artwork conservation: a laser based approach for process control and evaluation. Opt Lasers Eng 34:309–326

[42] Rode AV, Baldwin KGH, Wain A, Madsen NR, Freeman D, Delaporte P, Luther-Davies B (2008) Ultra-fast laser ablation for restoration of heritage objects. Appl Surf Sci 254:3137

[43] Lee JM, Watkins KG, Steen WM (2000) Angular laser cleaning for effective removal of particles from a solid surface. J Appl Phys A 71:671–674

[44] Watkins KG, Lee JM, Curran C (2002) Two new mechanisms for laser cleaning using Nd-YAG sources. J Cult Herit 4(Suppl 1):59–64

[45] Lee JM, Watkins KG (2001) Removal of small particles on silicon wafer by laser-induced airborne plasma shock waves. J Appl Phys 89:6496–6500

[46] Daurelia G, Chita G, Cinquepalmi M (1997) New laser cleaning treatments: cleaning, derusting, deoiling, depainting, deoxidising and de-greasing. In: Proceedings of the conference lasers and optics in manufacturing, Munich, June 1997, paper 3097-46

[47] Ploner L (1995) High power TEA laser available for paint stripping. Laser Focus World May 26–27

[48] Cooper MI, Emmony DC, Larson JH (1992) The use of laser energy to clean polluted stone sculpture. J Photogr Sci 40:55

[49] Cooper MI (1998) Laser cleaning in conservation: an introduction. Butterworth-Heinemann, Woburn

[50] Kolar J, Strlic M, Muller-Hess D, Grubner A, Troschke K, Pentzien S, Kautek W. Near-UV and visible pulsed laser interaction with paper. J Cult Herit 1:S221–S224

[51] Maravelaki PV, Zafiropolos V, Kilikoglou V, Kalaitzaki M, Fotakis C (1997) Laser induced breakdown spectroscopy as a diagnostic technique for the laser cleaning of marble. Spectrochem Acta B 52:41–53

[52] Lee JM, Watkins KG (2000) In-process monitoring techniques for laser cleaning. Opt Lasers Eng 34:429–442

[53] Lee JM, Watkins KG (2000) Chromatic modulation technique for in-line surface monitoring and diagnostic. J Cult Herit 1:S311–S316

[54] Pouli P, Bounos G, Georgiou S, Fotakis C (2007) Femtosecond laser cleaning of painted artefacts; is this the way forward? In: Nimmrichter J, Kautek W, Schreiner M (eds) Lasers in the conservation of artworks, LACONA VI proceedings, Vienna, Austria, September 21–25, 2005. Springer proceedings in physics, vol 116. Springer, Berlin, p 287

[55] Gaspard S, Oujja M, Moreno P, Méndez C, García A, Domingo C, Castillejo M (2008) Interaction of femtosecond laser pulses with tempera paints. Appl Surf Sci 254:2675

[56] Walczak M, Oujja M, Crespo-Arcá M, García A, Méndez C, Moreno P, Domingo C, Castillejo M. Evaluation of femtosecond laser pulse irradiation of ancient parchment. Appl Surf Sci 255:3179

[57] Pouli P, Nevin A, Andreotti A, Colombini P, Georgiou S, Fotakis C (2009) Laser assisted removal of synthetic painting-conservation materials using UV radiation of ns and fs pulse duration: morphological studies on model samples. Appl Surf Sci 255:4955–4960

[58] Lee JM (ed) Lasers and cleaning processes. Hanrimwon, Seoul

[59] Luk'yanchuk B (ed) (2003) Laser cleaning. World Scientific, Singapore

"No, Jonnie, you can't get cleaned in a flash – it's soap and water for you."

11 Biomedical Laser Processes and Equipment

Thou who didst come to bring

On thy redeeming wing

Healing and sight,

Health to the sick in mind,

Sight to the inly blind,

O now to all mankind

Let there be light.

J. Marriott (1780–1825), Hymns Ancient and Modern New Standard 180

for thyng that lightly cometh, lightly goeth

1475 Fortescue Works after Chaucer

11.1 Introduction

Light plays an important role in living processes. We rely on light to stimulate the formation of certain vitamins (*e.g.*, vitamin D from sunlight), we rely on light to perform the major manufacturing process of converting water and carbon dioxide into sugars, a process called photosynthesis, which is taking place all over our fields and forests. Without light our physical well-being would be greatly diminished if not destroyed. It is thus not surprising to find that lasers are beginning to play a significant role in medical science in more ways than simply as cutting tools.

The range of treatments currently practised is large and growing but includes the following generic areas:

* *Ophthalmology:* detached retinas (one of the first applications of lasers in medicine), glaucoma and laser-assisted *in situ* keratomileusis (LASIK) and photorefractive keratectomy (PRK).
* *Surgery:* the laser can be a self-cauterising knife or drill. Current uses include urology, gynaecology and otolaryngology.
* *Dermatology:* for the removal of hair, freckles, strawberry marks, tattoos and skin rejuvenation.
* *Cardiology:* angioplasty, transmyocardial laser revascularisation.

- *Orthopaedics*: for localised machining and smoothing, cartilage photothermolysis.
- *Dentistry*: painless drilling, cleaning, construction of implants.
- *Photodynamic therapy*: the use of optically switchable drugs to kill cancers cell by cell.
- *Biostimulation*: arthritis, rheumatism, enhanced healing.
- *Tissue welding*: rapid wound healing.
- *Diagnostics*: spectroscopy, fluorescence and visualisation.

Lasers of only a few watts are all that is needed for most of the processes in these different areas; the thought of a 2 kW laser operating on a person is not the image to hold in your mind. Biological tissue is not like metal; it does not have a clear melting point and if heated much above 60°C it undergoes minor chemical changes which make it dysfunctional and it dies, a process called "necrosis". The types of lasers used in medical processes are listed in Table 11.1. The main power intensity and interaction times for different process regimes are shown in Figure 11.1. This map shows the absorbed intensity. Whether or not the radiation is absorbed and at what depth is dependent on the wavelength of the beam and the absorptivity of the tissue or chromophores in the tissue at that wavelength. For penetration into the tissue there is a "diagnostic window" between 600 and 1,300 nm; for absorption near the surface then wavelengths shorter than 200 nm or longer than 1,500 nm should be used, the absorption peaks of water, which most tissues contain. Some values of the absorptivity of various chromophores and water are shown in Figure 11.2.

This chapter will discuss these different processes and some of the equipment that has been made to engineer light for medical purposes. But first we will discuss what happens when light interacts with biological tissue.

11.2 Interaction of Laser Radiation with Biological Tissue

What happens when radiation strikes biological tissue depends on the intensity and the interaction time, as illustrated in Figure 11.1, and wavelength, as shown by the different absorption ranges in Figure 11.2. Further effects occur both with high-energy photons, as with ultraviolet light, which can cause direct bond breaking resulting in strong absorption and very clean cuts (Figure 3.31, carved hair) and by nonlinear absorption due to multiphoton events at an intense focal spot. All these interactions depend to some extent on the optical, thermal and mechanical properties of the tissue.

11.2.1 Optical Properties of Biological Tissue

Biological tissue is essentially translucent. Light is more strongly scattered than absorbed, except in portions of the eye. This is due to refractive index gradients in organs and cells as well as discrete structures such as collagen fibrils and red blood cells. The scattering may be 10–100 times larger than the absorption, depending on the wavelength. Absorption is via chromophores, including water, various proteins,

Table 11.1 Types of laser used in medical applications

Laser type	Wavelength	Inventor/date	Typical uses	Comments
Er:YAG	1.53 or 2.94 μm		Medical applications (surgery, *etc.*)	Longer than 1.5 micron is "eye-safe": also as a fibre laser for telecommunications and power
Ho:YAG	2.12 μm		As Er:YAG	Advantages with absorption by water
Yb:YAG	1.03 mm		Telecommunications and power fibre lasers	Usual type of fibre or disc laser
Diode	0.5–1.55 μm	Robert Hall (1962)	Telecommunications, digital data reading; wide range of emerging medical applications	New wavelength versions being developed. Can be delivered via optical fibre
Alexandrite	700–820 nm		Selective photothermolysis	Can be CW, pulsed or Q-switched
Ruby	694 nm	Maiman (1960)	Holography, measurement of plasma properties, medical applications (skin treatments)	Can be Q-switched to create powerful pulses
Ti:sapphire	660–1,180 nm		IR spectroscopy, laser radar, remote sensing, tattoo removal	Can be tuned to required wavelength in the given range
Ar ion	409–686 and 275–363 nm	William Bridges (1964)	Phototherapy of the eye, cell cytometry, stereolithography, pumping dye lasers	Blue and green power beam and UV possible
Dye	320–1,200 nm	Sorokin and Lankard (1966)	Spectroscopy, physical research, medical applications (treatment of malignancies by selective absorption, shattering of kidney stones)	Over 200 dyes known to lase. Can select suitable dye to produce required wavelength in the given range, also tunable
Excimer	193 nm (ArF), 248 nm (KrF), 308 nm (XeCl) or 351 nm (XeF)	Basov and Ewing and Brau	Laser marking and drilling of microelectronic components, medical applications (eye surgery keratectomy)	Wavelength depends on type of rare halide filling gas
CO_2	9,000–11,000 nm	K. Patel (1964)	Skin rejuvenation, transmyocardial laser revascularisation	

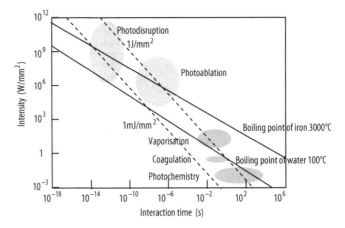

Figure 11.1 Laser–tissue interactions showing the general area for photodisruption, photoablation, vaporisation, coagulation and photochemistry

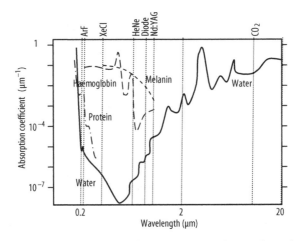

Figure 11.2 Absorption spectra for various tissue chromophores [62]

haemoglobin in blood and tissue pigments such as melanin. A graph of their wavelength-dependent absorption is shown in Figure 11.2. The low absorption of almost all the other constituents in the tissue in the 600–1,300 nm region leads to this wavelength region being known as the "diagnostic window" because of the resulting accessibility to deep tissue in that range of radiation; it also leads to many clinical possibilities.

Tissue is a continuous optical medium with a spatially varying complex refractive index. It is difficult to model except by assuming a distribution of discrete absorbers and scatterers. Usually in phototherapy it is necessary to know the optical dosage and hence it would be useful to calculate how light propagates within the tissue. This is difficult for the reasons given above. Approximate methods based on calculating the scattered and unscattered components, or using the Monte Carlo method of calculating each ray, have been developed. The unscattered radiation is accounted for by a form of Beer–

Table 11.2 Order of magnitude values of some tissue physical properties [62]

Property	Symbol	Soft tissue	Hard tissue
Absorption coefficient	μ_a	$0.2\,\text{cm}^{-1}$ at 635 nm (breast)	
Effective attenuation coefficient	μ_{eff}	$2.5\,\text{cm}^{-1}$	
Scattering coefficient	μ_s	$200\,\text{cm}^{-1}$	
Specific heat per unit volume	ρC_p	$3.96 \times 10^{-3}\,\text{J}\,^\circ\text{C}^{-1}\,\text{mm}^{-3}$	$0.88\,\text{J}\,^\circ\text{C}^{-1}\,\text{mm}^{-3}$ [65]
Thermal diffusivity	α	$0.106\,\text{mm}^2\,\text{s}^{-1}$	$0.4\,\text{mm}^2\,\text{s}^{-1}$ [66]
Speed of sound	C	$1{,}500\,\text{m}\,\text{s}^{-1}$	$2{,}800\,\text{m}\,\text{s}^{-1}$
Maximum elongation		25–100%	2.5%
Maximum stress	σ	600 bar	500–2,500 bar

Lambert absorption and the scattered portion is accounted for by a form of diffusion equation. This does not adequately account for the forward-scattered light which is a strong component of tissue scattering. Forward scattering is where the scattered light returns to the main direction of the original beam. Thus, at present much of the dosage analysis is empirical.

Some of the optical properties of biological tissue are listed in Table 11.2.

11.2.2 Thermal Properties of Tissue

Absorption of the shorter-wavelength visible and near-ultraviolet light in tissue occurs by exciting electronic transitions in the tissue molecules, whereas near- and mid-infrared radiation is absorbed via vibrational–rotational excitations. The lifetimes of these states are in the range of picoseconds, after which time thermal stabilisation occurs and is observed as heat, which disperses according to the normal diffusion principles and at a rate dependent on the thermal diffusivity. (The thermal diffusivity of tissue is approximately $0.106\,\text{mm}^2\,\text{s}^{-1}$, which is very similar to that of water – approximately $0.15\,\text{mm}^2\,\text{s}^{-1}$ – of which most of the tissue is composed). The optical absorption depth is inversely proportional to the concentration of chromophores times their molecular extinction coefficient (shown in Figure 11.2). The depth is found to vary from a few microns in the ultraviolet to several millimetres in the near-infrared "diagnostic window".

An estimate of the "thermal relaxation time", which is a characteristic time for thermal diffusion to significantly reach the penetration depth, can be found from the equation we have met for the HAZ (Fourier number is $1 = x^2/\alpha \Delta t$; see Section 5.3). In this case the Fourier number is usually taken as 4 to ensure that the thermal wave is encompassed. Hence, the thermal relaxation time is

$$\frac{1}{4\alpha}d_{eff}^2 = \frac{1}{4\alpha\mu_{eff}^2},$$

where μ_{eff} = is the effective attenuation coefficient, equivalent to the molar extinction coefficient, μm^{-1}) and d_{eff} is the depth of penetration at which the intensity has fallen by $1/e = \mu_{eff}^{-1}$.

This value is used to determine how short a pulse should be if collateral damage by heat conduction is to be avoided. For example if the laser has an optical penetration depth of 1 μm, (which is the reciprocal of the extinction coefficient), and the thermal diffusivity is 0.106 mm^2 s^{-1} then the relaxation time is

$$= \frac{1}{4 \times 0.106} 0.001^2 \, s = 2.3 \times 10^{-6} \, s \, .$$

Thus, to avoid thermal penetration beyond the optical penetration of 1 μm depth the pulse duration must be less than 2 μs. For longer pulses the depth of thermal penetration will be 23 μm for 1 ms and 0.72 mm for 1 s. It is often taken as a rule of thumb

Table 11.3 Summary of the thermal effects that can be caused by a laser-tissue interaction

Temperature (°C)	Thermal interaction	Description
37	Normal	Healthy body temperature
> 40	Hyperthermia	Body temperatures above 40 °C are life-threatening. This compares with normal body temperature of 36–37 °C. At 41 °C, brain death begins, and at 45 °C death is nearly certain. Internal temperatures above 50 °C will cause rigidity in the muscles and certain, immediate death if the whole body is above this temperature. In localised heating by a laser cells will die or be weakened if heated above 40 °C for long periods. See Figure 11.4
> 50	Low activity	Reduction in enzyme activity; cell immobility
> 60	Coagulation	Denaturing of proteins and collagen. Tissue coagulation
> 80	Cell breakdown	Cell membranes become permeable
> 100	Vaporisation	Vaporisation of water (particularly under the effect of pulsed Er:YAG laser treatment) can lead to photodisruption by the explosive expansion of water leading to tissue removal. This usually occurs at temperatures above the value of 100 °C
> 150	Carbonisation	At temperatures above 150 °C the organic molecules of the tissue may crack and release carbon, leading to a blackening in colour. This not only looks nasty but obscures the surgeon's view as well as destroying the tissue (necrosis)
> 300	Melting	For tissue that is capable of melting, such as fats
> 1,000	Ablation	At high intensities and very short pulses the material can be changed directly into vapour

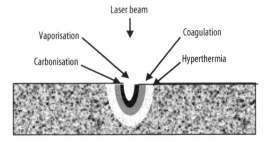

Figure 11.3 Possible effects within the irradiated zone

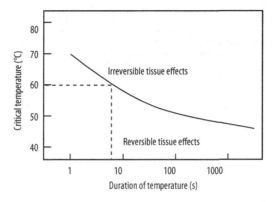

Figure 11.4 The dependence of cell necrosis on temperature and time [7]

that for pulses longer than 1 μs thermal damage should be expected beyond the optical penetration depth. Table 11.3 lists the various thermal interactions encountered in laser processing of biological tissue. In many procedures these effects occur together, illustrated in Figure 11.3; as in boiling, melting and HAZ when processing metal. Figure 11.4 shows the relationship between necrosis and time at a certain temperature; *e.g.* longer than 8 s above 60°C.

11.2.3 Mechanical Properties of Tissue

Tissue can be either hard or soft. These forms differ in their reaction to laser radiation. Soft tissue includes muscles, fat, skin, blood vessels and various internal organs. Hard tissue includes bones, teeth and calcified plaque. Soft tissue is mainly made of protein, water, electrolytes and complex organic molecules, some of which may be chromophores. Hard tissue consists of a composite of microcrystallites of calcium phosphate salts embedded in soft tissue. The salts provide compressive strength and the matrix tensile strength. In compact bone the salts are mainly hydroxyapatite deposited in a highly organised collagen matrix with a small amount of residual water. The composition of various hard tissues is given in Table 11.4. Kidney stones contain no soft material and in consequence are brittle and can be broken with ultrasound or by laser ablation, by which the plasma-generated shock wave cracks the stone so that the smaller fragments may then be passed naturally.

11.2.4 Tissue Heating Effects – Nonablative Heating

Only slight heating can cause cumulative damage in tissue owing to the breaking of the weak hydrogen or van der Waals bonds which hold the scaffold for the structure together. Such breakages may lead to a denaturation of proteins and enzymes, making them ineffective. The rate of this denaturing has been shown to follow a first-order Arrhenius type of reaction (one based on activation energy and the probability of ac-

Table 11.4 The composite nature of hard calcified tissue [62]

Tissue	Hard component	Soft component
Cardiovascular calcified deposits	75% mixed apatite salts (m.p. ≈ 1,500°C), granules 1–10 μm in size	20% lipids, cross-linked proteins (b.p. ≈ 300°C)
Bone	75% hydroxyapatite (m.p. ≈ 1,500°C), granules 1–10 μm	5% water (b.p. 100°C), 20% collagen (b.p. ≈ 300°C)
Tooth enamel	Hydroxyapatite	Keratin, 5% water (b.p. 100°C)
Kidney stones (urinary calculi)	Calcium oxalate	None

quiring it):

$$D = P \int_0^t \exp \frac{-E_a}{RT} dt \,,$$

where D is a measure of damage, P is a constant found to be approximately 10^{70} s^{-1}, E_a is the activation energy (found to be around 4×10^5 J mol^{-1} for liver tissue), R is the universal gas constant (8.314 J mol^{-1} K^{-1}) and T is absolute temperature (K).

This reaction rate is highly nonlinear. The rate can more than double for a 10°C rise in temperature:

$$\frac{D_T}{D_{T+\Delta T}} = \frac{e^{-\frac{E_a}{RT}}}{e^{-\frac{E_a}{R(T+\Delta T)}}} = e^{-\frac{E_a \Delta T}{RT^2}} \,.$$

11.2.5 Tissue Heating Effects – Ablation

The energy required for ablation is found to be an order of magnitude less than would be required for tissue vaporisation. This is thought to be due to inertial confinement. Soft tissue is essentially made of water. Water is normally heated at constant pressure, in which case it needs 300 J g^{-1} to start boiling as shown by the process line AB in the enthalpy–pressure diagram for water in Figure 11.5. But if the energy is provided so fast that the necessary expansion does not occur, then the heating will proceed more along the line of constant volume heating shown as line AD in Figure 11.5. Going this way, the same 300 J g^{-1} would heat the water to about the same temperature as before (approximately 98°C) but the pressure would now be huge at 900 bar. Thus, pulse heating soft tissue by laser may cause a mechanical breakdown of the soft tissue, which in turn will eject the salt components by the process of *photodisruption*. The energy has to be delivered in a time short compared with the time for the stress to propagate, which is determined by the time taken for a sound wave to traverse the optical penetration depth ($d_{eff} = \mu_{eff}^{-1}$):

$$\frac{t_{pulse} c_{sound}}{d_{eff}} = t_{pulse} \mu_{eff} c_{sound} \ll 1 \,,$$

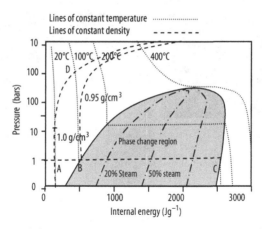

Figure 11.5 Enthalpy versus log pressure for water and steam. (After [62])

where t_{pulse} is the pulse length (s) and c_{sound} is the velocity of sound in the tissue (approximately 1,500 m s^{-1} for soft tissue and approximately 2,800 m s^{-1} for hard tissue).

For example, for a depth of penetration of 1 μm the pulse duration should be less than

$$\frac{d_{eff}}{c_{sound}} = \frac{0.001}{1\,500\,000} = 0.66\,\text{ns}.$$

For these very short pulses the tissue would rip under the pressures generated and fragments will be thrown away from the crater. The temperature, however, would remain quite low; in our example only reaching 98°C. There is a semblance of similarity to the behaviour of gun powder, which burns fast in air but explodes if confined. Figure 11.6 shows sections of skin that have been cut by laser and illustrate these various cutting mechanisms, from heating to photodisruption. The same inertial confinement applies to both soft and hard tissue. Experimentally, the phenomenon requires irradiances above 10^6 W cm^{-2} (around 100 W or 1 mJ for 10 μs, focused to a 100-μm spot), creating velocities of around 100 m s^{-1} to drag the salt particles away from the crater.

Vaporisation of hard or soft tissue is not precluded but requires considerably more power and will usually occur after the water has been vaporised. In the case of kidney stones, where they contain very little to no water, a laser spark of sufficient intensity will generate a plasma and shock wave (see Sections 6.19 and 10.2.2) which may be enough to break the stone into fragments which can be passed naturally.

Photoablation by energetic photons, such as photons of ultraviolet light, occurs when the energy of the photon is sufficient to break the chemical linkages in the tissue. In the case of corneal reshaping as with LASIK or PRK the corneal tissue is opaque to the very short wavelengths and also the C=C, C–C and C–H bonds can be directly broken, removing the material with very little collateral damage, which, of course, is very important in such operations (see Sections 3.3.3, 10.2.6, and 11.3.1.3). Table 11.5 shows the bond energy for various organic bonds and the energy of a single photon

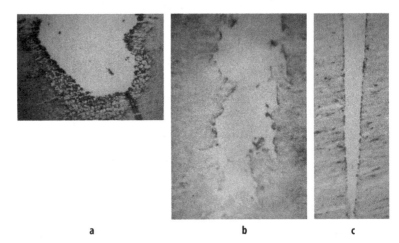

a   b   c

Figure 11.6 Examples of skin cut with different wavelengths and pulse lengths: **a** CW Nd:YAG, showing the heat-affected zone and coagulation heating, **b** 408-nm short-pulsed dye laser, showing photodisruption but little heat-affected zone, and **c** excimer laser, showing photoablation with short pulses of ultraviolet light (Courtesy of Lynton Laser Ltd., Crewe)

Table 11.5 Approximate energy of chemical bonds

Bond type	Energy (eV)	Bond type	Energy (eV)	Bond type	Energy (eV)
C–C	3.62	C=C	6.40	C≡C	8.44
C–H	4.30	C–N	3.04	C=S	4.96

Energy of photons			
Source	Wavelength (μm)	Frequency (Hz)	Energy (eV)
Cyclotron	0.1 (X-ray)	2.9×10^{15}	12.3
Excimer	0.248 (KrF)	1.2×10^{15}	4.9
Argon	0.488 (blue)	6.1×10^{14}	2.53
Nd:YAG	1.06 (IR)	2.8×10^{14}	1.16
CO_2	10.6 (IR)	2.8×10^{13}	0.12

from different lasers; but remember at the focus of a laser beam the photon density is such that there is a likelihood of multiphoton events.

11.2.6 Tissue Heating – Nonlinear Interactions with a Laser Beam

When light is brought to a tight focus, the electric field becomes very intense and multiphoton events will occur, possibly including avalanche ionisation (see Section 2.2). This leads to nonlinear absorption, *i.e.*, what was transparent becomes opaque. At the interaction zone a microplasma is formed, creating a cavity in fluid media and a strong acoustic signal. As the plasma is extinguished, a cavity is left behind which may grow or diminish depending on the nature of the tissue. The amount of collateral damage varies

Table 11.6 Typical parameters for photodisruption

Property	Pulse length		
	(ns)	(ps)	(fs)
Intensity (10^{12} W cm^{-2})	0.05 [3]	0.05–1 [3]	5–10 [3,8]
Fluence (J cm^{-2})	10–100 [4]	2–10 [4,5]	1–3 [4,5,8]
Pulse energy (μJ)	100–10,000 [3]	1–5 [3]	0.5–3 [3,8]
Amplitude of acoustic transient at 1-mm distance (bar)	100–500 [7]	10–100 [7]	1–5 [8]
Diameter of the cavitation bubble (μm)	1,000–2,000 [5]	200–500 [6]	< 30 [8]

with pulse energy above an intensity-dependent threshold. Shorter pulses produce less collateral damage and can create fine cuts. If the pulses are sufficiently short however (femtoseconds), self-focussing and dispersive pulse broadening may inhibit the process. Laser-induced breakdown is used in spectroscopy (Section 1.4.9) and in cutting transparent tissue in the eye, intraocular surgery, such as piercing the lens' posterior capsule if it becomes opaque after some months following a cataract operation. Femtosecond photodisruption is used in refractive corneal surgery where collateral damage is an important consideration.

Typical values for nonlinear photodisruption are shown in Table 11.6 [1–5].

11.3 Medical Applications of Lasers

The growing numbers of therapeutic applications of the laser were listed in Section 11.1. These will now be discussed in turn.

11.3.1 Ophthalmology

There are three basic ways in which the laser has been used so far in eye surgery. They are photocoagulation, photodisruption and photoablation.

11.3.1.1 Photocoagulation

This was one of the first uses of the laser in eye surgery. Parts of the eye are transparent to wavelengths between 400 and 1,200 nm, which is the visual range; hence, radiation of such wavelengths can be used to heat and probe within the eye, a far less intrusive process than using knives and tools. One of the first and continuing uses of the laser has been in the treatment of *detached retinas* caused by a variety of events, one being diabetes, which causes retinal haemorrhaging, microaneurysms and retinal neovascularisation and results in blindness in many people. If the disease is caught early, its progress can be halted by the technique of pan-retinal photocoagulation in which a series of small welds or scars are created with the blue light from an argon laser around

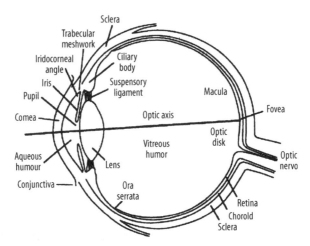

Figure 11.7 The structure of the human eye [62]

the macula region of the eye, which is the most sensitive zone, containing the colour receptors (the first operations used a ruby laser working with red light). The low-power light is absorbed in the pigmented epithelium of the retina and causes coagulation and hence a further bond between the retina and the holding material of the sclera (see Figure 11.7). The result is a slight loss of peripheral vision but a halt to the otherwise inevitable progression to total blindness. The previous treatment involved removing the eyeball and treating from behind!

11.3.1.2 Photodisruption

With the onset of aging or over exposure to ultraviolet or infrared light, the lens of the eye may become cloudy, causing impaired vision, a condition known as a "*cataract*". In treating cataracts, the normal procedure is to remove the crystalline lens and insert a replacement plastic lens into the same posterior capsule which held the original lens; such a procedure avoids possible infection and loss of the vitreous humour. In about one third of the cases this posterior capsule may cloud over and need to be removed. This used to be done in an intrusive way with a needle but now by using a Nd:YAG laser, the *posterior capsulotomy* is a simple outpatient operation.

Another use of photodisruption in ophthalmology is in the treatment of glaucoma. The eye has a stream of fluid passing across the lens generated in the ciliary body and passing between the iris and the lens into the cavity in front of the lens. From there the aqueous humour flows through the trabecular meshwork and so into the tear ducts. Blockages in this flow can occur either at the iris/lens passage (*closed-angle glaucoma*) or at the trabecular meshwork (*open-angle glaucoma*). In either case pressure builds up in the eye, causing some pain but also a possibility of blindness through stress on the retina. The treatment by laser for closed-angle glaucoma is to use a Nd:YAG laser to drill tiny holes in the iris in a procedure known as *iridectomy*. Open-angle glaucoma is treated at present by coagulation of the trabecular meshwork (*trabeculoplasty*) using an argon ion laser. It has also been treated by the photoablation of the trabecular mesh using a XeCl excimer laser with fibre-optic delivery. More recently, consideration has been given to using an Er:YAG fibre laser for a photodisruption of the trabecular mesh.

11.3.1.3 Photoablation

In 1983 Steve Trockel and R. Srinivasan proposed the idea of using photoablation to machine the eyeball to adjust the shape to cure myopic conditions. They used an ultraviolet excimer laser working at 193 nm to reshape the cornea. This enterprise resulted in two techniques: the LASIK and PRK procedures. The concept of corneal machining is illustrated in Figure 11.8. The two processes differ in the location of the material that is removed.

In PRK a thin 70-μm-thick "skin" cell layer called the "epithelium" is removed mechanically from the surface of the cornea and then the ultraviolet laser machines an appropriate thickness from the cornea of up to 10–100 μm. The superficial "skin" grows back rapidly within 24–48 h. This healing process removes machining irregularities. PRK is a safe and established process for low or medium myopia (near sightedness) of up to 6 dioptres; above that too much material would have to be machined and scar tissue may start to interfere.

In LASIK, developed in the early 1990s by Pallikaris, the inner part of the cornea is machined after the mechanical lifting of a surface flap of approximately 160-μm thickness is made from the corneal surface. The flap is replaced after the operation and it heals back without needing stitches. Myopic corrections of well over 6 dioptres can be made this way as well as corrections of more than 4 dioptres for astigmatism or hyperopia (far-sightedness). A more recent version of this process is IntraLASIK, in which the flap is cut by a femtosecond laser, which has the advantage over a knife of better precision in position and depth and in producing a more uniform flap thickness, which

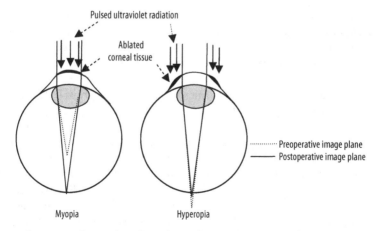

Figure 11.8 The areas to be machined in photorefractive keratectomy for myopia, in which the lens is too strong, and hyperopia, where it is too weak

avoids complications such as "buttonhole flaps" and thin and thick flaps. Femtosecond machining is also used in corneal transplants.

A more extraordinary development has been retinal prosthesis, in which an artificial retina is linked to the optic nerve when the retinal cells die owing to retinal pigmentosa or macula degeneration. The Space Vacuum Epitaxy Center at Houston, Texas, has made an artificial retina from ceramic photocells – 100,000 each 1/20th the diameter of a human hair. So small is the sensor that it is mounted in a soluble plastic sheet 1 mm × 1 mm for the surgeon to be able to handle it. The device is simply inserted into the retinal space. It requires no encapsulation and is porous to allow fluids to flow through it, thus avoiding atrophy. It may prove a partial cure for sufferers of macular degeneration or retinal pigmentosa; both are diseases of the retina sensors and not the optic nerve [6]. In other forms the artificial retina is a form of video camera [7].

11.3.2 Surgical Applications

Surgical applications are now fairly commonplace. The laser interacts with tissue as already noted by four routes: coagulation, photoablation, photodisruption and vaporisation (see Figure 11.1); all of these can lead to a severing or removal of the tissue material. The extraordinary advantages that the laser has as a surgical tool are:

- It cauterises as it cuts, except on the vaporisation route. This reduces bleeding and is sterile.
- It is a noncontact process; allowing cuts closer to tumour material without risk of cross-infection.
- No mechanical stress allows the surgeon to cut more accurately and with less trauma and tissue damage.
- The beam can be delivered by a variety of optical tooling, including endoscopes, transparent knives (made of sapphire) and by incisionless surgery for translucent material.

This last point has revolutionised keyhole surgery and led to many operations becoming outpatient events rather than something more serious. Surgical applications may benefit from the newly developed ultrafast pulse lasers (femtosecond to nanosecond pulses). The very short powerful pulses ionise the target faster than thermal energy can diffuse to the surrounding material, (as discussed in Section 11.2.5). This limits collateral damage and allows repeatable micron-sized cuts. There is little if any burning and tearing of the neighbouring tissue, unlike in all other surgical techniques. The ultrashort pulses do not cauterise like slower pulses and thus cuts made this way bleed. More extraordinary is that the pulsed beam may be focused beneath the skin, allowing some types of incisionless surgery without damage to the intervening tissue.

Ultrashort femtosecond pulses have also made possible a precision in surgery not previously possible. Some recent work by Garwe *et al.* [8] at the Friedrich Schiller University in Jena, Germany, has been able to dissect chromosomes using high numerical apertures and 170-fs pulses at 800 nm. They achieved cut sizes of 85 nm, less than the diffraction limit, by using the "sharpened pencil" effect whereby only the central power intensity is sufficient for the ablation. The implications of surgery at this scale are immense (see Section 11.5.5.1 on optical tweezers). Molecular biology is routinely capable of marking a section of a molecule with a dye and then breaking that section with a resonant frequency laser pulse (see Section 11.3.2.9). Some examples of surgical applications are given next.

11.3.2.1 Urology

The breaking of the brittle urinary calculi usually composed of calcium oxalate with very little water content, if any, can be achieved by inserting an optical fibre via the bladder into the urethra until it is in contact with the "*kidney stone*" and then applying pulses from a dye laser until the stone is fragmented. The fragments can then usually be passed naturally by the patient. An alternative technique is to smash the stone by ultrasound. This is not always possible if the stone is shielded by the pelvic bone. This is when the laser is often used.

Bladder tumours on the inner wall of the bladder can be destroyed by either coagulation of the tissue or photodynamic therapy (see Section 11.3.2.9).

Prostate gland problems, particularly enlargement, hypertrophy or benign prostatic hyperplasia affecting older men, concern the growth of the gland causing problems with urination. It can be treated by pharmaceutical medicines to control the growth but if they are insufficient, then surgery through the urethra may be required. Laser coagulation is an alternative which offers less bleeding and hence the reduced need for a blood transfusion. In the procedure a fibre optic is inserted up the penis into the prostate constriction. It is then pushed into the gland, where it heats a portion causing necrosis. As the dead tissue is absorbed over the next few days, the prostate will shrink. This procedure called *laser-induced interstitial thermotherapy* can be done as an outpatient appointment. The extent of the damage zone will depend on the laser power and wavelength, the exposure time, the geometry of the laser applicator and the nature of the tissue. The applicators are designed to give a uniform illumination over a specific area by having diffusive surfaces – roughened or frosted windows. Figure 11.9 illustrates the procedure.

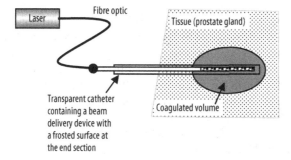

Figure 11.9 Laser-induced interstitial thermotherapy

11.3.2.2 Gastroenterology

Stomach ulcers have been a problem for many centuries, leading to strict diets and painful procedures. Today a bleeding ulcer can be coagulated by laser as an outpatient procedure. The patient is encouraged to swallow an endoscope (which is not much fun!) which contains an inflater, allowing the area to be seen, a light, a camera, and an optical fibre delivery system, which puts the surgeon in a position to operate. The patient still needs to follow some form of diet but not just bread and milk.

11.3.2.3 Gynaecology

Cervical cancer is a killer. It has a precondition of cervical dysplasia on the surface of the cervix. The cervix can be damaged by surgical treatment, which may affect the ability of the patient to bear children in the future by reducing the ability of the cervix to retain a fetus for the full term. The laser treatment to remove the diseased tissue can preserve the mechanical strength of the cervix and reduce blood loss.

In *endometriosis*, a common condition affecting middle-aged women, the endometrium (from *endo*, meaning "inside", and *metra*, meaning "womb") is found to be growing outside the uterus, on or in other areas of the body. Normally, the endometrium is shed each month during the menstrual cycle; however, in endometriosis, the misplaced endometrium is usually unable to exit the body. The endometriotic tissues still detach and bleed, but the result is far different: internal bleeding, degenerated blood and tissue shedding, inflammation of the surrounding areas, pain and formation of scar tissue may result. In addition, depending on the location of the growths, interference with the normal function of the bowel, bladder, small intestines and other organs within the pelvic cavity can occur. The laser forms part of the tooling for *laparoscopy* owing to its ability to coagulate and evaporate endometriotic implants very precisely. Laparoscopy is very useful not only to diagnose endometriosis, but also to treat it. With the use of scissors, cautery, lasers, hydrodissection or a sonic scalpel, endometriotic tissue can be ablated or removed in an attempt to restore normal anatomical structure. There are alternative treatments which include *hysterectomy* (removal of the uterus and surrounding tissue).

11.3.2.4 Otolaryngology

Hearing loss is frequently caused by the calcification of the tiny bones of the middle ear. These can be surgically removed by laser ablation and replaced with a cochlea implant, some of which are very sophisticated.

Growths on the vocal cords were removed from very early on in the application of lasers. The risk of a fire in the throat through the use of oxygen to keep the patient alive is real and has led to the use of special endotracheal tubes to keep the air to the patient's lungs separate from the operating area.

Sleep apnoea and snoring can be treated by reshaping the uvula in three or five 10-min sessions under local anaesthetic [9].

11.3.2.5 Dermatology

There are various dermatological laser applications, nearly all are cosmetic. They are based on the principle of selective photothermolysis, targeting a specific chromophore (colour centre) in the skin and ablating it. The applications include:

- removal of unwanted hair on any part of the body;
- treatment of acne;
- removal of unwanted pigment spots (sun spots, age spots, freckles, port wine stains, spider veins);
- tattoo removal; and
- Skin rejuvenation, wrinkle removal.

Hair removal is achieved by the destruction of the hair follicle by selectively targeting the melanin in the hair sheath and shaft with a diode laser operating at 800 nm and 10–40 J cm^{-2} and focused to a large spot of 9×9 mm^2 and with irradiation times of 5–30 ms. This is deemed to be able to heat the follicle structure sufficiently to deactivate it. People with grey, white or blond hair have insufficient melanin in the hair follicle for this procedure to work. They have some hope in that melanin can be infused over a 14-day treatment period using melanin encapsulated in a liposome that attaches the chromophore to the follicle. Laser hair removal systems have been around for years now and hair removal clinics have sprung up all over the world using alexandrite, diode or ruby lasers. The first generation of lasers were very ineffective, but now after years of improvements there are many lasers on the market that can be used for fast, effective, long-term hair removal. Some have very efficient cooling systems to cool the skin before the laser pulse. This limits oedema (swelling) and erythema (redness) of the skin [10]. The treatment is not as permanent as the advertisers would like one to believe. This treatment acts as an interesting example of how laser phototherapy works. Reference to Figure 11.2 shows that as the radiation penetrates tissue at certain wavelengths only

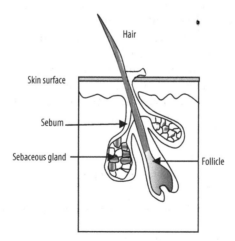

Figure 11.10 How hair is held in the skin. Only the melanin in the layer surrounding the follicle will strongly absorb 800 nm radiation

parts of that tissue (illustrated in Figure 11.10) may be absorbing the radiation significantly. In hair removal the wavelength is usually around 800 nm from a diode laser. The absorption of different parts of the skin is listed in Table 11.7.

If the dosage is 30 J cm^{-2}, then over 1 μm^2 it is 30×10^{-8} J μm^{-2}. If this is absorbed over a depth of 1 μm, the temperature rise can be calculated by a heat balance on the micron cube:

$$30 \times 10^{-8} = \rho C_p \Delta T = 3.96 \times 10^{-9} \Delta T,$$

where ρC_p is specific heat per unit volume (3.96×10^{-9} J °C^{-1} μm^{-3}).

Hence, $\Delta T = 75°C$.

This is sufficient to cause necrosis of the absorbing material, in this case melanin in the sheath of the hair follicle. Note that the more transparent parts will be unaffected, but any larger dosage may endanger them; also note how important the melanin concentration is for the success of the procedure.

Treatment of acne. Acne is caused by the clogging of the hair follicles in the skin. The clogged follicle becomes a fertile breeding ground for the bacterium *Propionibacterium acnes* – the main cause of acne. The *P. acnes* bacteria thrive when there is a lack of oxygen. New research has shown that by targeting the pore with light, it creates an oxygen-rich environment that destroys the *P. acnes* bateria. Understanding these principles has led to the development of acne clearance treatments. In one such treatment the area infected is irradiated first with red light then with green light to effect the breakdown of the bacteria and finally with heat to speed up the destruction of the *P. acnes* bacteria.

The removal of unwanted pigment spots. Many of these are caused by vascular lesions. They can be destroyed by a laser targeting the chromophore in haemoglobin (around 450 or 570 nm). One form of this is the *port wine stain* or *birthmark* caused by a number of abnormal blood vessels located in the dermis. Normal blood vessels are 90% empty, except when flushed as a result of exercise, cold, eating spicy food or embarrassment. In port wine stains they are normally full. The vessels are typically 40–100 μm in diameter. Thus, to avoid too much thermal diffusion and yet give sufficient time to heat the blood vessel, short pulses of around 1 μs are required. Successful treatment has been achieved using a dye laser tuned to 577 nm, an absorption peak for oxyhaemoglobin, and pulse lengths of 300 μs. Some clinics use multiple laser devices, short wavelength followed by longer wavelengths for deeper lesions, through an extended treatment protocol. The use of selective epidermal cooling, which permits the use of higher light dosages, is recommended to expedited lesion clearing.

Table 11.7 Absorption depth for different parts of tissue for 800-nm radiation

Material	Absorption coefficient at 800 nm (μm^{-1}) [a]	Depth of penetration
Melanin	$10^{-1.9}$	79 μm
Haemoglobin	$10^{-3.3}$	2 mm
Water	$10^{-5.5}$	3.1 m

[a] See Figure 11.2

Freckles and moles contain their own pigmentation. Skin is being constantly renewed by epidermal skin cells migrating upwards as they dehydrate to become skin, which is then sloughed off. In the generating basal layer there is a sprinkling of melanocyte cells, whose function is to generate the pigment melanin and inject it into the newly forming cells. Benign freckles and moles can be successfully treated by the laser destroying the melanocytes as with hair removal. The usual laser to be used is a Q-switched frequency-doubled Nd:YAG (green light) or ruby (red light) laser. The green light is particularly effective in destroying the melanocytes without too much collateral damage. However, if the pigmented lesion is suspected of being malignant, then this should be tested by biopsy and treated by conventional excision if it is found to be malignant.

Tattoos consist of artificial pigments that have been inserted in the skin layer. The pigments are insoluble (*e.g.*, carbon particles) and of sufficient size to prevent the normal bodily functions for their removal, such as phagocytosis. The laser can deliver short, sharp pulses that are sufficient to break the particles into smaller particles that can be digested but sufficiently quickly that collateral damage is minimal. Typically pulses would be around 1 ms from Ti:sapphire, alexandrite, dye or Q-switched Nd:YAG lasers, both fundamental and frequency-doubled. The frequency depends on the dye that is to be removed.

No single laser can remove all colours used in tattoos. A Q-switched Nd:YAG laser is used for black, grey, white and dark-blue inks; a Q-switched red ruby laser is used for black, light-blue and green inks; and a Q-switched frequency-doubled Nd:YAG laser (green) is used for red, purple, brown and orange inks. Figure 11.11 shows the stages of the removal of a tattoo mark using a Ti:sapphire laser.

Skin rejuvenation. The skin can be resurfaced by ablating the outer layers of the epidermis with a CO_2 (10.6-μm) or Er:YAG (2.94-μm) laser. This stimulates the regenerative potential and can remove sun damage and age-related changes such as small wrinkles. In this process a pulsed CO_2 laser of very precise power interacts with the collagen in the skin, causing up to 30% shrinkage. The laser must heat to a temperature

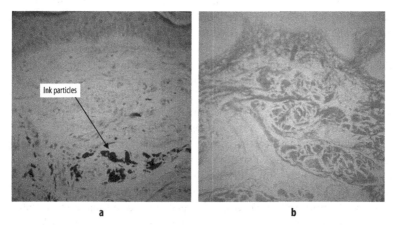

Ink particles

a

b

Figure 11.11 Subcutaneous tattoo ink removal by laser ablation: **a** before treatment, and **b** after treatment (Courtesy of Lynton Laser Ltd., Crewe)

of between 58 and 61°C to achieve the effect but avoid irreversible damage – hence the "precise" power.

Fat reduction. Still at the research stage is work in which a free-electron laser of the correct frequency to selectively interact with cellulite and body fats is used to heat the fatty cells and so destroy them by photothermolysis, after which they are removed by natural processes. The correct wavelength can be passed through the skin without breaking it [11].

11.3.2.6 Cardiology

One of the major areas of interest in cardiology is in how to keep the blood flowing to the heart. Narrowing of the arteries (stenosis) is caused by the build-up of plaque made of cholesterol, fat cells and in some cases calcified tissue. It was hoped in the early days of laser surgery that the laser would be able to clear this build-up. The application of an argon ion laser for *laser angioplasty* showed that it could ablate the soft plaque but not the calcified tissue and that there was a danger of perforating the artery. These problems have been overcome with pulsed ultraviolet lasers and special wire-guidance devices. However the renarrowing (restenosis) of the artery after treatment is no better than with the alternative treatments in which a balloon is used to expand the artery (balloon angioplasty). The restenosis can be reduced or even cured by the use of a stent. A stent is a flexible metal tube usually made by laser cutting (see Section 11.5.1.1).

Laser angioplasty was authorised by the US Food and Drug Administration in 1992. It is currently used in less than 1% of procedures owing to the advances of other techniques such as coronary artery bypass surgery and balloon angioplasty with or without stenting. The choice depends upon various factors, which include where the blockage is, how many blockages there are and the extent of the blockage(s). In laser angioplasty a thin plastic tube (a catheter) containing a wire and an optical fibre is passed into the femoral artery in the groin, from where it can be pushed up into the heart. The position of the catheter is monitored by X-ray or similar detection techniques. When the plaque is encountered, a thin guide wire is pushed through the plaque, followed by the optical fibre, which passes over the wire, ensuring that the laser pulse will not strike the artery wall. The laser may be used to clear the blockage by ablating the plaque or may be used to open a hole into which a balloon can be inserted. An alternative for hard deposits is atherectomy, in which a rotary shaver is used. Although the prognosis after laser angioplasty is generally good, there are a number of risks associated with the procedure. These risks include relatively high risk of reclosing (restenosis) of the artery, heart attack and a tear in the artery or heart tissues (coronary perforation). Other risks of the procedure are similar to those of any of the other methods of angioplasty and include stroke, abnormal heart rhythm (arrhythmia), emergency bypass surgery and allergic reaction to the dyes or stent material used in the procedure.

More recently a new procedure has been developed called *transmyocardial laser revascularisation* [12], in which a CO_2 laser is used to drill a hole directly through the heart muscle into the left ventricle. Oxygenated blood then perfuses into the heart muscle directly via this new passage. The outside hole is sealed by blocking until a clot forms. This is how a reptilian heart works and this drastic procedure has been shown to be effective in extreme cases. It is not clear why it works, but it has been suggested that the laser hole stimulates the rejuvenation of new blood vessels in the heart muscle. A version of this *percutaneous myocardial revascularisation* was first tried in 1997 at the Papworth Hospital in Cambridgeshire, UK. In this operation the laser is introduced by catheter through the groin and when the laser is in the left ventricle, powerful pulses drill around ten to 20 holes. The operation is conducted with the patient sedated but not anaesthetised so the patient can actually watch – if he or she wishes! [13]. A further development in 2006 [14] was to perform the operation robotically via a small plaster-size incision between the ribs, through which an optical fibre can be inserted to perform the heart-drilling procedure.

Many of us have heard of a "heart attack"; such a condition is often caused by a *thrombus* (blood clot) which gets stuck in an already partially blocked coronary artery, and so cuts off the blood flow, often quite suddenly. Emergency treatment is essential before serious damage can occur through lack of oxygen in various organs – such as the brain. There are some effective drugs, but many of these are unsuccessful or even dangerous to the patient. The thrombus type of clot is very soft compared with the artery wall or plaque and can be cleared easily by a laser working at a power well below that required to damage the artery wall. The laser beam can be delivered by using a liquid-filled catheter, in which the liquid acts as a waveguide.

11.3.2.7 Orthopaedics

The main applications for the laser in orthopaedics are in cartilage dissection in the knee or spine. It is used to cut, shrink or smooth the cartilage; for example, in meniscal cartilage smoothening or capsular shrinkage of the shoulder joint. The subject of orthopaedics was greatly enhanced by the invention of the arthroscope, with which the surgeon can look into a joint without cutting it open (percutaneous microsurgery). This device is often operated with a Ho:YAG laser at 2.1-μm wavelength, with the beam being delivered by a fibre. The beam can be passed a short way through the saline solution within the joint. The damaged section of torn herniated or bulging spinal discs can be reduced by heating the soft internal tissue of the cartilage, causing localised necrosis, as with laser-induced interstitial thermotherapy for prostate reduction. The reduction in size reduces the pressure on nerves with a consequent relief of back pain. Rough arthritic joints can be smoothed.

11.3.2.8 Dentistry

It is early days yet for any large-scale use of lasers in dentistry. Many ideas have been explored, from drilling out caries, sealing fissures in enamel, inhibiting demineralisation, etching the enamel, preparing retention pin holes to cleaning and preparing root canals. It is essential in laser drilling that the temperature of pulp of the tooth does not rise more than 5.5°C to avoid irreversible damage. A novel dental drill which passes an Er:YAG (2.9-μm wavelength) laser beam down a water spray, the Waterlase[1] (Figure 11.12), is proving successful in achieving quiet, painless drilling with very little

---

[1] Waterlase® is a registered trademark of BIOLASE Technology. http://www.biolase.com

Figure 11.12 The Waterlase® dental drill for painless drilling

heating. A wavelength of 2.9 μm is strongly absorbed by both water and hydroxyap-atite, which is the mineralised part of the tooth material.

11.3.2.9 Photodynamic Therapy

A more scientific use of the laser in medicine has been developed with the use of light-activated chemotherapies. Photodynamic therapy is a totally new form of medicine whereby drugs can be switched on and off by light – well, not so new since Neils Finson was awarded the Nobel prize in medicine in 1903 for using light treatment to cure lupus, but since the development of the laser this area of medicine has taken a significant stride forward. It should be noted that this is a photochemical process and does not necessarily involve any heating.

In the photodynamic therapy treatment of certain forms of cancer the patient is asked to take a porphyrin or tetracycline derivative (*e.g.*, Photofrin®[2]; the porphyrin family of chemicals include chlorophyll and haemoglobin oxygen carriers). Photofrin® is a light-sensitive nontoxic dye which is usually injected 48–72 h before phototherapy, although some recent work has shown that treatment can be performed in 20 min after injection. With skin cancers the drug is administered as a cream and the induction time is around 4–6 h. After injection or cream treatment the patient must stay in the dark and wear dark glasses until the natural body tissues have absorbed and excreted the dye. However, the higher metabolic rate of tumour tissue causes the dye to be accumulated in the tumour. Thus, there comes a time when the tumour is marked out specifically by the porphyrin dye. This can be checked by the surgeon, who illuminates the area of the tumour in a hollow organ with a specific frequency to cause the drug to fluoresce. In some cases two frequencies are used and a computer mixes the images to establish a very clear picture of where the tumour is for the surgeon to be able to treat it. The sur-

---

[2] Photofrin® is a registered trademark of Axcan Pharma PDT Inc. 22 Inverness Center Parkway Birmingham, Alabama 35242, USA. http://www.axcan.com

geon then changes the frequency of the irradiation to a specific value (around 800 nm for Photofrin®). The porphyrin will undergo a photochemical reaction by offloading an electron to any oxygen atom in the vicinity. This becomes a highly unstable free oxygen radical, with a half-life of a few tenths of a second, which will oxidise any molecule with which it comes into contact, thus destroying it. Since the dye was by now attached to or inside the cancer cell, the host is destroyed [15]. In those cases where the treatment has been successful the patients have recovered as if they had not had the cancer. There is no surgical damage or scar since the killed cells are absorbed naturally. The process is developing fast; there are now over 1.5 million publications on the subject! It is currently used on skin (but not for melanoma) and for cancers in organs where the light can be projected, such as throat (oesophagus), stomach, bladder, neck, bile duct, mouth and lungs. One of the more successful treatments has been for age-related macula deterioration of the eye in elderly people. In the later stages neovascularisation may cause bleeding leading to blindness.

With photodynamic therapy a "verteporphin photosensitiser" is injected intravenously to accumulate in the neovascular tissue. Irradiation at 690 nm leads to cytotoxicity and destroys the neovascular tissue. Development areas are centred on the search for light-sensitive nontoxic dyes which have good differentiation between normal and cancerous cells and for optical instrumentation which can deliver uniform dosages of radiation even in solid organs such as the liver or brain. Two devices for doing so are illustrated in Figure 11.13.

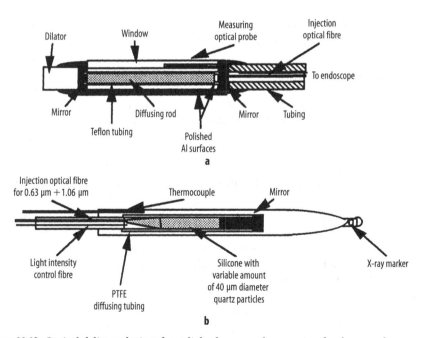

Figure 11.13 Optical delivery devices for: **a** light dosage and measuring for the oesophagus, and **b** light diffuser for simultaneous photodynamic therapy and hyperthermia in the oesophagus

Although Photofrin® selectively attaches itself to cancerous cells, up to 25% of it may attach to healthy cells, which will also be killed. A further problem was the lack of oxygen within a cancerous tumour, meaning the oxygen to form the singlet was not always available. The use of chlorophyll-based photosensitisers which carry their own oxygen has given the subject a boost (chlorophyll and haemoglobin are structurally similar but differ in haemoglobin having an iron atom while chlorophyll has a magnesium atom held by four nitrogen atoms in a flat chelate ring; vitamin B_{12}, which is similar, has cobalt in such a position). The Russians were the first to investigate algae and bacteriochlorophyll. Their red chlorophyll was sensitive to infrared light, which most significantly allowed treatment deeper into tissue. This stimulated the development of an improve chlorophyll-based sensitiser called "Photo Flora" based on spirulina (a blue–green alga of genus *Arthrospira*). This natural agent circulates easily in the bloodstream attached to lipoproteins, without toxic effect, because of its similarity to haemoglobin. Cancer cells produce their energy without oxygen in a fermentation process: they are acidic compared with normal cells, which are alkaline; they are also negatively charged and they need glucose to drive their activity. Photo Flora has two positively charged areas which energetically bond to the cancer cell; the acidity helps to detach it from the lipoprotein and it enters the cancer cell because it acts like glucose. The risk of binding to healthy cells is less than 3% compared with 25% for Photofrin®. The particularly interesting aspect of these latest agents is that they are sensitised by radiation that can penetrate more deeply than that required for Photofrin®, possibly allowing treatment of deep organs. It is not surprising that there is less cancer in countries where the inhabitants eat more vegetables and enjoy the sunshine, such as the Mediterranean or rural communities. The game is young but the method of play is becoming visible. The concept of a golden bullet for cancerous cells is a dream which could come true. In the meantime, eat vegetables and sunbathe – but not too much!

11.3.2.10 Biostimulation

This is a subject of great interest and controversy eliciting considerable scientific cynicism, yet one might wonder why. We are all familiar with a hot water bottle or an ice pack to allay the pain of muscle ache or bruising. For many years the beneficial effects of an infrared lamp have been enjoyed by those with lumbago and other aches and pains. So it is not so surprising that there are some 3,000 papers on the subject of low level laser light therapies suggesting that there is some interaction between this level of radiation and tissue such that some good things can happen. There is some parallel with acupuncture treatment, in fact in dentistry the low-power laser has been used instead of oral acupuncture. There appears to be a biological window of dosage in strength, duration and optical wavelength which is able to stimulate biological processes, such as gene expression, lymph flow, blood vessel dilation, cell membrane permeability and release of transmitter substances such as adenosine 5'-triphosphate production (adenosine 5'-triphosphate is a multifunctional nucleotide that has a role in transporting chemical energy within cells for cell metabolism). Such small events can radically affect the success of a healing process. The typical range of laser parameters is 50–500 mW from a diode laser operating in the near infrared for a few minutes per treatment. The way biostimulation has so far been studied includes [1, 16, 17]:

- enhanced wound healing – examples are venous leg ulcers, secondary spinal cord injuries in which the immune response is affected, pressure ulcers (bed sores) and open wounds;
- chronic joint disorders;
- degenerative osteoarthritis of the knee (treatment is of the form of 15 min twice a day for 10 days);
- rheumatoid arthritis;
- herpetic neuralgia;
- treatment of herpes simplex – removes the blisters within days, without side effects, repeat treatments will lower the incidence of recurrence;
- sinusitis, in which low level laser light therapy leads to a fast reduction of symptoms in many cases;
- postoperative pain particularly in dentistry;
- treatment of inflammation;
- oedema (swelling); and
- pain treatment.

The jury is still out on understanding what is happening just as it is on water divining, acupuncture and other practices that are hard to follow but appear to show results. Meanwhile many patients are being successfully treated, possibly psychosomatic or possibly real outcomes, we will find out one day; meanwhile carry on enjoying the sunshine!

11.3.2.11 Laser Tissue Welding

The early attempts at joining the severed ends of blood vessels by heating with a laser to make the surface proteins sticky was successful but needed sutures to hold them together while the tissue grew in strength. The result was no better than a good vascular surgeon could achieve by standard suturing. Control of the laser energy was critical, since if the heating is too strong, cauterising, or burning, occurs and the structure no longer has adequate internal openings to allow healing fluids to flow. Burns are notoriously difficult to heal. Recently, however, the concept of "laser soldering" biological tissue has been demonstrated with considerable success. In this process the joint is coated with a protein material, a human serum albumin, such as fibrin; the two ends are held together with a soluble scaffold, made of another albumin derivative; the joint is then heated by an 810-nm (for penetration of tissue and absorption by haemoglobin) diode laser to initiate the bonding.

This procedure has advanced way beyond simply postoperative bonding for joining severed blood vessels. It has successfully stopped bleeding in traumatised livers [18]. It shows promise of doing the same in kidneys, spleen and solid visceral organs. The hope is that one day it may be applicable to corneal grafts, middle ear repairs, arteries, bronchi, veins, urethra and skin closures. What is missing at the scientific level is a clear understanding of how electromagnetic radiation interacts with the complex chemistry of living organisms.

11.4 Medical Diagnostics

Considerable research is being devoted to laser-based medical diagnostics since the techniques for visualising tissue based on confocal microscopy, holography or tomography, and those for analysis based on high-resolution spectroscopy such as Raman spectroscopy and fluorescence are all nonionising, potentially inexpensive and noninvasive procedures compared with the expensive methods of magnetic resonance imaging and computed tomography.

11.4.1 Absorption Techniques

One of the simplest laser-based analytical systems is the oximeter, which is becoming prevalent in many hospitals. The pulsed laser oximeter uses infrared laser light at two wavelengths to determine the heart rate and blood oxygenation in clinics. This information is important in newborn babies and trauma patients. Oxygenated and deoxygenated haemoglobin have different absorption spectra. Thus, using two wavelengths, one can distinguish them. The sensor is clipped to the patient's finger or ear lobe and the infrared radiation is able to penetrate that thickness and give a continuous record of the goings on in the blood.

11.4.2 Spectral Techniques

11.4.2.1 Fluorescence

To ascertain whether there is a clinical problem, the physician may require a sample of the diseased tissue. This is usually obtained, with a certain level of pain and danger, by cutting out a biopsy sample. This sample is then analysed in the laboratory, which can be time-consuming. It is now possible to bypass this procedure to probe and quantitatively monitor a disease *in situ* using a laser fired through an endoscope, laparoscope, catheter or hypodermic needle to stimulate the natural fluorescence; most tissues can be analysed spectroscopically. Spectra obtained this way can, after careful analysis, yield data on the state of oxidation of the tissue, the nature of the plaque in an artery and whether the tissue in many organs is malignant or benign. For example, a comparison of the fluorescence spectrum from normal or adenomatous (benign tumour) colon tissue is shown in Figure 11.14. The fluorescence intensity and the ability to differentiate tissues can be greatly enhanced by using fluorescent dyes, as in photodynamic therapy. This allows the surgeon to gain a clear picture of the extent of a tumour or other tissue.

11.4.2.2 Raman Spectroscopy

Fluorescence is incapable of giving detailed biochemical information, which would have been obtained through standard histology. To obtain this information *in situ* can be done using the Raman spectrum obtained in a similar manner to the fluorescence spectrum. The weak Raman signals can be enhanced to allow analysis in a reasonably

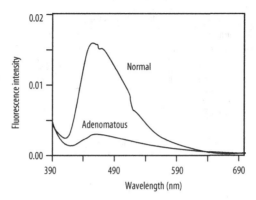

Figure 11.14 Laser-induced-fluorescence spectrum of normal and adenomatous colon tissue, illustrating the diagnostic capability of this technique. The excitation wavelength was 369.9 nm [62, 63]

short time with the latest CCD spectrometers. When the signal is deconvoluted by computer it yields data on the molecular vibrations of the tissue material. This can be interpreted accurately by a consultant to show what the structure of the observed material is. Thus, it is possible *in situ* to determine the nature of a deposit in an artery and to monitor its progress from one appointment to the next.

An exciting development is to observe the Raman spectrum from fluid in the eye, from which the blood sugar levels can be measured together with considerable further data. The Raman spectrum from tissue at the top of the nose is thought to give an early indication of Alzheimer's disease. Soon it may be that when you visit your physician, he or she will ask you to breathe into a machine like a breathalyser which will enable him or her to get diagnostic data on a variety of diseases simply through the Raman analysis of your breath; such a machine is being researched [19].

11.4.2.3 Cell Cytology

Biological fluids can have their cells tagged with exogenous fluorescent chromophores (*i.e.*, dyed) and then passed through a narrow tube through which several laser beams of differing wavelength may pass to cause fluorescence on the cells of interest, for example HIV, BSE, *etc*. When a cell is so detected, the cell cytometer is able to deflect the cell to a collection point. These machines can process approximately 100,000 cells per minute. It requires only 1,000 molecules per cell to make the identification. A diagram of a cell cytometer is shown in Figure 11.15.

11.4.3 Visualisation Techniques

11.4.3.1 Laser Scanning Confocal Microscopy

Confocal microscopy has similarities with scanning electron microscopy in that the image is built up by scanning a point source over the sample and analysing the reflected

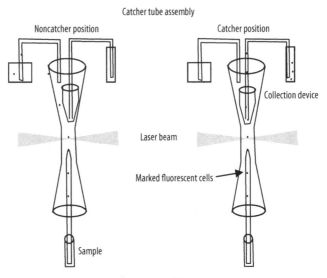

Figure 11.15 A cell cytometer

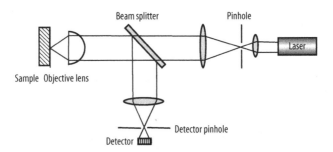

Figure 11.16 A confocal microscope [62]

radiation. In *laser scanning confocal microscopy* the point source is a tightly focused laser beam, which allows high brightness of the image; the reflected radiation is collected as shown in Figure 11.16 and refocused onto a pinhole in front of the detector. Only points that are in focus at both the object and image planes (hence the "con" in confocal) will be illuminated, insufficient light coming from any other location. This technique gives a higher resolution than an optical microscope, allows living tissue to be examined in its natural environment and can be used to optically section tissue by adjusting the position of the object plane to different depths within the sample separated only by microns. It has been used extensively in ophthalmology to create images of the retina.

11.4.3.2 Optical Coherence Tomography

This is a new confocal scanning technique using low-coherence radiation usually from a diode laser. The word "tomography" comes from the Greek word *tomos* meaning

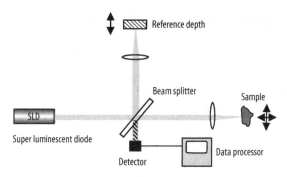

Figure 11.17 Single-point optical coherence tomography (After [62])

"slice" or "section"; tomography is a way of imaging a slice through an object. The arrangement is shown in Figure 11.17. The scanning optic is focused on a point in the sample and the reference arm is moved to create an interference with that image, thus establishing its depth position, which is recorded in the computer. The scanning optic moves to the next spot and the process is repeated until a full image of the sample in the *XY* and *Z* planes (up to the coherence length of the radiation) is obtained.

The internal structure of living cells can be imaged by optical coherence tomography to give a three-dimensional image of cell structure, such as nucleoli (clusters within a cell's nucleus) with a resolution of 2 μm. The system uses broadband pulses of less than 10 fs from a Ti:sapphire laser source [20].

11.4.3.3 Laser Tomography

This is another laser-based diagnostic tool, known as diffuse optical tomography [21–23]. In this technique a number of fibre-optic sources, arranged around the part to be examined, transmit intensity-modulated radiation through it. The emergent beams, identified by the intensity modulation, are analysed by a number of sensors also arranged around the part. The computer-generated image of the part calculated from the variations in absorption of the object when it is probed from a number of directions gives the shape of the object, as in standard tomography using X-rays or magnetic resonance, but in the case of diffuse optical tomography it also measures both scattering and absorption effects. This gives additional tissue information, such as the concentration of haemoglobin in the sample (from tissue absorption) or the stage of inflammation (given by the scattering). Light pulses passing into diffusing tissue will take several paths; some will be almost direct, known as "ballistic" signals, some will pass with minor meandering, known as "snake" signals, and the bulk will be diffused within the sample, known as "diffuse" signals. Working with the ballistic and snake signals, which can be identified by time-resolved techniques, one can build up an image.

11.5 Laser Manufacture of Medical Devices

11.5.1 Laser Cutting

The subject of laser cutting was discussed in Chapter 3. There are numerous bits of metal and plastic that are being cut by laser for the manufacture of medical devices, such as:

- Bone reamer: three-dimensional cutting of the reamer head and marking.
- Endoscopic instruments: radial cutting of the guide tubes.
- Implantable plating system: high-precision cutting together with marking. There must be no burr on plates, which can be up to 2 mm thick; variable cut angles for perfect screw fitting; smooth surfaces; good biocompatibility.
- Stent cutting: this has become an industry almost on its own and hence deserves its own section.

11.5.1.1 Stents

The growth in demand for stents and other microimplants is forecast to grow significantly in both the developed and the emerging markets of the world. By 2020 it is forecast to have doubled from the market in 2007 in Germany alone.

Stents are expandable metal grid structures used to stabilise weakened blood vessels, for example, aneurysms and blockages (Figure 11.18) [24]. The clearing of a blockage was discussed in Section 11.3.2.6. After the blockage has been cleared, a stent and a balloon are inserted into the catheter which was used for the clearance of the blockage. The stent is placed where needed as seen by X-ray and the balloon is expanded within it to extend the stent to form a scaffold for the blood vessel. The balloon and catheter are then removed and the stent stays in place. The stent is usually made of some form of biocompatible strong material such as stainless steel, cobalt chromium alloys or a memory alloy such as nitinol (51% Ni, 49% Ti); in the latter case the collapsed stent is passed

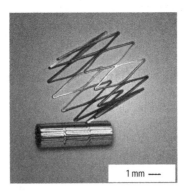

Figure 11.18 A stent shown before and after expansion. (Courtesy of AILU, http://www.designforlasermanufacture.com, photograph courtesy of Rofin-Baasel, UK)

down the catheter and when it is in the right location, it is heated to make it revert to its original shape. Stents are produced from thin-walled tubes (hypertube diameter 0.25 mm with wall thickness up to 400 μm). Filigrees, intricate grid structures that are a prerequisite for achieving their impressive diameter expansion ratios, are produced using precision laser cutting technology. This technology has advanced considerably to meet the growing sophistication of the stent itself. Any foreign body will cause a reaction in the host; to minimise this effect stents are made with drug-eluting coatings. The control of the drug requires a strict control over the surface area of the stent and hence the accuracy of the cut has increased from a scatter band on area of 15% to 6–8%. Strut dimensions have been reduced from 110 to 60–85 μm [25]. This accuracy has altered the design of the stent-making machines. They are now fitted with high brightness disc or fibre lasers with spot sizes of the order of 10–12 μm, giving a kerf width of that size. The whole machine is mounted on a solid granite vibration absorbing base and uses a linear drive X axis with automated focus control and CCD inspection. The lasers operate at 50–100 W and cut at speeds up to 500–600 mm min^{-1}, a threefold increase over what was being used in 2005. Further speed increases up to 2,000 mm min^{-1} have been attained when using internal water cooling for the tube. For greater accuracy, temperature stabilisation of the whole machine will be required.

After cutting, the stent is inspected 100% by an automatic optical system; it then goes through electroplating, etching, micro sandblasting and heat treatment prior to a final optical inspection. There is currently a shortage of stents.

11.5.2 Marking

Laser marking was discussed in Section 6.18; so here it is only necessary to mention some of the applications for medical equipment. There is a need for traceability and identification of all medical equipment. The mark needs to include the part number, size, compliance standard (often this is held in a bar code or a two-dimensional data matrix code), company name and logo, direction of rotation and scale, if appropriate. Almost all equipment is now marked by one technique or another; competing processes are electrochemical, dot printing and ink jet. The laser makes a durable mark which can be sterilised. High-frequency pulsed laser marking reduces metallurgical damage; diathermic coatings (transparent to infrared radiation) can be marked without breaking through the layer as can anodised aluminium [26].

Laser surface texturing of prostheses is being investigated to enhance the growth of protein on the surface of a bone implant as is surface cladding with hydroxyapatite.

11.5.3 Wire Stripping

Wire stripping for ultrasound equipment, endoscopes and heart pacemakers poses a problem owing to the extreme fineness of the wire (*e.g.*, 0.2 mm outside diameter of coated wire); mechanical stripping may damage the wire. Radiation of 10.6 μm from a CO_2 laser is absorbed strongly in the plastic insulator, which can be vaporised down to the wire conductor that reflects the radiation and is thus unaffected. Many of the

a   b

Figure 11.19 Examples of laser wire stripping: **a** the coiled electrode from a heart pacemaker whose insulation has been removed by a high-powered pulse from a CO_2 laser without damaging either the metal wire or the shape of the coil, and **b** the laser stripping of a microcoaxial ribbon used for video data transfer to LCD screens. Each wire has an outer diameter of 0.22 mm. The wires were processed with a combination of CO_2 and YAG shield cutting systems. (Photographs courtesy of Spectrum Technologies Ltd)

wires are now ribbons which are not easily manipulated mechanically, but which can be manipulated straightforwardly by laser stripping. Microcoaxial cables used to transmit high-frequency image signals from ultrasound equipment may contain 100–200 core wires. Endoscope cables, also containing many wires, have to be as small as possible and currently use microcoaxial individual wires of approximately 0.16 mm diameter. Each microcoaxial wire has to have its insulation jacket, earth shield and dielectric cut back to the lengths required for the termination to a transducer head or circuit board (Figure 11.19). Heart pacemakers have a lead from the power supply to the point of stimulation; this is usually a 1 mm-diameter wire containing two insulated coils for the cathode and anode. Stripping this mechanically would damage the wire and require the uncoiling of the coils. The laser can do this successfully. Laser stripping is fast becoming an industry standard [27].

11.5.4 Laser Welding

The subject of laser welding was discussed in Chapter 4. It is essential for implanted devices that the weld is strong and ductile, not liable to corrosion, biocompatible, cosmetically pleasing and does not change the dimensions of the part. The laser is used extensively in the medical device industry because of the quality of the welds and the controlled HAZ. The coming of the high-brightness lasers (disc and fibre) has created a further improvement in the market since the width of the weld can be in scale with the trend to miniaturise equipment. Examples of the areas of application are [28]:

- Cardiac pacemakers (Figure 11.20): seam welding and marking of the titanium housing and spot welding of the inner electronics.
- Neurosurgical aneurysm clips: spot welding of the clip jaw and marking for identification and traceability.

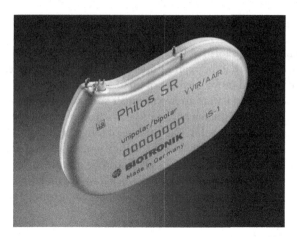

Figure 11.20 A heart pacemaker with a titanium housing that has been hermetically welded using a pulsed Nd:YAG laser. It has also had the contacts inside the housing welded and the housing marked by a pulsed Nd:YAG laser (photo: TRUMPF)

- Welding hubs for catheters.
- CCD endocam: seam welding and marking of the camera housing.
- Retrieval basket in minimally invasive surgery: spot welding of nitinol wires.
- Endoscope instrument: simultaneous nonradial cut-outs and welding of the guiding tubes to the shaft.
- Endoscope circumferential butt weld of the tip on a bronchoscope.
- Wound spreader: cutting and welding of the tip shape.
- Polymer welding, particularly for fluidic devices,
- Sealing glass ampoules.
- Balloon catheters.
- Hearing aids: spot welding.

11.5.5 Nanomedicine

Medical devices are becoming smaller and smaller and more mobile for use in many different locations. This thrust was discussed in greater detail in Chapter 8. A short list of the topics that are covered by nanomedicine include:

- drug delivery on demand;
- heart assistance;
- devices capable of crossing biological barriers; and
- nanostructured scaffolds.

The growth in microsystems in the coming years is expected to be of the order of 17% compared with 11% for macrosystems (source Optech Consulting, 2008).

11.5.5.1 Optical Tweezers

The laser can be used as a form of tweezers to hold or move nanoparticles such as cells and large molecules so that they can be operated upon. Optical tweezers use the pressure of light to hold and move particles of up to a few microns in size. The instrument focuses a few watts of radiation to a very small spot size; a transparent object at the focus will then have the refracted and scattered radiation from it exerting a small force to drive it towards the focal point. The balance of forces is shown in Figure 11.21. This "tractor beam" can be used by attaching a cell or molecule of interest to a small glass bead and moving the bead so that the sample can be examined. Working in this way, one can measure the force from muscle cells, separate sperm cells and adjust chromosomes [29]. See also section 8.5.1 referring to optical traps using Bessel beams.

11.5.5.2 Artificial Lungs

One of the more exciting possibilities is the development of artificial lungs, which is made possible by the extreme accuracy of ultrafast laser lithography. The need for artificial lungs is great. Tens of millions of people in the USA currently suffer from a lung disease such as chronic obstructive pulmonary disease, lung cancer, cystic fibrosis, or tuberculosis [30]. Many of these diseases lead to death, but all are unpleasant to the sufferer. Lung transplants are possible but are only done for extreme cases, partly through lack of donors. Current devices for aiding breathing are not of use for temporary or permanent replacement of the lung [31]. Extracorporeal membrane oxygenation [32–37] is a standard treatment for newborn babies with acute reversible breathing problems. It does, however, incur intensive clinical management, large priming volumes of blood, massive anticoagulation and multiple transfusions. Alternatives are interventional lung assistance [32, 34–40], which relies on a low-pressure gas exchange with a membrane integrated in a passive arteriovenous shunt, and intravascular lung-assist devices, such

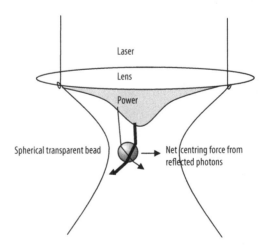

Figure 11.21 The principles of optical tweezers

as the Hattler catheter [31, 34, 41–44]. This latter device is a bundle of twisted microfibres placed within the vena cava or right atrium. Gas passes through the fibres to elute CO_2 and bring oxygen; thus, their gas exchange is intrinsically limited by surface area. Currently, the devices are intended for partial respiratory support, up to 50% basal requirement, but are not expected to serve as long-term support or as a bridge to transplant. Total artificial lungs [31–34, 45–52] have been developed with the goal of complete respiratory support for up to several months as a rescue therapy and as a bridge to transplant or recovery. The devices are designed to have low impedance, and can thus be driven by the right ventricle connecting to the pulmonary circulation, avoiding external pumps.

The gas-exchanger modules of the current generation of lung-assist devices are composed of bundles of microporous hollow-fibre membranes as shown in Figure 11.22. In structural terms, the inefficiency of current hollow-fibre blood oxygenators is attributable to their nonphysiological features: long gas diffusion distance and vulnerability to blood damage and blood activation [31, 53].

To mimic the highly complex micromesh of blood capillaries and air sacs as in a lung requires uniform blood flow through a low-impedance structure. To achieve this the size of the branching flow channels should follow Murray's law [54], which states that the cube of the radius of a parent vessel equals the sum of the cubes of the radii of the daughter vessels,

$$r_p^3 = \sum_{i=1}^{n} r_{di}^3 ,$$

where r_p is the radius of the parent channel, r_{di} is the radius of a daughter channel, and the summation includes n daughter channels.

The laser can make etched tracks according to these rules by mask-based lithography [55–57]. This, however, is slow and costly and the variation in the depth of the track is limited. Laser direct writing, on the other hand, does not need a mask and does have the flexibility to vary the machining depth. Branching networks have been prepared this way in polydimethylsiloxane using the laser-machined silicon structures as

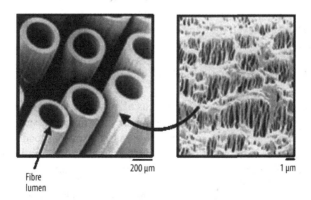

Fibre
lumen

200 µm   1 µm

Figure 11.22 Microporous hollow-fibre membranes [64]

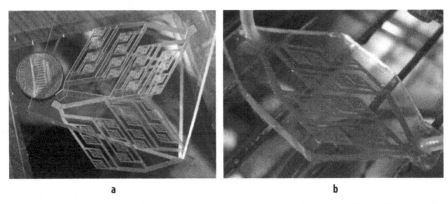

Figure 11.23 **a** Laser-machined silicon master and polydimethylsiloxane replica, and **b** blood flow in the microchannel network [55–57]

moulds, shown in Figure 11.23. The uniform flow, pressure and stress distributions in such designed structures are good as shown in Figure 11.24.

It is, however, important for the stability of the blood that the flow channels are smooth. This limits the type of laser that can make such fine smooth flow paths. A nanosecond laser will cause roughness when machining silicon owing to the strong thermal reaction during ablation. The explosive boiling for each pulse and the overlap of pulses are the main causes, but then there is also some splash debris. On the other hand, a femtosecond laser with ultrashort pulses will sublimate the silicon and not so much debris formation is not observed. The track can be made sufficiently smooth by taking care with the galvanometer scanning speed and the repetition rate (Figure 11.25a). The remaining problem is to obtain a sufficiently high productivity. This is being addressed by the team at MC3 working on the Biolung®[3] (MC3, Ann Arbor, MI, USA) by controlling the scanning speed. Some results are shown in Figure 11.25b.

The Biolung®, based on sheets of these etched panels, as the oxygenator has been mounted in a capsule that can be implanted or mounted externally. Trials on sheep have demonstrated feasibility and safety in a number of tests. So far the maximum survival time has been 30 days. For longer uses and clinical trials, improvements in biocompatibility and control of coagulation are required. The future for these artificial organs looks very promising, even though the challenge is daunting.

11.5.6 Scaffolds for Tissue Engineering

Tissue engineering is one of the growth areas of medicine (if the reader will forgive the pun). It is a promising alternative for the repair of damaged tissues or organs. In the case of bone regeneration, one approach is to insert a scaffold loaded with osteogenic cells which will grow into bone tissue *in situ*, eventually making the scaffold redundant. Cells from a biopsy are cultured on the three-dimensional scaffold before being

---

[3] Biolung® is a registered trademark of MC3 Inc, 3550 West Liberty, Suite 3, Ann Arbor, MI 48103. http://www.mc3corp.com

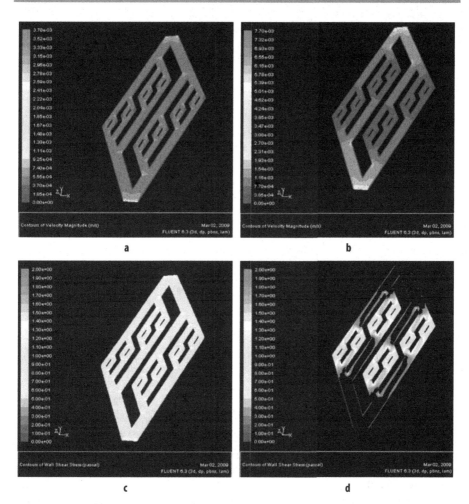

Figure 11.24 Blood flows in the microchannel branching network are characterised with computational fluid dynamics simulation; velocity distribution in **a** multiple-depth channel and **b** 20-μm-depth channel; shear stress distribution in **c** multiple-depth channel and **d** 20-μm-depth channel [55–57]

implanted. Bone is one of the tissues with the highest demand for tissue reconstruction or replacement. For this purpose the scaffold needs to be strong enough to take the load and not biologically harmful. The primary roles of a scaffold are (1) to serve as an adhesion substrate for the cell, facilitating the localisation and delivery of cells when they are implanted, (2) to provide temporary mechanical support to the newly grown tissue and (3) to guide the development of new tissues with the appropriate function. Different biomaterials have been used as scaffolds, including bioactive metallic, ceramics and polymers for bone tissue engineering. Many techniques have been developed to fabricate three-dimensional porous biodegradable scaffolds, such as particle leaching, phase separation, laser sintering of polymers and self-assembly.

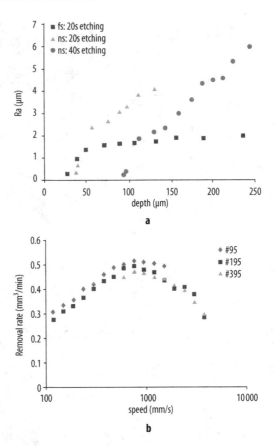

Figure 11.25 a Average roughness (Ra) versus depth for femtosecond and nanosecond laser-machined channels after acid etching, and **b** the material removal rate of femtosecond ablation versus scan speed

Recently, the laser-aided Direct Metal Deposition DMD process (see Chapter 7) has proven to be a promising method to produce three-dimensional metallic scaffolds [58, 59]. The DMD technique [60, 61] allows good control of composition, geometry and surface finish. An illustration of the equipment is shown in Figure 11.26. There are five primary components of the DMD assembly: the laser system, the powder delivery system, sensors for the feedback loop for dimension control, the controlled environment and the CAD-driven motion control system. In one example a 6 kW CO_2 laser unit was used to fabricate different Ti–6Al–4V scaffolds. The laser processing parameters used were as follows: laser scanning speed 36.6 mm min^{-1}, powder feed rate 5 g min^{-1} and Z increment 0.25 mm. The scaffold was fabricated on a substrate of Ti–6Al–4V rolled plate. During DMD processing, a 300 W laser beam was focused onto the substrate to create a melt pool into which the powder feed was delivered in an inert gas (helium) jet flowing through a special nozzle, where the powder streams converged at the same point as the focused laser beam (beam diameter 0.5 mm). An inert gas

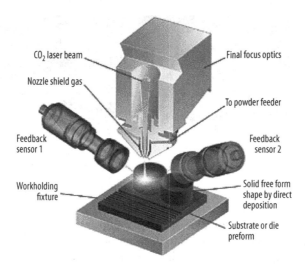

Figure 11.26 The direct metal deposition system [60]

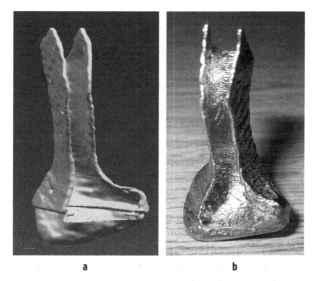

a   b

Figure 11.27 Example of a solid scaffold and its CAD design: **a** the CAD image of a condylar ramus of a Yucatan minipig, and **b** an identical Ti–6Al–4V scaffold produced by the direct metal deposition process

shroud containing a mixture of helium and argon was used as a protective atmosphere for preventing oxidation during deposition.

Figure 11.27 shows an example of a solid scaffold and its CAD for a condylar ramus of a Yucatan minipig produced by the DMD process. The surface of the as-fabricated piece needed sandblasting and chemical polishing in Kroll's reagent for 5 min to make it smooth enough to avoid causing inflammations due to macrophase activation. This left an average surface roughness of approximately 8 μm. The tensile and yield strengths

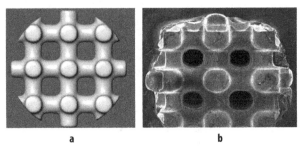

a   b

Figure 11.28 **a** The CAD design of a porous scaffold cylinder, and **b** the fabricated porous scaffold

of the as-deposited structure were $1,163 \pm 22$ and $1,105 \pm 19$ MPa, respectively, which are higher than the ASTM limit for Ti–6Al–4V implants (896 and 827 MPa).

A development of this process is to form porous titanium scaffolds with designed interconnecting cylindrical pores to hold bioactive materials such as cells or specific drugs to aid growth and avoid rejection. Figure 11.28 illustrates the CAD and the fabricated scaffold of a porous cylinder 5 mm in diameter and 3 mm high with 700-μm-diameter holes. Such devices have been tested *in vivo* on mice and have been shown to generate bone growth. The *in vivo* gene therapy approach consists in delivering the virus within fibrin clots held in the scaffold or directly drying virus onto the surface of the scaffold pores.

This approach to organ regeneration appears to work in cooperation with nature and hence should be the source of some major medical breakthroughs in the near future, with the laser via the DMD or similar techniques offering the necessary flexibility to the designs needed.

11.6 Conclusion

Living tissue only survives in a small temperature range, unlike engineering materials. It is also highly nonhomogeneous. Radiation can interact with the electric field within a molecule and can resonate with specific chemical bonds, resulting in a fluorescent response, spectral emission, breakage or heating of the whole molecule. A laser beam can thus be seen as a high-precision surgical tool or as a general heater or ablator. It has been illustrated through the examples in this chapter that by choosing the right wavelength of radiation the depth of penetration that the molecules to be heated can be controlled to some extent. Table 11.8 summarises the procedures discussed in this chapter.

It seems that phototherapy in its many current forms and those yet to be found will cause massive changes in our diagnostic and curing capabilities. Large areas of research have yet to be explored; the possibility of selectively activating specific chemicals or cells with the use of mixed wavelengths, as with isotope separation, is an area totally unexplored to date. The potential for particular frequencies to bypass some cells and interact with others is a tool no surgeon has yet enjoyed. The future for lasers in medicine looks very exciting.

Table 11.8 Summary of laser medical procedures

Main activity	Procedure	Example
Photochemistry	PDT	Optically switchable drugs and dyes: cancer cure
	Fluorescence	Diagnostic method for evaluating molecular species, cell cytology
	Raman spectroscopy	Diagnostic method for evaluating molecular structure
	Biostimulation	Relief of pain and enhanced healing
	Thermolysis	Hair removal, acne treatment, removal of freckles and birthmarks, skin rejuvenation
	Optical	Oximeter, laser scanning confocal microscopy, OCT, DOT
Photocoagulation	Eye surgery	Detached retinas
	LLLT	Prostate or cartilage shrinkage
		Wound healing and tissue welding
		General surgery
Photodisruption	Eye surgery	Glaucoma
		Stomach ulcers
		Endometriosis and cervical cancer surgery
		General surgery
Photoablation	LASIK, PRK, IntraLASIK	Correcting the eye focus
	Plasma generation	Kidney stone shattering
		Spectroscopy
		Tattoo removal
		TMLR
		Dentistry drilling
		Laser angioplasty

PDT photodynamic therapy, *OCT* optical coherence tomography, *DOT* diffuse optical tomography, *LLLT* low level laser light therapy, *LASIK* laser-assisted *in situ* keratomileusis, *PRK* photorefractive keratectomy, *TMLR* transmyocardial laser revascularisation

Questions

1. What are the main ways in which radiation can interact with tissue; give the approximate range of laser power and interaction time?
2. When would it be appropriate to consider using photodynamic therapy?
3. What is optical coherence tomography?
4. What are the advantages and disadvantages of the laser as a surgical cutting tool?
5. What is the mechanism for laser–tissue interaction for:

   a. photodynamic therapy;
   b. laser-induced interstitial thermotherapy; and
   c. low level laser light therapy

6. What pulse length is required to allow thermal diffusion to significantly reach the penetration depth of the laser treatment radiation, given that the thermal diffusivity of tissue is $0.106 \, \text{mm}^2 \, \text{s}^{-1}$ and the wavelength used has a penetration depth of $2.3 \, \mu\text{m}$?

References

[1] Niemz MH, Hoppeler T, Juhasz T (1993) Intrastromal ablations for refractive corneal surgery using... infrared laser pulses. Laser Light Ophthalmol 5(3):149

[2] Vogel A, Noack J, Nahen K, Theisen D, Birngruber R, Hammer DX, Noojin GD, Rockwell BA (1998) Laser-induced breakdown in the eye at pulse durations from 80 ns to 100 fs. Proc SPIE 3255:34–49

[3] Loesel FH, Niemz MH, Horvath C, Juhasz T, Bille JF (1997) Experimental and theoretical investigations on threshold parameters of laser-induced optical breakdown on tissues. Proc SPIE 2923:118–126

[4] Vogel A, Busch S, Asiyo-Vogel M (1993) Time-resolved measurements of shock-wave emission and cavitation-bubble generation in intraocular laser surgery with ps- and ns-pulses and related tissue effects. Proc SPIE 1877:312–322

[5] Vogel A, Busch S (1996) Shock wave emission and cavitation bubble generation by picosecond and nanosecond optical breakdown in water. J Acoust Soc Am 100(1):148–165

[6] Ignatiev A (2002) Bionic eyes-ceramic micro detectors may cure blindness. Mater World 10(7):31–32

[7] Loudin JD et al. (2007) Optoelectronic retinal prosthesis: system design and performance. J Neural Eng 4:s72–s84

[8] Garwe F et al. (2008) Optically controlled thermal management on the nanometer length scale. Nanotechnology 19:055207

[9] Kincade K (1993) New treatment may bring better night's sleep. Med Laser Rep 7(6):1

[10] Body4Real.co.uk (2001–2010) Laser hair removal – does it really work? http://www.body4real.co.uk/pages.php?pagecode=Laser_Hair_Removal&js=y

[11] BBC News (2006) Hope over laser 'that melts fat'. http://news.bbc.co.uk/2/hi/health4895148.stm

[12] Anderson JJ (2000) Transmyocardial laser revascularization. Prog Cardiovasc Nurs 15(3):76–81

[13] Hall C (1998) Pioneer watches laser fired inside his heart. Electronic Telegraph 908, 18 Nov 1998

[14] Summa Health System (2006) Summa performs first robotic laser heart surgery in Ohio. http://www.summahealth.org/common/templates/article.asp?ID=11992

[15] Vrouenraets MB, Visser GWM, Snow GB, van Dongen GAMS (2003) Basic principles, applications in oncology and improved selectivity of photodynamic therapy. Anticancer Res 23:505–522

[16] Tunér J, Christensen PH (2009) Low level lasers in dentistry. http://www.laser.nu/lllt/Laser_therapy_%20in_dentistry.htm

[17] Neimz MH (1996) Laser tissue interactions. Springer, Berlin

[18] Wadia Y, Xie H, Kajitani M (2000) Liver repair and hemorrhage control by using laser soldering of liquid albumin in a porcine model. Lasers Surg Med 27:319–328

[19] Wallace J (2008) Analysis of human breath holds key to disease. Laser Focus World Apr 43–44

[20] Wojtkowski M, Leitgeb R, Kowalczyk A, Bairaszewski T (2002) In vivo human retinal imaging by Fourier domain optical coherence tomography. J Biomed Opt 7:457

[21] Opto & Laser Europe (2005) Tomography beats micron resolution. Opto & Laser Europe Nov 15

[22] Ferguson B, Wang S, Gray D, Abbot D (2002) T-Ray computed tomography. Opt Lett 27(15):1312–1314

[23] Jiang H (2001) Tomography captures bone damage. Opto & Laser Europe May 19

[24] Albrecht GJ (2007) A field report of stent cutting machines with disk laser technology. In: Proceedings of AILU workshop on laser in medical device manufacture, West Bromwich, 7 Nov 2007

[25] Vollarth K (2007) Cutting intricate microstructures. Industrial Laser Solutions Jun 28–32

[26] Ogden CC (2007) Laser marking for the identification and traceability of medical devices. In: Proceedings of AILU workshop on laser in medical device manufacture, West Bromwich, 7 Nov 2007.

[27] Dickson P (2007) Laser enabling wire stripping for ultra sound equipment and heart pacemakers. In: Proceedings of AILU workshop on laser in medical device manufacture, West Bromwich, 7 Nov 2007

[28] Knitsch A (2007) Made to measure medical device manufacture: the laser advantage. In: Proceedings of AILU workshop on laser in medical device manufacture, West Bromwich, 7 Nov 2007

[29] Ashkin A (2000) History of optical trapping and manipulation of small-neutral particle, atoms, and molecules. IEEE J Select Top Quantum Electron 6(6):841–856

[30] American Lung Association (2010) Lung disease. http://www.lungusa.org/lung-disease

[31] US Organ Procurement and Transplantation Network and the Scientific Registry of Transplant Recipients (2006) OPTN/SRTR 2006 annual report: transplant data 1996–2005. Available via http://www.ustransplant.org/annual_Reports/archives/2006/default.htm

[32] Zwischenberger JB, Alpard SK (2002) Artificial lungs: a new inspiration. Perfusion 17:253–268

[33] Haft JW, Griffith BP, Hirschl RB, Bartlett RH (2002) Results of an artificial-lung survey to lung transplant program directors. J Heart Lung Transplant 21:467–473

[34] Matheis G (2003) New technologies for respiratory assist. Perfusion 18:245–251

[35] Fischer S, Simon AR, Welte T, Hoeper MM, Meyer A, Tessmann R, Gohrbandt B, Gottlieb J, Haverich A, Strueber M (2006) Bridge to lung transplantation with the novel pumpless interventional lung assist device Novalung. J Thorac Cardiovasc Surg 131:719–723

[36] Kopp R, Dembinski R, Kuhlen R (2006) Role of extracorporeal lung assist in the treatment of acute respiratory failure. Minerva Anestestiol 72:587–595

[37] Iglesias M, Jungebluth P, Petit C, Matute MP, Rovira I, Martinez E, Catalan M, Ramirea J, Macchiarini P (2008) Extracorporeal lung membrane provides better lung protechtion than conventional treatment for severe postpneumonectomy noncardiogenic acute respiratory distress syndrome. J Thorac Cardiovasc Surg 135:1362–1371

[38] Iglesias M, Martinez E, Badia JR, Macchiarini P (2008) Extrapulmonary ventilation for unresponsive severe acute respiratory distress syndrome after pulmonary resection. Ann Thorac Surg 85:237–244

[39] Kjaergaard B, Christensen T, Neumann PB, Nürnberg B (2007) Aero-medical evacuation with interventional lung assist in lung failure patients. Resuscitation 72:280–285

[40] McKinlay J, Chapman G, Elliot S, Mallick A (2008) Pre-emptive Novalung-assist carbon dioxide removal in a patient with chest, head and abdominal injury. Anaesthesia 63:767–770

[41] Hewitt TJ, Hattler BG, Federspiel WJ (1998) A mathematical model of gas exchange in an intravenous membrane oxygenator. Ann Biomed Eng 26:166–178

[42] Hattler BG, Lund LW, Golob J, Russian H, Lann MF, Merrill TL, Frankowski B, Federspiel WJ (2002) A respiratory gas exchange catheter: in vitro and in vivo tests in large animals. J Thorac Cardiovasc Surg 124:520–530

[43] Eash HJ, Frankowski BJ, Litwak K, Wagner WR, Hattler BG, Federspiel WJ (2003) Acute in vivo testing of a respiratory assist catheter: implants in calves versus sheep. ASAIO J 49(4):370–377

[44] Snyder TA, Eash HJ, Litwak KN, Frankowski BJ, Hattler BG, Federspiel WJ, Wagner WR (2006) Blood biocompatibility assessment of an intravenous gas exchange device. Artif Organs 30(9):657–664

[45] Lynch WR, Montoya JP, Brant DO, Schreiner RJ, Iannettoni MD, Bartlett RH (2000) Hemodynamic effect of a low-resistance artificial lung in series with the native lungs of sheep. Ann Thorac Surg 69:351–356

[46] Haft JW, Montoya P, Alnjjar O, Posner SR, Bull JL, Iannettoni MD, Bartlett RH, Hirschl RB (2001) An artificial lung reduces pulmonary impedance and improves right ventricular efficiency in pulmonary hypertension. J Thorac Cardiovasc Surg 122:1094–1100

[47] Lick SD, Zwischenberger JB, Wang D, Deyo DJ, Alpard SK, Chambers SD (2001) Improved right heart function with a compliant inflow artificial lung in series with the pulmonary circulation. Ann Thorac Surg 72:899–904

[48] Lick SD, Zwischenberger JB, Alpard SK, Witt SA, Deyo DM, Merz SI (2001) Development of an ambulatory artificial lung in an ovine survival model. ASAIO J 47:486–491

[49] Zwischenberger JB, Wang D, Lick SD, Deyo DJ, Alpard SK, Chambers SD (2002) The paracorporeal artificial lung improves 5-day outcomes from lethal smoke/burn-induced acute respiratory distress syndrome in sheep. Ann Thorac Surg 74:1011–1018

[50] Haft JW, Alnajjar O, Bull JL, Bartlett RH, Hirschl RB (2005) Effect of artificial lung compliance on right ventricular load. ASAIO J 51:769–772

[51] Sato H, Griffith GW, Hall CM, Toomasian JM, Hirschl RB, Bartlett RH, Cook KE (2007) Seven-day artificial lung testing in an in-parallel configuration. Ann Thorac Surg 84:988–994

[52] Sato H, Hall CM, Lafayette NG, Pohlmann JR, Padiyar N, Toomasian JM, Haft JW, Cook KE (2007) Thirty-day in-parallel artificial lung testing in sheep. Ann Thorac Surg 84:1136–1143

[53] Wegner JA (1997) Oxygenator anatomy and function. J Cardiothorac Vasc Anesth 11(3):275–281

[54] Murray CD (1926) The physiological principle of minimum work applied to the angle of branching of arteries. Proc Natl Acad Sci xii(207):835–841

[55] Kam DH, Mazumder J (2008) Three-dimensional biomimetic microchannel network by laser direct writing. J Laser Appl 20(3):185–191

[56] Kam DH, Shah L, Mazumder J (2008) Laser micromachining of branching networks. Proc SPIE 6880:68800S

[57] Kam DH, Shah L, Mazumder J (2008) Laser micromachining of microchannel branching networks into silicon with a femtosecond fiber laser. In: ICALEO 2008 proceedings. LIA, Orlando, paper M506

[58] Dinda G, Song L, Mazumder J (2008) Laser deposited Ti-6Al-4V scaffolds for patient specific bone tissue engineering. Metall Trans A 39:2914

[59] Hollander DA, von Walter M, Wirtz T, Sellei R, Schmidt-Rohlfing B, Paar O, Erli HJ (2006) Structural, mechanical and in vitro characterization of individually structured Ti–6Al–4V produced by direct laser forming. Biomaterials 27:955–963

[60] Koch JL, Mazumder J (1993) Rapid prototyping by laser cladding. In: Denney IMP, Mordike BL (eds), ICALEO'93 proceedings. LIA, Orlando, pp 556–565

[61] Mazumder J, Dutta D, Kikuchi N, Ghosh A (2000) Closed loop direct metal deposition, art to part. J Opt Laser Eng 34:397–414

[62] Itzkhan I, Izatt JA, Lubatschowski H (2003) Lasers medical use of. In: Brown TG et al. (2003) The optics encyclopedia, vol 2. Wiley-VCH, Weinheim, pp 1211–1250

[63] Richards-Kortum RR (1990) Fluorescence spectroscopy as a technique for diagnosis in human arterial, urinary bladder and gastrointestinal disease. PhD thesis, MIT, p 200

[64] Federspiel WJ, Hewitt T, Hout MS et al (2004) Recent progress in engineering the Pittsburgh intra-venous membrane oxygenator. In: Bowlin GL, Wnek G (eds) Encyclopedia of biomaterials and biomedical engineering. Dekker, New York, pp 910–921

[65] The Engineering ToolBox (2005) Densities of various solids. http://www.engineeringtoolbox.com/density-solids-d_1265.html

[66] Rodriguez GP, Arenas AC, Hernandez RAM, Stolik S, Orea AC, Sinencio FS (2001) Measurement of thermal diffusivity of bone and hydroxyapatite and metal for biomedical applications. Anal Sci 17:357–360

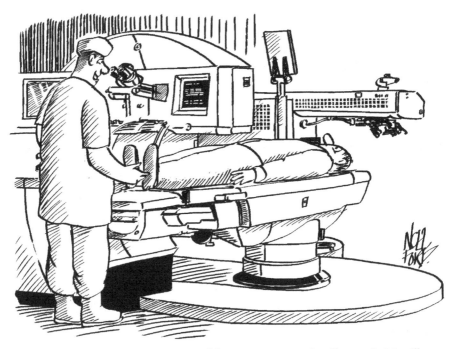

"Could I interest you in our new fish-eye conversion for all-round vision?"

12 Laser Automation and In-process Sensing

> Everything must be like something, so what is this like?
>
> *E.M. Forster (1879–1970), Abinger Harvest (1936)*
>
> To govern is to make choices
>
> *Duc de Levis (1764–1830), Politique xix*

12.1 Automation Principles

The recent developments in industry, particularly through the activities of Ford Motor Company, where the word "automation" was first used in the 1940s, have sketched a progression through "mechanisation" – the use of machines which enhanced speed, force or reach, but where the control was human, to "automatic" machinery – in which the machine will go through its programmed movements without human intervention and the machine is self-regulating, until today when we have "automation" – in which there is usually a sequence of machines all controlling themselves under some overall control. In the future there is the prospect of "adaptive control" or "intelligent" machines – in which the machine can be set a task and it teaches itself to do the task better and better according to some preset criteria. The drive towards automation is powered by the possibility of cost reductions, increased productivity, increased accuracy, saving of labour, greater production reliability, longer production hours, better working conditions for the human staff, increased flexibility of production to meet the needs of changing markets and improved quality. This list is a formidable argument for automation but it is only justified for certain production volumes. Figure 12.1 gives an idea of the stages which are most economical in setting up an automatic production facility. If very few pieces are needed, then it is cheapest to make them by hand. If a very large number of pieces are needed, then it is cheapest to make them on a purpose -built production line – "hard automation". In-between there is the relatively new area of flexible manufacturing, possibly using robots and linked machines. This middle zone in production size is growing owing to the manufacturing market becoming more fashion-conscious and pandering to the human appetite for novelty and change. In this manufacturing pattern the laser has a place as another tool, but one with some significant advantages.

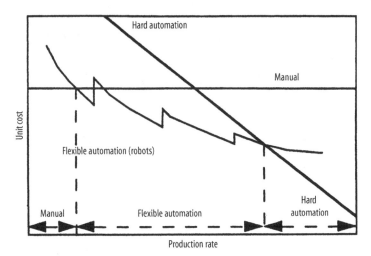

Figure 12.1 Variation of unit costs with production rate for different manufacturing strategies

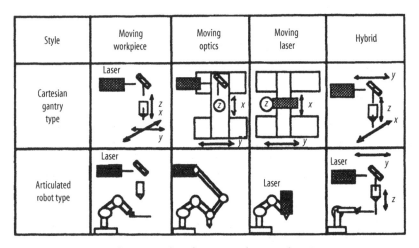

Figure 12.2 Examples of automatic laser workstations

Firstly, it is very *flexible* in the way it can be programmed to direct the optical energy. Figure 12.2 illustrates some of the options. One of the most flexible forms of laser beam guidance is via a robotic beam-delivery system. However, the accuracy and neatness of the laser, particularly as a cutting tool, shows up the poor line-following capability of current robots. This low level of accuracy in the robot precludes its use for many applications. Figure 12.3 gives an idea of the growth of applications which would result from a successful development of an accurate robot. Whatever workstation is used, it must be as vibration-free as possible. This is becoming more important as higher accelerations put larger loads on the structure and machine space means impact machines may be close by. Linear motors are capable of 7–8G. In consequence, some manufacturers

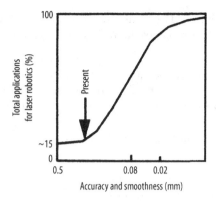

Figure 12.3 Market for laser robotics versus accuracy

build their tables with very heavy bases, such as 6 tons or more. There is surprisingly little effort being devoted to considering robot control algorithms for the drive mechanisms. Some work is discussed in Craig [1] amongst others. The main effort in robotic research seems to be spent on controlling the sequencing of robots rather than their accuracy.

Secondly, there is very little environmental disturbance in delivering optical energy. For example, there is no electric field, no magnetic field (except in the electromagnetic radiation itself), no sound, no light or other optical signal (except at the frequency of the laser beam), no heat and no mechanical stress. Thus, any signal in these areas will probably have come from the process itself. This gives a *wide, open window for in-process diagnostics*, which is unique to the laser.

To have a self-regulating system for a laser or any machine it will have either an open-loop or a closed-loop controller. The open-loop controller might be a clock if the sequencing is done by time rather than events. In this case the machine actions are taken automatically but without reference to the state of the process or the position of

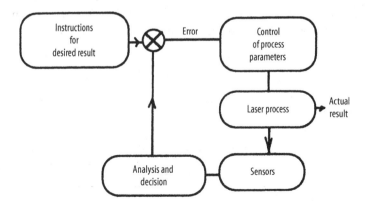

Figure 12.4 The structure of a closed-looped control circuit

the machine. If it is controlled by events, then it would be a closed-loop control system. A schematic of a closed-loop controller is shown in Figure 12.4. It can be seen that the control sequence is:

1. A process variable or product quality is measured.
2. The signal is compared with the desired value and an error is detected.
3. This error initiates a change in the process manipulators or drives, thus affecting the process.
4. What is changed and by how much is decided by the controller.

The start – if not the heart – of this process is to be able to take the signal from the process while the process is running, and to do so sufficiently quickly that the error detection can be made and the machine corrected before there is any considerable product waste. In-process signalling should be, and is fast becoming, one of the strengths of the laser. This we will discuss next and then, in Sections 12.3 and 12.4, we will discuss the possibility of in-process control and "intelligent" control systems.

12.2 In-process Monitoring

For the control or monitoring of laser material processing, the in-process signals for the variables listed in Table 12.1 are required. Many ideas are being and have been devised for these tasks. Table 12.2 lists some of the main concepts being investigated or which have been engineered. This is not a complete list, but nearly so. It does, I hope, illustrate the wide number of sensing options open for laser process monitoring. Some of these techniques we will now discuss in more detail.

12.2.1 Monitoring Beam Characteristics

The beam characteristics of interest during processing are as listed in Table 12.1: beam power, diameter, mode and location. These characteristics can be measured using beam-blocking methods or for in-process measurements by using scanning samplers, camera-based arrangements or possibly acoustic signals from mirrors (Section 12.2.3.1).

Table 12.1 Principal variables which characterise a laser process

Beam	Workstation	Workpiece
Power	Traverse speed	Surface absorptivity
Diameter	Vibration, stability	Seam location
Mode structure	Focal position	Temperature
Location	Shroud velocity and direction	"Quality" of product

Table 12.2 Some in-process sensors currently under investigation or use

Signal	Type of sensor	Principle	Speed fast or slow	Directional or omnidirectional
Acoustic	Mirror	Surface stress waves	F	O
	Probe	Keyhole shock waves	F	D
	Nozzle	Keyhole shock waves	F	O
	Workpiece	Stress waves in work piece	F	D
Radiation emission	Photoelectric	From plasma or melt pool viewed above below or through the mirror	F	D/O
	Pyroelectric	Emissions over wide range of wavelengths	F	D/O
	CCD camera	Direct viewing usually uses filters	S	D/O
	IR camera		S	D/O
Space charge	Plasma charge sensor nozzle or ring, with or without external voltage	Diffusion of electrons gives space charge field	F	O
	Langmuir probe	Conductivity of plasma	F	D
Radiation injection	Laser beam	Reflected and transmitted	F	D

F fast, *S* slow, *D* directional, *O* omnidirectional

12.2.1.1 Monitors that Block the Beam

The laser power is normally measured when the beam is not being used by some technique which totally blocks the beam. For example, most lasers are fitted with a beam dump which doubles as a calorimeter. A conical beam dump is illustrated in Figure 12.5. The power is measured as the rise in temperature of the flowing water. An absolute black-body calorimeter for measuring the power is illustrated in Figure 12.6. This simple device is highly mobile, but does require a lens to focus the beam into the spherical absorbing chamber, which is usually coated black on the inside. This need for a lens is not usually a disadvantage since the most interesting place to know the power is after

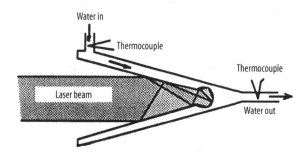

Figure 12.5 A standard cone calorimeter power meter/beam dump

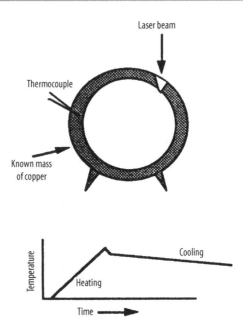

Figure 12.6 An absolute black-body calorimeter

the lens at the point where it is to be used. Care must be exercised that all the beam enters the calorimeter. The power can then be read off from a chart recorder reading of the thermocouple output. The equation for the rate of heating of the calorimeter is

$$P = MC(dT/dt)_{\text{heating}} - MC(dT/dt)_{\text{cooling}}.$$

If the calorimeter is made of copper, then the specific heat, C, is known to be 4,300 J kg^{-1} °C^{-1}. The value of the mass of the calorimeter, M, is easily found by weighing. The slopes for heating and cooling come from the chart recorder read-out. There is usually some surprise at how much power is lost between the laser and the workpiece by mirrors and lenses (approximately 1–2% per optical element). Beam-blocking devices are of no use for in-process sensing. It is amazing that some production machines, fitted only with these methods of power sensing, have the laser on for considerable periods of time and therefore have no way of being monitored – they are simply running on faith that the manufacturer has made a stable product.

Polarisation, which was discussed in Section 2.7.4, can be measured by reflecting a high-powered beam at an angle, near the Brewster angle, from a cone or rotating mirror and measuring the power reflected in different directions. This parameter, though important in processes involving reflection at glancing angles, is rarely measured in engineering applications but is of great importance in signal transmission down fibres. For these low-powered signals a rotating polarising filter, such as a Polaroid® sheet (a molecular analogue of a wire grid), is often used. For the sophisticated measurement of polarisation mode dispersion within a long fibre, coherent interferometry is used, whereby the delay of different wavelengths and states of polarisation can be precisely measured. At present, in engineering applications this level of information is not needed.

12.2.1.2 In-process Beam Monitors

Table 12.3 lists a number of the techniques that have been patented or developed for in-process monitoring of the laser beam – that is, monitoring while the beam is being used. The main ones currently of interest consist of scanning systems and camera systems.

Scanning systems: the ALL laser beam analyser (LBA) [2]. The LBA consists of a reflecting molybdenum rod which is rotated fast through the beam. The reflections off the rod are measured by two pyroelectric detectors placed as shown in Figure 12.7a. The two detectors pick up signals proportional to the power on two simultaneous orthogonal passes of the beam, as illustrated. It is this ability of the instrument to collect the power distribution within around 0.01 s in two dimensions simultaneously with only 0.1% beam interference that has made it one of the more popular beam measuring instruments in a laser facility. It is often fitted after the output window and before the shutter assembly so that the beam can be monitored even when it is not being used. The arrangement is illustrated in Figure 12.7b. The signals from the instrument can be displayed on an oscilloscope or passed to a computer for further analysis. The data which can be gained from the signals, which are illustrated in Figure 12.7c, are:

1. Overall power, measured from the integral under the curve or the root mean square (RMS) value.
2. Beam diameter, measured from the $1/e^2$ position of the power rise.
3. Beam wander, measured from any variation in the relative rise positions A and B of the two traces.

Table 12.3 Methods available for the monitoring of beam characteristics

Instrument	Refs.	Characteristic					
		Power	Diameter	Mode	Wander	Dirt[a]	Response time
Laser beam analyser	[2]	✓	✓	✓	✓	✓	Fast
Perforated mirror	[50]	✓	✓	✓			Fast
Chopper devices	[51, 52]	✓	✓	✓			Fast
Heating mirrors	[53]	✓					Slow
	[54]	✓	✓	✓	✓		Slow
	[55]	✓	✓	✓	✓		Fast
Heating wire	[56]	✓	✓	✓			Slow
Photon drag in Ge[b]	[57]	✓	✓	✓			Fast
Piezoelectric[b]	[58]	✓	✓	✓			Fast
Heating gas	[59]	✓					Slow
	[60]	✓					Slow
	[61]	✓					Fast
Optical scattering	[62]	✓				✓	Fast
Acoustic signals	[22]	✓	✓	✓	✓	✓	Fast

[a] Mirror or lens fouling
[b] In-process only if used with a beam splitter

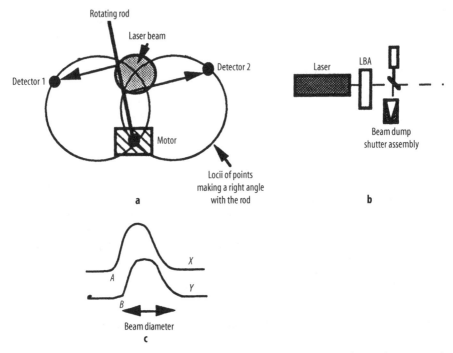

Figure 12.7 The laser beam analyser (*LBA*): **a** the principle whereby two orthogonal passes can be made simultaneously, **b** an arrangement for continuous viewing of the beam, and **c** an example of the oscilloscope output

4. Mode structure, measured from the shape of the curves, in particular comparison with previous mode structures can be made to check on cavity tuning or fouling of cavity mirrors.
5. Any instantaneous variations in the above properties. It is quite surprising how much power vibration and beam dilation there is in the beam from some lasers.

The instrument can be used to inspect the focus of the beam but that, of course, cannot be done in-process. Detailed analysis of the signals from an LBA is given in Lim and Steen [3].

Instruments which only measure power have measured only half the story. The beam diameter is equally or more important. For example, it should be remembered that the cutting and melting ability can be correlated with the group P/VD, the penetration in welding with P/VD^2 and the depth of hardness with $P/(VD)^{1/2}$. So the versatility of this instrument and the small beam interference are its main strengths.

The UM profiler. An improved version of the ALL LBA has been designed by O'Neill and Sparkes [4]. They used the basic principles of the ALL LBA, but redesigned the optical system and the signal analysis to produce a smaller machine that can be inserted into the beam path with no interference with the beam except when the rotated wand flicks through the beam. The "UM profiler", as they call it, need only be operated when beam data are required. It gives instantaneous data on the beam diameter, position and

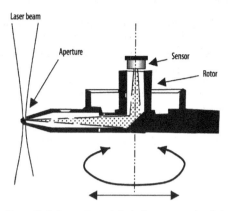

Figure 12.8 Scanning pinhole beam sampler [9]

$x–y$ power profiles. One of the more interesting aspects of data of this sort is that they can be stored and compared over periods of time, thus measuring mode stability over months or beam wander during processing. Beam wander has been noted owing to fumes in the flight tubes or changing beam paths in large gantries. This latter problem may well lead in the future to all large gantries being fitted with steerable mirrors, controlled by beam profilers of this sort. A study along this line was made by Sparkes [5] in his Ph.D. work.

The Promotec Laserscope or PRIMES FocusMonitor. These instruments collect the beam data from a rotating aperture (approximately 23 μm diameter, approximately twice the wavelength) that then reflects the signal down a waveguide (a hollow needle) to a central detector (Figure 12.8). Mapping a complete beam requires a traversing system – as with all scanning profilers, but with a single aperture only one signal is obtained at any given instant compared with the $x–y$ signals from the rotating wand reflectors. The complete profile usually takes around 5 s to accumulate. It is more of a diagnostic tool than an in-process control instrument. It can be used to measure the focal spot size at low powers and has been used to measure the profile of pulsed beams by careful synchronisation of the sweep with the pulse [6]. In Feurschbach and Norris's paper [6] a comparison between the laser beam spot size from the Promotec Laserscope and the size of a burn print in Kapton[®][1] showed that the Kapton[®] indicated 30% greater diameters for larger spots sizes and hence a greater beam divergence. This is a difference that questions the definition of beam diameter.

Camera-based beam monitors [6–9]. Camera-based systems using CCD, infrared (usually a pyroelectric sensor array) or complementary metal oxide semiconductor (CMOS) cameras offer the possibility of an almost instantaneous picture of the complete beam profile and position. Such data can be collected at over 48 frames per second or faster for CCD or CMOS systems. This appears to be the ideal beam monitor, but for three problems:

---

[1] Kapton[®] is a registered trademark of E.I. du Pont de Nemours and Company. http://www.dupont.com

1. how to get a reliable sample without destroying the camera or the beam image;
2. how to get the signal size down to the size of the detector array in the camera (typically 12.5 mm square) without any distortion of the beam image; and
3. how to do this at a competitive price.

Camera systems are based on a detector array of pyroelectric, CCD or CMOS sensors that convert the infrared energy falling on it into an electric signal or charged area that can be electronically interrogated later. This signal is digitised and sent to a computer for analysis and display. Some 16,384 discrete intensity levels can usually be identified, with good spatial resolution. This makes camera systems far better than burn prints or even single beam profiles as from a scanning system. However, the problems noted above have to be answered.

The first problem, that of attenuation, is not straightforward. The options are beam splitters, diffraction gratings or diffuse reflecting surfaces. The CCD, CMOS or pyroelectric array can only stand a few watts of incident radiation. Both CCD and CMOS, in particular, are very sensitive; their quantum efficiency in the visible range of 400–700 nm is reckoned to be 50% of collected electrons per photon but can be as high as 90% with antireflection coatings and special design [9]. Hence, if the beam splitter option were considered, then even a 99.7% beam splitter would not reduce the beam sufficiently for a 2–10 kW beam to be analysed. It would require two beam splitters to reduce the intensity sufficiently and a telescope to reduce the image to the size of the array. This implies perfect optics in all this equipment and hence this is not an attractive route. It is, however, a route taken by some instruments.

If the diffraction grating is considered, then a large grating is required for the 70 mm-diameter beams in current use. These gratings are usually made by photolithography onto a polymeric resin mounted on a flat surface and gold-coated. This is not suitable for high-powered laser beams and hence the grating would need to be specially made, which may fall foul of the cost problem. However, that is not always the case; Armstrong Optical [10] has adapted a single-point diamond-turned mirror by observing that such mirrors have a very faint diffractive image that can be used to observe the beam. A relatively cheap quadrant thermopile was used as the sensor to give in-process data on beam position, power and mode. The angle of diffraction can be customised to suit any optical system. One ingenious arrangement based on this principle is shown in Figure 12.9. This allows separate, yet simultaneous, monitoring of the beam and process.

Diffuse reflecting surfaces have been used in the latest Pyrocam®2 III from Spiricon. In essence the camera, with focusing optics, images the diffuse reflection from the incident spot on any of the beam-guidance mirrors. This achieves not only the correct intensity level but also the correct size. The image distortion, if any, is caused only by irregularities of the diffuse reflection from the guidance mirror surface and the focusing optics. This device is totally nonintrusive and can be used while the beam is in use, with a caveat on back-reflected light affecting the reading.

One can get mesmerised when chasing after a clear target, like beam characteristics. Devices such as those described give all the data asked for, which breaks the spell by asking the question what was it needed for in the first place! As seen in Chapter 2 most

---

[2] Pyrocam® is a registered trademark of Ophir-Spiricon Inc. http://www.ophiropt.com

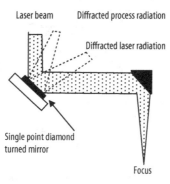

Figure 12.9 Beam sampling via a diffractive mirror, which allows signals from both the beam and the process [10]

of the behaviour of the beam as it passes through optical components can be described by λM^2 (see Section 2.8.1.1). So the characteristic that would solve most problems is M^2 together with the beam diameter and beam divergence. A camera-based technique capable of measuring M^2 in-process was developed by Dearden *et al.* [8] The arrangement is shown in Figure 12.10. The attenuation is achieved with a beam splitter and attenuation plates (diffuse reflection would not work in this case since the beam structure has to be measured at known distances along the beam path and diffuse reflection could only be used if the reflecting mirror was moved precisely). The beam quality M^2 is independent of the optical elements placed in the path of the beam, so the value found on the sampled part of the beam will be the same as that for the working beam. In Section 2.8, it was shown that $D_0 \Theta_0 / M^2 = 4\lambda/\pi$ = constant. Thus, measuring the divergence, Θ_0, and knowing D_0 from the laser geometry, one can infer M^2. This follows the ISO 11146 method [11]. The entire measurement of the beam at five set locations

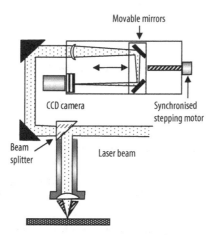

Figure 12.10 A beam monitor for measuring M^2 – the YAG:MAX [8]

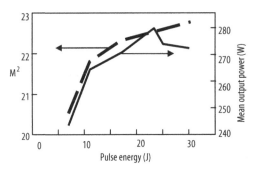

Figure 12.11 The variation of M^2 and mean output power with pulse energy and constant input (pump) [8]

takes less than 5 min. Some results are shown in Figure 12.11 on the performance of a JK704 TR Nd:YAG laser in pulse mode illustrating the change in beam focusability with power.

A CMOS camera microscope has been developed by PRIMES [9] to address the problem of ultrashort pulses of short-wavelength radiation at extreme intensity levels. It relies on heavy attenuation but is faster than CCD technology, although it has lower resolution. For very short events frame grabbing technology is used, whereby the picture is seized and held as a charged array that is subsequently interrogated electronically.

12.2.2 Monitoring Worktable Characteristics

The worktable variables are fairly straightforward to measure, as in the control of position (encoders), traverse speed (encoders) or nozzle gas velocity (pressure). However, the focal position and seam location are also crucial yet subtler to measure. Their measurement will now be discussed in more detail.

12.2.2.1 Monitoring Focal Position

The focal position needs to be measured and controlled in real time because the workpiece may warp slightly during processing, or the part may be contoured, making programming tedious. There are several signals that are used for sensing and controlling the height of the nozzle above the workpiece. They are shown in Figure 12.12 including electrical, mechanical and optical sensors.

Inductance [12] *and capacitance* [13] *devices* can be made to look very neat and can be built as part of the nozzle. Their problems arise from being only applicable to processing metals and the fact that they suffer signal changes near edges or in the presence of surface water or debris, particularly oxides. The edge problem may be acute when cutting nested articles. Plasma generated in welding may also affect the electrical environment (see Section 12.2.3.3), causing some judder on the focus. This is a particular problem when piercing.

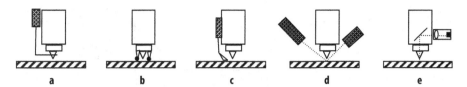

Figure 12.12 Examples of various ways of sensing the focal position: **a** capacitance, inductance, **b** skids, **c** feelers, **d** optical sensors, and **e** chromatic focal device

Skid devices have the nozzle riding on the workpiece. To avoid bouncing and recoil due to the air-cushion effects from the nozzle gas when cutting without full penetration, the nozzle is loaded with a force of 1 kg or so. This means that when cutting soft material, it may cause a scar on the surface. Also cutting small parts may be difficult by this method, owing to the movement of the part. The nozzle is usually made with a slotted piece of tungsten carbide or a set of roller balls. This system does have the advantage of allowing the nozzle to be very close to the cut surface and hence gives good aerodynamic coupling between the jet and the cut slot.

Feeler devices are sensitive, lightly loaded, lever systems that operate a piston in an inductive pot. They are fast-acting and, with properly shaped feeler skids, can be built to work close to edges. The advantages of this system and the skid system are that they are not sensitive to the material being processed nor water, debris on the surface or plasma effects.

Optical height sensors have been built based on a He–Ne or diode laser beam illuminating a spot very near to the interaction point. The image of this spot is focused by a telescope onto an array of detectors capable of sensing any variation in the image location (see Figure 12.12d and section 1.4.3.4 and Fig. 1.39). The probe beam can be modulated and the detector fitted with a filter to only allow the probe frequency to be measured. By this means the signal can be separated from the light emitted by the process. An alternative to this elaborate system was one used by Li [14] using an infrared transceiver. The intensity of the reflected signal varies strongly with height. However, this small electronic component has to be mounted away from intense heat and hence further from the interaction zone than in the probe beam method. Both optical devices have the advantage that they are suitable for all materials, are not sensitive to edges and are relatively insensitive to surface debris and water.

Dual-wavelength optical sensor. An ingenious focus control system was invented by the Laser Engineering for Manufacturing Applications (LEMA) team (EPSRC programme) [15, 16]. The arrangement is as for the optical weld monitor described in Section 12.2.3.4 and shown in Figures 12.12e and 12.13. The back radiation from the process, which the sensor detects, contains all the colours of the spectrum, being essentially from a black-body radiant source. Hence, provided optics that are transparent to the ultraviolet and infrared radiation are being used, the back-reflected beam can be imaged. The ultraviolet or visible end of the spectrum will be focused shorter than the infrared end when the signal is focused onto the detector or collecting optic. Thus, the intensities of the two signals from the ultraviolet and infrared components vary with focal position as shown in Figures 12.14 and 12.15. The optimal position is eas-

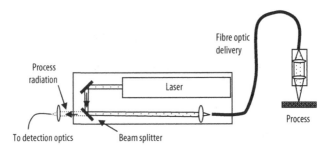

Figure 12.13 The optical sampling for dual-wavelength focus control

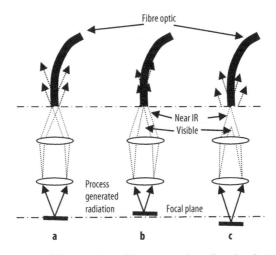

Figure 12.14 The variation of the coupling of the process broadband radiation into the delivery fibre optic: **a** in focus; **b** in front of the focal plane; **c** lower than the focal plane. *IR* infrared

ily determined owing to the asymmetry of the signal rise and fall for the two regimes (Figure 12.15). This system has been successfully used to weld wrinkled sheets of steel and to control the stand-off distance during laser direct casting. It has the significant advantage over all the other height sensors in being the only noncontact method that works best when the process is operating with a keyhole – plasma and all. Most other techniques find the plasma affects the capacitance or inductance or it blinds the optical sensors. The response time is a function of the height control gear. Interestingly, it is possible to mount this sensor on the laser as shown in Figure 12.13. It may therefore become a standard add-on to a Nd:YAG laser in the future.

12.2.2.2 Monitoring Seam Location

In butt welding there is a need to follow the joint line. In laser welding, because the fusion zone is very narrow there is a need for a seam-following system that is accurate and fast. Several systems have been suggested. There is the optical system of Lucas and Smith [17] and Goldberg's inductive sensors [12] which have been adapted to look

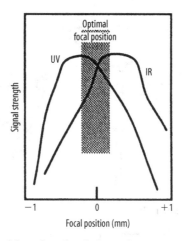

Figure 12.15 Variation of the infrared and ultraviolet intensities with focal position

at variations of the magnetic field around a joint. In the method of Lucas and Smith, a laser line from a He–Ne laser or diode laser is projected, by cylindrical optics, across the joint and the shape of this line is detected by a CCD camera. The straightness of this line is analysed by a computer and the location of the seam is identified within microseconds. The control system has been tested and proved to work for TIG welding [18]. An alternative is to scan the beam as with Oomen's method [19].

12.2.2.3 Monitoring the Effectiveness of the Shroud Gas

The spectrum of the plasma radiation can be used to detect any failure of the shroud gas during welding [20]. Ocean Optics (Dunedin, FL, USA) has made a miniature spectrometer fed by a 600 μm step-index fibre to receive the radiation from the process and analyse it for the telltale 426-nm peak for oxygen.

12.2.3 Monitoring Process Characteristics

Process characteristics are witnessed mainly through integral signals observing the radiation, noise or electric field from the whole event or spatially resolved versions of these signals. The sensors are either used separately or as a multiple-sensor fusion, and the signals collected are analysed for total intensity, frequency or wavelength. A review of sensors for laser weld quality is given in Sun and Kannatey-Asibu [21]. A number of sensors that have been studied are discussed next.

12.2.3.1 Acoustic Mirror Beam and Process Monitor

This instrument [22] may serve to illustrate the surprises in store for laser engineers. It was found by Weerasinghe and Steen in 1984 that high-frequency acoustic signals

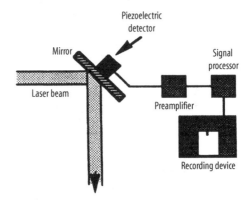

Figure 12.16 The acoustic mirror arrangement

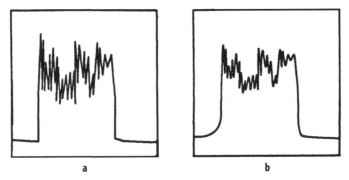

Figure 12.17 a The form of the raw signal from the acoustic mirror, and **b** the form of the signal after passing through a low-pass filter [22]

were generated in mirrors which were reflecting laser radiation. The arrangement is shown in Figure 12.16. A typical signal is shown in Figures 12.17 and 12.18. Perhaps the strangest part was the long-term oscillation in the RMS signal shown in Figure 12.19. The signal responds to the power, beam diameter, position on the mirror, state of tuning of the laser and even the gas composition in the laser cavity! What is considered to be happening is that radiation falls on the mirror and, in the action of reflecting the power, some power is absorbed, which instantaneously heats the surface atoms. This causes an expansion and hence a stress wave which passes through the mirror (and water cooling at the back if necessary) to the piezoelectric detector. The concept of the stress being caused by photon pressure is not considered likely since, as seen in Chapter 2, the stress would be relatively very low and would be expected to increase with the reflectivity of the mirror rather than what happens, which is the opposite. Thus, the instrument is recording only the variation in power – not the absolute power. With this picture in mind, the phenomena observed with this instrument can be understood.

Firstly, an increase in laser power is usually associated with an increase in the power variation. Some lasers are more stable than others. For example, a slow flow laser will give a smaller signal than a fast axial flow laser for the same power. A beam of larger

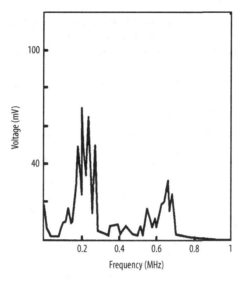

Figure 12.18 Frequency spectrum of the raw signal from the acoustic mirror showing two peaks, one in the 100-kHz region and one towards the megahertz region [49]

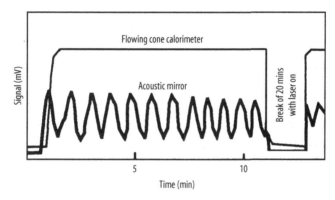

Figure 12.19 Example of the long-term signal variation from the acoustic mirror. (From Steen and Weerasinghe [22])

diameter will have a higher surface thermal stress than one of smaller diameter and hence the signal would rise if the diameter increased with the same power fluctuations. So the question becomes why should there be a fluctuation in the power of a laser beam? Figure 12.18 shows that the signal is made of two frequency components, one at around 100 kHz and one around 1 MHz. The low-frequency component varies with the gas mixture and state of tuning of the cavity. In fact for certain gas mixtures it will disappear altogether. This low-frequency element correlates with the frequency observed for the brightness of the plasma in the laser cavity and is considered to be due to the plasma attachment frequency at the cathodes. The higher-frequency component is thought to be a function of the cavity design and corresponds to the photon oscillations in the cavity. The time to exhaust the inverted population and the time to rebuild it results in the laser virtually spitting power rather than being truly continuous.

Figure 12.19 shows that the RMS signal with a 10 ms time constant has a slowly oscillating intensity not detected by the flowing cone calorimeter with a 10 s time constant. Since the oscillations in signal strength take over 1 min, this should have been detected by both instruments if it had been due to a power variation. The LBA also showed no power variation, but by very careful observation of the LBA signal a small oscillation in beam diameter was observed. The acoustic mirror picks up this beam dilation – or rather power variation – strongly. It is believed, but has not yet been proven, that as the laser warms up, so the cavity expands and the optical oscillations in the cavity form standing waves on the optic axis when the cavity is an exact number of wavelengths long and slightly off-axis when it is not an exact number of wavelengths long. This very small variation would cause a variation in beam size but more importantly a variation in the beam stability. It does mean this instrument, with no beam interference, could be a tool to identify automatically when a laser is warmed up and properly tuned.

There is a variation in signal with beam position on the mirror as shown in Figure 12.20. This is due to the different distances the signal must travel within the mirror to get to the detector and is also possibly due to some of the beam missing the mirror at the edges. This feature has been studied as a means for automatically sensing the location of a beam on a mirror, such as for automatic alignment [23].

Thus, the acoustic mirror is capable of measuring instantaneously and without any beam interference variations in the following:

1. power;
2. beam diameter;
3. beam position on mirror;
4. state of laser tuning;
5. approximate gas composition in the cavity; and
6. back reflections from the workpiece.

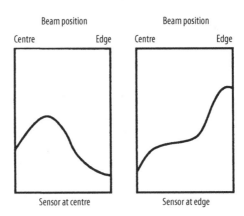

Figure 12.20 Variation of root mean square (RMS) signal from the acoustic mirror with beam position on the mirror

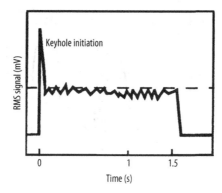

Figure 12.21 Acoustic signal obtained from a good weld with a steady keyhole

This formidable collection of data is unfortunately too much of a cornucopia of data. It is difficult to unravel one variable from another when the only analytical variables are signal strength and frequency. It seems to be a case waiting for high-speed pattern recognition, which may possibly come with neural logic.

The last item on the above list is the detection of back-reflection signals. If the acoustic mirror is mounted near the workpiece, it will receive back reflections from the workpiece, which should be diagnostic of the process condition. These back reflections are strongly fluctuating signals and on the whole they cover the whole mirror surface, so their thermal stress variations are very large compared with the relatively stable, but far more powerful, laser beam. A 5% variation in a 2 kW beam is 100 W, whereas a back reflection of 200 W may have a 100% variation. Since the reflectivity of many materials is over 60%, it can be appreciated that this calculation is modest and that the acoustic sensor would be expected to measure back reflections more strongly than the main beam. Some of the results are illustrated in Figures 12.21 and 12.22. In these figures the use of the acoustic mirror as a keyhole sensor is illustrated. In Figure 12.23 the signals from a stationary spot being melted by a laser are shown. The first pulse on a flat plate has a fairly uniform signal with a possible 15 Hz variation in it. The second pulse on the same spot has a high initial peak due to the strong reflection from the resolidified smooth concave cavity left from the first pulse. It appears that this took around 53 ms to remelt and then the signal again showed a faint 15 Hz variation. Calculations by Postacioglu *et al.* [24] indicated that the expected natural wave frequency on pools of this size would be around 15 Hz. Are we looking at (listening to) the waves on the pool? Certainly the acoustic mirror would be a very significant in-process tool if such intimate data could be obtained without any process interference.

12.2.3.2 Acoustic Nozzle

A variation on this acoustic theme is to place the piezoelectric sensor on the nozzle assembly and shield it from direct radiation as shown in Figure 12.24 [25]. The signals in this case are almost the reverse of those from the acoustic mirror. When there is poor back reflection, there is a stronger signal and *vice versa*. This detector is thought to be

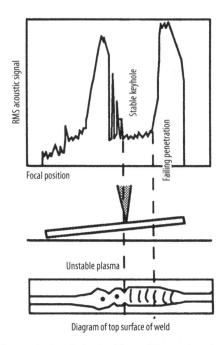

Figure 12.22 Variation of acoustic signal obtained by welding a sloped sample. Very high signals are obtained when the keyhole fails and there is a large back reflection

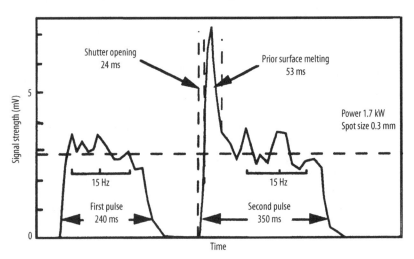

Figure 12.23 The acoustic signal obtained from two laser pulses on the same spot

responding to shock waves coming from the keyhole. It is thus weakest when there is no keyhole, whereas the acoustic mirror is strongest since the back reflection is then at its greatest.

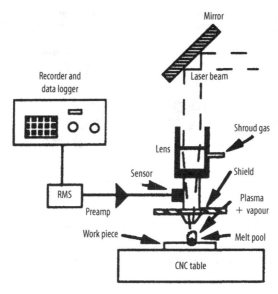

Figure 12.24 Arrangement for sensing with an acoustic nozzle [25]

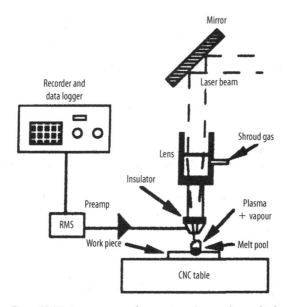

Figure 12.25 Arrangement for sensing plasma charge [26]

12.2.3.3 Plasma Charge Sensor

This is another surprise. Li and Steen [26] measured a space charge voltage with an insulated nozzle assembly. The arrangement is shown in Figure 12.25. The space charge is only present when there is a plasma and the signal varies with the process as indicated

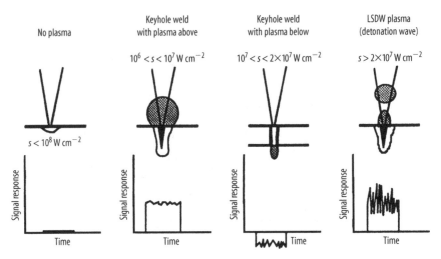

Figure 12.26 Variation in plasma charge sensor signal with the keyhole dynamics. *LSDW* laser-supported detonation wave

in Figure 12.26. The signal arises from the different diffusion rates and velocities of the electrons and ions in a plasma. This causes a momentary imbalance in the space charge. The electron velocities for the temperatures being considered would be of the order of 5×10^5 m s^{-1} and those of the ions would be of the order of 10^4 m s^{-1}. On the basis of this theory Li *et al.* [27] built a model from which they derived that the plasma charge sensor signal, V_s, is

$$V_s \approx 2.5 k T / e,$$

where V_s is the signal strength (V), k is the Boltzmann constant (J K^{-1}), T is the absolute temperature (K) and e is the charge of an electron (C).

It has been found by Li *et al.* that the signal increases with depth of penetration, corresponding to the higher temperatures with depth. They have also shown a distinct variation in the signal quality with shroud gas composition. Argon gives a stronger signal than nitrogen or helium. The signal gives detectable variations for most weld faults, from keyhole failure, humping and perforations to fainter signals for depth of penetration and mistracking.

The instrument itself causes no beam interference and has a very fast response time, measured in picoseconds. One variation of Li *et al.* was to have four plasma charge sensors (*i.e.*, a quadrant nozzle), which they found could give an indication of the correct centring of the beam within the nozzle.

12.2.3.4 Radiation Emissions

Optical signals from the process can be obtained by direct viewing, collecting the signal via a fibre mounted close to the process or peering through the mirror, as shown in Figure 12.27. These signals have an intensity and wavelength distribution in line with that predicted by Planck's law. The keyhole acts like a black-body emitter, whereas the

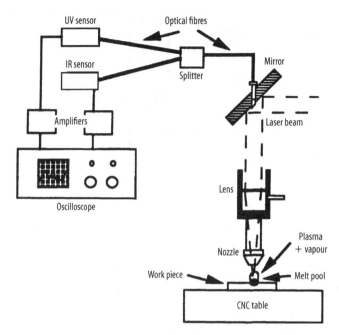

Figure 12.27 Dual-wavelength optical sensing

plasma has its own spectral emissions. In one version of a detector based on this principle the fibre is divided and the two parts are separately analysed for their infrared and ultraviolet components – not forgetting to allow for the different transmissivities with wavelength of the fibre and the different detector sensitivities across the spectrum. The result is a fast-response signal which varies with most weld faults. One interesting feature noticed by Chen *et al.* [28] is that when the fault occurs relatively slowly, for example, a drift out of focus, the ultraviolet signal "forecasts" the coming failure before it actually happens! This potentially predictive signal could be the dream of a control engineer interested in adaptive control. The explanation is that with any cooling of the keyhole the ultraviolet component is the first to change significantly. This same arrangement can be adapted as an automatic focus control (see Section 12.2.2.1).

The directly viewed spectral signals from the plasma have also been used to control the laser cleaning of art work and other artefacts, as in the processes of laser-induced-breakdown spectroscopy and laser-induced-fluorescence spectroscopy (see Section 1.4.9, Chapter 10) [29].

Kaierle *et al.* [30] used a camera in place of the fibre shown in Figure 12.27. This gave a spatially resolvable image of the cut front or weld pool. Figure 12.28 shows an outline of the hot spot they observed while cutting at varying speed around a corner. It illustrates the new layer of information that will be available from spatially resolvable signals – plus a new layer of computing to extract useful data. The images do show, however, the variation in the size of the melt pool and HAZ with speed. Note the extra side burning on slowing and also the increase in the radiation passing through the kerf as full speed is reduced as seen by the reduced size of the melt pool.

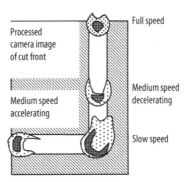

Figure 12.28 Outline of a processed camera image of the cut front as seen through the cutting optics. Two intensity levels have been outlined. (From Anglos *et al.* [29])

12.2.3.5 Temperature

The temperature of the workpiece is important in determining the extent of a transformation hardening process or the level of dilution expected during laser cladding. There are several methods for examining temperature. There is a straightforward pyrometer looking at the interaction zone, as in Bergmann's method [31] for controlling the transformation hardening process. There are CCD cameras which can look at the welding process and with appropriate software compute the size of the weld pool, from which the penetration might be determinable [32]. Infrared scanning pyrometers have been used to measure the thermal profile around the event.

Optical intensity has been viewed through the mirror by Zheng *et al.* [33] using a fibre mounted in the last beam-guidance mirror. Their signals indicated dross adhesion and the formation of striations during laser cutting. Olsen [34] and Kaierle *et al.* [30] using a beam splitter were able to witness these detailed events during cutting. Miyamoto *et al.* [35] had a system of viewing along the kerf to the cut face during cutting. Their equipment was used to take a very informative film and could be adapted to be less intrusive using fibres. Mazumder [36] used the signal from three photodiodes to sense the temperature during DMD cladding. These data were used to control the height of the clad.

12.2.3.6 Keyhole Monitoring

The monitoring of the keyhole by an acoustic mirror was discussed in Section 12.2.3.3. It can also be achieved using the "see-through" mirror of Zheng *et al.* [33]. There is also an electric signal, identified by Li *et al.* [27] (the plasma charge sensor), which can diagnose the general health of a laser weld while it is occurring. The plasma intensity is also an indicator and has been observed by many using optoelectric sensors, for example, the work of Beyer [37]. Others have simply listened to the welding process by microphone.

12.2.3.7 Dilution Monitoring

There are many new variables introduced during laser cladding; for example, there is the powder feed rate, the height of the clad track, the temperature of the substrate and the dilution of the clad. Li *et al.* [38] invented a device for in-process monitoring the dilution of a clad layer based on a microinductance sensor. The signal varied with the level of dilution in certain alloy systems, but it also varied with change of height and temperature. Thus, for the signal to be useful there has to be a computation between several signals. This is a greater order of complexity than for the other sensors just considered, but indicates the way the control of laser processing may probably progress.

12.2.3.8 Spark Discharge Monitoring

The angle at which the sparks leave a cut and the cone angle of the discharge is indicative of the health of the cutting process. ETCA [39] in Paris has developed a TV camera system to look at the spark discharge from the underside of the cut. The viewing angle is transverse to the cutting direction. The image is passed to an image processor, from which control data are elicited and used to control the process. The process being a multivariable, multioption one means that some form of "intelligent" processing had

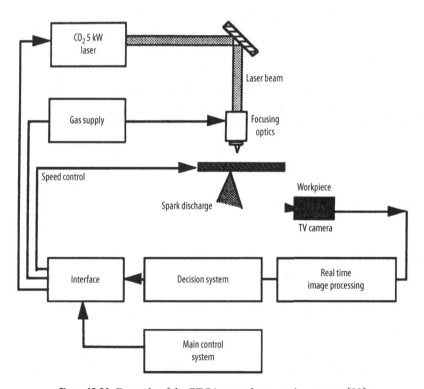

Figure 12.29 Example of the ETCA smart laser cutting system [39]

to be developed by ETCA. The general arrangement is shown in Figure 12.29. Its weakness is that the instrumentation has to be able to see beneath the workpiece, which is not always very convenient.

12.2.3.9 Clad Bond Condition Sensor

A photoelectric sensor was used by Li *et al.* [40] to look at the radiation from the melt pool during cladding. The signal was around 3–4 times higher than normal when the clad was not bonding correctly.

12.2.3.10 Ultrasonic Sensors for Measuring Welding Defects and Microstructure

Ultrasonics has been used for a long time as a method for inspecting welds by reflected signals from defects. It can also be used to measure the velocity of sound, which can be modulated by the microstructure (compression, shear, Rayleigh and Lamb waves; also grain size can be estimated by the spectrum of scattered waves). The problem with standard ultrasonic nondestructive testing inspection is that the ultrasonic generator and receiver (piezoelectric or electromagnetic acoustic transducers) have to be in good sonic contact with the workpiece. This has basically stopped the process from being used on hot material or in-process. A pulsed laser (typically 10 ns of intense infrared radiation) can generate an ultrasonic pulse within a workpiece and the resulting sound waves can be inspected by another laser whose reflected beam is being examined by an interferometer. Minute vibrations in the surface will cause a Doppler shift in the reflected laser light. Both lasers can be sited up to 2 m from the surface, making this process of laser ultrasonic technology [41] particularly applicable for high-temperature and high-speed applications. The arrangement being used in continuous microstructure measurements on a steel strip mill [42] is shown in Figure 12.30. This diagram also includes an arrangement for a single spot analysis for deep defects. The laser ultrasonic technology method using a twin-beam analyser can give data on thickness by measuring the time of flight of the compression and surface waves, for thicknesses up to 20 mm in steel or aluminium and with an accuracy of 0.2%, which is some 4 times better than with X-ray gauges.

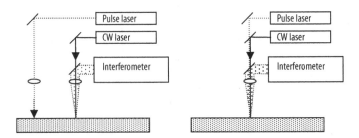

Figure 12.30 Two arrangements for laser ultrasonic inspection of microstructure or defects

A similar arrangement using an ultrasonic signal generated from a laser pulse and a laser Doppler velocimeter to measure the resulting surface vibrations can be used to measure laser weld defects in-process [43]. Since these detectors are only interrogating very high frequencies they are not sensitive to lower frequencies caused by workpiece vibration and by welder-induced turbulence. This technique, also used by spies to listen to conversations by reading the vibration of windowpanes, is likely to feature strongly in the future.

Acoustic emission technology has been used to listen to martensitic changes during laser transformation hardening [44] and cracking during laser cladding trials. Laser cladding onto a workpiece that is ultrasonically vibrated has been shown to reduce cracking [45]. The intensity of the acoustic signal made by a laser pulse used for cleaning has also been used to control the depth of cleaning. This signal can also be measured by a broadband microphone operating around 10–15 kHz [46] (see Chapter 10). These techniques could be revived with the use of noncontact laser Doppler velocimeters.

12.2.3.11 Spectral Analysis of Reflected Radiation

This technique [46] has been investigated for laser cleaning processes. It is based on the standard cleaning inspection technique of "Does it look clean?" In this process white light is shone on the surface being treated and the reflected light is analysed spectrally for the desired change (see Chapter 10).

12.3 In-process Control

It is one step to gain the diagnostic signals, the next is to use them in a closed-loop control system. These control systems come in two basic varieties. There are those which are one to one, that is, for example, the power is measured and the power is controlled. It is obvious what to do. If the power is too low, raise the current to the discharge tubes; if it is too high, do the opposite. The alternative control systems occur where there are many diagnostic signals and many interrelated operating conditions to be adjusted. This latter type of control system requires decision-making software, which could constitute "intelligent" processing as discussed in Section 12.4. For the straight control loops of the first type which do not require a decision as to what has to be altered but only by how much, we will discuss the in-process control of power and temperature, by way of example.

12.3.1 In-process Power Control

Li *et al.* [40] were able to monitor the beam power using the LBA. They showed that the RMS signal with a 10 ms time constant was linearly proportional to the observed

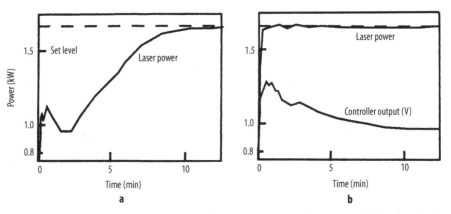

Figure 12.31 The effect of in-process control of the laser power: **a** the uncontrolled power during the warm-up of the Control Laser 2 kW machine, and **b** the same process but using feedback control of the power [40]

power as detected by a flowing cone calorimeter with a 10 s time constant. This signal was passed into a proportional, integral and differential (PID) controller and the PID constants were adjusted to give fast damping. The PID-generated control signal was used to control the potential to the cavity tubes and thus the current and so the laser power. The results are shown in Figure 12.31. The warm-up time of the Control Laser 2 kW CO_2 laser used was around 12 min, during which time the power varied sufficiently that the laser could not be used reliably. For the same warm-up period, but with use of the control system, the power was attained within 1 min and was then stable to within ±5 W for 1.6 kW. What was surprising in these results was that the mode structure during this period was also stabilised. Working with ultrastable beams is something new to material processing engineers, though becoming more usual with the use of diode and fibre lasers. The weakness of this system is that the controlled power must be below the maximum power so that the control can call on higher currents without causing an overload situation.

12.3.2 In-process Temperature Control

In transformation hardening and laser cladding it is important to know or control the temperature of the substrate since this determines the extent of hardening or the level of dilution. Drenker *et al.* [47] using a pyrometer to measure the temperature at the heated zone were able, via a control circuit, to control the speed of the table and so control the temperature of the workpiece. The depth of hardness achieved was uniform. This is a simple and reliable control system for a real industrial problem. It could have problems when there is a burning of the graphite or other coating used to aid absorption, which one might expect to interfere with the signal. This was overcome by using an adaptive temperature control algorithm [48].

12.4 "Intelligent" In-process Control

If one wishes to control a whole process such as welding, cutting or cladding, then the control system will have to handle many different in-process sensing signals and will be faced with several control options. For example, in cladding there will be signals from the height, temperature, dilution, powder feed rate, laser power, traverse speed, height of clad and possibly others. The control options are to vary the power, powder feed, focal position and traverse speed. Thus the process will need straight control loops on power, powder feed, traverse speed, *etc.*, to have control over these variables and to know they will not wander. Moreover, on top of this an overall diagnostic package is required to determine what is wrong, if anything, and what to alter. This decision is then passed to another package to determine by how much that parameter should be altered. The "intelligent" part is for this decision-making process to have its own feedback so that it is capable of making faster and more accurate diagnoses and adjustments. The logical frame for all this is illustrated in Figure 12.32. The wider view has been seen in the arrangement of the ETCA cutting system (Figure 12.29).

The diagnostics could be arranged as in Li's method [14] as a probability matrix for the solution of fuzzy logic problems. An alternative is the use of neural logic. As an example of a fuzzy logic matrix, some of the diagnostic options for the sensor signals and responses are shown in Table 12.4.

These options will be seen by the computer as a probability matrix, illustrated in Table 12.5. The sensing logic will have indicated a "true" or "false" situation regarding the state of each symptom. This matrix example, which is only a partial matrix for the full problem, indicates that the most probable diagnoses are d1.1, d1.2 and d1.5.

The diagnoses are selected when their probability ratings exceed a threshold amount, say, 0.5. They are then fed into a decision probability matrix, illustrated in Table 12.6.

Table 12.4 Examples of signals, diagnosis and responses for probability matrix

Sensed signal		Possible diagnosis		Possible responses	
s1	Clad bead too rough	d1.1	Powder flow rate too high	r1	Decrease powder flow rate
s2	Clad bead discontinuous	d1.2	Powder particle velocity too low	r2	Increase powder delivery gas pressure
s3	Bad laser mode structure	d1.3	Bad mode structure	r3	Decrease nozzle gas pressure
s4	Nozzle temperature too high	d1.4	Lens cracked	r4	Nozzle gas pressure too high: set alarm
s5	Clad bead too narrow	d1.5	Nozzle gas pressure too high	r5	Powder delivery pressure too high: set alarm
s6	Clad is too thin	d1.6	Traverse speed too high	r6	Laser power too high
		d1.7	Power too high	r7	Traverse speed too high
		d2.1[a]	Powder delivery gas		
		d2.2[a]	pressure cannot be altered online		
			Nozzle gas pressure cannot be altered online		

[a] Some control restriction affecting the diagnosis

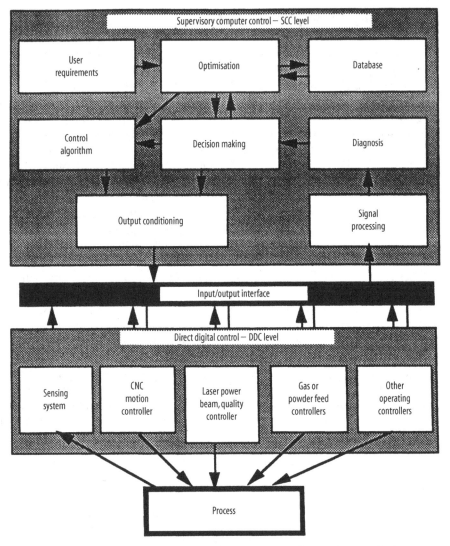

Figure 12.32 The logical framework for an "intelligent" control system [38]

Thus, the response should most probably be r1, or failing that r5 or failing that r4. It may be that r1 cannot be altered any more in a given direction.

The probability values in the probability matrices are variables which could be altered by the decision-making program. Thus, if r1 is altered and it is found not to work, then the probability of 1 would be reduced by a calculated amount. In this way the program would "learn" a better probability table and therefore a swifter and more accurate response will be the result. The subject is developed in greater depth by Li [14].

That is only half the problem. The next half is to decide by how much r1 should be adjusted; in fact by how much the powder flow rate should be decreased. This is stan-

Table 12.5 Diagnosis probability matrix

Signal State	s1 true	s2 true	s3 false	s4 false	s5 false	s6 true	sum d1.x
d1.1	0.5	0.5	0	0	0	0	1.0
d1.2	0.4	0.5	0	0	0	0	0.9
d1.3	0.2	0.1	0.3	1	0	0	0.3
d1.4	0.2	0.2	1	0.3	0	0	0.4
d1.5	0.3	0.4	0	0	0	0	0.7
d1.6	0	0.3	0	0	0.5	0	0.3

Table 12.6 Response probability matrix

Diagnosis Probability State	d1.1 (1) True	d1.2 (0.9) True	d1.5 (0.7) True	dn.1 (1) True	dn.2 (1) True	Sum rx
r1	1	0	0	0	0	1.0
r2	0	0.5 × 0.9	0	−1	0	−0.55
r3	0	0	0.5 × 0.7	0	−1	−0.65
r4	0	0	0.5 × 0.7	0	0.5	0.85
r5	0	0.5 × 0.9	0	0.5	0	0.95

dard control theory, which is designed to make these alterations without introducing instability into the control system. A rule-based adaptive control strategy is given in Li [14].

12.5 Conclusions

The laser is an ideal partner for automation owing to the ease of gaining in-process signals regarding the state of the process. The gains to be made by having a fully automatic self-correcting and improving process are currently only partly perceived. For factories to work through the night with the lights out and the heating off, for the factory owner to play golf all day and the machines to work all day and night and for the product quality to be uniformly high may seem a Utopia. The laser is becoming well suited to help bring this about. Then gentle reader, what will you do? There is the challenge of positive use of leisure – one could always start a laser job shop or something!

Questions

1. What is automation?
2. What methods can be used to monitor the laser power while the process is running?
3. Sketch the concept behind neural logic control systems.
4. What is an acoustic nozzle and why does it work?
5. Where does the signal for a plasma charge sensor come from?
6. Describe an optical autofocus device.

References

[1]   Craig JJ (1988) Adaptive control of mechanical manipulators. Addison-Wesley, Reading
[2]   Lim GC, Steen WM (1982) The measurement of the temporal and spatial power distribution of a high powered CO_2 laser beam. Opt Laser Technol Jun 149–153
[3]   Lim GC, Steen WM (1984) Instrument for the instantaneous in-situ analysis of the mode structure of a high power laser beam. J Phys E Sci Instrum 17:999–1007
[4]   Sparkes M, O'Neill W, Gabzdyl J (2002) In process laser beam diagnostics. In: ICALEO'02 proceedings, Phoenix, October 2002. LIA, Orlando, paper 403
[5]   Sparkes MR (1996) Automatic CO_2 laser beam alignment systems. PhD thesis, Liverpool University
[6]   Feurschbach PW, Norris JT (2002) Beam characterisation for Nd:YAG spot welding lasers. In: ICALEO'02 proceedings, Phoenix, October 2002. LIA, Orlando, paper 407
[7]   Green LI (2002) New methods for beam profiling high power CO2 lasers with an IR camera-based system. In: ICALEO'02 proceedings, Phoenix, October 2002. LIA, Orlando, paper 401
[8]   Dearden G, Sharp M, French PW, Watkins KG, Green LI (2001) Initial studies of laser beam performance monitoring using a novel camera-based in-line beam monitoring system. In: ICALEO'01 proceedings, Jacksonville, October 2001. LIA, Orlando, paper 543
[9]   Kramer R, Hansel K, Schwede H, Brandl V, Klos M (2002) Logarithmic CMOS camera microscope for beam diagnostics – profiling of high power Q-switch lasers. In: ICALEO'02 proceedings, Phoenix, October 2002. LIA, Orlando, paper 404
[10] Johnstone I (2000) Beam sampling and process monitoring in laser material applications. Ind Laser User 20:34–35
[11] International Organization for Standardization (1993) Test methods for laser beam parameters: beam widths, divergence angle and beam propagation factor. Document ISO11146. International Organization for Standardization, Geneva
[12] Goldberg F (1985) Inductance seam tracking improves mechanisation and robotic welding. In: Proceedings of automation and robotisation of welding, Strasbourg, France
[13] Hanicke L (1987) Laser technology within the Volvo Car Corp. In: Proceedings of the 4th international conference laser in manufacturing (LIM4), Birmingham, UK, May 1987. IFS, Kempston/Springer, Berlin, pp 49–58
[14] Li L (1989) Intelligent laser cladding control system design and construction. PhD thesis, University of London
[15] Morgan SA, Fox MDT, McLean MA, Hand DP, Haran FM, Su D, Steen WM, Jones JDC (1997) Real time process control in CO2 laser welding and direct casting focus and temperature. In: ICALEO'97 proceedings, San Diego, October 1997. LIA, Orlando, pp 290–299
[16] Hand DP, Fox MDT, Haran FM, Peters C, Morgan SA, McLean MA, Steen WM, Jones JDC (2000) Optical focus control system for laser welding and direct casting. J Opt Lasers Eng 34:415–427
[17] Lucas J, Smith JS (1988) Seam following for automatic welding. Proc SPIE 952:559–564
[18] Sloan K, Lucas J (1982) Microprocessor control of TIG welding systems. IEE Proc Part D.1 1–8
[19] Oomen G, Verbeek W (1984) Real time optical profile sensor for robot arc welding. In: Proceedings of intelligent robots ROVISEC 3, Cambridge, MA
[20] Boas G (2001) Welding system monitors gas shield. Appl Opt 20:6606–6610
[21] Sun A, Kannatey-Asibu E (1999) Sensor systems for real time monitoring of laser weld quality. J Laser Appl 11(4):153–168
[22] Steen WM, Weerasinghe VM (1986) Monitoring of laser material processing. Proc SPIE 650:16–166
[23] Tashiro H, Suetsugu Y (1991) Localisation of incident laser beam in the optical element by on-site photo-acoustic detection. J Appl Phys 69(9):6741–6743
[24] Postacioglu N, Kapadia P, Dowden J (1988) Capillary waves on the weld pool in production welding with a laser. J Phys D Appl Phys 22:1050–1061
[25] Li L, Steen WM (1992) Non contact acoustic emission monitoring during laser welding. In: Farson D, Steen WM, Miyamoto I (eds) ICALEO'92 proceedings, Orlando, October 1992. LIA, Orlando, pp 719–728
[26] Li L, Brookfield DJ, Steen WM (1996) In-process laser weld monitoring. In: Proceedings of IIW Asian Pacific welding congress, February 1996, pp 119–136

[27] Li L, Qi N, Brookfield DJ, Steen WM (1990) Laser weld quality monitoring and fault diagnosis. In: Proceedings of the conference laser system and applications in industry, Turin, Italy, November 1990, pp 165–178

[28] Chen HB, Li L, Steen WM, Brookfield DJ (1993) Multi-frequency fibre optic sensors for in-process laser welding quality monitoring. J Non Destruct Test Eval 26(2):67–73

[29] Anglos D, Couris S, Mavromanolakis A et al (1995) Artwork diagnostics: laser induced breakdown spectroscopy (LIBS) and laser induced fluorescence (LIF) spectroscopy. In: Proceedings of LACONA I (1st international conference on lasers in conservation of artworks), Greece, 1995, pp 113–118

[30] Kaierle S, Abels P, Kapper G, Kratzsch C, Michel J, Schulz W, Poprawe R (2001) State of the art and new advances in process control for laser materials processing. In: ICALEO'01 proceedings, Jacksonville, October 2001. LIA, Orlando, paper 805

[31] Rubruck V, Geisler E, Bergmann HW (1990) Case depth control for laser treated materials. In: Proceedings of the 3rd European conference on laser treatment of materials, ECLAT'90, Erlangen, Germany, September 1990. Sprechsaal, Coburg, pp 207–216

[32] Juvin D, de Prunelle D, Lerat B. SAO par Imagerie. In: Proceedings of aut des procedes de soudage, 1986, Grenoble, France

[33] Zheng HY, Brookfield DJ, Steen WM (1989) The use of fibre optics for in-process monitoring of laser cutting. In: ICALEO'89 proceedings, Orlando, 12–22 October 1989. LIA, Orlando, pp 140–154

[34] Olsen F (1988) Investigations in methods for adaptive control of laser processing. Opto Electron Mag 4(2):168

[35] Miyamoto I, Ohie T, Maruo H (1988) Fundamental study of in-process monitoring in laser cutting. In: Proceedings of CISFFEL 4, Cannes, France, pp 683–692

[36] Mazumder J, Dutta D, Kikuchi N, Ghosh A (2000) Closed loop direct metal deposition: art to part. J Opt Lasers Eng 34:397–414

[37] Beyer E (1988) Plasma fluctuation in laser welding with CW CO2 laser. In: ICALEO'87 proceedings, San Diego, CA, May 1987. IFS, Kempston, and Springer, Berlin, in association with LIA, Toledo, pp 17–23

[38] Li L, Steen WM, Hibberd R, Brookfield DJ (1990) In-process monitoring of clad quality using optical method. Proc SPIE 1279:89–100

[39] Burg B (1986) Smart laser cutter. Proc SPIE 650:27–278

[40] Li L, Hibberd R, Steen WM (1987) In-process laser power monitoring and feedback control. In: Steen WM (ed) Proceedings of the 4th international conference on lasers in manufacturing (LIM4), Birmingham, UK, May 1987. IFS, Kempston, pp 165–175

[41] Scruby CB, Drain LE (1990) Laser ultrasonics, techniques and applications. Hilger, Bristol

[42] Baker G (2003) Laser inspection of materials online. Materials World Jan 22–23

[43] Klein MB, Pouet B, Kercel S, Kisner R (2001) In-process detection of weld defects using laser ultrasonics. In: ICALEO'01 proceedings, Jacksonville, October 2001. LIA, Orlando, paper P534

[44] Rawlings RD, Steen WM (1981) Acoustic emission monitoring of surface hardening by laser. Opt Lasers Eng Nov 173–187

[45] Powell J, Steen WM (1981) Vibro laser cladding. In: Mukherjee K, Mazumder J (eds) Proceedings of the symposium lasers in metallurgy. Metallurgical Society of AIME, Warrendale, pp 93–104

[46] Lee JM, Watkins KG (2000) In-process monitoring techniques for laser cleaning. J Opt Lasers Eng 34:329–442

[47] Drenker A, Beyer E, Boggering L, Kramer R, Wissenbach K (1990) Adaptive temperature control in laser transformation hardening. In: Proceedings of the 3rd European conference on laser treatment of materials ECLAT'90, Erlangen, Germany, September 1990. Sprechsaal, Coburg, pp 283–290

[48] Li L, Steen WM, Hibberd RD, Weerasinghe VM (1988) Real time expert system for supervisory control of laser cladding. In: ICALEO'87 proceedings, San Diego, May 1987. IFS, Kempston, and Springer, Berlin, in association with LIA, Toledo, pp 9–16

[49] Willmott NFF, Hibberd R, Steen WM (2008) Keyhole/plasma sensing system for laser welding control system. In: ICALEO'88 proceedings, Santa Clara, October–November 1988. LIA, Orlando, pp 109–118

[50] Spalding IJ (1986) High power laser beam diagnostics, part I. In: Proceedings of the 6th international gas flow and chemical laser conference (GLC6), Jerusalem, Israel, September 1986. Springer, Berlin, pp 314–322

[51] Sepold G, Juptner G, Rothe R (1980) Remarks on deep penetration cutting with a CO_2 laser. In: Proceedings of the international conference on welding research, Osaka, Japan, 1980. JWRI, paper A-29

[52] Oakley PJ (1983) Measurement of laser beam parameters. IIW DOC IV-350-83. International Institute of Welding, Roissy Charles de Gaulle Airport

[53] Rasmussen AL (1973) Double plate calorimeter for measuring the reflectivity of the plates and energy in beam of radiation. US Patent 3,622,245, December 1971

[54] Mansell DN (1973) Laser beam scanning device. US Patent 3,738,168, 12 June 1973

[55] Davis JM, Peter PH (1971) Calorimeter with a high reflective surface for measuring intense thermal radiation. Appl Opt 10(8):1959–1960

[56] Shrakura T et al (1984) Methods and apparatus for measuring laser beam. US Patent 4,474,468, 2 October 1984

[57] Gibson AF, Kimitt MF, Walker AC (1970) Photon drag radiation monitors for use with pulsed CO_2 lasers. Appl Phys Lett 17:75–77

[58] Satheesshkumar MK, Vallabhan CPG (1985) Use of a photo-acoustic cell as a scientific laser power meter. J Phys E Sci Instrum 18:435–436

[59] Ulrich PB (1983) Power meter for high energy lasers. US Patent 4,481,148, 26 April 1983

[60] Miller TG (1982) Power measuring device for pulsed lasers. US Patent 4,325,252, 20 April 1982

[61] Shifrin GA (1970) Absorption radiometer. US Patent 3,487,685, 6 January 1970

[62] Crow TG (1980) Laser energy monitor. US Patent 4,424,581, 30 December 1980

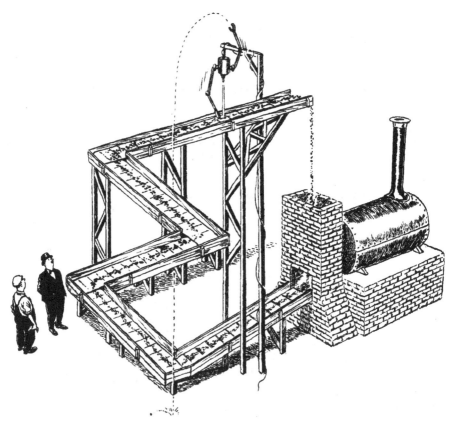

"Self-referencing loops can be self-deluding when applied to machinery and people."

13

Laser Safety

He saw; but blinded with excess of light, closed his eyes in endless night

Thomas Grey (1716–1771), The Progress of Poesy (1757) iii.2 (Milton)

13.1 The Dangers

All energy is dangerous, even gaining potential energy walking up stairs is dangerous! The laser is no exception, but it poses an unfamiliar hazard in the form of an optical beam. Fortunately, to date, the accident record for lasers is very good, but there have been accidents. The risk is reduced if the danger is perceived.

The main dangers from a laser are:

- damage to the eye;
- damage to the skin;
- electrical hazards; and
- hazards from fume.

13.2 The Standards

These risks can be minimised by following standards which have been laid down by various authorities. Most countries have their own set of standards, but recently the laser community has started coming together on a single set of principles. Safety regulations are of two kinds: manufacturer requirements and user requirements. The regulations can be legally binding or voluntary standards, which could be used in court in evaluating liabilities in the event of an accident.

The Technical Committee No. 76 of the International Electrotechnical Commission (IEC) drew up the basic standard from which most others have developed; this is IEC 825-1 (1993), the original version was written in 1984 and amended in 1990. This covers manufacturers and users and applies to both lasers and LEDs. It is a bit conservative and has been amended [1,2].

The European Committee for Electrotechnical Standardization (CENELEC) adopted the IEC 825 standard in 1992 as European Norm EN 60825. EN 60825-1, which is identical to IEC 825-1, was approved in 1994 and amended in 1996 to correct for the too-conservative approach which was affecting the LED manufacturers and users. Now only EN 60825-1 can be used for product certification in Europe [3]. It supersedes any other standards on laser safety in EU and EFTA countries [4].

In the USA the American National Standards Institute (ANSI) issued ANSI Z136.1-1993 [5]. This differs from IEC requirements primarily in labelling, class 1 limits, interlocks, measurement criteria and collateral radiation.

These standards evolved into the next generation in which there is a closer fit between Europe and the USA in EN 60825-1:2001 [1] and ANSI Z136.1-2007 [2] and a major redefinition of the class 3 lasers. They are applicable for all lasers and LED sources.

Laser material processing is specifically addressed by the European Committee for Standardization (CEN) and the ISO. This is part of a mandatary requirement for machinery in general – the machinery directive [ISO 11553, "Safety of machinery – laser processing machines – safety requirements" (1996), published by ISO, Geneva. Also EN 292 parts 1 and 2 "Safety of machinery" and ISO/Tr12100:1992 parts 1 and 2 "Safety of machinery"].

In the multitude of counsellors there is safety
  Proverbs 11 v14.

These standards give guidance and rules concerning engineering controls, advice on personal protective equipment, administrative and procedural controls and special controls. Class 4 laser installations, in which category nearly all material processing systems fall, should also have a laser safety officer (LSO), who should see that these guidelines are observed. He or she is also responsible for evaluating laser hazards and establishing appropriate control measures. It is common these days for the LSO to have attended a specialist course. If these rules are followed, the laser installation will be safe. Breaking the rules may result in an accident.

13.3 The Safety Limits

13.3.1 Damage to the Eye

The ocular fluid has its own spectral transmissivity as shown in Figure 13.1. It indicates that there are two types of problem with radiation falling on the eye. There is potential damage to the retina at the back of the eye and potential damage to the cornea at the front of the eye. Radiation which falls on the retina will be focused by the eye's lens to give an amplification of the power density by a factor of around 10^5. This means that lasers with wavelengths in the visible or near-visible waveband (Ar, He–Ne, Nd:YAG, Nd:glass) are far more dangerous than those with wavelengths outside that band (CO$_2$, excimer, Er:YAG). The nature of the threat from different lasers is listed in Table 13.1.

Safe exposure limits have been found by experiment and they are listed as the maximum permissible exposure (MPE) levels. These levels are plotted in Figures 13.2 and 13.3 for retinal and corneal damage. At power density and times greater than these safe limits damage may occur owing to cooking or boiling or at higher levels

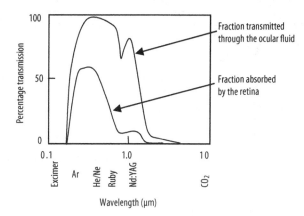

Figure 13.1 Spectral transmissivity of the ocular fluid and the absorptivity of the retina

Table 13.1 Basic laser biological hazards

Laser type	Wavelength (μm)	Biological effects	Skin	Cornea	Lens	Retina
CO_2	10.6	Thermal	X	X		
H_2F_2	2.7	Thermal	X	X		
Er:YAG	1.54	Thermal	X	X		
Nd:YAG	1.33	Thermal	X	X	X	X
Nd:YAG	1.06	Thermal	X			X
GaAs diode	0.78–0.84	Thermal	_a		X	
He–Ne	0.633	Thermal	_a		X	
Ar	0.488–0.514	Thermal, photochemical	X			X
Excimer:						
XeF	0.351	Photochemical	X	X	X	
XeCl	0.308	Photochemical	X	X		
KrF	0.254	Photochemical	X	X		

[a] Insufficient power

owing to explosive evaporation. The cooking/boiling limit is the reason for the very low levels of power which the eye can tolerate. For example, a 1 mW He–Ne laser with a 3 mm-diameter beam would have a power density in the beam of $(0.001 \times 4)/(3.14 \times 0.3 \times 0.3) = 0.014\,\text{W cm}^{-2}$. On the retina this would be amplified by 100,000 to be $0.014 \times 10^5\,\text{W cm}^{-2} = 1,400\,\text{W cm}^{-2}$. A blink reflex at this level would only allow a 0.25 s exposure, which is the MPE level for a class 2 laser. Notice that the calculation assumes that all the radiation can enter the pupil of the eye. Thus, it is common practice to ensure that working areas around lasers are painted with light colours and are brightly illuminated – not so with holographic laboratories and others involved with photography, of course!

The hazard zone around a laser is that in which radiant intensities exceed the MPE level. These zones are known as the nominal hazard zone [6]. The size of the zone can be calculated on the basis of the beam expansion from the cavity, lens or fibre, or from diffuse or specular reflection from a workpiece. For example, consider a 2 kW CO_2 laser

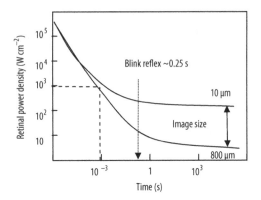

Figure 13.2 Approximate exposure limits for the retina

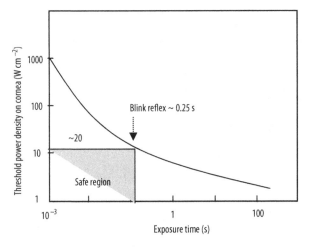

Figure 13.3 Approximate damage threshold power densities for the cornea of the eye

beam with a 1 mrad divergence. The MPE level for safe direct continuous viewing of the beam (not that much would be seen with infrared radiation!) is 0.01 W cm^{-2}. This would occur when the beam has expanded to 504 cm diameter – a distance of 5,020 m away, around 3 miles! This means that precautions must be taken to avoid the beam escaping from the area of the laser by installing proper beam stops, screens for exits and enclosed beam paths. Similar calculations for a 500 W CO_2 laser give a nominal hazard zone for diffuse reflections of 0.4 m. Therefore, it is necessary to wear goggles when near a working laser. As a general rule, *never look at a laser beam directly* – it is like looking down a gun barrel.

13.3.2 Damage to the Skin

There are also MPE levels for skin damage. These are far less severe than for the eye and so are essentially irrelevant. The laser is capable of penetrating the body at speeds as fast as that for steel and so the focused beam needs to be seriously respected. Without meaning to trivialise the problem with skin effects, the damage done is usually blistering or cutting, neither is pleasant but the wound is clean and will heal – unlike some eye damage. Incidentally, a vein or artery cut by laser will bleed even though it is cauterised!

As a general rule, *never put parts of your body in the path of a laser beam.* If an adjustment has to be made to the beam path, do it by holding the edges of mirrors, *etc.*

13.4 Laser Classification

Lasers are classified in EN 60825-1:2001 and ANSI Z136.1:2007 according to their relative hazard. All lasers of interest to material processing will be classified as class 4, except some which are totally built into a machine in which there is no human access possible without the machine being switched off; so the list is somewhat academic for this book. Table 13.2 provides a summary of the classification, based on accessible emission limits.

Table 13.2 Classification of lasers and LED sources [8]

Class	Definition	Warning label[a]
1	Intrinsically safe for continuous viewing. Includes embedded products that totally enclose a higher classification of laser, *e.g.*, CD players, laser printers and most production industrial laser material processing machines	None
1M	Low risk to eyes. No risk to skin. Safe, provided binoculars, *etc.*, are not used for viewing	"Laser radiation. Do not view directly with optical instruments"
2	Low risk to the eyes. No risk to the skin. Visible radiation in which protection is by blink reflex (0.25 s), <1 mW CW laser	"Do not stare into the beam"
2M	Low risk to the eyes. No risk to the skin. Same as for class 2 except binoculars and telescopes are not to be used to directly view the beam	"Do not stare into the beam or view directly with optical instruments"
3R	Low risk to eyes. Low risk to skin. Protection by blink reflex and beam size. The output accessible emission is up by a factor of 5 on that for class 1 or class 2 lasers	0.4-1.4-mm wavelengths: "avoid direct eye "exposure". Other wavelengths: "Avoid exposure to the beam"
3B	Medium risk to the eyes. Low risk to the skin. Direct or specular reflection exposure of the eyes is hazardous, even allowing for the blink reflex. Skin damage is prevented by natural aversion. Not a diffuse reflection or a fire hazard	"Avoid exposure to the beam"
4	High risk to the eyes and skin. May cause a fire. Standard safety precautions must be observed	"Avoid eye or skin exposure to direct or scattered radiation"

[a] All lasers of class 2 or above need to be labelled with a laser warning triangle

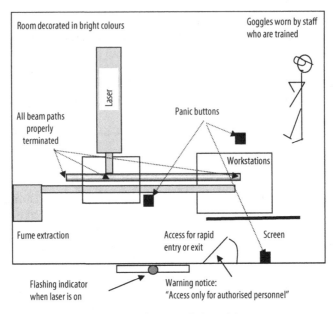

Figure 13.4 Safety features of a laser laboratory

13.5 Typical Class 4 Safety Arrangements

The following precautions are advised:

1. all beam paths must be terminated with material capable of withstanding the beam for several minutes;
2. stray specular reflections must be contained;
3. all personnel in the nominal hazard zone must wear safety goggles; for CO_2 radiation they can be made of glass or perspex, in fact normal spectacles may do if the lenses are large enough;
4. noninvolved personnel must have approval for entry;
5. there should be warning lights and hazard notices so that it is difficult (impossible) to enter the area without realising that it is being entered;
6. extra care should be taken when aligning the beam; and
7. there should be a laser safety officer to check that these guidelines are followed.

These guidelines are summarised in Figure 13.4 for a typical laser material processing arrangement.

13.6 Where Are the Risks in a Properly Set Up Facility?

If the facility is properly designed, then the beam is enclosed and all beam paths are terminated so that the beam cannot escape to do damage. A standard set-up would have the beam focused and pointing downwards. As the beam expands after the lens the nominal hazard zone is considerably reduced. For robotic beam-steering systems this nominal hazard zone is the distance to which the operators can approach the robot. This system is thus safe except in certain unlikely events. These events can be classified in a risk analysis tree. They would include breaking of the lens and total removal with loss of the nozzle, failure of a mirror mount and the mirror swinging free, *etc.* The essence of such an analysis is to devise a system for rapidly identifying an errant beam. This can be achieved by beam monitoring and/or enclosure monitoring. If the beam is monitored as leaving the laser but not arriving at the expected target, then the system should immediately shut down. If a hot spot appears within the enclosure, then again the system should shut down.

13.7 Electrical Hazards

Nearly all the serious or fatal accidents with lasers have been to do with the electric supply. A typical CO_2 laser may have a power supply capable of firing the tubes with 30,000 V with 400 mA. This is a dangerous power supply and when working on it the standard procedures for electric supplies should be followed. The smoothing circuit contains large capacitors and so even when the power is switched off a fatal charge is still available, and proper precautions to earth the system before working on it are essential. Panic buttons must be available at the laser and at the main exit. Access to the high-tension circuit should be protected by interlocks. As a general rule, *do not enter the high-voltage supplies without first carefully earthing the system.*

13.8 Fume Hazards

The very high temperatures associated with laser processing are able to volatilise most materials and thus form a fine fume, some of which can be poisonous. With organic materials, in particular, the plasma acts as a sort of dice shaker and a wide variety of radical groups may reform into new chemicals. Some of these chemicals are highly dangerous, such as the cyanides, and some are potential carcinogens. It is necessary, as a general rule, to have a well-ventilated area around the laser processing position as for standard welding. Some of the problems with cutting nonmetallic materials have been identified by Lyon *et al.* [7]. These are shown in Table 13.3, but it should be remembered that the volumes per second are not very large and represent a hazard only if much work is being done over an extended time. As a general rule, *the laser processing zone should be adequately ventilated.*

13.9 Conclusions

The laser is as safe as any other high-energy tool and should be properly handled. It is the responsibility of the user to learn how to handle it correctly.

Table 13.3 Main decomposition products from laser-cut nonmetallic materials [2] (an approximate percentage in fume)

Decomposition products	Material Polyester	Leather	PVC	Kevlar®	Kevlar®/epoxy
Acetylene	0.3–0.9	4.0	0.1–0.2	0.5	1.0
Carbon monoxide	1.4–4.8	6.7	0.5–0.6	3.7	5.0
Hydrogen chloride			9.7–10.9		
Hydrogen cyanide				1.0	1.3
Benzene	3.0–7.2	2.2	1.0–1.5	4.8	1.8
Nitrogen dioxide				0.6	0.5
Phenylacetylene	0.2–0.4			0.1	
Styrene	0.1–1.1	0.3	0.05	0.3	
Toluene	0.3–0.9	0.1	0.06	0.2	0.2

Questions

1. What are the main points separating the definitions of Class 1,2, 3 and 4 lasers?
2. What is the nominal hazard zone?
3. How are MPE levels calculated?
4. In what ways can the laser be dangerous?

References

[1] British Standards Institute (2001) EN60825-1:2001. Radiation safety of laser products and systems. British Standards Institute, London
[2] American National Standards Institute (2007) ANSI Z136.1-2007. American National Standard for safe use of lasers. LIA, Orlando
[3] British Standards Institute (1994) BS EN 60825-1. Radiation safety of laser products and systems. British Standards Institute, London
[4] Weiner R (1997) Status of laser safety requirements. Lasers and Optronics Euro Summer 21–23
[5] American National Standards Institute (1993) ANSI Z136.1-1993. ANSI Standard for the safe use of lasers. LIA, Orlando
[6] Rockwell RJ (1990) Fundamentals of industrial laser safety. In: Industrial laser annual handbook 1990. PennWell Books, Tulsa, pp 131–148
[7] Lyon TL, Wood RL, Sliney DH (2001) Hazards and safety considerations. In: Ready JF, Farson DF (eds) LIA handbook of laser and material processing. LIA, Orlando, pp 209–210
[8] Green M (2002) The new laser hazard classification scheme. Ind Laser User 26:38

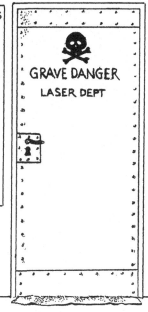

INTERNATIONAL SAFETY CODES
ON OPERATION OF LASERS.

1. GOGGLES MUST BE WORN AT ALL TIMES.
2. DO NOT USE THE LASER AS A
   CIGARETTE LIGHTER.
3. DO NOT DIRECT LASER BEAM AT
   CANS OF TINNED BEANS.
4. IN THE EVENT OF SERIOUS INJURY
   TRY NOT TO MAKE A FUSS.
5. DO NOT FIDDLE ABOUT WITH
   THE LASER - IT'S EXPENSIVE!
6. ON NO ACCOUNT MUST THE
   LASER BE LENT TO POP GROUPS.
7. ALL DOGS TO BE KEPT ON LEADS
8. WHEN YOU'VE FINISHED USING
   THE LASER ROOM PLEASE LOCK
   UP AND REPLACE KEY UNDER
   THE MAT.
9. DO NOT FORGET TO BLOW
   OUT THE CANDLES.
10. DO NOT USE THE LASER AS A
    SUN LAMP.

GRAVE DANGER
LASER DEPT

14 Epilogue

They are ill discoverers that think there is no land when they can see
nothing but sea

Francis Bacon (1561–1626) Advancement of Learning book I vii

This book started with the extravagant claim that optical energy was a new form of energy and therefore should lead to a major advance in our quality of life, as has been the case with the mastery of other forms of energy. Thus, the expectation was high. Hopefully the patient reader has seen in the chapters of this book that some of this expectation is beginning to shape up. It seems appropriate, therefore, to finish with a little thoughtful wander through the future possibilities for laser material processing, which it must be remembered is only one small part of the impact of optical energy.

The major developments are likely to be centred on the differences of optical energy from other forms of energy as well as any equipment developments which will alter the capital cost or processing capabilities.

Some of the principal differences between optical energy and other forms of industrial energy are:

- That current laser power densities are amongst the highest available to industry today.
- That optical energy is amongst the easiest forms of energy to direct and shape.
- That this power can be delivered with very little signal noise, allowing a unique window into the process for automation and adaptive control.
- That the laser beam contains properties as yet unexplored by material processing engineers but which are contributing to the rapidly growing "optical" applications of lasers. These may spill over into material processing.
- That multiphoton events and interactions with radiation can transform transparent material into opaque material or *vice versa* and if of very pure spectral frequency may allow chemical processing which is unexpected thermodynamically.

Finally it is very likely that there are a wealth of applications so novel that we have not seen the possibilities yet – this was certainly the case with the development of applications for fire, steam, electricity and oil. Consider these points in turn.

14.1 Power Intensity

Quick and accurate control of the power intensity would allow precise thermal histories to be engineered. This would allow control of cooling rates, stirring action, time above certain temperatures and weld bead profile. New materials for surfacing applications can be expected based on increased solid solubility, *in situ* alloy formation and composites containing thermally sensitive material such as diamond, Si_3N_4 and others.

14.2 Power Transmission

Optical energy is one of the few forms of energy which can be transmitted intact through air or space. Currently there is a great thrust for fibre delivery systems for laser beams. This is throwing away one of the distinctive advantages of this form of energy. The ease of placing optical energy should lead to developments in time-sharing of beams over considerable distances within factories. The placement of optical energy within existing tooling and machinery may lead to the saving of entire unit operations. The application of fibres, particularly monomode fibres, has enhanced the opportunities for beam insertion into the production line. Some early work is emerging on remote welding and cutting using the long depth of focus of high-brightness beams. Fibres will always be haunted by the limited power which can be delivered without destroying the fibre or the focusability, but the limit is currently adequately high for most material processing uses.

14.3 Power Shaping

No other form of energy can be shaped with such precision as optical energy. Holographic and phase-conjugated mirrors have been invented and developments are promising. Scanning systems are becoming subtler. The future appears to lead to complex shaping of high-powered beams, allowing single-shot processing of areas for surfacing, cladding, drilling or even cutting. The development of mode-matching optics could lead to the conversion of poor modes to high-quality Gaussian modes. Processing with two or more wavelengths giving simultaneously different images, as in a colour picture, is just emerging.

14.4 Automation

Laser power has so many ways of being monitored. The process is wide open to in-process sensing and the development to "intelligent" processing is only just out of reach. Is such a dream a chimera or a real processing advantage?

14.5 Beam Coherence

This is a property as yet unused by material processing engineers, yet it has the property of measuring distance, even possibly penetration distance in-process. It has the property of allowing interference effects, which would in turn allow unusually detailed surface heat sources such as fringe systems to be used. Such ideas are being developed.

14.6 Beam Spectral Purity

This is another property hardly used in material processing except in isotope separation. A challenge or a red herring?

14.7 Multiphoton Events

This is something the laser has brought to the world which we had never contemplated before. The excimer laser with its ability to cut "cold" has shown the extraordinary activity of photons in quantity. There are many nonlinear events demonstrating that photons in bulk lead to unusual events; consider, Brillouin, Rayleigh and Raman scattering. The ubiquitous internally etched glass blocks in tourist shops illustrate how multiphoton events can transform transparent material into opaque material and *vice versa* as in saturable Bragg reflectors. This area of science involving immense waves of electric force is in its infancy.

14.8 Frequency-related Events

Isotope separation, photodynamic therapy and laser enhanced plating all indicate a possible mine of new techniques awaiting the wit of the inventor.

14.9 Equipment Developments

If someone were to build a mass-produced laser for half the current price, then firstly he or she would still make a large profit per laser, but secondly a whole new range of applications would become cost-effective. In fact the laser market is an elastic market, one which would grow faster than the reduced profit from each laser if prices were reduced – but it requires mass production, not the current hand-built systems.

The invention of the fibre and disc lasers has created a high-brightness laser that is compact, lightweight and efficient. The intrinsic cost of such lasers is low since there is relatively little construction work in their designs, which are based mainly on solid-state diodes and doped fibres. As the price of such lasers comes down, so the market will expand. One might wonder if there is an analogy with electric motors, which started as huge devices for lathes and lifts and have developed into many markets, including driving toothbrushes.

Hand-held laser devices for the art and do-it-yourself markets seem an obvious development: desktop laser erasers, and microcladding, welding or cleaning lasers would sporn a surge in small businesses.

Diode-pumped solid-state lasers and phase-matched diode arrays are among other contenders for the crown of laser of choice, with the diode laser being the one with the greatest potential future. Diodes are efficient, compact and simple to operate; understanding their cavity optics and controlling their mode output could make them the device of the future.

Thus, it seems the market is waiting for a wealthy laser manufacturer with a good design, clear vision and a cool nerve to make the substantial investment a mass-production line requires.

Apart from the laser, the development of optics for high-powered beams has a long way to go yet and one might expect some substantial surprises in novel beam shaping and controlling systems. For example, conversion of poor modes or even Gaussian modes into beams that can be focused to near theoretical diffraction-limited spot sizes, whereas phase-conjugated mirrors for power applications would allow beam shaping on the fly.

Automatic beam-guidance systems, which are currently receiving some research interest, will lead to the simple control of laser beams over long distances, allowing a new field of applications to develop.

Finally, the development of beam and workpiece translation equipment, including the possibility of remote processing using long depths of focus, with high-speed software and user-friendly programming is an ongoing vision for many engineers.

14.10 Unthought-of Concepts

Analogy with the introduction of all other forms of energy leads us to believe with some confidence that there are some remarkable things awaiting us based on laser technology. One thing is certain, this book is finished but the story of the laser has just begun.

"An early example of collaborative research with high energy density beams.
Let's hope we can do better in the future!"

Index